Pattern Classifiers and Trainable Machines

Jack Sklansky
Gustav N. Wassel

Pattern Classifiers and Trainable Machines

With 117 Illustrations

Springer-Verlag
New York Heidelberg Berlin

Jack Sklansky
Department of Electrical Engineering
University of California at Irvine
Irvine, CA 92717
U.S.A.

Gustav N. Wassel
Department of Electronic and Electrical Engineering
California Polytechnic State University
San Luis Obispo, CA 93407
U.S.A.

Library of Congress Cataloging in Publication Data
Sklansky, Jack.
Pattern classifiers and trainable machines.

1. Pattern recognition systems. I. Wassel, Gustav N. II. Title.
III. Title: Trainable machines.
TK7882.P3S57 621.3819'598 81-5814
AACR2

Softcover reprint of the hardcover 1st edition 1981

9 8 7 6 5 4 3 2 1

ISBN-13:978-1-4612-5840-7 e-ISBN-13:978-1-4612-5838-4
DOI: 10.1007/978-1-4612-5838-4

To
Gloria and Ruth

Preface

This book is the outgrowth of both a research program and a graduate course at the University of California, Irvine (UCI) since 1966, as well as a graduate course at the California State Polytechnic University, Pomona (Cal Poly Pomona). The research program, part of the UCI Pattern Recognition Project, was concerned with the design of trainable classifiers; the graduate courses were broader in scope, including subjects such as feature selection, cluster analysis, choice of data set, and estimates of probability densities.

In the interest of minimizing overlap with other books on pattern recognition or classifier theory, we have selected a few topics of special interest for this book, and treated them in some depth. Some of this material has not been previously published. The book is intended for use as a guide to the designer of pattern classifiers, or as a text in a graduate course in an engineering or computer science curriculum. Although this book is directed primarily to engineers and computer scientists, it may also be of interest to psychologists, biologists, medical scientists, and social scientists.

We give special attention in this book to linear trainable classifiers and the extensions of these "linear machines" to nonlinear classifiers. The techniques for designing such classifiers were first developed in the late 1950s and early 1960s. Several inadequacies of these techniques moved us to the discovery of new design concepts—such as window training and close opposed pairs of prototypes—and to the development of the continuous-state model of Markov training processes. These results, as well as several other heretofore inadequately published topics within the scope of pattern classifiers and trainable machines, are included in this book.

A graph of the interchapter dependencies is shown below. In this graph, the circle at each node encloses a chapter number. A branch directed from node i to node j indicates that Chapter j depends on Chapter i.

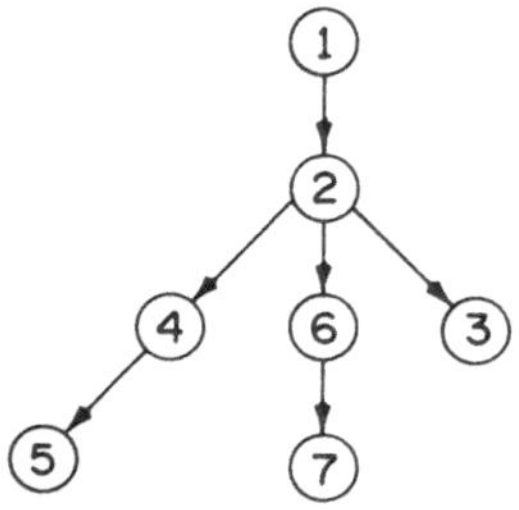

Thus for a short course the reading may be restricted, for example, to Chapters 1, 2, 4, and 5. Further shortening is possible by deletion of portions of chapters. For example Sections 2.2, 2.3, and 2.4 may be omitted in chapter sequence 1, 2, 4, 5.

Exercises for the student are placed at the end of each chapter.

We thank three successive deans of the UCI School of Engineering, Dr. Robert M. Saunders, Dr. James H. Mulligan, Jr., and Dr. Allen R. Stubberud; as well as Dr. Beamont Davison, Dean of Engineering at Cal Poly Pomona, for their support and encouragement of the research and course development leading to this book.

We acknowledge with thanks the contributions of Dr. Leo Michelotti to the locally trained piecewise linear classifier described in Chapter 3; and we thank him for his criticisms of the manuscript—particularly Chapter 2. We are indebted to Dr. Phil Merryman for his doctoral thesis on continuous-state models, upon which most of Chapter 7 is based. We thank Lee Kilday for discussions leading to our procedure for choosing the minimal-cost operating point. We thank Ramalingam Chellappa for his careful reading and constructive criticisms of the manuscript; in particular for his contributions to our discussion of Bayesian learning and sufficient statistics. We thank Carolyn Kimme-Smith, Walter Blume, and Dr. Gloria Frankl for their contributions to the design and programming of the breast tissue classifier described in Chapter 5.

We are grateful to our students and research colleagues for their criticisms of the text and constructive suggestions.

We are indebted to the University of California at Irvine, the National Science Foundation, the Air Force Office of Scientific Research, the National Institute of General Medical Sciences, and Spectra Research Systems for support of the research leading to the results included in this book. Much of this support was provided under the following grants: NSF Grant No. GK-4226, NSF Science Faculty Fellowship 60196, U.S. Air Force Grant No. 69-1813, and U.S. Public Service Grant GM-17632.

We thank Audrey Bennett, Frances Candelori, Deborah S. Germain, Edna Nemetz, Mary Phillips, Dorothy Shearer, Martha Sue Spence Campbell, and Suzanne Costelloe for typing the manuscript, and Arlene Sanders for drawing the figures.

Contents

CHAPTER 1

Introduction and Overview

Just as the successful invention of the airplane stimulated the study of aerodynamics, the modern digital computer has stimulated the study of intelligence and learning.

A frequently occurring form of intelligent behavior is the sorting or classification of data. The process of distinguishing poisonous objects from food is an example of such behavior. Extreme forms of the classification process are scientists' transformations of observations of nature into "laws" of nature.

Because of the difficulty of many practical classification tasks, it is no surprise that this form of intelligent behavior often depends on a learning process. For example, the accuracy with which a radiologist classifies the image of a lesion in a radiograph as either benign or malignant is highly dependent on an extended period of training. Since the late 1950s the growth of the technology of digital computers has spurred both a technology of pattern classification machines and a mathematical theory of simple forms of human and machine learning.

Pattern classification is an information-transformation process. That is, a classifier transforms a relatively large set of mysterious data into a smaller set of useful data. It is not surprising, therefore, that computing machines, as well as living organisms, exhibit the ability to detect and classify patterns. Examples of such machines that have been constructed and used effectively—including a few commercial successes—are blood cell classifiers, chromosome analyzers, analyzers of aerial photographs, speech analyzers, postal zone readers, fingerprint analyzers, and radiograph analyzers.

In this book we describe several mathematical methods for the design of artificial pattern classifiers, with special attention given to training techniques. We also describe the mathematics of continuous-state and Markov-chain learning models—which are applicable to both artificially constructed classifiers and human decision processes.

In this first chapter we develop a few preliminary concepts and notation, and discuss the utility of machine learning in a broad sense. We start with a few basic definitions.

1.1 Basic Definitions

A *classifier* is a device or a process that sorts data into categories or classes.

A *trainable* classifier is a classifier that can improve its performance in response to information it receives as a function of time. Let $\mathscr{C}$ denote such a classifier and let I denote the received information. $\mathscr{C}$ may be a machine, a biological organism or human being, a biological species, a man–machine system, a business organization, or a nation's economic system. I is often a mathematical function of $\mathscr{C}$'s past performance; sometimes it is just a special sequence of observations and correct classifications; usually it is a combination of all three: a special set of observations, associated correct classifications, and the value of a function of some or all of the past performance of $\mathscr{C}$.

Training is the process by which the parameters of $\mathscr{C}$ are adjusted in response to I. (If $\mathscr{C}$ is a human, these parameters are usually parts of a psychological model of the learning process.) A *training procedure* is an algorithm—often a computer algorithm—that implements the training process.

Learning is the motion of a system's effectiveness (or performance) from one level to another. Learning is positive if the motion is in the direction of increased effectiveness. (Implicit in this definition is the existence of a procedure for computing effectiveness or performance quantitatively. We will take up the subject of these performance measures later.) Learning is often associated with feedback, and provides a means by which humans may control their technologies, machines may adapt automatically to changing environments, and species may survive.

The social utility of learning in intelligent machines is particularly evident in the following ways:

1. Models of learning can lead to the construction of machines that learn and relearn the users' goals, even when these goals change over a period of time. It is only by such a feedback process that the human users can have assurance that these machines will serve the users' purposes, and avoid the sorcerer's apprentice syndrome.
2. Models of learning can lead to the construction of machines that learn to overcome the inadequacies and failures of the machine's own

structures and parts. These self-organizing and self-repairing properties may contribute significantly to the economy of designing and operating a complex intelligent machine.

3. Models of human learning provide a basis for the construction of machines that can share with humans the task of learning. We envision here a learning loop consisting of two learn–teach sequences. In one sequence the machine learns those aspects of a task that it is best suited for, and teaches some of what it has learned to the human user. In the second learn–teach sequence, the human learns a task at which he or she is efficient, and teaches the machine a task based on what he or she has learned.

1.2 Trainable Classifiers and Training Theory

In this section we refine our view of pattern classifiers and define and discuss training theory. Recall that we defined a classifier as a device that sorts data into categories. These data are often structured as vectors in *feature space*. Every point in this space is called a *feature vector*. Each component x_i of the feature vector $\mathbf{x}$ is usually a feature, attribute, or property of an object under analysis. For example, the classifier may be analyzing the chromosomes of a single human cell, and sorting these chromosomes into 23 pairs (a karyotype). The feature vector of each chromosome may have components consisting of, for example, the width of its centromere (its "waist"), the average length of its lobes, the distance of its convex hull's* centroid to its mass centroid, etc. [3].

It is often assumed that the feature vectors of a given class are in some sense nearer to all feature vectors in that class than to all or most of the feature vectors in other classes. This is the *compactness hypothesis* [4].

The feature vectors in a given class occupy a region in feature space which we call a *class region*. It is often assumed that every class region is bounded. Another common assumption is that none of the class regions overlap. (However, in most practical problems, some overlap exists.) When the class regions don't overlap, the classes are said to be *separable*, and to have the property of *separability*. If, for every class region, a hyperplane can be placed so that it separates that region from all other class regions, the classes are said to be *linearly separable*. Much of the early work on the theory of pattern classification was concerned with linearly separable classes. Recently, however, a significant amount of work has been concerned with *nonseparable* (i.e., overlapping) classes and classes which are not linearly separable.

* The convex hull of a plane region $\mathscr{R}$ is the smallest convex region that contains $\mathscr{R}$. One can visualize the convex hull by imagining a rigid board cut in the exact of $\mathscr{R}$ and a stretched elastic band placed around the edge of the board. The band lies on the boundary of the convex hull.

The classifier assigns every feature vector to a particular *decision region* $\mathscr{R}_j$ in feature space by means of a set of *decision hypersurfaces* (Figure 1.1). Each such assignment may or may not correspond to a correct or desirable classification. A *trainable classifier* is a classifier which attempts to make the number of incorrect classifications small or zero by adjusting the set of decision regions $\{\mathscr{R}_j\}$ in response to observations on a sequence of feature vectors, $\{\mathbf{x}(n)\}$. This sequence of observations is said to take place during a *learning* or *training* phase.

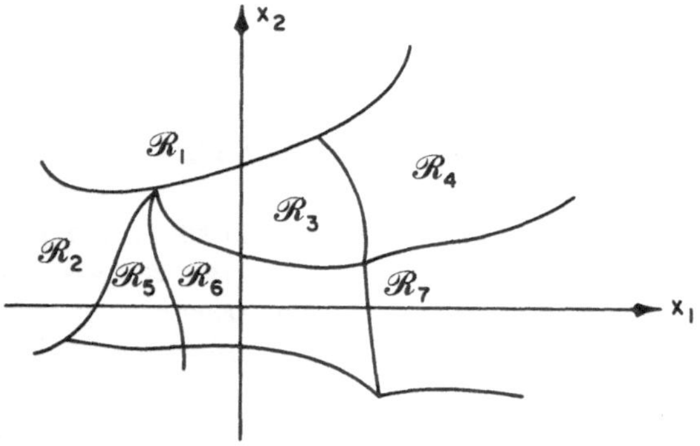

Figure 1.1. An example of a set of decision regions $\{\mathscr{R}_j\}$ in two-dimensional feature space.

Along with the feature vectors, the observations may include information that correctly classifies these feature vectors. If the observations include the correct-classification data, the training is said to be *supervised* or *with a teacher.* These correct-classification data are sometimes referred to as *reinforcements.* If no reinforcements are included in the data, the training is said to be *unsupervised* or *without a teacher.* The process of forming a computer program that classifies various forms of heart enlargements in radiographs in response to radiologists' diagnoses is an example of supervised training [5]. The process by which a botanist learns to classify new forms of plant life into families, genera, species, etc., is an example of unsupervised training.

The procedure or algorithm by which the members of $\{\mathscr{R}_j\}$ are adjusted in response to the observed feature vectors is called a *training procedure.* Each adjustment of $\{\mathscr{R}_j\}$ takes place in response to one or more feature vectors. Each such adjustment, together with the associated observations and reinforcements is called a *trial*, following the terminology of psychologists and psychophysicists. The number of trials is an index of the length of training.

After a classifier is trained, it is usually presented with input data whose classes are unknown. This mode of operation of the classifier is referred to as the *working phase*, while the mode during which training takes place is known as the *training phase.* The set of feature vectors or observations used as input data during the training phase is referred to as the *training set.* Sometimes the training phase and working phase can coincide or overlap. This is often the case when the training is unsupervised.

Often the decision hypersurfaces of a classifier are determined by a set of *discriminant functions* $\{g_j(\mathbf{x})\}$ as follows:

$$\mathscr{R}_j = \{\mathbf{x} \mid g_j(\mathbf{x}) \geqslant g_i(\mathbf{x}) \text{ for all } i\}. \tag{1.1}$$

When $g_j(\mathbf{x})$ is of the form

$$g_j(\mathbf{x}) = \mathbf{w}_j^T\mathbf{x} + w_{j0}, \tag{1.2}$$

where $\mathbf{w}_j^T$ denotes the transpose of a *weight vector* $\mathbf{w}_j$, then $g_j(\mathbf{x})$ is a *linear discriminant function* and the classifier is referred to as a *linear classifier.* An advantage of linear over nonlinear classifiers is that the available training procedures for linear classifiers are relatively simple and well understood, and the number of weights w_{ji} to be adjusted during the training phase is relatively small.

We define *training theory* as a body of insights into the relationship between training procedures and learning. The types of insights provided by training theory include relationships among expected performance [6], length of training, stability in the large [7], and the dimensionality of feature space. For example, certain parts of this theory permit us to predict the average performance of trainable classifiers as a function of the length of training [8].

1.3 Assumptions and Notation

It is usually assumed in this book that the data enters the classifier as a sequence of statistically independent d-dimensional random vectors $\{\mathbf{X}(n) \mid n = 0, 1, \ldots, T-1\}$, each component of $\mathbf{X}(n)$ denoting a feature, property, or measurement of an object under observation. The quantity T denotes the number of trials in the training process. Often we use a lowercase form, namely $\mathbf{x}(n)$ or $\mathbf{x}$, for this vector when the indication of a random process is not important. Unless stated otherwise, it is assumed that the objects under observation change at random as n increases, and that the sequence of classes to which these objects belong is a statistically independent random sequence.

Let $\omega(n)$ denote the class to which $\mathbf{X}(n)$ belongs. In this book the discussion is usually restricted to two classes: ω_1 and ω_2, so that the range of $\omega(n)$ has just two elements. However, two-category classification techniques can often be extended to multiclass cases. One way in which such cases may be handled is by assigning a weight vector to each class, and training each weight vector of the classifier to distinguish between members of the assigned class and members of all other classes. Another technique, devised by Kesler, constructs a single two-category problem from the multiclass problem. For a description of the Kesler technique see Section 2.6.

1.4 Illustrative Training Process

The fictitious scenario described below illustrates a few of the types of tasks involved in the design of a simple pattern classifier. Real design processes are usually more complex than this one.

Scenario of a Simplified Design Process. A medical instrumentation manufacturer (MIM) wants to construct a pilot model of an artificial classifier that, using microscopic images of blood cells as the only input data, can distinguish malignant cells from benign cells with an accuracy comparable to or better than that of human pathologists. MIM wants the classifier to be capable of being trained by examples of malignant and benign cells, with each example labeled M or B, according as it is malignant or benign, respectively. To obtain these examples MIM enlists the services of a competent pathologist, PA, and a person skilled in the design of automatic pattern classifiers, KL.

Let $\mathscr{M}$ denote the class of all malignant cells, and let $\mathscr{B}$ denote the class of all benign cells. The pathologist analyzes N cells, and labels the ith cell M_i or B_i, denoting membership in $\mathscr{M}$ or $\mathscr{B}$, respectively. (The choice of N is made under the advice of KL.) Let us assume that all of these labels are correct, so that $\{M_i\} \subset \mathscr{M}$ and $\{B_j\} \subset \mathscr{B}$. Let N_M = number of members of $\{M_i\}$, N_B = number of members of $\{B_j\}$. Thus $N_M + N_B = N$. The quantity N_M is chosen equal to the expected number of members of $\mathscr{M}$ in a sample of N cells, and N_B is the expected number of members of $\mathscr{B}$ in a sample of N cells.

PA and KL discuss the problem of assigning physical measurements to the components of a feature space. They decide on the following two-dimensional feature space. (In practice the dimensionality of the feature space is likely to be much higher than two). Let c_n denote a closed plane curve lying on the boundary of the image of the nucleus of the nth cell. Let r_n denote the region enclosed by c_n. Let $X_1(n)$ denote the ratio of the area of r_n to the square of the length of c_n. Let $X_2(n)$ denote the ratio of the area of r_n to the area of the convex hull of r_n. The feature vector $\mathbf{X}(n)$ is

$$\mathbf{X}(n) = [X_1(n), X_2(n)]^T.$$

A *training input sequence* is then formed as follows. Using a table of random numbers or the random number generator of a computer, $\mathscr{M}$ or $\mathscr{B}$ is chosen at random in accordance with the estimated probability of occurrence of $\mathscr{M}$ or $\mathscr{B}$, respectively. For each choice of $\mathscr{M}$ or $\mathscr{B}$ at trial n, an example $\mathbf{X}(n)$ from $\mathscr{M}$ or $\mathscr{B}$, respectively, is selected at random. Let

$\omega(n)$ denote a function of n whose value is the label $\mathscr{M}$ or $\mathscr{B}$ at trial n. For the nth example, the components of $\mathbf{X}(n)$ are measured and recorded. In this way KL obtains the sequence of pairs $\{[\mathbf{X}(n), \omega(n)]\}$ $(n = 0, 1, 2, 3, \ldots, T - 1)$. This is the training input sequence obtained for the artificial classifier.

KL decides on the following form of classifier and training procedure (many others could have been chosen). The classifier is of the form whose output at trial n, namely $R(n)$, is determined by

$$R(n) = \begin{cases} \mathscr{M}, & \text{if } \mathbf{W}(n)^T\mathbf{X}(n) + W_0(n) \geqslant 0, \\ \mathscr{B}, & \text{if } \mathbf{W}(n)^T\mathbf{X}(n) + W_0(n) < 0, \end{cases} \tag{1.3}$$

where $\mathbf{W}(n) = [W_1(n), W_2(n)]^T$ = weight vector of a linear classifier. If we define

$$\mathbf{V}(n) = \begin{bmatrix} W_0(n) \\ W_1(n) \\ W_2(n) \end{bmatrix}$$

and

$$\mathbf{Y}(n) = \begin{bmatrix} 1 \\ X_1(n) \\ X_2(n) \end{bmatrix},$$

then Equation (1.3) may be simplified to

$$R(n) = \begin{cases} \mathscr{M}, & \text{if } \mathbf{V}(n)^T\mathbf{Y}(n) \geqslant 0, \\ \mathscr{B}, & \text{if } \mathbf{V}(n)^T\mathbf{Y}(n) < 0. \end{cases}$$

The chosen training procedure is the form in which the weight vector and $W_0(n)$ are adjusted in accordance with the following rule:

$$\mathbf{V}(n+1) = \begin{cases} \mathbf{V}(n) - \rho\mathbf{Y}(n), & \text{if } R(n) = \mathscr{M}, \omega(n) = \mathscr{B}, \\ \mathbf{V}(n) + \rho\mathbf{Y}(n), & \text{if } R(n) = \mathscr{B}, \omega(n) = \mathscr{M}, \\ \mathbf{V}(n), & \text{if } R(n) = \omega(n), \end{cases} \tag{1.4}$$

where the *step size* ρ is any positive number. This is the *proportional increment* training procedure—one of the most popular training procedures in the early development of the technology of trainable classifiers [9]. The initial values $W_0(0)$, $W_1(0)$, $W_2(0)$ are chosen on the best available evidence for the shapes of the class regions in feature space. Further discussion of this training procedure appears in Section 2.4.

During the training process the augmented weight vector $\mathbf{V}(n)$ moves toward an asymptotic value with a random sequence superimposed on its expected motion. Thus $\mathbf{V}(n)$ does not necessarily achieve a final value with certainty. It may interminably move at random in a neighborhood of the limit of the expected value of $\mathbf{V}(n)$ as $n > \infty$. This phenomenon is discussed in detail in Chapters 4–7.

The training process is terminated by a *stopping rule*, and classifier performance is evaluated on new data obtained from an additional N cells. See Section 1.10 for further discussion of these matters.

1.5 Linear Discriminant Functions

A discriminant function g_j is a mapping from the set of feature vectors $\mathscr{X}$ to the real numbers, each value of j associated with a single decision region $\mathscr{R}_j$. Thus for each $\mathbf{x} \in \mathscr{X}$ there is a set of values $\{g_j(\mathbf{x})\}$.

Usually, but not always, each $\mathscr{R}_j$ is associated with a single class region ω_j. Let p denote the number of decision regions, let c denote the number of classes, and let d denote the number of features. In most applications, either $p = c$ or $p = c + 1$. (The cases in which $p = c + 1$ usually include a rejection region, representing the set of feature vectors that the classifier cannot classify reliably.) Since p is usually much less than d, each $g_j(\mathbf{x})$ may be viewed as a macrofeature formed from the features $x_i, \ldots, x_d$. The index of the largest of $\{g_j(\mathbf{x})\}$ is the classifier's assignment of $\mathbf{x}$ to a class. Let $r(\mathbf{x})$ denote this assignment. The determination of $r(\mathbf{x})$ from the discriminant functions $\{g_i(\mathbf{x})\}$ is specified by

$$r(\mathbf{x}) = j \quad \text{whenever } g_i(\mathbf{x}) \leqslant g_j(\mathbf{x}) \text{ for all } i.$$

In a linear classifier the discriminants are linear:

$$g_j(\mathbf{x}) = \omega_{j0} + \mathbf{w}_j^T\mathbf{x}. \tag{1.5}$$

The d-dimensional vector

$$\mathbf{w}_j = (w_{j1}, \ldots, \omega_{jd})^T \tag{1.6}$$

is called a *weight vector*. The elements of this vector are called *weights*. The manner in which the decision regions of a linear classifier are formed by the discriminant functions is illustrated in Figure 1.2.

It is often convenient to express linear discriminant functions in terms of augmented vectors. An *augmented feature vector* $\mathbf{y}$ is defined by

$$\mathbf{y} = (x_0, x_1, x_2, \ldots, x_d)^T, \tag{1.7}$$

where in most cases $x_0 = 1$.

An *augmented weight vector* $\mathbf{v}$ is defined by

$$\mathbf{v} = (w_0, w_1, \ldots, w_d)^T. \tag{1.8}$$

For convenience we define $g_j(\mathbf{y})$ and $r(\mathbf{y})$ as the value of $g_j(\mathbf{x})$ and $r(\mathbf{x})$ respectively, when $\mathbf{x}$ is mapped into $\mathbf{y}$ by Equation (1.7). Thus in augmented feature space, Equation (1.5) becomes

$$g_j(\mathbf{y}) = \mathbf{v}_j^T\mathbf{y}.$$

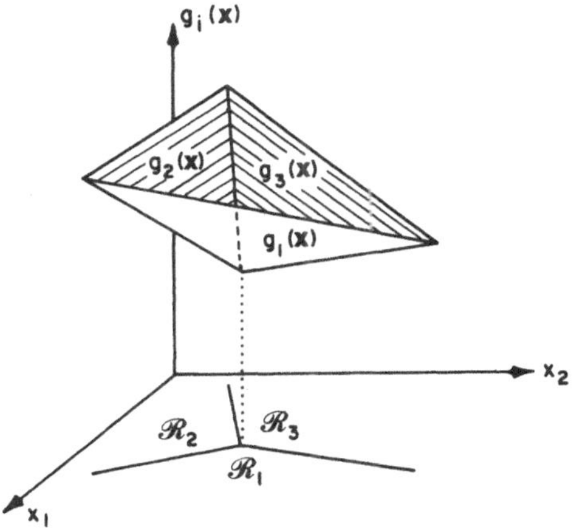

Figure 1.2. Relationship between linear discriminant functions and decision regions.

In a p-class discriminant-function classifier, with $p = c$, the output of the classifier is

$$r(\mathbf{y}) = j \quad \text{whenever } g_i(\mathbf{y}) \leqslant g_j(\mathbf{y}) \text{ for all } i.$$

A diagram of this form of classifier is shown in Figure 1.3. The output $r(\mathbf{y})$ of this classifier may be formed by a sequence of $p - 1$ binary comparisons of the form

$$\text{Is } g_j(\mathbf{y}) \geqslant g_i(\mathbf{y})?$$

At each comparison, the larger of the compared quantities is retained in memory for use in the next comparison.

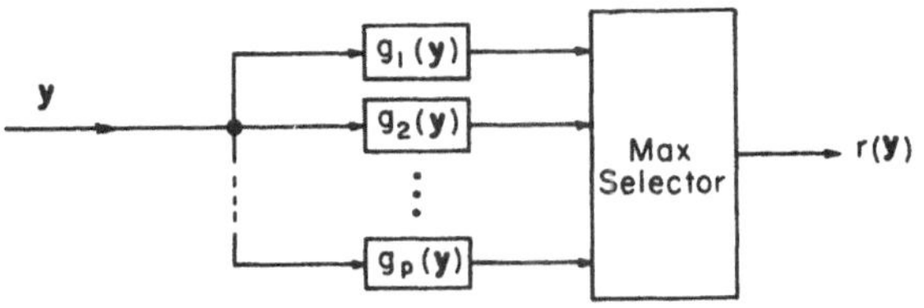

Figure 1.3. A discriminant function classifier.

In a two-class classifier, let

$$g(\mathbf{y}) = g_2(\mathbf{y}) - g_1(\mathbf{y}), \tag{1.9}$$

which we sometimes refer to as a two-class discriminant function. Then the two discriminant functions and the maximum selector may be replaced by a single function $g(\mathbf{y})$ followed by a sign detector:

$g(\mathbf{y}) \geqslant 0$ implies $r(\mathbf{y}) = 2$, corresponding to decision region $\mathscr{R}_2$.
$g(\mathbf{y}) < 0$ implies $r(\mathbf{y}) = 1$, corresponding to decision region $\mathscr{R}_1$.

This may also be expressed by

$$r(\mathbf{y}) = \tfrac{1}{2}\{\operatorname{sgn}[g(\mathbf{y})] + 3\} \tag{1.10}$$

where sgn(z) is defined as 1 if $z \geqslant 0$, -1 if $z < 0$. A diagram of this form of classifier is shown in Figure 1.4.

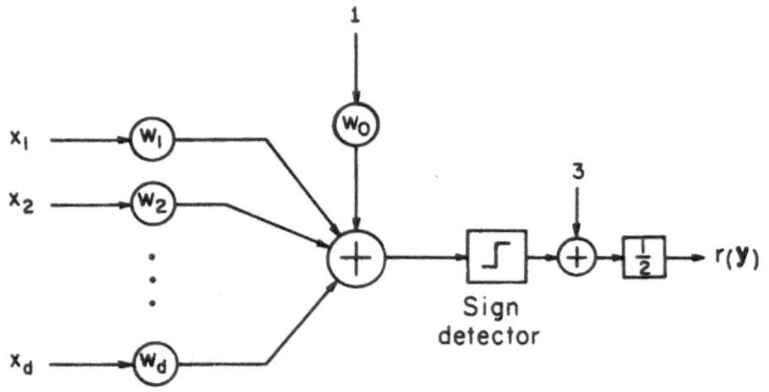

Figure 1.4. A two-class linear classifier.

In a linear classifier, the decision regions are bounded by hyperplanes or portions of hyperplanes. Thus the decision surface between decision regions j and k must satisfy

$$(\mathbf{v}_j - \mathbf{v}_k)^T\mathbf{y} = 0, \tag{1.11}$$

which is the equation of a hyperplane. In the case of a two-class linear classifier, there is just one decision surface:

$$g(\mathbf{y}) = (\mathbf{v}_2 - \mathbf{v}_1)^T\mathbf{y} = 0 \tag{1.12}$$

or

$$g(\mathbf{y}) = \mathbf{v}^T\mathbf{y} = 0, \tag{1.13}$$

where

$$\mathbf{v} = \mathbf{v}_2 - \mathbf{v}_1.$$

Note that all of these hyperplanes pass through the origin of augmented feature space.

To help visualize the relationship between augmented feature space and nonaugmented feature space, consider the case $d = 2$. Then Equation (1.13) can be written in the form

$$g(\mathbf{y}) = w_0x_0 + w_1x_1 + w_2x_2 = 0.$$

This determines the hyperplane shown as a shaded region in Figure 1.5. Nonaugmented feature space is the plane $x_0 = 1$. In this space, the above equation becomes

$$g(\mathbf{x}) = g(\mathbf{y}\,|\,x_0 = 1) = w_0 + \mathbf{w}^T\mathbf{x} = 0.$$

This equation is represented by the line AB in Figure 1.5. Thus line AB in Figure 1.5 is the hyperplane in nonaugmented feature space corresponding to the augmented-feature-space hyperplane shown shaded in that figure.

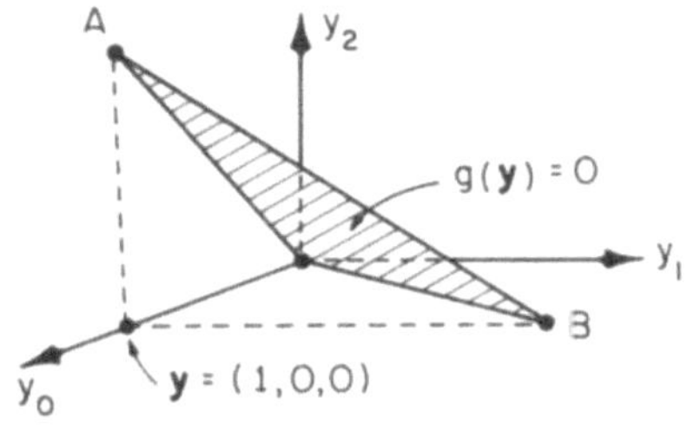

Figure 1.5. Relationship between a hyperplane in augmented feature space and its corresponding hyperplane in nonaugmented feature space.

EXAMPLE 1.1. Plot, in $\mathbf{x}$ space, the decision hyperplane specified by the augmented weight vector $\mathbf{v}^T = (2, 2, -1)$. Also indicate on this plot the weight vector $\mathbf{w}$ and the decision regions $\mathscr{R}_1$ and $\mathscr{R}_2$.

Solution.

$$\mathbf{v}^T\mathbf{y} = (2, 2, -1)\begin{pmatrix} 1 \\ x_1 \\ x_2 \end{pmatrix} = 2 + 2x_1 - x_2.$$

Therefore, $\mathbf{v}^T\mathbf{y} = 0$ gives $2 + 2x_1 - x_2 = 0$, which is the equation of the desired hyperplane (i.e., a line in this two-dimensional example) and is plotted on Figure 1.6. The components of $\mathbf{v}$ other than the v_0 component determine $\mathbf{w}$. Therefore, $\mathbf{w}^T = (2, -1)$ as indicated on the figure. From expressions (1.1) and (1.13), region $\mathscr{R}_1$ is defined as the set $\{\mathbf{x} | \mathbf{v}^T\mathbf{y} < 0\}$. Similarly $\mathscr{R}_2 = \{\mathbf{x} | \mathbf{v}^T\mathbf{y} > 0\}$. These regions are also illustrated on the figure. Note that the vector $\mathbf{w}$ will always point away from the region-$\mathscr{R}_1$ half-space.

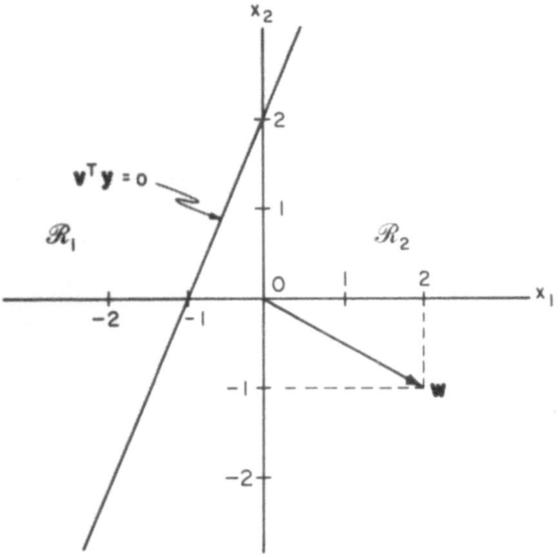

Figure 1.6. Solution to Example 1.1.

Figure 1.7 displays some distance relationships in two-dimensional feature space. Note that $\mathbf{w}^T\mathbf{x}/\|\mathbf{w}\|$, $-w_0/\|\mathbf{w}\|$, and $\mathbf{v}^T\mathbf{y}/\|\mathbf{w}\|$ represent signed distances with a positive or negative value indicating, respectively, a direction

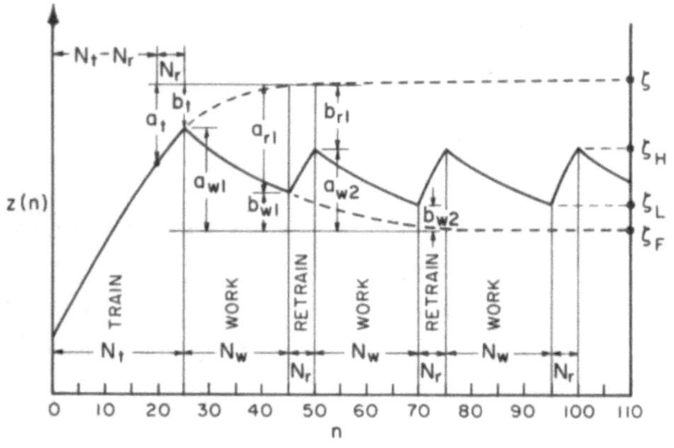

Figure 1.7. Feature vector and decision hyperplane distance relationships.

equal to or opposite to the direction of **w**. The demonstration that $\mathbf{v}^T\mathbf{y}/\|\mathbf{w}\|$ is the signed distance of a sample **x** from the hyperplane is left as an exercise.

It is sometimes convenient to represent the augmented feature vector **y** by

$$\boldsymbol{\eta} = \begin{cases} -\mathbf{y}, & \text{if } \mathbf{x} \in \omega_1, \\ \mathbf{y}, & \text{if } \mathbf{x} \in \omega_2. \end{cases} \tag{1.14}$$

We sometimes refer to $\boldsymbol{\eta}$ as a *sign-normalized* augmented feature vector. Note that with this notation, correct classification is accomplished whenever

$$\mathbf{v}^T\boldsymbol{\eta} > 0$$

for an $\boldsymbol{\eta}$ which is determined by feature vectors from either class ω_1 or ω_2.

1.6 Expanding the Feature Space

Let the training set $\mathscr{X}$ be partitioned into two subsets $\mathscr{X}_1$, $\mathscr{X}_2$ such that $\mathscr{X}_1 \subseteq \omega_1$ and $\mathscr{X}_2 \subseteq \omega_2$. In some situations the sets $\mathscr{X}_1$, $\mathscr{X}_2$ are not linearly separable in **x**-space, but are linearly separable in $\boldsymbol{\xi}$-space, where $\boldsymbol{\xi}$ is a function of **x** and the dimensionality of $\boldsymbol{\xi}$ is larger than that of **x**. To be specific, suppose d is the dimensionality of **x**, and suppose

$$\boldsymbol{\xi} = \boldsymbol{\Phi}(\mathbf{x}) \tag{1.15}$$

where

$$\boldsymbol{\Phi}(\mathbf{x}) = \begin{bmatrix} \Phi_1(\mathbf{x}) \\ \vdots \\ \Phi_r(\mathbf{x}) \end{bmatrix}, \qquad r \geqslant d.$$

Then a separating hypersurface may be linear in $\boldsymbol{\xi}$-space but nonlinear in $\mathbf{x}$-space.

For example, suppose

$$\xi_i = \Phi_i(\mathbf{x}) = \begin{cases} x_i, & \text{for } 1 \leqslant i \leqslant d, \\ x_{i-d}^2, & \text{for } d+1 \leqslant i \leqslant 2d, \\ x_1 x_2, & \text{for } i = 2d+1, \\ x_1 x_3, & \text{for } i = 2d+2, \\ \cdots & \\ x_{d-1} x_d, & \text{for } i = d(d+3)/2. \end{cases} \tag{1.16}$$

(The $\Phi_i(\mathbf{x})$'s are sometimes called Φ-functions [10].) In this case the set of all hyperplanes in $\boldsymbol{\xi}$-space is equivalent to the set of all quadratic hypersurfaces in $\mathbf{x}$-space.

This concept permits the theory of linear classifiers to be applicable to nonlinear classifiers consisting of the transformation

$$\boldsymbol{\xi} = \boldsymbol{\Phi}(\mathbf{x})$$

followed by a linear classifier in $\boldsymbol{\xi}$-space. This class of nonlinear classifiers, which we refer to as *Φ-classifiers*, is illustrated in Figure 1.8.

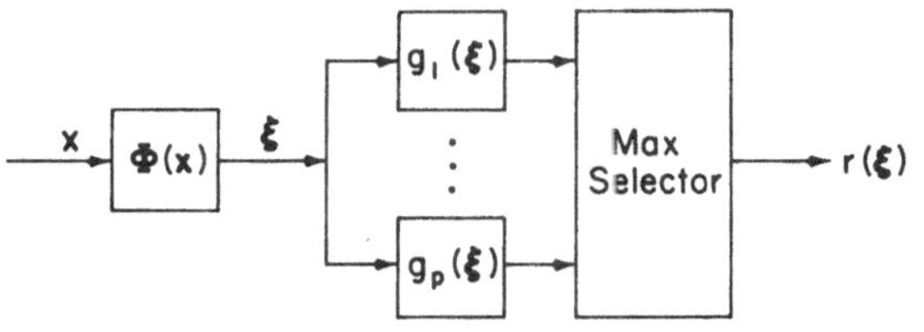

Figure 1.8. A Φ-classifier.

An advantage of the use of Φ-classifiers arises from the convenience of finding the $g_i(\mathbf{x})$'s when the $g_i(\mathbf{x})$'s are linear. A disadvantage of the use of Φ-classifiers is the need for an enlarged training set, since the number of elements in the set must be of the order of r. (This requirement is a consequence of a capacity property discussed in the next section.) Another disadvantage is the need for computing many more features than the original d features: for example, in a quadratic Φ-classifier, $r = d(d+3)/2$, as indicated in Equation (1.16).

1.7 Binary-Input Classifiers

In certain applications of pattern classification, the features x_i, $i = 1, \ldots, d$ are binary-valued, i.e., x_i takes on only one of two values. Often these values are denoted by 0 and 1, -1 and $+1$, or T and F (for True and False). For example we may have

$$x_i = \begin{cases} 1, & \text{if the observed object is convex,} \\ 0, & \text{if the object is not convex.} \end{cases}$$

We say that a vector is binary-valued if each of its elements is binary-valued.

We refer to classifiers operating on binary-valued vectors as *binary-input classifiers*. If such a classifier separates its inputs into two decision regions, we refer to it as a *two-class binary-input classifier*. The mapping performed by such a classifier is a *switching function*. Common examples of two-class binary-input classifiers are the AND gates, OR gates, NAND gates, and NOR gates appearing in computer circuits.

A linear two-class binary-input classifier, i.e., a two-class binary-input classifier implemented by linear discriminant functions, is a *threshold gate* or *threshold logic unit*. A block diagram of a threshold gate is shown in Figure 1.9. The mapping performed by a threshold gate is usually called a *threshold function* [11].

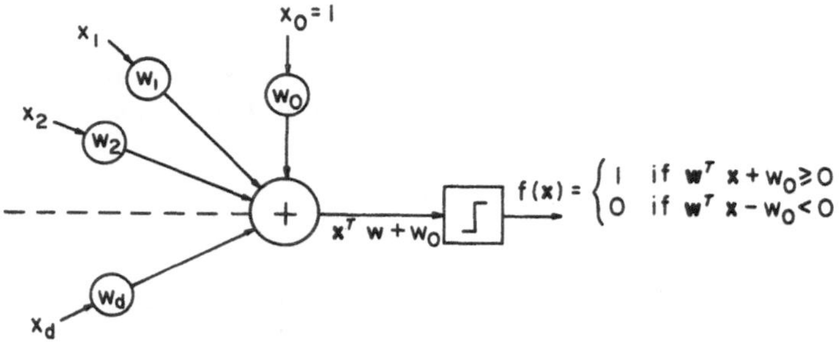

Figure 1.9. Block diagram of a threshold gate.

It is sometimes useful to visualize the operation of a threshold gate in terms of the intersection of a hyperplane in d-space and the vertices of a d-cube, where d is the dimensionality of $\mathbf{x}$. Figure 1.10 illustrates this for the case $d = 3$. Here vertex $(1, 1, 1)$ lies in one decision region and the remaining vertices lie in the other decision region.

The threshold gate has also been used as an elementary model of the neuron. A number of properties of the nervous systems of animals, including humans, have been simulated by networks of threshold gates [12, 13]. Among these properties are the conditioned reflex and the reliability or robustness of the network in the face of changes in the signal processing characteristics of the component neurons.

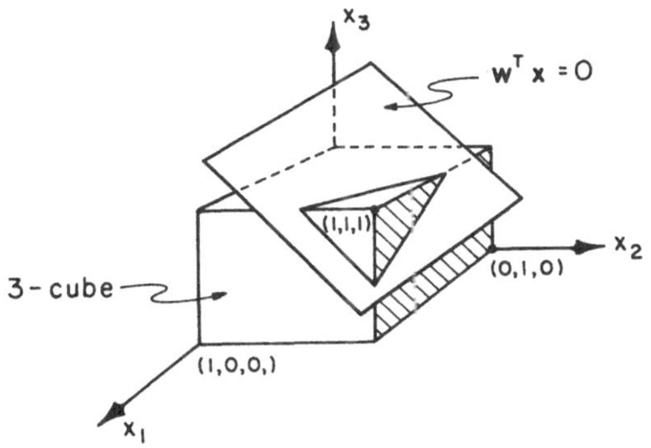

Figure 1.10. Geometric interpretation of the operation of a three-feature binary classifier (threshold gate).

The threshold gate thus has three distinct roles: (a) it serves as a binary device in computer circuits; (b) it is a binary classifier; and (c) it serves as an elementary model of a neuron. As a result, threshold gates have been investigated by researchers in computer logic, pattern recognition, and mathematical neurophysiology.

Note that the set of all possible binary-valued vectors in d-space coincides with the set of 2^d vertices of a d-cube. In the design of threshold gates for computer circuits, d is usually small, say $d \leqslant 7$, and usually the training set occupies all or almost all of the 2^d vertices. Under these conditions the threshold gate is said to be completely or almost completely *specified.* For this type of application a large number of properties of threshold gates have been derived.

These properties, unfortunately, are usually not helpful in the design of typical binary classifiers, because (a) the value of d in typical applications of classifiers is usually large, say $d \geqslant 20$; and (b) the training set usually occupies a small fraction of the 2^d vertices. In fact Cover [14] has shown that only about $2d$ points out of the possible 2^d are needed in the training set, on average, assuming that the points are in general position (i.e., the rank of the matrix formed by the vectors in the training set is d) and that the position of the separating hyperplane is chosen at random. Suppose, for example, that $d = 30$. Then the number of vertices on the d-cube is 2^{30}, which is approximately 10^9. Cover's theory states that of these vertices only 60 randomly chosen vertices in general position are very likely to specify the space adequately for a linear classification.

Widrow [15] refers to $2d$ as the *capacity* of a threshold gate, since a random specification of more than $2d$ feature vectors in the training set is unlikely to be linearly separable, while a specification of fewer than $2d$ feature vectors is likely to be linearly separable.

Thus we need a set of techniques or a mathematical theory suited to the case where d is large and the training set is small, perhaps of the order of $2d$. The concept of asummability, described in the next chapter, fits this need.

1.8 Weight Space Versus Feature Space

It is often helpful to visualize the training process in augmented weight space, rather than in augmented feature space.

An example of a training process in augmented feature space is shown in Figure 1.11(a). In this figure the training set consists of the four sign-normalized feature vectors $\boldsymbol{\eta}(0)$, $\boldsymbol{\eta}(1)$, $\boldsymbol{\eta}(2)$, $\boldsymbol{\eta}(3)$. In this feature space, the hyperplane solution region $\mathscr{F}$ consists of that portion of $\boldsymbol{\eta}$-space containing all hyperplanes which pass through the origin and which are contained in the exterior of the smallest polyhedral cone, with its vertex at the origin, containing all of the given $\boldsymbol{\eta}(n)$'s. (In other words, $\mathscr{F}$ is contained in the exterior of the polyhedral cone with its vertex at the origin subtending the convex hull of the $\boldsymbol{\eta}(n)$'s.) The solution region $\mathscr{F}$ is shown shaded in the figure. Any hyperplane decision surface passing through the origin, contained in $\mathscr{F}$, and with all $\boldsymbol{\eta}(n)$'s in its positive half-space separates the pair of classes in the given training set. The training process in this space is indicated by a sequence of hyperplanes, as illustrated by the dashed lines in Figure 1.11(a), each hyperplane corresponding to an equation of the form $\mathbf{v}^T(n)\boldsymbol{\eta} = 0$.

The same training process appears in augmented weight space as a sequence of points. This form of display of the training process is often easier to visualize than the sequence of hyperplanes in feature space. In weight space the solution region $\mathscr{V}^*$ is defined as the set of all vectors $\mathbf{v}$ each of which satisfies $\mathbf{v}^T\boldsymbol{\eta}(n) > 0$ for all n. This region is the intersection of all the positive half-spaces bounded by the hyperplanes $\{\mathbf{v}^T\boldsymbol{\eta}(n) = 0\}$. Thus in all cases where the training set is finite, $\mathscr{V}^*$ is a polyhedral cone with its vertex at the origin.

The training process is the motion of $\mathbf{v}(n)$ in a search for any point in $\mathscr{V}^*$. In augmented weight space this motion is displayed by the endpoints

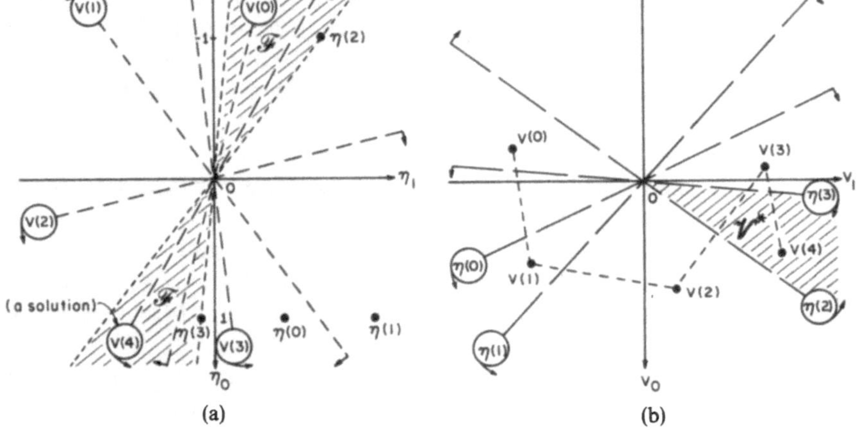

Figure 1.11. Training process (a) in feature space and (b) in weight space.

of the sequence of vectors $\{\mathbf{v}(n)\}$, while in augmented feature space the motion is displayed by a sequence of directed hyperplanes (i.e., hyperplanes with a positive direction toward one of its half-spaces).

Figure 1.11(b) illustrates a training process in weight space. This training process is the same as that illustrated in the feature space of Figure 1.11(a). The motion of $\mathbf{v}(n)$ in Figure 1.11(b) is indicated by a dashed line.

A hyperplane $\mathbf{v}^T\boldsymbol{\eta}(n) = 0$ appearing in augmented feature space, as illustrated in Figure 1.11(a), is sometimes called a *pattern hyperplane*, a *feature-vector hyperplane*, or an *$\boldsymbol{\eta}(n)$-hyperplane*. We prefer $\boldsymbol{\eta}(n)$-hyperplane. Each of the $\boldsymbol{\eta}(n)$-hyperplanes in Figure 1.11(b) is labeled by an $\boldsymbol{\eta}$ in a circle attached to the line representing each hyperplane. The arrow on each $\boldsymbol{\eta}(n)$-hyperplane indicates the direction of the associated positive half-space.

1.9 Statistical Models

Often the observed feature vectors of a set of classes are effectively modeled by a set of probability densities rather than a set of class regions. This statistical model allows us to deal effectively with the substantial overlap that sometimes occurs among the class regions. In the statistical model, the classes $\{\omega_j\}$ are represented by a set of a priori probabilities $\{P(\omega_j)\}$ and a set of conditional densities $\{p(\mathbf{x}|\omega_j)\}$. $P(\omega_j)$ is defined as the a priori probability of the event $\omega = \omega_j$; and $p(\mathbf{x}|\omega_j)$ is defined as the conditional probability of the random vector variable $\mathbf{X}$, given that $\omega = \omega_j$.

For convenience we also define the *constituent density* $f_j(\mathbf{x})$ as the product of $p(\mathbf{x}|\omega_j)$ and $P(\omega_j)$:

$$\begin{aligned} f_j(\mathbf{x}) &= P(\omega_j)p(\mathbf{x}|\omega_j) \\ &\equiv p(\mathbf{x}, \omega_j). \end{aligned}$$

Thus, in the single feature case, $f_j(x)dx$ is the joint probability of the events $X \in (x, x + dx)$ and $\omega = \omega_j$. Let

$$p(\mathbf{x}) = f_1(\mathbf{x}) + \cdots + f_c(\mathbf{x}),$$

where c is the number of classes. The function $p(\mathbf{x})$ is the probability density of $\mathbf{X}$. Note that $p(\mathbf{x})$ satisfies

$$\int_{-\infty}^{\infty} \cdots \int_{-\infty}^{\infty} p(\mathbf{x})\, dx_1 \cdots dx_d = 1.$$

The constituent densities $\{f_j(\mathbf{x})\}$ determine the *Bayes optimum* decision regions:

$$\mathscr{R}_j = \{\mathbf{x} | f_j(\mathbf{x}) \geqslant f_i(\mathbf{x}) \text{ for all } i\}.$$

We illustrate these concepts for a one-dimensional feature space and two classes. For this case

$$f_1(x) = P(\omega = \omega_1)p(x|\omega = \omega_1) \equiv P(\omega_1)p(x|\omega_1) \tag{1.17}$$

$$f_2(x) = P(\omega = \omega_2)p(x|\omega = \omega_2) \equiv P(\omega_2)p(x|\omega_2). \tag{1.18}$$

These densities are illustrated in Figure 1.12. Note that

$$\int_{-\infty}^{\infty} p(x)\,dx \equiv \int_{-\infty}^{\infty} [f_1(x) + f_2(x)]\,dx = 1$$

and that $p(x)\,dx$ is the probability of the event $X \in (x, x + dx)$. Let θ denote a decision threshold such that $w_0 + w_1\theta = 0$. Then $\theta = -w_0/w_1$, $\mathscr{R}_1 = \{x|x < \theta\}$, and $\mathscr{R}_2 = \{x|x > \theta\}$. In Figure 1.12, the value of the threshold θ is marked by the dashed line. The decision regions $\mathscr{R}_1$ and $\mathscr{R}_2$ are, respectively, the portions of the x-axis to the left and right of θ.

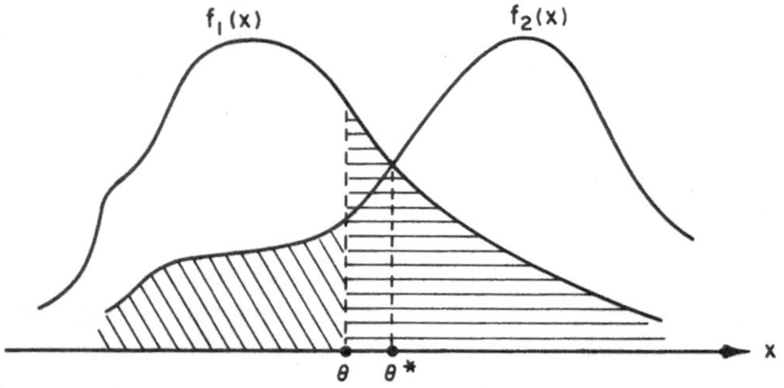

Figure 1.12. Constituent densities of a single-feature classifier.

We now can compute $P(\text{error}|\theta)$, the conditional error probability for a given θ. The threshold θ partitions the x-axis into the two decision regions $\mathscr{R}_1$ and $\mathscr{R}_2$. If $x \in \mathscr{R}_j$, then the classifier assigns x to ω_j. Let $P(\omega = \omega_2, \mathscr{R}_1) =$ probability that the classifier assigns x to ω_1 and that $x \in \omega_2$. Let $P(\omega = \omega_1, \mathscr{R}_2) =$ probability that the classifier assigns x to ω_2 and that $x \in \omega_1$. Note that

$$P(\omega = \omega_2, \mathscr{R}_1) = \int_{-\infty}^{\theta} f_2(x)\,dx \tag{1.19}$$

$$P(\omega = \omega_1, \mathscr{R}_2) = \int_{\theta}^{\infty} f_1(x)\,dx. \tag{1.20}$$

Thus

$$P(\text{error}|\theta) = P(\omega = \omega_2, \mathscr{R}_1) + P(\omega = \omega_1, \mathscr{R}_2) \tag{1.21}$$

$$= \int_{-\infty}^{\theta} f_2(x)\,dx + \int_{\theta}^{\infty} f_1(x)\,dx. \tag{1.22}$$

The error tail probabilities in Equations (1.19) and (1.20) are indicated by the diagonally and horizontally shaded regions, respectively, in Figure 1.12.

From Equation (1.22) one can find that value of θ which minimizes $P(\text{error}|\theta)$. To do this, one first differentiates both members of Equation (1.22) and sets the derivative of $P(\text{error}|\theta)$ equal to zero. This yields

$$f_1(\theta^*) = f_2(\theta^*), \tag{1.23}$$

where θ^* is that value of θ for which $P(\text{error}|\theta)$ is a local minimum or a local maximum. In particular if $f_1(x)$ and $f_2(x)$ have a single point of intersection at $x = \theta^*$, then the minimum error probability is achieved when $\theta = \theta^*$. This is often referred to as the *Bayes optimum* value of θ.

It can be shown (see Exercise 6.2) that $P(\text{error}|\theta)$ is a local minimum at $x = \theta^*$ even if $P(\text{error}|\theta)$ has a finite discontinuity at $\theta = \theta^*$, provided

$$(x - \theta^*)[f_1(x) - f_2(x)] < 0 \quad \text{for all } x \neq \theta^*. \tag{1.24}$$

If there is just one value of θ^* that satisfies Equations (1.23) and (1.24), it is clearly the Bayes optimum value of θ. If there is more than one such value, say $\{\theta_i^*\}$, an optimum value of θ, say $\theta = \theta_m^*$, is one for which

$$P(\text{error}|\theta_m^*) \leqslant P(\text{error}|\theta_i^*) \text{ for all } \theta_i^* \neq \theta_m^*.$$

EXAMPLE 1.2. Given the constituent densities

$$f_1(\mathbf{x}) = \begin{cases} \frac{1}{8}, & \text{if } -1 < x_1 < +1 \text{ and } -1 < x_2 < +1, \\ 0, & \text{otherwise;} \end{cases}$$

$$f_2(\mathbf{x}) = \begin{cases} \frac{1}{8}, & \text{if } 0 < x_1 < 2 \text{ and } 0 < x_2 < 2, \\ 0, & \text{otherwise,} \end{cases}$$

find the probability of error for a linear classifier with

$$\mathbf{v} = \begin{bmatrix} -0.75 \\ 0.75 \\ 0.75 \end{bmatrix}.$$

Solution. The contours of the constituent densities are illustrated in Figure 1.13. Also illustrated are the decision hyperplane $\mathbf{v}^T\mathbf{y} = 0$, the weight vector $\mathbf{w}$, and the decision regions $\mathscr{R}_1$ and $\mathscr{R}_2$. These were found following the methods of Section 1.5. The error

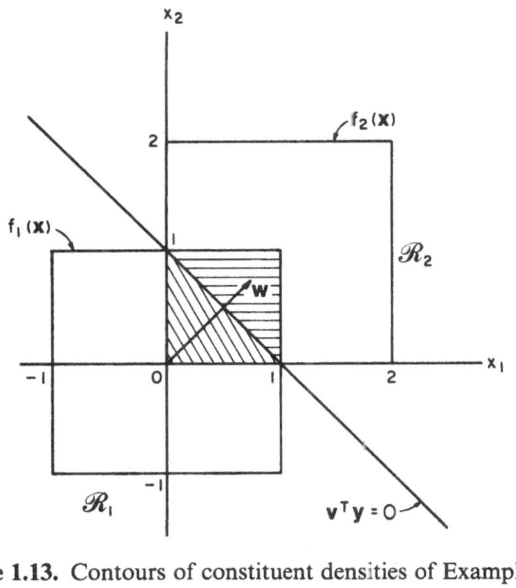

Figure 1.13. Contours of constituent densities of Example 1.2.

tail regions are shown shaded in the figure. Since the constitutent densities have a value of $\frac{1}{8}$ in these regions, the error tail probabilities are given by

$$P(\omega = \omega_2, \mathscr{R}_1) = \tfrac{1}{8}[(\tfrac{1}{2})(1)(1)] = \tfrac{1}{16}$$
$$P(\omega = \omega_1, \mathscr{R}_2) = \tfrac{1}{8}[(\tfrac{1}{2})(1)(1)] = \tfrac{1}{16}.$$

Therefore

$$P(\text{error}|\mathbf{v}) = P(\omega = \omega_2, \mathscr{R}_1) + P(\omega = \omega_1, \mathscr{R}_2) = \tfrac{1}{8}.$$

1.10 Evaluation of Performance

The evaluation of the performance of an automatic classifier is an important part of the process of designing that classifier. The available data, in the form of a correctly classified set of feature vectors, must be partitioned into two parts: a *design set* (or *training set*) and a *test set*. The design set is partitioned into several *design subsets*, one design subset for each class. Each design subset provides exemplars of the class with which it is associated. It is not good practice to test the classifier on the design set, since the classifier design tends to be biased in favor of the design set. If the test set is too small, the evaluation of the classifier's performance is not sufficiently reliable. If the design set is too small, the classifier's performance is likely to be too far from optimum.

If the size of the available set of data is large—say 2000 or more feature vectors for each class—then dividing the data into two equal-size sets, one for design and one for test, is effective. If the size of the data set is small, then unequal partitioning is often necessary. Strategies for unequal partitioning are discussed in subsection 1.10.3.

1.10.1 Stopping Rules for Large Data Bases

Some applications provide large data bases at relatively low cost. An example of such an application is the segmentation of aerial photographs into labeled regions, such as *road*, *forest*, *corn*, *urban area*, etc. In this application each *pixel* (picture element)—usually in the form of multispectral data—provides a feature vector for the data base. Under these conditions the size of the design set is determined primarily by the available time and money for the design process, while the cost of the data is not a significant parameter. If a trainable classifier is used in such an application, part of the evaluation of the classifier can take place during the training phase in the form of a *stopping rule*.

Most stopping rules are semiempirical in nature. A well designed rule will stop the training process when the error probability lies within a specified interval with a confidence (probability) greater than or equal to some specified value. A step toward the construction of such rules appears in

[16], where a stopping rule for the training of one-dimensional classifiers is derived.

If the size of the available training set is limited or if the acquisition of training data is expensive, one must make an economical division of the data into design and test sets, and one must account for the finiteness of the data in the estimate of the classifier's performance. These considerations are discussed in the remainder of this section. In the discussion it is assumed that every feature vector in both the test set and the design set is obtained independently of all the other feature vectors in these sets.

1.10.2 Operating Characteristics and Confidence Intervals

When the a priori class probabilities $\{P(\omega_j)\}$ of a two-class classifier are unknown, or when one of the parameters of a two-class classifier is particularly difficult to choose, the performance of the classifier is conveniently described by an *operating characteristic* (similar to the receiver operating characteristics used for communication systems). We define an operating characteristic as a curve of b versus a, where

$$b = \text{probability of guessing that } \mathbf{x} \in \omega_1, \text{ when actually } \mathbf{x} \in \omega_2;$$
$$a = \text{probability of guessing that } \mathbf{x} \in \omega_2, \text{ when actually } \mathbf{x} \in \omega_1.$$

Both b and a are functions of the a priori probability $P(\omega_1)$ or the parameter in question. Recall that in a two-class classifier, $P(\omega_2) = -P(\omega_1)$. An operating characteristic is illustrated by the solid curve in Figure 1.14.

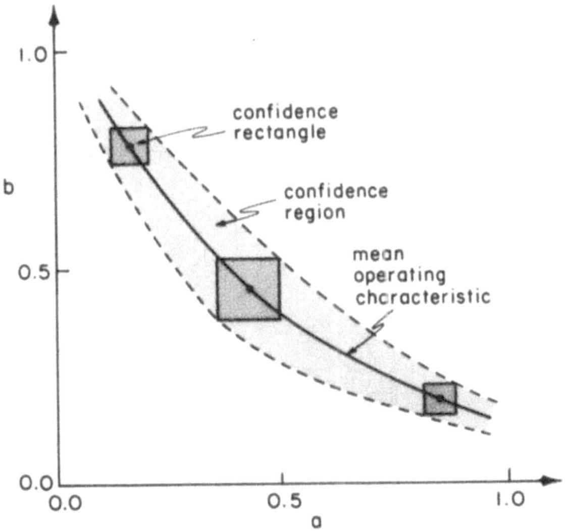

Figure 1.14. Operating characteristic and confidence region.

In the literature on communication engineering and psychophysics, a receiver operating characteristic is usually defined as the curve of $1 - b$ versus a, rather than b versus a. For example, see Swets [17] or Van Trees [18]. We have modified the definition in order to facilitate symmetry in our analyses of the operating characteristics.

For a given set of design data, with $P(\omega_1)$ unknown, $P(\omega_1)$ can be simulated by adjusting the step sizes in the training procedure, or equivalently, by repeating the occurrence of the feature vector in ω_j a number of times proportional to $P(\omega_j)$. For example, if $P(\omega_1) = 0.2$, then $P(\omega_2) = 1 - 0.2 = 0.8$. If the sizes of the design subsets for the two classes are equal, then the step size for class ω_2 would be four times the step size for class ω_1. By varying $P(\omega_1)$, computing $1 - P(\omega_1)$, and adjusting the step sizes appropriately, one obtains a decision surface $\mathscr{S}$ as a function of $P(\omega_1)$. Each such decision surface would yield a point on the operating characteristic if a and b could be estimated accurately.

Since only a finite set of test data is available, we cannot compute a and b exactly. Let $\hat{a}$, $\hat{b}$ denote estimates of a, b respectively, computed as follows:

$$\hat{a} = \frac{A}{Z}, \qquad \hat{b} = \frac{B}{R}$$

where

A = number of misclassified members of ω_1 in the test set,
Z = number of members of ω_1 in the test set,
B = number of misclassified members of ω_2 in the test set,
R = number of members of ω_2 in the test set.

For given values of R, Z, and for a given decision surface $\mathscr{S}$, the quantities A and B are random variables. The probability distributions of A and B are binomial:

$$\begin{aligned} q(A) &= \binom{Z}{A} a^A (1-a)^{Z-A} = \binom{Z}{\hat{a}Z} a^{\hat{a}Z}(1-a)^{(1-\hat{a})Z} \\ q(B) &= \binom{R}{B} b^B (1-b)^{R-B} = \binom{R}{\hat{b}R} b^{\hat{b}R}(1-b)^{(1-\hat{b})R} \end{aligned} \tag{1.25}$$

Both of these distributions have the following form:

$$q(\hat{p}N) = \binom{N}{\hat{p}N} p^{\hat{p}N}(1-p)^{(1-\hat{p})N}. \tag{1.26}$$

From this equation one can compute, for every $p \in (0, 1)$, an interval I such that

$$\sum_{\hat{p} \in I} q(\hat{p}N) = 0.95. \tag{1.27}$$

This is the 95-*percent confidence interval* of p, when the number of observations entering the estimate $\hat{p}$ is N. Figure 1.15 shows the curves of the endpoints of the 95-percent confidence intervals of p, each curve plotted for a constant value of N. (For further discussion of these matters, see References [1] *and* [2]). These curves yield confidence intervals for p as a function of $\hat{p}$. For example, if $\hat{p} = 0.45$ and if $N = 100$, then the probability that p lies in $(0.35, 0.55)$ is 0.95. In a similar fashion one may obtain confidence intervals for a and b, based on Equations (1.25) and Figure 1.15.

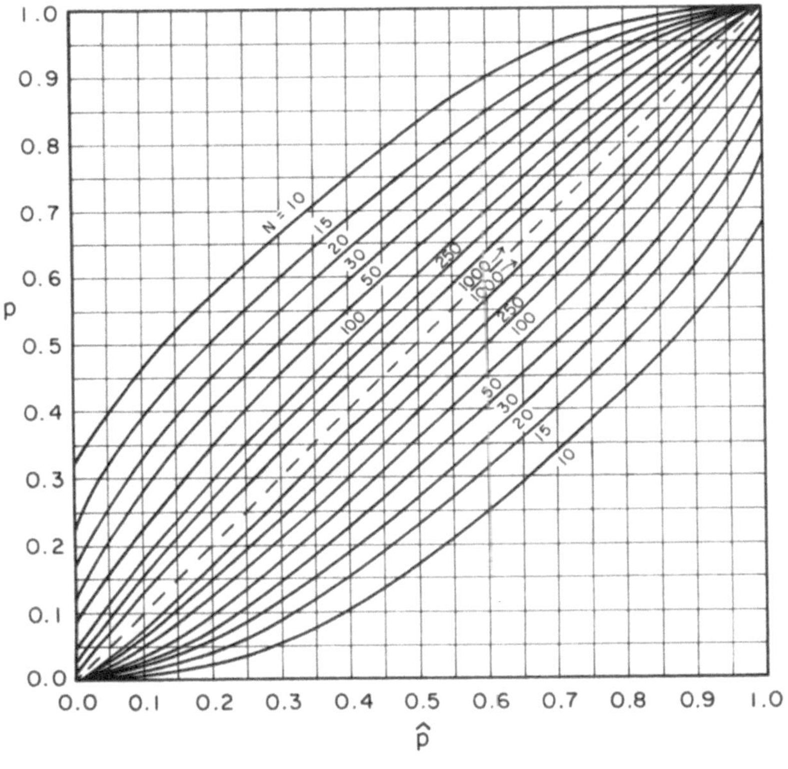

Figure 1.15. 95-percent confidence intervals based on binomial distribution of $\hat{p}$. (Copyright 1962, American Telephone and Telegraph Company. Reprinted by permission.)

For each value of $P(\omega_1)$, the pair of confidence intervals for (a, b) forms a *confidence rectangle in a, b*-space. This is illustrated in Figure 1.14 for three values of $P(\omega_1)$. The envelope of these confidence rectangles forms two curves that enclose the *confidence region* of the operating curves.

In a situation where a two-class classifier has a parameter θ which is difficult to choose, one many obtain an operating characteristic by plotting (a, b) for various values of θ. The resulting locus of (a, b) in the a,b-plane is the operating characteristic of the classifier with respect to θ. If $P(\omega_1)$ is known, one may use this operating characteristic to find the value of θ that minimizes the cost of misclassification. Below we describe how to do this.

Note that $aP(\omega_1)$ is the joint probability that the classifier assigns $\mathbf{x}$ to $\mathscr{R}_2$ and that $\mathbf{x} \in \omega_1$—i.e., $aP(\omega_1)$ is the probability of wrongly assigning $\mathbf{x}$ to $\mathscr{R}_1$. (Recall that $P(\omega_2) = 1 - P(\omega_1)$.) For each wrong assignment or *misclassification*, the classifier incurs a cost c_{21} for wrongly assigning $\mathbf{x}$ to $\mathscr{R}_2$, and c_{12} for wrongly assigning $\mathbf{x}$ to $\mathscr{R}_1$. Thus the expected cost C is

$$\begin{aligned} C &= c_{21}P(\omega_1)a + c_{12}P(\omega_2)b \\ &\equiv c_{21}P(\omega_1)a + c_{12}(1 - P(\omega_1))b. \end{aligned} \tag{1.28}$$

From this equation it follows that a curve of constant cost in the a,b-plane is a straight line, with a slope of $-c_{21}P(\omega_1)/[c_{12}(1 - P(\omega_1))]$.

For example, suppose ω_1 is the class of forged endorsements of bank checks in a specific bank, and ω_2 is the class of valid check endorsements in this bank. Suppose this bank analyzes its check-honoring operations, and finds that its costs of honoring forged checks and investigating wrongly suspected checks are \$1,000 per forged check and \$5 per wrongly suspected check. Then $c_{21} = \$1{,}000$ and $c_{12} = \$5$. If, on the average, three out of every 100 checks are forged, then the expected cost is

$$\begin{aligned} C &= (1000)(0.03)a + 5(0.97)b \text{ dollars} \\ &= 30a + 4.85b \text{ dollars.} \end{aligned} \tag{1.29}$$

In other applications the cost may be other than money—e.g. time, subjective pain, fuel, etc.

By plotting lines of constant cost in the a,b-plane, one sees that a point of minimum cost on any operating characteristic must be a point of contact of the operating characteristic with a constant-cost line such that all the remaining points of the operating characteristic lie above the constant-cost line. Most operating characteristics are everywhere differentiable and concave upward. For such curves the minimum cost is achieved at a point of tangency to the constant-cost line, as illustrated in Figure 1.16. This occurs where the slope of the operating characteristic is

$$\frac{db}{da} = -\frac{c_{21}P(\omega_1)}{c_{12}P(\omega_2)} \quad -\frac{c_{21}P(\omega_1)}{c_{12}[1 - P(\omega_1)]}. \tag{1.30}$$

We refer to this as the *minimum-cost operating point.*

Figure 1.16 illustrates how we may account for inaccuracies in the finite-data estimates of a and b. To do this we find the costs corresponding to the constant-cost lines that support the confidence rectangle at the minimum-cost operating point. These are the lower and upper endpoints of the range of costs associated with the inaccuracies in estimating a and b. In Figure 1.16, this range is $(3, 5)$.

In this way we may set a parameter of the classifier to a value that mini-imizes the expected (or average) cost of misclassifications, and we may

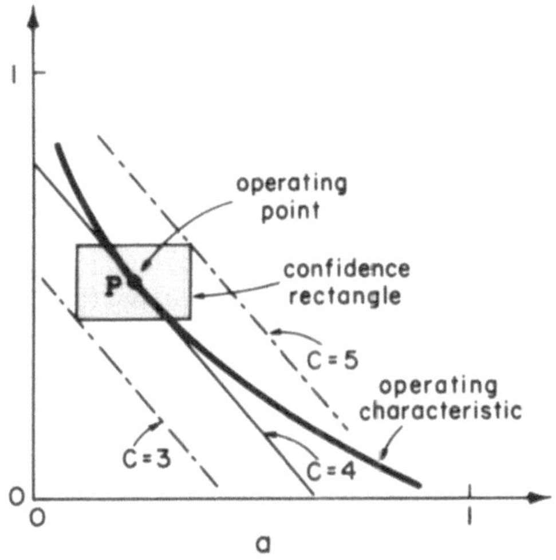

Figure 1.16. Minimum-cost operating point is at *P*.

estimate the changes in this expected cost incurred by inaccuracies in our estimates of a and b.

1.10.3 Partitioning the Data into a Test Set and a Design Set

In many applications, the amount of data available for designing the classifier is so limited that special partitioning procedures are needed to maximize the exploitation of the data. Such procedures are often necessary when m_i, the number of feature vectors in the data for class ω_i, is less than 1000 for every i and d, the dimensionality of feature space, is 4 or more.

If $100 < m_i < 1000$ for every i, and if $d \geqslant 4$, a *rotation* (or "leave-k-out" or "jack-knife") procedure is likely to prove useful. (These numbers are just guidelines for typical applications; they are not accurate.) In this procedure, k out of the m feature vectors in the data are chosen at random for testing, where $m = \sum_i m_i$. This forms a rotation test set set of size k. The remaining $m - k$ feature vectors are used for designing the classifier. To obtain an adequate set of test feature vectors, the random selection of k feature vectors is repeated, yielding a new rotation set, preferably excluding the feature vectors in the preceding rotation test sets. A new design is obtained for each rotation test set, until the tested feature vectors cover the entire data set.

We suggest that the computation of the operating characteristics and the confidence intervals under these conditions be estimated as follows:

A = number of misclassified members of ω_1 among all the k-vector tests,

B = number of misclassified members of ω_2 among all the k-vector tests,

Z = number of members of ω_1 in the full set of data,

R = number of members of ω_2 in the full set of data.

When the number of feature vectors, m, is less than 100, and the number of features (i.e., the dimensionality of the feature space) is 4 or more, the "leave-one-out" or U procedure is likely to be economical. This is equivalent to a rotation procedure in which $k = 1$ [19]. In this procedure, all but one of the data vectors are used in the design set, and only the single withheld feature vector in the test set. The design process is repeated for every possible partition of the data into $m - 1$ vectors for design and one vector for test.

The rotation method is inherently less accurate in its estimate of the error probability than the U method. On the other hand, the computational cost of the rotation method is $1/k$ of that of the U method. An effective way of exploiting the rotation method to yield an estimate nearly as accurate as that of the U method is sketched below.

If the entire set of data is used for the design of a classifier, the resulting classifier is called a *resubstitution* classifier. The error probability of this classifier is the resubstitution error probability, $P_e(R)$. The error rate of a resubstitution classifier on the data used for the design of that classifier yields an estimate $\hat{P}_e(R)$ of the resubstitution error probability $P_e(R)$. The average of $\hat{P}_e(R)$ over an ensemble of equal-size data sets from a statistically defined population is $E[\hat{P}_e(R)]$. $E[\hat{P}_e(R)]$ is a lower bound for the error probability of any particular resubstitution classifier: i.e., $E[\hat{P}_e(R)] \leqslant P_e(R)$.

Let $E[\hat{P}_e(\pi)]$ denote the average of the error probabilities of the leave-k-out classifiers. $E[\hat{P}_e(\pi)]$ is an upper bound for the error probability of a resubstitution classifier: i.e., $P_e(R) \leqslant E[\hat{P}_e(\pi)]$. Toussaint [19] and Foley [20] have suggested that an improved estimate of the error probability of a resubstitution classifier may be obtained by the formula

$$\hat{P}_e^* = \alpha E[\hat{P}_e(\pi)] + (1 - \alpha)\hat{P}_e(R), \tag{1.31}$$

where α satisfies $0 \leqslant \alpha \leqslant 1$. The value of α depends on the sample size m, the dimensionality d of feature space, and the size k of the test set. Toussaint and Sharpe [21] reported that $\alpha = \frac{1}{2}$ yielded excellent results in an experiment in which $m = 300$ and $k = 30$. In their results, $\hat{P}_e^*$ by the above formula was essentially equal to the estimate of error probability obtained by a leave-one-out analysis. The leave-one-out analysis would have required 300 training processes, while the computation of $\hat{P}_e^*$ by the above formula required only 11 training processes: one training process for $\hat{P}_e(R)$ and ten training processes for $E[\hat{P}_e(\pi)]$.

An extensive overview of these and related matters appears in an article by Kanal [22].

Exercises

For typographical reasons, a column vector will sometimes be written as the transpose of a row vector.

1.1. Demonstrate by an example that $\mathbf{w}$ points to region $\mathscr{R}_2$ if $\mathbf{v}^T\mathbf{Y} > 0$ for augmented feature vectors which are determined by an $\mathbf{X}$ from class ω_2.

1.2. Given the following column vectors

$$\mathbf{w} = \begin{bmatrix} 4 \\ 3 \end{bmatrix}, \qquad \mathbf{x} = \begin{bmatrix} x_1 \\ x_2 \end{bmatrix}.$$

(a) Find $\|\mathbf{w}\|^2$ and $\|\mathbf{w}\|$.
(b) Plot the equation $\mathbf{w}^T\mathbf{x} = -2$ in $\mathbf{x}$ space.

1.3. Given the augmented weight vector $\mathbf{v} = [-1, 1, -1]^T$, find the equation of the hyperplane determined by the intersection of $\mathbf{v}^T\mathbf{y} = 0$ and the plane $y_0 = 1$.

1.4. Given the following column vectors

$$\mathbf{v} = \begin{bmatrix} 2 \\ 4 \\ 3 \end{bmatrix}, \qquad \mathbf{y} = \begin{bmatrix} y_0 \\ y_1 \\ y_2 \end{bmatrix} = \begin{bmatrix} 1 \\ x_1 \\ x_2 \end{bmatrix},$$

plot the intersection of the hyperplane $\mathbf{v}^T\mathbf{y} = 0$ with the plane $y_0 = 1$. (Note that $y_1 = x_1$ and $y_2 = x_2$ and that the intersection will take place in $\mathbf{x} = [x_1, x_2]^T$ space.)

1.5. Demonstrate analytically that $l = -w_0/\|\mathbf{w}\|$, where l denotes the shortest Euclidean distance of the hyperplane $\mathbf{v}^T\mathbf{y} = 0$ from the origin. (See Figure 1.7. Do not use the distance $\mathbf{v}^T\mathbf{y}/\|\mathbf{w}\|$, since this is derived from l.)

1.6. Find expressions for the intersection of the hyperplane $\mathbf{v}^T\mathbf{y} = 0$ with each of the coordinate axes x_1, x_2, and x_3 when $\mathbf{x}$ is a three-component feature vector.

1.7. Given the conditional probability densities

$$p(x|\omega_1) = \begin{cases} \frac{1}{4}, & \text{if } -3 < x < 1, \\ 0, & \text{otherwise}, \end{cases} \qquad p(x|\omega_2) = \begin{cases} \frac{1}{3}, & \text{if } 0 < x < 3, \\ 0, & \text{otherwise}, \end{cases}$$

and the class probabilities $P(\omega_1) = 0.4$ and $P(\omega_2) = 0.6$, plot the constituent densities $f_1(x)$ and $f_2(x)$ on the same axis.

1.8. Given the following constituent densities

$$f_1(\mathbf{x}) = \begin{cases} \frac{1}{8}, & \text{if } -1 < x_1 < 1 \text{ and } 0 < x_2 < 2, \\ 0, & \text{otherwise}, \end{cases}$$

$$f_2(\mathbf{x}) = \begin{cases} \frac{1}{8}, & \text{if } 0 < x_1 < 2 \text{ and } 1 < x_2 < 3, \\ 0, & \text{otherwise}, \end{cases}$$

and the augmented weight vector

$$\mathbf{v} = \begin{bmatrix} -\frac{3}{2} \\ 0 \\ 1 \end{bmatrix}$$

(a) Will a feature vector $\mathbf{x} = [\frac{1}{2}, 2]^T$ from class ω_2 be correctly classified by this $\mathbf{v}$? Show work or clearly state why.
(b) Find the probability of misclassification for the given $\mathbf{v}$.

1.9. Given the constituent densities shown in Figure 1.17.
(a) Find $E(X| -1 < X < 0)$.

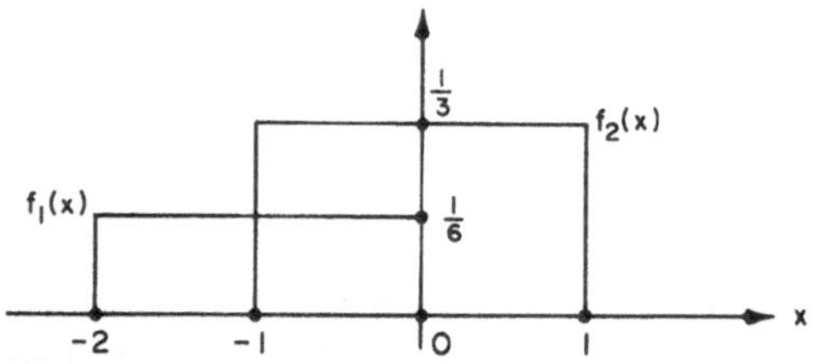

Figure 1.17. Illustration for Exercise 1.9.

(b) Find $E(X \mid -2 < X < 0)$.
(c) Find $P(\text{error} \mid \mathbf{V})$ for a linear classifier when $\mathbf{V} = [-\frac{1}{2}, 1]^T$.

1.10. Given

$$f_2(\mathbf{x}) = \begin{cases} \dfrac{1}{2\pi}, & \text{if } \|\mathbf{x}\| \le 1, \\ 0, & \text{otherwise,} \end{cases} \qquad f_1(\mathbf{x}) = \begin{cases} \dfrac{1}{2\pi}, & \text{if } \left\| \mathbf{x} - \begin{bmatrix} -0.6 \\ 0.8 \end{bmatrix} \right\| \le 1, \\ 0, & \text{otherwise,} \end{cases}$$

$$\mathbf{V} = \begin{bmatrix} 0 \\ -4 \\ -3 \end{bmatrix}$$

(a) Sketch the class regions, the weight vector $\mathbf{w}$, and the hyperplane specified by $\mathbf{v}$.
(b) Find the probability of classification error for the given $\mathbf{v}$.

1.11. Given the following separable, but not linearly separable, constituent densities

$$f_1(\mathbf{x}) = \begin{cases} \frac{1}{8}, & \text{if } |x_1| < 1 \text{ and } |x_2| < 1, \\ 0, & \text{otherwise,} \end{cases}$$

$$f_2(\mathbf{x}) = \begin{cases} \dfrac{1}{10\pi}, & \text{if } 2 < \sqrt{x_1^2 + x_2^2} < 3, \\ 0, & \text{otherwise,} \end{cases}$$

(a) Plot the regions in feature space where class ω_1 and class ω_2 feature vectors can occur.
(b) Indicate a $\boldsymbol{\Phi}(\mathbf{x})$ mapping to ξ-space which will allow for linear separability in ξ-space.
(c) Find a solution vector $\mathbf{V}^*$ for your $\boldsymbol{\Phi}(\mathbf{x})$ mapping.

1.12. Show that the expected value of the error estimate $\hat{b}$, defined in Section 1.10, is exactly probability b.

1.13. Given the following test set:

$\mathscr{X}_1$: (1, 2), (0, 2), (−1, 2), (−2, 2), (0, 1),
(−1, 1), (−3, 1), (−1, 0), (−1, −2), (0, −2)
$\mathscr{X}_2$: (−1, −2), (0, −2), (1, −2), (2, −2), (0, −1),
(1, −1), (3, −1), (1, 0), (2, 1), (1, 2),

the decision surface in x_1, x_2-space is

$$x_1 = x_2.$$

(a) Suppose $P(\omega_1) = P(\omega_2) = \frac{1}{2}$. Find an estimate of the error rate of the decision surface for class ω_1, based on Section 1 10. Find a similar estimate for class ω_2.
(b) What is the 95-percent confidence interval for each of the error rates found in part (a)?
(c) Suppose $P(\omega_1)$ is unknown. Plot the mean operating characteristic for the error rates for class ω_1 and class ω_2. Assume that the decision surface is optimized while lying parallel to its original position as $P(\omega_1)$ changes.
(d) Plot the confidence region for the mean operating characteristic found in part (c).

References

1. R. O. Duda and P. E. Hart, *Pattern Classification and Scene Analysis*. John Wiley & Sons, New York, 1973.
2. W. H. Highleyman, The design and analysis of pattern recognition experiments. *Bell System Technical*, **41** (March): 723–744 (1962).
3. M. A. B. Brazier and W. J. Dixon, Imagery in analytical methods. In *Image Processing in Biological Science*, D. M. Ramsey (ed.), University of California Press, 1968.
4. A. G. Arkadev and E. M. Braverman, *Computers and Pattern Recognition*. Thompson Book Co., Inc., Washington, D.C., 1966.
5. R. P. Kruger, J. R. Townes, D. L. Hall, S. J. Dwyer III, and G. S. Lodwick, Automated radiographic diagnosis via feature extraction and classification of cardiac size and shape descriptors. *IEEE Transactions on Biomedical Engineering*, **BME-19** (3): 174–186 (1972).
6. N. J. Bershad and J. Sklansky, Threshold learning and Brownian motion. *IEEE Trans. on Information Theory*, **IT-17**(3): 350–352 (1971).
7. J. Sklansky, Stability of threshold learning. In *Proc. 1971 IEEE Conference on Decision and Control*, Institute of Electrical and Electronics Engineers, NY, pp. 348–350.
8. J. Sklansky and N. J. Bershad, The dynamics of time-varying threshold learning. *Information and Control*, **15**(6): 455–486 (1969).
9. J. Sklansky, ed., *Pattern Recognition: Introduction and Foundation*, Part I. Dowden, Hutchinson & Ross, Stroudsburg, PA, 1973.
10. N. J. Nilsson, *Learning Machines*. McGraw-Hill, New York, 1965.
11. C. L. Sheng, *Threshold Logic*, Academic Press, New York, 1969.
12. W. S. McCulloch and W. H. Pitts, A logical calculus of the ideas immanent in nervous activity. *Bulletin of Mathematical Biophysics*, **5**: 115–133 (1943).
13. M. L. Minsky, *Computation: Finite and Infinite Machines*. Prentice-Hall, Englewood Cliffs, NJ, 1967.
14. T. M. Cover, Geometrical and statistical properties of systems of linear inequalities with applications in pattern recognition. *IEEE Transactions on Electronic Computers*, **EC-14**(3): 326–334 (1965).
15. B. Widrow, Pattern recognition and adaptive control. *IEEE Transactions on Applications and Industry*, **83**(74): 269–277 (1964).

16. J. Sklansky and H. R. Ramanujam, A stopping rule for threshold learning. *International Journal of Systems Science*, **4**(1): 129–148 (1973).

17. J. A. Swets, *Signal Detection and Recognition by Human Observers*. John Wiley & Sons., New York, 1964, pp. 13–15.

18. H. L. Van Trees, *Detection, Estimation and Modulation Theory, Part I*. John Wiley & Sons, New York, 1968, 36–37.

19. G. T. Toussaint, Bibliography on estimation of misclassification. *IEEE Transactions on Information Theory*, **IT-20** (July): 472–479 (1974).

20. D. H. Foley, Considerations of sample and feature size. *IEEE Transactions on Information Theory*, **IT-18** (Sept.): 618–626 (1972).

21. G. T. Toussaint and P. M. Sharpe, An efficient method for estimating the probability of misclassification applied to a problem in medical diagnosis. School of Computer Science, McGill University, Montreal, P.Q., Canada, Nov. 1973.

22. L. Kanal, Patterns in pattern recognition: 1968–1974. *IEEE Transactions on Information Theory*, **IT-20** (Nov.): 697–722 (1974).

CHAPTER 2

Linearly Separable Classes

2.1 Introduction

Much of the analysis of trainable classifiers is simplified if it is known that every pair of class regions encountered by the classifier can be separated by a linear hypersurface. In augmented feature space such a surface is determined by an equation of the form

$$\mathbf{v}^T\mathbf{y} = 0. \tag{2.1}$$

In this case we say that the pairs of classes are *linearly separable*. More generally, if all the class regions can be separated by a linear classifier (using linear discriminant functions) then the entire set of classes is said to be linearly separable. An augmented weight vector $\mathbf{v}^*$ which separates a given pair of classes is known as a *solution vector*.

One of the early significant results on the subject of training procedures for linearly separable classes was Rosenblatt's error-correction training theorem. This theorem states that if each of two class regions is composed of a finite set of points (feature vectors), and if these two regions are linearly separable, then the proportional increment training procedure will yield a solution in a finite number of trials—i.e., it will yield a vector $\mathbf{v} = \mathbf{v}^*$ satisfying Equation (2.1). (This theorem is discussed in detail in Section 2.4). Thus in the design of training procedures it is useful to determine at the outset whether or not the given classes are linearly separable.

In this chapter we describe the proportional increment training procedure and other related training procedures, along with demonstrations that these training procedures converge to solutions when solutions exist. We also describe iterative testing procedures that determine whether or not a pair of finite classes (i.e. classes consisting of a finite number of elements) is linearly separable, and—if linear separability is confirmed—for finding a solution weight vector $\mathbf{v}^*$.

A central problem of the theory of linear classifiers is that of finding algorithms for choosing the weights of the discriminant functions so as to nearly maximize or completely maximize the expected number of correctly classified patterns. We refer to this problem as the problem of *synthesizing* the classifiers, and the solution to this problem as a *synthesis technique*. When a pair of classes is linear separable, an acceptable synthesis technique is usually one that finds a solution vector.

Many of the known synthesis techniques are forms of training procedures, as a result of the fact that the required size of the training set depends on the shapes and interrelationships of the class regions in feature space, and because these shapes and interrelationships are initially unknown or very imprecisely specified. In subsequent sections of this chapter we derive several of these training procedures, starting from the basic necessary and sufficient conditions for linear separability of the classes in the training sets.

To facilitate our demonstrations of the validity of the various training and testing procedures in this chapter, we derive in the next two sections some of the basic mathematics of linear separability.

2.2 Convex Sets, Summability, and Linear Separability

In this section we derive relationships among the following concepts: convex sets, summability, and linear separability. These relationships contribute to our ability to determine whether a given training set on two classes is linearly separable.

2.2.1 Convex Sets and Linear Separability

A set of vectors $\mathscr{X}$ is said to be *convex* if a straight line segment joining any pair of vectors in $\mathscr{X}$ is entirely contained in $\mathscr{X}$. Thus $\mathscr{X}$ is convex if and only if for $\mathbf{x}_1, \mathbf{x}_2 \in \mathscr{X}$ and $0 \leqslant b \leqslant 1$ we have

$$b\mathbf{x}_1 + (1 - b)\mathbf{x}_2 \in \mathscr{X}$$

The *convex hull* of $\mathscr{X}$, which we denote by $\mathscr{C}\mathscr{X}$, is the intersection of all convex sets containing $\mathscr{X}$ [1]. Since any intersection of convex sets is convex, $\mathscr{C}\mathscr{X}$ is convex. If $\mathscr{X}$ is a finite set of k vectors, $\{\mathbf{x}(1), \ldots, \mathbf{x}(k)\}$, then $\mathscr{C}\mathscr{X}$ is the set of all vectors $\boldsymbol{\xi}$ such that

$$\boldsymbol{\xi} = \sum_{i=1}^{k} b_i\mathbf{x}(i), \qquad \mathbf{b} \geqslant \mathbf{0}, \qquad \sum_{i=1}^{k} b_i = 1, \tag{2.2}$$

where $\mathbf{b}$ and $\mathbf{0}$ denote the vectors $[b_1, \ldots, b_k]^T$ and $[0, \ldots, 0]^T$, respectively, and $\mathbf{b} \geqslant \mathbf{0}$ means that every component of $\mathbf{b}$ is nonnegative. The linear sum above is called a *convex sum* of the vectors in $\mathscr{X}$.

Caratheodory noted that if the dimensionality of each $\mathbf{x}(i)$ in the above set of equations is d, then every ξ in $\mathscr{C}\mathscr{X}$ may be expressed as a convex sum of $d + 1$ or fewer members of $\mathscr{X}$, regardless of whether $|\mathscr{X}|$, the number of members of $\mathscr{X}$, is finite or infinite (see page 201, Theorem 6 of [1]). Thus the number of nonzero terms in the convex sum in Equation 2.2 need never exceed $d + 1$. We express this observation in the following theorem.

Theorem 2.1. *If $\mathscr{X}$ is any set (not necessarily finite) of position vectors in R^d, then every vector in $\mathscr{C}\mathscr{X}$ is a convex sum of $d + 1$ or fewer position vectors in $\mathscr{X}$.*

It is useful to note that if $\mathscr{X}$ is a finite set of points (position vectors) in the Euclidean plane R^2 (the set of all pairs of real numbers), then $\mathscr{C}\mathscr{X}$ is a convex polygon whose vertices are points in $\mathscr{X}$. The angle between two adjacent edges of $\mathscr{C}\mathscr{X}$ is less than or equal to π radians.

These concepts are illustrated in Figure 2.1. The small circles in this figure denote the members of $\mathscr{X}_1$, and the dots denote members of $\mathscr{X}_2$. The polygonal curves represent the boundaries of $\mathscr{C}\mathscr{X}_1$ and $\mathscr{C}\mathscr{X}_2$. This figure suggests that a hyperplane can separate two finite sets $\mathscr{X}_1$ and $\mathscr{X}_2$ if and only if their convex hulls do not intersect. We prove this below (see Corollary 2.2.1).

If $\mathscr{H}$ is a $(d - 1)$-dimensional hyperplane in R^d, then $R^d - \mathscr{H}$ is the disjoint union of two convex sets. Let $\mathscr{A}$, $\mathscr{B}$ denote these two convex sets. If a set $\mathscr{X}_1$ is contained in $\mathscr{A}$, then $\mathscr{C}\mathscr{X}_1 \subseteq \mathscr{A}$. Similarly if $\mathscr{X}_2 \subseteq \mathscr{B}$, then $\mathscr{C}\mathscr{X}_2 \subseteq \mathscr{B}$. Thus if $\mathscr{X}_1$ and $\mathscr{X}_2$ are linearly separable, then $\mathscr{C}\mathscr{X}_1 \cap \mathscr{C}\mathscr{X}_2 = \varnothing$. The next theorem is a partial converse of this result. To prepare the reader for that theorem, we define closed and bounded sets of points.

A set of points is *closed* if it contains the limit of every convergent sequence of points in the set. Thus a closed set contains its boundary. A set of points is *bounded* if there exists a positive number M such that $\|\mathbf{x}\| < M$ for every point $\mathbf{x}$ in the set.

Examples. The region $\|\mathbf{x}\| < 5$ is bounded but not closed. The region in R^d specified by $x_1 \geqslant 0$ is closed but not bounded. The region $x_1 > 0$ is neither bounded nor closed. A set $\mathscr{U}$ is closed and bounded if it is the convex hull of a finite set contained in R^d.

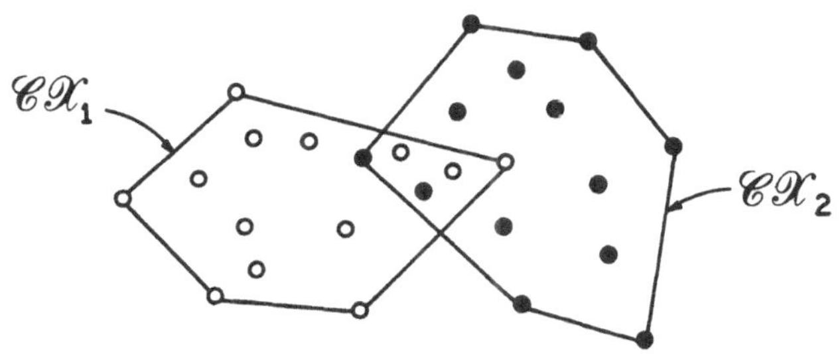

Figure 2.1. A pair of intersecting convex hulls of class regions in feature space.

Theorem 2.2. *If $\mathscr{U}$ and $\mathscr{V}$ are closed and bounded convex sets in R^d, and $\mathscr{U} \cap \mathscr{V} = \varnothing$, then $\mathscr{U}$ and $\mathscr{V}$ are linearly separable.*

PROOF. Since $\mathscr{U}$ and $\mathscr{V}$ are closed and bounded, there are vectors $\mathbf{u} \in \mathscr{U}$ and $\mathbf{v} \in \mathscr{V}$ such that

$$0 < \|\mathbf{u} - \mathbf{v}\| = \inf\{\|\mathbf{u}' - \mathbf{v}'\| \,|\, \mathbf{u}' \in \mathscr{U} \text{ and } \mathbf{v}' \in \mathscr{V}\}. \tag{2.3}$$

Thus $\mathbf{u}$ and $\mathbf{v}$ are not further apart than any other pair $(\mathbf{u}', \mathbf{v}' \,|\, \mathbf{u} \in \mathscr{U}, \mathbf{v}' \in \mathscr{V})$. Let $\mathscr{H}$ denote the $(d-1)$-dimensional hyperplane that is the perpendicular bisector of the line segment joining $\mathbf{u}$ to $\mathbf{v}$, as illustrated in Figure 2.2, and let $\mathbf{p}$ denote the intersection of $\mathscr{H}$ with that line segment. We prove below that $\mathscr{H}$ separates $\mathscr{U}$ and $\mathscr{V}$ by showing that if $\mathscr{H} \cap \mathscr{U} \neq \varnothing$ or $\mathscr{H} \cap \mathscr{V} \neq \varnothing$ then we would have a contradiction of our hypothesis, Inequality (2.3).

Since $\mathscr{H}$ separates $\mathbf{u}$ and $\mathbf{v}$, $\mathscr{U}$ and $\mathscr{V}$ cannot both be on the same side of $\mathscr{H}$. Consequently, if $\mathscr{H}$ did not separate $\mathscr{U}$ and $\mathscr{V}$, then either $\mathscr{U}$ or $\mathscr{V}$, say $\mathscr{U}$, would need to intersect $\mathscr{H}$. Assuming $\mathscr{U} \cap \mathscr{H} \neq \varnothing$, let $\mathbf{q}$ denote a vector belonging to $\mathscr{U} \cap \mathscr{H}$. Since $\mathscr{U}$ is convex, the line segment $b\mathbf{u} + (1-b)\mathbf{q}$, $0 \leqslant b \leqslant 1$, is contained in $\mathscr{U}$. We now show that we can find b_0 such that for $\mathbf{r} = b_0\mathbf{u} + (1-b_0)q$, $0 \leqslant b_0 < 1$, we will have $\|\mathbf{r} - \mathbf{p}\| < \|\mathbf{u} - \mathbf{p}\|$. (See Figure 2.2.) To find $\mathbf{r}$ we have to show that there are values of b, $0 \leqslant b < 1$, satisfying the inequality

$$\|b\mathbf{u} + (1-b)\mathbf{q} - \mathbf{p}\| < \|\mathbf{u} - \mathbf{p}\|. \tag{2.4}$$

Since a translation does not change lengths of vectors or orthogonality relations among directed line segments, we can replace every vector $\mathbf{w}$ by $\mathbf{w}' = \mathbf{w} - \mathbf{p}$ without affecting the validity of the above inequality. Since $\mathbf{u}'$ and $\mathbf{q}'$ are orthogonal, we have, by the Pythagorean theorem,

$$\|b\mathbf{u}' + (1-b)\mathbf{q}'\|^2 = b^2\|\mathbf{u}'\|^2 + (1-b)^2\|\mathbf{q}'\|^2.$$

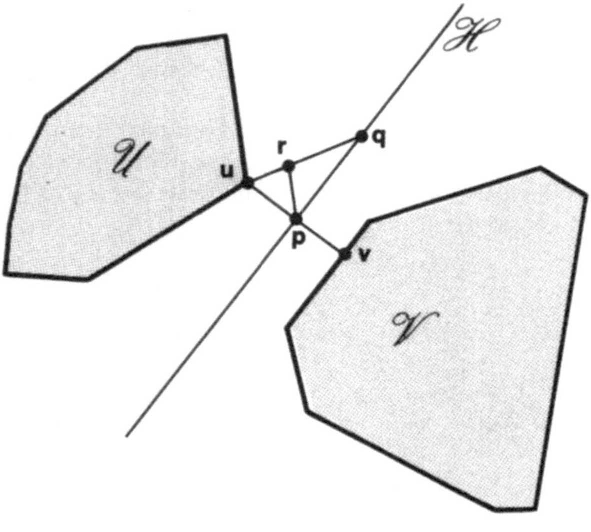

Figure 2.2. Illustration for proof of Theorem 2.2.

Thus there are values of b, $0 \leqslant b < 1$, satisfying (2.4) if and only if there are values of b, $0 \leqslant b < 1$, satisfying

$$b^2\|\mathbf{u}'\|^2 + (1-b)^2\|\mathbf{q}'\|^2 < \|\mathbf{u}'\|^2$$

or

$$\frac{\|\mathbf{q}'\|^2}{\|\mathbf{u}'\|^2} < \frac{1+b}{1-b}. \tag{2.5}$$

With $b > 0$, the right-hand side of the Inequality (2.5) becomes indefinitely large as b approaches 1. Consequently, there are values of b satisfying both $0 < b < 1$ and Inequality (2.5). For these values of b Inequality (2.4) is also satisfied. Recalling that $\mathbf{u}$, $\mathbf{p}$, and $\mathbf{v}$ are collinear, it follows that

$$\begin{aligned}\|\mathbf{u}-\mathbf{v}\| &= \|\mathbf{u}-\mathbf{p}\| + \|\mathbf{p}-\mathbf{v}\| \\ &> \|\mathbf{r}-\mathbf{p}\| + \|\mathbf{p}-\mathbf{v}\| \\ &\geqslant \|\mathbf{r}-\mathbf{v}\|.\end{aligned}$$

That is,

$$\|\mathbf{u}-\mathbf{v}\| > \|\mathbf{r}-\mathbf{v}\|,$$

which contradicts our hypothesis, Inequality (2.3). Thus $\mathscr{U} \cap \mathscr{H}$ must be empty. Similarly, $\mathscr{V} \cap \mathscr{H}$ must be empty. Therefore $\mathscr{H}$ separates $\mathscr{U}$ and $\mathscr{V}$. □

Since $\mathscr{C}\mathscr{X}$ is closed and bounded if $\mathscr{X}$ is closed and bounded, Theorem 2.2 implies the following corollary.

Corollary 2.2.1. *If $\mathscr{X}_1$ and $\mathscr{X}_2$ are closed and bounded sets in R^d, and $\mathscr{C}\mathscr{X}_1 \cap \mathscr{C}\mathscr{X}_2 = \varnothing$, then $\mathscr{X}_1$ and $\mathscr{X}_2$ are linearly separable.*

EXAMPLE 2.1. If $\mathscr{X}_1 = \{[x_1, x_2]^T \mid x_1^2 + x_2^2 < 1\}$, and $\mathscr{X}_2 = \{[1,0]^T\}$, then $\mathscr{X}_1$ is not closed, $\mathscr{C}\mathscr{X}_1 = \mathscr{X}_1$, $\mathscr{C}\mathscr{X}_2 = \mathscr{X}_2$, and $\mathscr{C}\mathscr{X}_1 \cap \mathscr{C}\mathscr{X}_2 = \varnothing$, but $\mathscr{X}_1$ and $\mathscr{X}_2$ are not linearly separable.

EXAMPLE 2.2. If $\mathscr{X}_1 = \{[x_1, x_2]^T \mid x_1 x_2 \geqslant 1,\ x_1 > 0\}$ and $\mathscr{X}_2 = \{[0, x_2]^T \mid x_2 \geqslant 0\}$, then $\mathscr{X}_1$ and $\mathscr{X}_2$ are both closed but not bounded, $\mathscr{C}\mathscr{X}_1 = \mathscr{X}_1$, $\mathscr{C}\mathscr{X}_2 = \mathscr{X}_2$, $\mathscr{C}\mathscr{X}_1 \cap \mathscr{C}\mathscr{X}_2 = \varnothing$, but $\mathscr{X}_1$ and $\mathscr{X}_2$ are not linearly separable.

2.2.2 Summability and Linear Separability

For a specified positive integer value of k, two sets of vectors, $\mathscr{X}_1$ and $\mathscr{X}_2$, are said to be *k-summable* if there are k vectors, not necessarily distinct, in $\mathscr{X}_1$ and another k vectors, not necessarily distinct, in $\mathscr{X}_2$ such that

$$\sum_{i=1}^{k} \mathbf{x}^{(1)}(i) = \sum_{j=1}^{k} \mathbf{x}^{(2)}(j).$$

To indicate the allowed repetitions of vectors in these sums this definition may be restated as follows. $\mathscr{X}_1$ and $\mathscr{X}_2$ are *k-summable* if there are positive integers $m_1, \ldots, m_r, n_1, \ldots, n_s$, and vectors $\mathbf{x}^{(1)}(1), \ldots, \mathbf{x}^{(1)}(r) \in \mathscr{X}_1$,

$\mathbf{x}^{(2)}(1), \ldots, \mathbf{x}^{(2)}(s) \in \mathscr{X}_2$ such that

$$\sum_{i=1}^{r} m_i \mathbf{x}^{(1)}(i) = \sum_{j=1}^{s} n_j \mathbf{x}^{(2)}(j)$$

and

$$k = \sum_{i=1}^{r} m_i = \sum_{j=1}^{s} n_j.$$

Two sets of vectors are *summable* if there is some integer $k > 0$ for which the two sets are k-summable. The two sets of vectors are said to be *k-asummable* if they are not k-summable. The sets are *asummable* if they are not summable.

The following theorem relates asummability to linear separability.

Theorem 2.3. *Let $\mathscr{X}_1$ and $\mathscr{X}_2$ denote two sets of vectors in the Euclidean d-dimensional space of real numbers R^d. Suppose every component of each vector in $\mathscr{X}_1$ and $\mathscr{X}_2$ is rational. Then $\mathscr{X}_1$ and $\mathscr{X}_2$ are summable if and only if $\mathscr{C}\mathscr{X}_1 \cap \mathscr{C}\mathscr{X}_2 \neq \varnothing$.*

PROOF. If $\mathscr{X}_1$ and $\mathscr{X}_2$ are summable then there are k vectors $\mathbf{x}^{(1)}(1), \ldots, \mathbf{x}^{(1)}(k)$ in $\mathscr{X}_1$ and k vectors $\mathbf{x}^{(2)}(1), \ldots, \mathbf{x}^{(2)}(k)$ in $\mathscr{X}_2$ such that

$$\sum_{i=1}^{k} \mathbf{x}^{(1)}(i) = \sum_{j=1}^{k} \mathbf{x}^{(2)}(j).$$

Then

$$\sum_{i=1}^{k} k^{-1} \mathbf{x}^{(1)}(i) = \sum_{j=1}^{k} k^{-1} \mathbf{x}^{(2)}(j).$$

Note that $\sum_{i=1}^{k} k^{-1} = 1$. Thus

$$\sum_{i=1}^{k} k^{-1} \mathbf{x}^{(1)}(i), \qquad \sum_{j=1}^{k} k^{-1} \mathbf{x}^{(2)}(j) \in \mathscr{C}\mathscr{X}_1 \cap \mathscr{C}\mathscr{X}_2,$$

which proves that if $\mathscr{X}_1$ and $\mathscr{X}_2$ are summable then $\mathscr{C}\mathscr{X}_1 \cap \mathscr{C}\mathscr{X}_2 \neq \varnothing$.

Suppose $\mathscr{C}\mathscr{X}_1 \cap \mathscr{C}\mathscr{X}_2 \neq \varnothing$. Then there are vectors $\{\mathbf{x}^{(1)}(i)\}$ in $\mathscr{X}_1$ and $\{\mathbf{x}^{(2)}(j)\}$ in $\mathscr{X}_2$ such that

(i) $$\sum_{i=1}^{r} b_i \mathbf{x}^{(1)}(i) = \sum_{j=1}^{s} c_j \mathbf{x}^{(2)}(j)$$

(ii) $$\sum_{i=1}^{r} b_i = \sum_{j=1}^{s} c_j = 1.$$

(iii) $$0 < b_i, c_j \quad \text{for } 1 \leqslant i \leqslant r \text{ and } 1 \leqslant j \leqslant s.$$

If the numbers b_i, c_j are rational for all i, j, let k denote their least common denominator. Then kb_i, kc_j are positive integers for all i, j. Let $m_i = kb_i$, $n_j = kc_j$. Multiplying (i) and (ii) by k, we obtain

$$\sum_{i=1}^{r} m_i \mathbf{x}^{(1)}(i) = \sum_{j=1}^{s} n_j \mathbf{x}^{(2)}(j)$$

or

$$\sum_{i=1}^{r} m_i = \sum_{j=1}^{s} n_j = k.$$

Hence $\mathscr{X}_1$ and $\mathscr{X}_2$ are k-summable.

If not all of the numbers b_i, c_j are rational, there are rational numbers b_i^*, c_j^* arbitrarily close to b_i, c_j such that (i), (ii), (iii) are satisfied. We show this below.

Let

$$\mathbf{C} \triangleq \begin{bmatrix} -1 & \cdots & -1 & 1 & \cdots & 1 \\ -\mathbf{x}^{(1)}(1) & \cdots & -\mathbf{x}^{(1)}(r) & \mathbf{x}^{(2)}(1) & \cdots & \mathbf{x}^{(2)}(s) \end{bmatrix}$$

$$\mathbf{d} \triangleq \begin{bmatrix} d_1^{(1)} \\ \vdots \\ d_r^{(1)} \\ d_1^{(2)} \\ \vdots \\ d_s^{(2)} \end{bmatrix}.$$

Using the above notation, (i), (ii), (iii) are equivalent to

$$\mathbf{Cd} = \mathbf{0} \quad \text{and} \quad \mathbf{d} > \mathbf{0} \tag{2.6}$$

where $\mathbf{d} > \mathbf{0}$ means that every component of $\mathbf{d}$ exceeds 0. Now we derive a vector $\mathbf{d}^*$ all of whose components are rational and which satisfies

$$\mathbf{Cd}^* = \mathbf{0} \quad \text{and} \quad \mathbf{d}^* > \mathbf{0}.$$

By the Gauss–Jordan elimination method [2] one obtains matrices $\mathbf{F}$, $\mathbf{P}$, and $\mathbf{R}$ such that either

$$\mathbf{FCP} = \begin{bmatrix} \mathbf{I} & \mathbf{R} \\ \mathbf{0} & \mathbf{0} \end{bmatrix}$$

or

$$\mathbf{FCP} = [\mathbf{I} \quad \mathbf{R}],$$

where $\mathbf{F}$ is a $(d + 1) \times (d + 1)$ invertible matrix, $\mathbf{P}$ is an $(r + s) \times (r + s)$ invertible matrix obtained by permuting the columns of the $(r - s) \times (r + s)$ identity matrix, and $\mathbf{I}$ is an identity matrix.

A derivation of the Gauss–Jordan elimination procedure is given in Section A.2 of Appendix A. In that derivation we show that the elements of $\mathbf{F}$ are rational if the elements of $\mathbf{C}$ are rational, and that the elements of $\mathbf{P}$ are 0's and 1's. By our hypothesis, all the elements of $\mathbf{C}$ are rational. Hence, all the elements of $\mathbf{R}$ are rational.

Since $\mathbf{Cd} = \mathbf{0}$, it follows that

$$\mathbf{FCPP}^{-1}\mathbf{d} = \mathbf{0}.$$

From the discussion in Section A.2, we see that $\mathbf{P}^{-1}\mathbf{d}$ is a vector obtained by permuting the elements of $\mathbf{d}$. Hence $\mathbf{P}^{-1}\mathbf{d} > \mathbf{0}$. Let $\mathbf{g}$ denote an $(r + s)$-vector

$$\mathbf{g} = [g_1, \ldots, g_{r+s}]^T.$$

Let k denote the number of columns in $\mathbf{I}$. If

$$\mathbf{FCPg} = \begin{bmatrix} \mathbf{I} & \mathbf{R} \\ \mathbf{0} & \mathbf{0} \end{bmatrix} \mathbf{g} = \mathbf{0}$$

or

$$\mathbf{FCPg} = [\mathbf{I} \quad \mathbf{R}]\mathbf{g} = \mathbf{0}$$

and if $g_{k+1}, \ldots, g_{r+s}$ are specified arbitrarily, then $g_1, \ldots, g_k$ are uniquely determined. Suppose $g_{k+1}, \ldots, g_{r+s}$ are rational. Then since the elements of $\mathbf{R}$ are rational, it follows that $g_1, \ldots, g_k$ are rational.

The continuity of the arithmetic operations, the fact that irrational numbers can be approximated arbitrarily closely by rational numbers, and the fact that $\mathbf{P}^{-1}\mathbf{d} > \mathbf{0}$ imply that we can choose rational numbers $g_{k+1}, \ldots, g_{r+s}$ sufficiently close to the last $r + s - k$ elements of $\mathbf{P}^{-1}\mathbf{d}$ so that not only are $g_{k+1}, \ldots, g_{r+s}$ positive but so also are $g_1, \ldots, g_k$. Thus, choosing $\mathbf{g}$ in this way, we have

$$\mathbf{g} > \mathbf{0}.$$

Let $\mathbf{d}^* = \mathbf{Pg}$. Since $\mathbf{P}$ permutes the elements of $\mathbf{g}$, it follows that $\mathbf{d}^* > \mathbf{0}$. Since $\mathbf{FCPg} = \mathbf{0}$, it follows that

$$\mathbf{FCd}^* = \mathbf{0}.$$

Hence $\mathbf{F}^{-1}(\mathbf{FCd}^*) = \mathbf{Cd}^* = \mathbf{0}$. Thus we found $\mathbf{d}^* > \mathbf{0}$ such that $\mathbf{Cd}^* = \mathbf{0}$. Hence $\mathscr{X}_1$ and $\mathscr{X}_2$ are summable. □

The following example demonstrates that the restriction in the above theorem on the rationality of the components of $\mathscr{X}_1$ and $\mathscr{X}_2$ cannot be removed without invalidating the theorem.

EXAMPLE 2.3. Let $\mathscr{X}_1 = \{[1]\}$, and $\mathscr{X}_2 = \{[\pi], [0]\}$. Then $\mathscr{C}\mathscr{X}_1 \cap \mathscr{C}\mathscr{X}_2 = \{[1]\}$. But $\mathscr{X}_1$ and $\mathscr{X}_2$ cannot be summable because

$$a[1] = b[\pi] + c[0]$$

with a and b rational would imply that π is rational.

EXAMPLE 2.4. Consider a 3×3 array of integers $x_1, \ldots, x_9$ arranged according to Figure 2.3. Let the vector

$$\mathbf{x} = [x_1, \ldots, x_9]^T$$

represent this array.

Let $\mathbf{x}_H(i)$ denote an instance of $\mathbf{x}$ representing an array consisting of 1's along one row and 0's everywhere else. Note that there are only three such arrays. We represent them by $\mathbf{x}_H(1)$, $\mathbf{x}_H(2)$, $\mathbf{x}_H(3)$. Let $\mathscr{H}$ denote the set $\{\mathbf{x}_H(i) | i = 1, 2, 3\}$.

Let $\mathbf{x}_V(i)$ denote an instance of $\mathbf{x}$ representing an array consisting of 1's along one column and 0's everywhere else. There are three such vectors: $\mathbf{x}_V(1)$, $\mathbf{x}_V(2)$, $\mathbf{x}_V(3)$. Let $\mathscr{V}$ denote this set of three vectors.

x_1	x_2	x_3
x_4	x_5	x_6
x_7	x_8	x_9

Figure 2.3. Illustration for Example 2.4.

Table 2.1.

$\mathscr{H}$	$\mathscr{V}$
$[1,1,1,0,0,0,0,0,0]^T$	$[1,0,0,1,0,0,1,0,0]^T$
$[0,0,0,1,1,1,0,0,0]^T$	$[0,1,0,0,1,0,0,1,0]^T$
$[0,0,0,0,0,0,1,1,1]^T$	$[0,0,1,0,0,1,0,0,1]^T$

Problem: Are $\mathscr{H}$ and $\mathscr{V}$ linearly separable?

To see the solution to this problem, arrange the vectors in $\mathscr{H}$ and $\mathscr{V}$ as in Table 2.1. By examining the sums of all possible pairs of vectors from $\mathscr{H}$ and pairs of vectors from $\mathscr{V}$, we find that $\mathscr{H}$ and $\mathscr{V}$ are not 2-summable. But the sum of all the members of $\mathscr{H}$ equals the sum of all the members of $\mathscr{V}$, namely $[1,1,1,1,1,1,1,1,1]^T$. Hence $\mathscr{H}$ and $\mathscr{V}$ are 3-summable. This proves that $\mathscr{H}$ and $\mathscr{V}$ are not linearly separable.

When the dimensionality of feature space is seven or less, then the test for asummability reduces to a test for 2-asummability [3, 4]. The 2-asummability test consists of examining the columns of **C** four at a time to see whether they sum to zero.

When the number of features is large, however, a more general k-asummability test is required. The direct application of this test in this situation is often tedious, since it involves an exhaustive examination of all possible k-term sums. Alternative tests that in many cases overcome this inconvenience are described in Section 2.7.

The theorems of this section may be combined as follows.

Theorem 2.4. *If $\mathscr{X}_1$, $\mathscr{X}_2$ are finite sets of rational vectors (i.e., vectors all of whose components are rational), then the following properties of $\mathscr{X}_1$ and $\mathscr{X}_2$ are equivalent.*

(a) *$\mathscr{X}_1$ and $\mathscr{X}_2$ are linearly separable.*

(b) $\mathscr{C}\mathscr{X}_1 \cap \mathscr{C}\mathscr{X}_2 = \varnothing$.

(c) *$\mathscr{X}_1$ and $\mathscr{X}_2$ are asummable.*

(d) *There is no rational positive nonzero vector* $\mathbf{d}$ *that satisfies* $\mathbf{C}^T\mathbf{d} = \mathbf{0}$.

If the vectors in $\mathscr{X}_1$, $\mathscr{X}_2$ are not necessarily rational then the following properties of $\mathscr{X}_1$ and $\mathscr{X}_2$ are equivalent.

(a) *$\mathscr{X}_1$ and $\mathscr{X}_2$ are linearly separable.*

(b) $\mathscr{C}\mathscr{X}_1 \cap \mathscr{C}\mathscr{X}_2 = \varnothing$.

2.3 Notation and Terminology

In this section we recapitalate our vector notation for linear classifiers. We introduce new terminology—matrices and switching functions—and relate this terminology to linear separability and asummability.

We use the notation

$$\mathbf{x} = (x_1, \ldots, x_d)^T$$

for a d-dimensional feature vector, and

$$\mathbf{w} = (w_1, \ldots, w_d)^T$$

for a d-dimensional weight vector. For mathematical convenience we define the *augmented feature vector*

$$\mathbf{y} = [1, x_1, \ldots, x_d]^T$$

and the *augmented weight vector*

$$\mathbf{v} = [w_0, \ldots, w_d]^T.$$

If two classes ω_1, ω_2 are linearly separable, there is an augmented weight vector $\mathbf{v}$ such that

$$\mathbf{v}^T\mathbf{y}\begin{cases} <0 & \text{for } \mathbf{x} \in \omega_1, \\ >0 & \text{for } \mathbf{x} \in \omega_2. \end{cases} \tag{2.7}$$

Here the equation

$$\mathbf{v}^T\mathbf{y} = 0$$

represents a hyperplane that separates ω_1 and ω_2. Using the sign-normalized augmented feature vector, $\boldsymbol{\eta}$, defined as

$$\boldsymbol{\eta} = \begin{cases} -\mathbf{y} & \text{for } \mathbf{x} \in \omega_1, \\ \mathbf{y} & \text{for } \mathbf{x} \in \omega_2, \end{cases}$$

we may represent the pair of inequalities (2.7) by the single inequality

$$\mathbf{v}^T\boldsymbol{\eta} > 0.$$

Suppose $(\mathscr{X}_1, \mathscr{X}_2)$ is a finite training set such that $\mathscr{X}_1 \subseteq \omega_1$ and $\mathscr{X}_2 \subseteq \omega_2$. Then the linear separability of $\mathscr{X}_1$ and $\mathscr{X}_2$ is equivalent to the existence of an augmented weight vector $\mathbf{v}$ such that

$$\mathbf{v}^T\boldsymbol{\eta} > 0 \quad \text{for all } \mathbf{x} \in \mathscr{X}_1 \cup \mathscr{X}_2. \tag{2.8}$$

If Equation (2.7) is satisfied we say that the vectors

$$\{\boldsymbol{\eta} \,|\, \mathbf{x} \in \mathscr{X}_1 \cup \mathscr{X}_2\}$$

are *linearly contained.* Inequality (2.8) is identical to

$$\boldsymbol{\eta}^T\mathbf{v} > 0 \quad \text{for all } \mathbf{x} \in \mathscr{X}_1 \cup \mathscr{X}_2. \tag{2.9}$$

If the training set $\mathscr{X}_1 \cup \mathscr{X}_2$ consists of m vectors $\mathbf{x}(1), \ldots, \mathbf{x}(m)$, the first k of which belong to $\mathscr{X}_1$ and the remaining $m - k$ belong to $\mathscr{X}_2$, then Inequality (2.9) may be expressed as follows.

$$\begin{bmatrix} -1, & -\mathbf{x}^T(1) \\ \vdots & \vdots \\ -1, & -\mathbf{x}^T(k) \\ 1, & \mathbf{x}^T(k+1) \\ \vdots & \vdots \\ 1, & \mathbf{x}^T(m) \end{bmatrix} \mathbf{v} = \begin{bmatrix} \boldsymbol{\eta}^T(1) \\ \vdots \\ \boldsymbol{\eta}^T(m) \end{bmatrix} \mathbf{v} > \begin{bmatrix} 0 \\ \vdots \\ 0 \end{bmatrix}.$$

Let

$$\mathbf{A} \triangleq \begin{bmatrix} \boldsymbol{\eta}^T(1) \\ \vdots \\ \boldsymbol{\eta}^T(m) \end{bmatrix} \quad \text{and} \quad \mathbf{t} \triangleq \begin{bmatrix} t \\ \vdots \\ t \end{bmatrix}.$$

By dividing the elements of $\mathbf{v}$ by a suitably small positive number we obtain, for any $t > 0$,

$$\mathbf{A}\mathbf{v} \geqslant \mathbf{t} > \mathbf{0}.$$

Thus $\mathscr{X}_1$ and $\mathscr{X}_2$ are linearly separable (i.e., $\mathscr{C}\mathscr{X}_1 \cap \mathscr{C}\mathscr{X}_2 = \varnothing$) if and only if there exists a vector $\mathbf{v}$ satisfying

$$\mathbf{A}\mathbf{v} \geqslant \mathbf{1},$$

where $\mathbf{1} \triangleq [1, \ldots, 1]^T$. The equation $\mathbf{v}^T\mathbf{y} = 0$ represents a hyperplane in $\mathbf{x}$-space that separates $\mathscr{X}_1$ and $\mathscr{X}_2$. Recalling Equation (2.6), we note that if a vector $\mathbf{b}$ satisfies

$$\mathbf{A}^T\mathbf{b} = \mathbf{0}, \qquad \mathbf{b} \geqslant \mathbf{0}, \qquad \mathbf{b} \neq \mathbf{0}, \tag{2.10}$$

then $\mathscr{C}\mathscr{X}_1 \cap \mathscr{C}\mathscr{X}_2 \neq \varnothing$.

Thus to test whether $\mathscr{X}_1$ and $\mathscr{X}_2$ are linearly separable, one may search for a nonnegative nonzero vector $\mathbf{b}$ satisfying $\mathbf{A}^T\mathbf{b} = \mathbf{0}$. If such a $\mathbf{b}$ can be found then $\mathscr{X}_1$ and $\mathscr{X}_2$ are not linearly separable. If no such $\mathbf{b}$ exists, then $\mathscr{X}_1$ and $\mathscr{X}_2$ are linearly separable.

From a practical standpoint it is realistic to assume that whenever ω_1 and ω_2 are linearly separable some variation in $\mathbf{v}$ can take place while satisfying Equation (2.9). This is illustrated in Figure 2.4. In this figure the two straight lines, one solid and one dashed, represent two distinct hyperplanes both of which separate ω_1 and ω_2 perfectly. It follows from Theorem 2.3 that if all the elements of $\mathbf{A}$ and $\mathbf{b}$ in Equation (2.10) are rational then $\mathscr{X}_1$

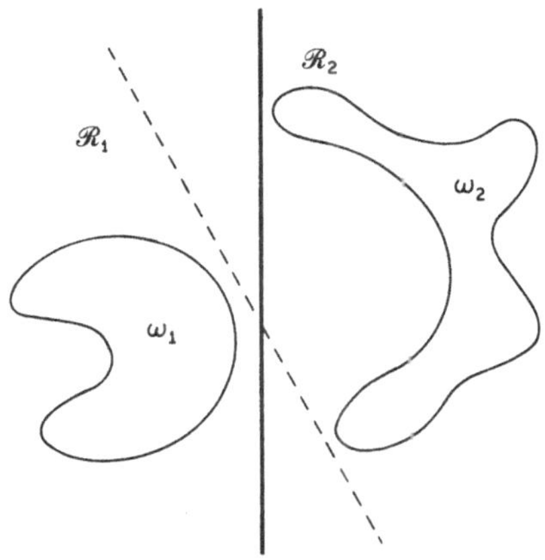

Figure 2.4. Two straight lines separating the classes ω_1 and ω_2.

and $\mathscr{X}_2$ are summable, i.e., $\mathbf{b}$ can be chosen so that all its elements are non-negative integers and satisfy Equation (2.10).

Let $U(\mathbf{x})$ denote the following characteristic function on $\mathscr{X}_1 \cup \mathscr{X}_2$:

$$U(\mathbf{x}) = \begin{cases} 0 & \text{for } \mathbf{x} \in \mathscr{X}_1, \\ 1 & \text{for } \mathbf{x} \in \mathscr{X}_2, \end{cases}$$

assuming that $\mathscr{X}_1 \cap \mathscr{X}_2 = \varnothing$. We say that $U(\mathbf{x})$ is summable, asummable, or k-summable if $\mathscr{X}_1$ and $\mathscr{X}_2$ are summable, asummable or k-summable, respectively.

If the elements of $\mathbf{x}$ are restricted to 0's and 1's then $U(\mathbf{x})$ is a *switching function*, representing a binary-input classifier (see Section 1.7). In many such classifiers $U(\mathbf{x})$ is a *partially specified* switching function—i.e., a switching function defined over a proper subset of $\{0, 1\}^d$, the set of all d-dimensional binary vectors. Example 2.4 illustrates a partially specified switching function $U(\mathbf{x})$ if we define $U(\mathbf{x}) = 1$ for $\mathbf{x} \in \mathscr{H}$, 0 for $\mathbf{x} \in \mathscr{V}$. The test for asummability often is applicable to partially specified switching functions, as in Example 2.4.

We say a switching function $U(\mathbf{x})$ is realizable by a *threshold gate* (or is *threshold realizable*) if there is an augmented weight vector $\mathbf{v}$ such that

$$U(\mathbf{x}) = \begin{cases} 0 & \text{for } \mathbf{v}^T\mathbf{y} < 0, \\ 1 & \text{for } \mathbf{v}^T\mathbf{y} > 0. \end{cases}$$

Thus $U(\mathbf{x})$ is asummable if and only if $U(\mathbf{x})$ is threshold realizable. We use the same terminology if $U(\mathbf{x})$ is a characteristic function on a domain $\mathscr{X}_1 \cup \mathscr{X}_2$ where all the coordinates of each $\in \mathscr{X}_1 \cup \mathscr{X}_2$ are rational.

2.4 The Perceptron and the Proportional Increment Training Procedure

One of the earliest techniques for finding a solution vector for a pair of linearly separable classes is Rosenblatt's error-correcting training procedure. In this book we refer to this as the *proportional increment* training procedure. This training procedure, as well as the associated termination theorem, lie at the core of much of the theory of trainable pattern classifiers.

The basic version of Rosenblatt's perceptron is illustrated in Figure 2.5. In this figure the object observed by the perceptron is the letter X. This object is sensed by an array of sensor units, each of which emits a 0 or a 1, depending on whether or not it lies on the image of the object. Randomly generated subsets of the sensor units are fed to a large set of associator units $\{A_1, \ldots, A_d\}$. It is assumed that although d is large it is significantly smaller than the number of sensor units, by analogy with the way in which the optic nerve of the human visual system transforms the relatively large number of signals in the rods and cones of the retina to a much smaller number of

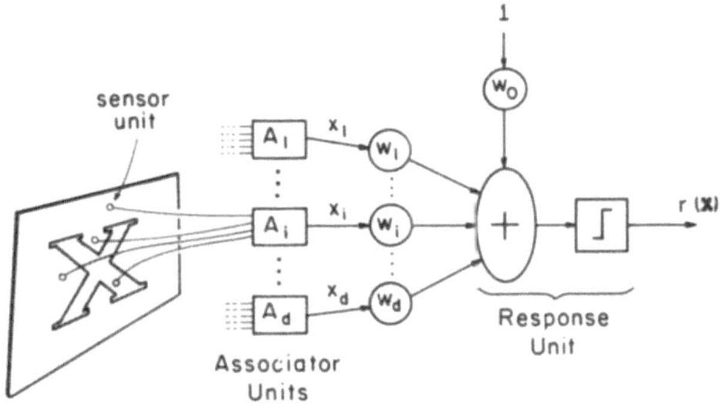

Figure 2.5. Basic version of perceptron.

features in a corresponding set of neural fibers. (These neural fibers carry the visual signals through various stages of preprocessing to the cortex).

The output x_i of each associator unit is a fixed linear sum of the outputs of the sensor units. A linear sum $\mathbf{w}^T\mathbf{x}$ of the outputs of the associator units is fed to the response unit through a set of adjustable weights $\{w_i\}$. The response unit emits $r(\mathbf{x}) = 2$ or $r(\mathbf{x}) = 1$, depending on whether the sign of $w_0 + \mathbf{w}^T\mathbf{x}$ is positive or negative, respectively.

It is convenient to describe this decision process in augmented feature space. In this space, $r(\mathbf{x}) = 1$ or 2 whenever $\mathbf{v}^T\mathbf{y} < 0$ or >0, respectively. Let $(\mathscr{Y}_1, \mathscr{Y}_2)$ denote a training set consisting of a finite number of augmented feature vectors in classes ω_1 and ω_2, respectively. The training procedure seeks a solution vector $\mathbf{v}$ such that

$$\mathbf{v}^T\mathbf{y}\begin{cases} <0, & \text{if } \mathbf{y} \in \mathscr{Y}_1, \\ >0, & \text{if } \mathbf{y} \in \mathscr{Y}_2. \end{cases} \tag{2.11}$$

Equivalently, using the notation of Equation (2.8) the training procedure seeks a vector $\mathbf{v}$ such that

$$\mathbf{v}^T\boldsymbol{\eta} > 0 \quad \text{for all } \boldsymbol{\eta}. \tag{2.12}$$

The proportional increment training procedure searches for such a $\mathbf{v}$ by starting with $\mathbf{v}(0)$ as a guess at trial 0. (A trial, following the usage in mathematical psychology, consists of an observation $\mathbf{y}$, a response r and a reinforcement containing information about the class to which $\mathbf{y}$ belongs. Trial n denotes the nth member of a sequence of trials.) Then the procedure applies the following adjustment of $\mathbf{v}$ at trial n:

$$\mathbf{v}(n+1) = \begin{cases} \mathbf{v}(n) + \rho\boldsymbol{\eta}(n), & \rho > 0, \quad \text{if } \mathbf{v}^T(n)\boldsymbol{\eta}(n) \leqslant 0, \\ \mathbf{v}(n), & \text{otherwise.} \end{cases} \tag{2.13}$$

This equation is an example of an error correction procedure, so called because $\mathbf{v}(n)$ is adjusted only when the feature vector $\mathbf{x}(n)$ is misclassified.

A computationally convenient alternative way of expressing Equation (2.13) is

$$\mathbf{v}(n+1) = \mathbf{v}(n) + \rho[\Omega(n) - r(n)]\mathbf{y}(n), \tag{2.14}$$

where $\Omega(n) = j$ when $\omega(n) = \omega_j$, and $r(n)$ is the response of the classifier at trial n.

In the termination theorem stated and proved below, we show that this training procedure converges to a solution vector in a finite number of trials whenever a solution exists. On the other hand, when no solution exists, the procedure produces a weight vector which remains bounded for any size of the training set [5].

Theorem 2.5. (Proportional Increment Termination Theorem). *Suppose $\mathscr{X}_1$ and $\mathscr{X}_2$ are finite, linearly separable sets of feature vectors. (These sets would necessarily be finite if $\mathscr{X}_1$ and $\mathscr{X}_2$ were derived from digitizations of black objects on a finite white background, since there is a finite upper bound on the number of such digitizations.) Construct a training sequence $\{\mathbf{x}(n)\}$ in which every member of $\mathscr{X}_1 \cup \mathscr{X}_2$ appears an infinite number of times. Then a proportional increment training procedure on $\{\mathbf{x}(n)\}$, as described in Equation* (2.14), *yields a sequence $\{v(n)\}$ such that, for some finite positive number k,*

$$\mathbf{v}(k+j) = \mathbf{v}(k) \quad \text{for all } j \geq 0.$$

Proof. (After Ridgway [6].) Let $\mathscr{V}^*$ denote the set of solution vectors of Equation (2.12). Let $\mathbf{v}^*$ denote any member of $\mathscr{V}^*$. Let

$$B = \min_{\boldsymbol{\eta}} \frac{\mathbf{v}^{*T}\boldsymbol{\eta}}{\rho}. \tag{2.15}$$

(Note that $B > 0$.) Let

$$T = \max_{\boldsymbol{\eta}} \|\boldsymbol{\eta}\| \tag{2.16}$$

$$\mathbf{u}^* = \frac{T^2}{\rho B}\mathbf{v}^*. \tag{2.17}$$

Let

$$D(n) = \text{distance from } \mathbf{u}^* \text{ to } \mathbf{v}(n)/\rho$$

$$= \left\| \mathbf{u}^* - \frac{\mathbf{v}(n)}{\rho} \right\|.$$

Assume $\mathbf{v}(n)$ is not a solution vector. Thus there exists $m \geq n$ such that $\mathbf{v}^T(m)\boldsymbol{\eta}(m) \leq 0$. For convenience we assume that $m = n$, since this assumption has no effect on the convergence or nonconvergence of $\mathbf{v}(n)$ as $n \to \infty$. Then

$$D^2(n+1) = \left\| \mathbf{u}^* - \frac{\mathbf{v}(n+1)}{\rho} \right\|^2 = \left\| \mathbf{u}^* - \frac{\mathbf{v}(n)}{\rho} - \boldsymbol{\eta}(n) \right\|^2$$

$$= \left\| \mathbf{u}^* - \frac{\mathbf{v}(n)}{\rho} \right\|^2 - 2\left(\mathbf{u}^* - \frac{\mathbf{v}(n)}{\rho}\right)^T \boldsymbol{\eta}(n) + \|\boldsymbol{\eta}(n)\|^2.$$

Using Equation (2.16) and the inequality $\mathbf{v}^T(n)\boldsymbol{\eta}(n) \leqslant 0$, we deduce that

$$\begin{aligned} D^2(n+1) &\leqslant D^2(n) - 2\mathbf{u}^{*T}\boldsymbol{\eta}(n) + T^2 \\ &= D^2(n) - \frac{2T^2}{B}\frac{\mathbf{v}^{*T}\boldsymbol{\eta}(n)}{\rho} + T^2 \quad \text{by Equation (2.17)} \\ &\leqslant D^2(n) - \frac{2T^2}{B}B + T^2 \quad \text{by Equation (2.15).} \end{aligned}$$

Thus

$$D^2(n+1) \leqslant D^2(n) - T^2. \tag{2.18}$$

Suppose that $\{\mathbf{v}(n)\}$ does not terminate in a solution. Since every member of $\mathscr{X}_1 \cap \mathscr{X}_2$ appears an infinite number of times in $\{\mathbf{x}(n)\}$, it follows from Equation (2.18) that $D^2(n)$ becomes negative for sufficiently large n. This is impossible, since $D^2(n) \geqslant 0$. Hence $\{D(n)\}$ must terminate at a value of $n = k$ such that

$$D^2(k) = D^2(k+j) \quad \text{for all } j \geqslant 0. \tag{2.19}$$

Hence $\mathbf{v}(n)$ must eventually reach $\mathscr{V}^*$. □

2.5 The Fixed Fraction Training Procedure

After the appearance of the proportional increment training procedure in the technical literature, several other error-correcting training procedures and their convergence properties were established. Among those procedures is the *fixed fraction* training procedure. In this training procedure the augmented weight vector $\mathbf{v}(n)$ is adjusted by a fixed fraction λ of the distance of $\mathbf{v}(n)$ from the $\boldsymbol{\eta}(n)$-hyperplane, namely the hyperplane determined by $\mathbf{v}^T\boldsymbol{\eta}(n) = 0$, in augmented weight space, provided the classifier's guess at trial n is incorrect. We assume here that $0 < \lambda < 2$.

To derive the equation for this adjustment rule, note that the distance of the hyperplane $\mathbf{v}^T\boldsymbol{\eta}(n) = 0$ from $\mathbf{v}(n)$ in weight space is

$$\pm\frac{\mathbf{v}(n)^T\boldsymbol{\eta}(n)}{\|\boldsymbol{\eta}(n)\|} \tag{2.20}$$

as illustrated in Figure 2.6. Also, note that $\mathbf{v}(n)^T\boldsymbol{\eta}(n) < 0$ when the classifier's guess is incorrect. Hence the adjustment rule is given by

$$\mathbf{v}(n+1) = \begin{cases} \mathbf{v}(n) - \lambda\dfrac{\mathbf{v}(n)^T\boldsymbol{\eta}(n)}{\|\boldsymbol{\eta}(n)\|^2}\boldsymbol{\eta}(n), & \text{for } \mathbf{v}(n)^T\boldsymbol{\eta}(n) \leqslant 0, \\ \mathbf{v}(n), & \text{otherwise,} \end{cases} \tag{2.21}$$

where $0 < \lambda < 2$.

We now demonstrate the convergence of this training procedure. By reasoning in a manner similar to the derivation of Equation (2.18) one can show that

$$\|\mathbf{v}(n+1) - \mathbf{v}^*\| \leqslant \|\mathbf{v}(n) - \mathbf{v}^*\|, \tag{2.22}$$

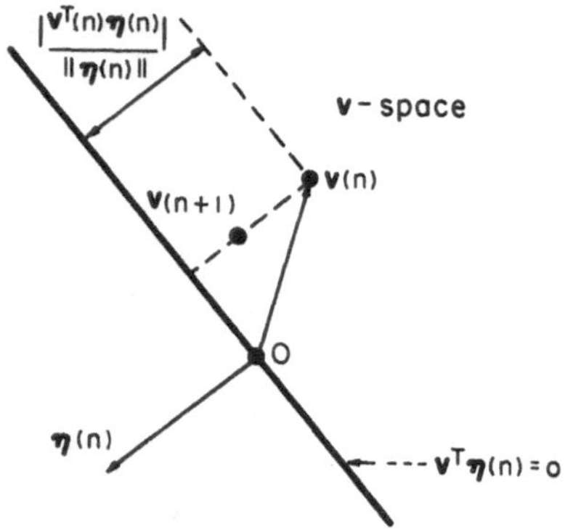

Figure 2.6. Distance of v(n) from η(n)-hyperplane.

whenever $0 < \lambda \leqslant 2$, where $\mathbf{v}^*$ is any solution vector, i.e., $\mathbf{v}^*$ is any member of the set of solution vectors $\mathscr{V}^*$. (Again we assume that $\mathbf{v}(n)^T\boldsymbol{\eta}(n) \leqslant 0$ at all successive values of n whenever $\mathbf{v}(n)$ is not in $\mathscr{V}^*$.) In this case, however, we cannot show that $\|\mathbf{v}(n) - v^*\|^2 - \|\mathbf{v}(n+1) - \mathbf{v}^*\|^2$ necessarily exceeds a positive constant, as it did in Equation (2.18), because the size of the adjustment approaches zero as $n \to \infty$ if $\mathbf{v}(n)^T\boldsymbol{\eta}(n)$ approaches zero as $n \to \infty$. Under these circumstances $\mathbf{v}(n)$ either terminates in $\mathscr{V}^*$ (where $\mathscr{V}^*$ includes the boundary of $\mathscr{V}^*$) or converges to a point on the boundary of $\mathscr{V}^*$ whenever $\mathscr{V}^*$ is not empty and the training set is finite.

To show this, note from Equation (2.22) that $\mathbf{v}(n)$ must converge to a hypersphere defined by

$$\begin{aligned}\|\mathbf{v} - \mathbf{v}^*\| &= l(\mathbf{v}^*, \mathbf{v}(0)) \\ &= \text{a nonnegative function of } \mathbf{v}^* \text{ and } \mathbf{v}(0),\end{aligned} \tag{2.23}$$

where $\mathbf{v}$ is any augmented weight vector on the surface of this hypersphere. Since $\mathbf{v}$ is independent of $\mathbf{v}^*$, $\mathbf{v}(n)$ must reach one or more points that lie simultaneously on the set of all hyperspheres given by Equation (2.23) over all $\mathbf{v}^*$ for a fixed value of $\mathbf{v}(0)$. This is impossible unless the intersection of all the hyperspheres is a single point. Hence all the hyperspheres have a radius of zero. Hence $\mathbf{v}(n)$ must converge to $\mathbf{v}^*$. Recall that $\mathbf{v}^*$ belongs to $\mathscr{V}^*$.

If $\mathbf{v}^*$ lies in the interior of $\mathscr{V}^*$ then $\mathbf{v}^*$ must be a termination point, by the following reasoning. If $\mathbf{v}^*$ were a limit point, the sequence $\{\mathbf{v}(n)\}$ would contain at least one element, say $\hat{\mathbf{v}}$, lying in an ε-neighborhood of $\mathbf{v}^*$ inside $\mathscr{V}^*$. In this case $\mathbf{v}(n)$ would terminate at $\hat{\mathbf{v}}$, contradicting the assumption that $\mathbf{v}^*$ is a limit point. Hence $\mathbf{v}^*$ is a termination point.

Hence if $\mathbf{v}^*$ lies in $\mathscr{V}^*$ and is a limit point, $\mathbf{v}^*$ must be on the boundary of $\mathscr{V}^*$.

Thus we have the following convergence theorem for the fixed fraction training procedure.

Theorem 2.6. (Fixed Fraction Convergence Theorem). *If $\mathscr{X}_1 \cup \mathscr{X}_2$ is a finite linearly separable training set and $\mathscr{V}^*$ is the set of augmented solution vectors, then the fixed-fraction training procedure, as described by Equation* (2.21), *yields a sequence of augmented weight vectors $\{\mathbf{v}(n)\}$ that either terminates in $\mathscr{V}^*$ or on the boundary of $\mathscr{V}^*$ or converges to a point on the boundary of $\mathscr{V}^*$.*

2.6 A Multiclass Training Procedure

It is possible to extend the basic concepts of a two-class linear classifier to a p-class linear classifier as follows. Suppose we are given a training sequence of pairs $(\mathbf{x}(n), \mathscr{X}(n))$. Each $\mathscr{X}(n)$ is a member of the set of p sets $\{\mathscr{X}_1, \ldots, \mathscr{X}_p\}$, and serves to indicate the class to which $\mathbf{x}(n)$ belongs. Each $\mathscr{X}_i$ is a finite subset of the class region ω_1 (i.e., we assume that the number of classes c equals the number of decision regions p). Construct a linear classifier composed of p linear discriminant functions:

$$\{\mathbf{v}_i^T\mathbf{y} \,|\, i = 1, \ldots, p\}.$$

These discriminant functions in a linear classifier map the feature space into p decision regions $\{\mathscr{R}_1, \ldots, \mathscr{R}_p\}$. The desired operation of the training process of this classifier is to converge to or terminate on a set of weight vectors $\{\mathbf{v}_i^*\}$ such that $\mathscr{R}_i \supseteq \omega_i$ for every i. With these weight vectors the classifier assigns $\mathbf{x}$ to $\mathscr{R}_i$ if $\mathbf{v}_i^T\mathbf{y} > \mathbf{v}_j^T\mathbf{y}$ for all j such that $i \neq j$.

The training procedure we will describe here—an extended form of proportional increment training—finds $\{\mathbf{v}_i^*\}$ such that $\mathscr{R}_i \supseteq \mathscr{X}_i$ for every i if at least one such set $\{\mathbf{v}_i^*\}$ exists.

Suppose $\mathbf{x}(n) \in \mathscr{X}_i$ and suppose there are b values of j, $b \geqslant 1$, for which $\mathbf{v}_j^T\mathbf{y}(n) > \mathbf{v}_i^T\mathbf{y}(n)$, so that the classifier misclassifies $\mathbf{x}(n)$. Then an extended proportional increment training procedure is as follows.

$$\begin{aligned} &\mathbf{v}_i(n+1) = \mathbf{v}_i(n) + b\rho\mathbf{y}(n) \\ &\mathbf{v}_j(n+1) = \mathbf{v}_j(n) - \rho\mathbf{y}(n) \quad \text{if } \mathbf{v}_j^T\mathbf{y}(n) > \mathbf{v}_i^T\mathbf{y}(n),\ i \neq j \\ &\mathbf{v}_k(n+1) = \mathbf{v}_k(n) \quad \text{if } \mathbf{v}_k^T\mathbf{y}(n) \leqslant \mathbf{v}_i^T\mathbf{y}(n),\ k \neq i. \end{aligned} \tag{2.24}$$

Thus at every trial involving a misclassification of a member of $\mathscr{X}_i$ the weight vectors $\{\mathbf{v}_k\}$ are adjusted in accordance with Equation (2.24).

One can show that this training procedure converges to a set of solution vectors $\{\mathbf{v}_i^*\}$ in a finite number of trials, provided there exists a finite number $M > 0$ such that each member of each $\mathscr{X}_i$ occurs at least once in every sequence of M successive trials. To prove this, convert the p-class classifier to an equivalent 2-class classifier by means of Kesler's construction [7], as follows.

For each $\mathbf{x} \in \mathscr{X}_i$, define the following set of $p(d+1)$-dimensional vectors:

$$\boldsymbol{\eta}_{ij}(\mathbf{x}) = \begin{bmatrix} \boldsymbol{\eta}_{ij}^{(1)} \\ \vdots \\ \boldsymbol{\eta}_{ij}^{(p)} \end{bmatrix}, \qquad j = 1, \ldots, p; j \neq i.$$

where each of the p vectors $\boldsymbol{\eta}_{ij}^{(k)}$ is defined by

$$\begin{aligned} \boldsymbol{\eta}_{ij}^{(k)} &= \mathbf{0} \qquad \text{for } k \neq i, k \neq j; \\ \boldsymbol{\eta}_{ij}^{(j)} &= -\mathbf{y} \\ \boldsymbol{\eta}_{ij}^{(i)} &= \mathbf{y}. \end{aligned}$$

Replace every occurrence of $(\mathbf{x}, \mathscr{X}_i)$ in the training sequence $\{\mathbf{x}(n), \omega(n)\}$ by a sequence

$$\{\boldsymbol{\eta}_{ij}(\mathbf{x}) \mid j = 1, \ldots, p; i \neq j\}$$

Let $\{\boldsymbol{\eta}(n)\}$ denote the resulting sequence of $\boldsymbol{\eta}_{ij}(\mathbf{x})$'s. Let

$$\mathbf{v} = \begin{bmatrix} \mathbf{v}_1 \\ \vdots \\ \mathbf{v}_p \end{bmatrix}.$$

Notes. (a) Every member of $\{\boldsymbol{\eta}_{ij}(\mathbf{x}) \mid \mathbf{x} \in \bigcup_i \mathscr{X}_i, i \in \{1, \ldots, p\}, j \in \{1, \ldots, p\}, i \neq j\}$ occurs at least once in every subsequence of length $M(p-1)$ in $\boldsymbol{\eta}(n)$. (b) $\mathbf{v}_T \boldsymbol{\eta}_{ij} < \mathbf{0}$ for $j = 1, \ldots, p\,(i \neq j)$ is equivalent to

$$\mathbf{v}_i^T \mathbf{y} < \mathbf{v}_j^T \mathbf{y}, \qquad \mathbf{x} \in \mathscr{X}_i, i \neq j.$$

(c) $\mathbf{v}(n)^T \boldsymbol{\eta}_{ij} > 0$ for all i, j and all $\mathbf{x} \in \bigcup_i \mathscr{X}_i$ is equivalent to a perfect classification of the training set $\{\mathbf{x}(n)\}$. Thus $\mathbf{v}(n)^T \boldsymbol{\eta}_{ij} < 0$ for one or more values of j and for some $\mathbf{x} \in \mathscr{X}_i$ whenever the classifier misclassifies a member of $\mathscr{X}_i$.

Thus one can see that the p-class linear classifier can be transformed into an equivalent 2-class linear classifier in which $\boldsymbol{\eta}(n)$ is a $p(d+1)$-dimensional sign-normalized augmented feature vector, and $\mathbf{v}(n)$ is a $p(d+1)$-dimensional augmented weight vector. In this case, however, the absolute value of the first element of $\boldsymbol{\eta}(n)$ is not always unity—in contrast to the usual situation for augmented feature vectors for 2-class classifiers.

Using the above notation the p-class proportional increment training procedure described by Equation (2.24) becomes

$$\mathbf{v}(n+1) = \begin{cases} \mathbf{v}(n) + \rho \boldsymbol{\eta}(n), & \text{if } \mathbf{v}(n)^T \boldsymbol{\eta}(n) < 0, \\ \mathbf{v}(n), & \text{otherwise.} \end{cases} \tag{2.25}$$

This equation is equivalent to Equation (2.24) and is the same as Equation (2.13).

Thus the behavior of $\mathbf{v}(n)$ in the p-class proportional increment training process of Equation (2.25) is completely analogous to that of the sequence of augmented weight vectors in a 2-class proportional increment training procedure. Hence the p-class proportional increment training procedure of

Equations (2.24) and (2.25) produces a sequence $\{\mathbf{v}(n)\}$ that terminates to a solution vector $\mathbf{v}^*$ (equivalent to a set of solution vectors $\{\mathbf{v}_i^*\}$) within a finite number of trials.

Other forms of p-class proportional increment training procedures are possible, and termination of the weight sequence during the training process can be proved in a similar manner. One such training procedure is given in Exercise 2.12.

2.7 Synthesis by Game Theory

The proportional increment and the fixed-fraction training procedures have two major weaknesses:

1. If it is not known whether or not the classes are linearly separable, and if the classes are not linearly separable, these training procedures do not notify the user that the classes are not linearly separable.
2. If the classes are linearly separable, these training procedures usually notify the user that $\mathbf{v}(n)$ is a solution vector only after several tests of all members of the training set on the vector $\mathbf{v}(n)$.

The game theoretic approach [8, 9], described below, overcomes both of these weaknesses at the expense of additional complexity in the processing of the training set. It is particularly well suited to cases in which the number of features d is relatively large; say $d \geqslant 20$.

The mathematical theory of games—or game theory, for short—is concerned with the properties of products of the form $\mathbf{R}^T\mathbf{q}$, where $\mathbf{R}$ is a matrix, $\mathbf{q}$ is a vector, $\mathbf{q} \geqslant \mathbf{0}$ and $\mathbf{R}^T\mathbf{q} > \mathbf{0}$. To exploit these properties for the synthesis of linear classifiers, Singleton derived Theorem 2.7, presented below [10]. This theorem is based in part on the fact that for every matrix $\mathbf{A}$, defined in Section 2.3, there is a permutation matrix $\mathbf{P}$ and a nonsingular row-operation matrix $\mathbf{F}$ such that $\mathbf{FA}^T\mathbf{P}$ transforms $\mathbf{A}^T$ to one of the following four forms:

$$\mathbf{FA}^T\mathbf{P} = \begin{bmatrix} \mathbf{I} & \mathbf{R} \\ \mathbf{0} & \mathbf{0} \end{bmatrix}, \quad \begin{bmatrix} \mathbf{I} \\ \mathbf{0} \end{bmatrix}, \quad [\mathbf{I} \quad \mathbf{R}], \quad \text{or } \mathbf{I}, \tag{2.26}$$

where $\mathbf{I}$ is an identity matrix. This result is derived in Appendix A.

Before discussing Singleton's theorem we first examine the conditions under which $\mathbf{FA}^T\mathbf{P}$ may have the forms in Equation (2.26). We assume here that every row of $\mathbf{A}$ has at least one nonzero element. Let N denote the number of rows in $\mathbf{A}$ (i.e., N = the cardinality of $\mathscr{X}_1 \cup \mathscr{X}_2$). Let d denote the number of coordinates in nonaugmented feature space. Thus $\mathbf{A}$ has N rows and $d + 1$ columns.

Suppose $\mathbf{A}$ is of rank N. Then, necessarily, $N \leqslant d + 1$. Also, it follows that the endpoints of the row vectors $\{\mathbf{a}_i^T\}$ of $\mathbf{A}$ determine an $(N - 1)$-dimensional hyperplane $\mathscr{H}_A$ which does not contain the origin. (If $\mathscr{H}_A$ were to contain the origin, then $\{\mathbf{a}_i^T\}$ would be linearly dependent—since $N-1$ linearly independent vectors form the basis of an $(N-1)$-space—and $\mathbf{A}$ would be of rank less than N.) It follows that there exists a vector $\mathbf{v}$—e.g., a vector orthogonal to $\mathscr{H}_A$—such that $\mathbf{a}_i^T \mathbf{v} > 0$ for every i. Hence if $\mathbf{A}$ is of rank N then $(\mathscr{X}_1, \mathscr{X}_2)$ is linearly separable. Note that if $\mathbf{A}$ is of rank N, then $N \leqslant d+1$ and hence $\mathbf{FA}^T\mathbf{P}$ is of the form

$$\begin{bmatrix} \mathbf{I} \\ \mathbf{0} \end{bmatrix} \quad \text{or} \quad \mathbf{I}.$$

Now suppose $\mathbf{A}$ is of rank less than N. Then $\mathbf{FA}^T\mathbf{P}$ will be of the form

$$\begin{bmatrix} \mathbf{I} & \mathbf{R} \\ \mathbf{0} & \mathbf{0} \end{bmatrix} \quad \text{or} \quad [\mathbf{I} \quad \mathbf{R}], \tag{2.27}$$

with $\mathbf{R} \neq \mathbf{0}$, where $\mathbf{0}$ denotes a matrix all of whose elements are zeros. To see that $\mathbf{R} \neq \mathbf{0}$, note that every column of $\mathbf{A}^T\mathbf{P}$, say $\mathbf{a}_i$, is not composed solely of zeros, i.e., $\mathbf{a}_i \neq \mathbf{0}$. Hence the rank of $\mathbf{a}_i$ is 1 for every i. Let $\mathbf{Fa}_i = \mathbf{b}_i$. Since $\mathbf{F}$ is nonsingular, $\mathbf{a}_i = \mathbf{F}^{-1}\mathbf{b}_i \neq \mathbf{0}$. Hence $\mathbf{b}_i \neq \mathbf{0}$, which proves that $\mathbf{R} \neq \mathbf{0}$.

Thus if $\mathbf{A}$ is of rank less than N, then $\mathbf{FA}^T\mathbf{P}$ is of the form of (2.27), with $\mathbf{R} \neq \mathbf{0}$. In the subsequent discussion we assume that the rank of $\mathbf{A}$ is less than N.

Theorem 2.7. (Singleton's Theorem). *For any given matrix $\mathbf{A} \neq \mathbf{0}$, suppose $\mathbf{R}$ denotes a residual matrix obtained from $\mathbf{A}^T$ by means of the permutation matrix $\mathbf{P}$ and the nonsingular row operation matrix $\mathbf{F}$ in the transformation*

$$\mathbf{FA}^T\mathbf{P} = \begin{bmatrix} \mathbf{I} & \mathbf{R} \\ \mathbf{0} & \mathbf{0} \end{bmatrix} \quad \text{or} \quad [\mathbf{I} \quad \mathbf{R}]. \tag{2.28}$$

Then there exists $\mathbf{v}$ such that $\mathbf{Av} > \mathbf{0}$ (i.e., $\mathscr{X}_1$ and $\mathscr{X}_2$ are linearly separable) if and only if there exists $\mathbf{q}$ such that $\mathbf{q} \geqslant \mathbf{0}$, and $\mathbf{R}^T\mathbf{q} > \mathbf{0}$.

Note that one can restrict $\mathbf{q}$ to satisfy $\mathbf{q}^T\mathbf{1} = 1$ without affecting the validity or nonvalidity of this theorem. A vector $\mathbf{q}$ that satisfies $\mathbf{q}^T\mathbf{1} = 1$, $\mathbf{q} \geqslant \mathbf{0}$, $\mathbf{q} \neq \mathbf{0}$ is known as a *stochastic vector*. Thus the theorem can be restated as follows:

$\mathscr{X}_1$, $\mathscr{X}_2$ are linearly separable if and only if there exists a stochastic vector $\mathbf{q}$ satisfying $\mathbf{R}^T\mathbf{q} > \mathbf{0}$.

PROOF. (a) Suppose there exists $\mathbf{v}$ such that $\mathbf{Av} > \mathbf{0}$. Then $\mathbf{P}^T\mathbf{Av} > \mathbf{0}$, since $\mathbf{P}^T\mathbf{A}$ is a matrix obtained by permuting rows of $\mathbf{A}$. Hence

$$\mathbf{P}^T\mathbf{AF}^T(\mathbf{F}^T)^{-1}\mathbf{v} > \mathbf{0}$$

and by Equation (2.28),

$$\mathbf{P}^T\mathbf{AF}^T = (\mathbf{FA}^T\mathbf{P})^T = \begin{bmatrix} \mathbf{I} & \mathbf{0} \\ \mathbf{R}^T & \mathbf{0} \end{bmatrix}.$$

Hence

$$\begin{bmatrix} \mathbf{I} & \mathbf{0} \\ \mathbf{R}^T & \mathbf{0} \end{bmatrix} \mathbf{q}' > \mathbf{0}, \quad \text{where } \mathbf{q}' = (\mathbf{F}^T)^{-1}\mathbf{v}.$$

Delete the portion of $\mathbf{q}'$ corresponding to the columns of 0's in $\mathbf{P}^T\mathbf{A}\mathbf{F}^T$. Call the resulting vector $\mathbf{q}$. Thus

$$\begin{bmatrix} \mathbf{I} \\ \mathbf{R}^T \end{bmatrix} \mathbf{q} > \mathbf{0}.$$

The upper half of this equation yields $\mathbf{q} > \mathbf{0}$. The lower half yields $\mathbf{R}^T\mathbf{q} > \mathbf{0}$. Hence $\mathbf{q} \geqslant \mathbf{0}$ and $\mathbf{R}^T\mathbf{q} > \mathbf{0}$.

(b) Suppose there exists $\mathbf{q} \geqslant \mathbf{0}$ such that $\mathbf{R}^T\mathbf{q} > \mathbf{0}$. Let (2.29)

$$\mathbf{R} = (\boldsymbol{\rho}_1, \boldsymbol{\rho}_2, \ldots, \boldsymbol{\rho}_s) \tag{2.30}$$

$$\hat{\mathbf{q}} = \mathbf{q} + \varepsilon\mathbf{1}, \tag{2.31}$$

where

$$\mathbf{1} = \begin{bmatrix} 1 \\ 1 \\ \vdots \\ 1 \end{bmatrix}, \qquad \varepsilon > 0.$$

With $\hat{\mathbf{q}}$ as defined in Equation (2.31), we have $\hat{\mathbf{q}} > \mathbf{0}$, $\mathbf{R}^T\hat{\mathbf{q}} > \mathbf{0}$. Hence

$$\begin{bmatrix} \mathbf{I} \\ \mathbf{R}^T \end{bmatrix} \hat{\mathbf{q}} > \mathbf{0}.$$

Recall that

$$\mathbf{P}^T\mathbf{A}\mathbf{F}^T = \begin{bmatrix} \mathbf{I} & \mathbf{0} \\ \mathbf{R}^T & \mathbf{0} \end{bmatrix}.$$

Let M denote the number of columns of zeros in $\mathbf{P}^T\mathbf{A}\mathbf{F}^T$. Append any M positive numbers, say M 1's, to $\hat{\mathbf{q}}$, forming $\mathbf{q}'$. Then

$$\mathbf{P}^T\mathbf{A}\mathbf{F}^T\mathbf{q}' > \mathbf{0}, \qquad \mathbf{q}' > \mathbf{0}. \tag{2.32}$$

Let

$$\mathbf{v} = \mathbf{F}^T\mathbf{q}'. \tag{2.33}$$

Substituting $\mathbf{v}$ in Equation (2.32), we obtain

$$\mathbf{P}^T\mathbf{A}\mathbf{v} > \mathbf{0}. \tag{2.34}$$

Thus, $\mathbf{A}\mathbf{v} > \mathbf{0}$. □

By this theorem the synthesis of linear classifiers is reduced to searching for a nonnegative nonzero $\mathbf{q}$ satisfying $\mathbf{R}^T\mathbf{q} > \mathbf{0}$. The game-theoretic technique of *iterative fictitious play*—a form of training procedure—provides, in a finite number of iterations, a means of finding such a $\mathbf{q}$, if it exists, or determining that no such $\mathbf{q}$ exists.

Suppose such a $\mathbf{q}$ exists, and it has been found by this game-theoretic technique. One may then obtain $\mathbf{v}$ by

$$\mathbf{v} = \mathbf{F}^T\mathbf{q}' \tag{2.35}$$

as indicated in Equation (2.33).

In the game-theoretic technique of searching for $\mathbf{q}$, the matrix $\mathbf{R}$ is viewed as the *payoff matrix* of a two-person zero-sum game, with two opposing players, P and Q. In each game P and Q choose *pure strategies* P_i and Q_j, simultaneously. Each element r_{ij} of $\mathbf{R}$ is the payoff to player Q when player Q uses pure strategy Q_i and player P uses pure strategy P_j. In a *zero-sum* game the payoff to Q is equal to the loss to P, i.e., by representing losses as negative payoffs, the sum of the payoffs to the two players is zero, regardless of the strategies used by the players. Thus hereafter we use "payoff" to mean the payoff to Q or the negative of the payoff to P. In playing the game the players may or may not know each other's strategies, each player may or may not know whether the other player knows the first player's strategy, etc., and the payoff matrix may or may not be known to one or both players. The payoff matrix is merely a model of a possible interaction between two entities (people, machines, organizations, species, etc.)

As an example, consider the following payoff matrix.

$$\mathbf{R} = \begin{bmatrix} 3 & 2 & 4 & 5 \\ -8 & 4 & 5 & -3 \\ 3 & 1 & 9 & 4 \\ 7 & 8 & -3 & 6 \end{bmatrix}. \tag{2.36}$$

For this payoff matrix the payoff to Q is 5 when Q's strategy is Q_2, corresponding to the second row of $\mathbf{R}$, and P's strategy is P_3, corresponding to the third column of $\mathbf{R}$.

Sometimes one considers the case where a player assumes that his opponent will always use the pure strategy that yields the highest payoff to the opponent. With this pessimistic attitude, using the payoff matrix in Equation (2.36), and assuming the payoff matrix is known to both players, Q would assume that if he used strategy Q_1 then P would use strategy P_2, if Q used strategy Q_2 then P would use strategy P_1, etc. Thus for Q's strategies $\{Q_i | i = 1, 2, 3, 4\}$ the payoffs are 2, -8, 1, -3, respectively. The largest of these payoffs is 2, corresponding to Q_1. Thus, using this pessimistic approach, Q chooses Q_1, which is the strategy yielding the largest of the minimum possible payoffs. This is the *maximin strategy* for Q. The associated *maximin point* is $(i, j) = (1, 2)$. The analogous strategy for P, if P is a pessimist, is the *minimax strategy*, P_4. The *minimax point* is $(i, j) = (4, 4)$. Note that the maximin point and minimax point are different—although in some cases they may be the same.

A more sophisticated form of play involves the use of a *mixed strategy*, i.e., a scheme in which pure strategies are determined by a random variable. Thus suppose P chooses strategies $\{P_j | j = 1, 2, \ldots, s\}$ with probabilities $p_1, \ldots, p_s$. Then the expected payoff to Q when Q chooses pure strategy Q_i is

$$\begin{aligned} \pi_i &= r_{i1}p_1 + \cdots + r_{is}p_s \\ &= \mathbf{r}_i^T \mathbf{p}, \end{aligned} \tag{2.37}$$

where $\mathbf{r}_i^T$ is the ith row of $\mathbf{R}$, and $p = [p_1, \ldots, p_s]^T$. We refer to $\mathbf{p}$ as the mixed strategy of player P.

If Q uses a mixed strategy $\mathbf{q} = [q_1, \ldots, q_t]^T$, then the expected payoff is

$$\begin{aligned}\pi &= q_1\mathbf{r}_1^T\mathbf{p} + \cdots + q_t\mathbf{r}_t^T\mathbf{p} \\ &= \mathbf{q}^T\mathbf{R}\mathbf{p}.\end{aligned} \tag{2.38}$$

The mixed strategies $\mathbf{p}$, $\mathbf{q}$ are forms of stochastic vectors.

For any given mixed strategy $\mathbf{p}$ of player P, there is a mixed strategy $\mathbf{q}_p$ such that π is a maximum. Imagine plotting $(\mathbf{p}, \mathbf{q}_p)$ as a trajectory in **p**,**q**-space. Somewhere on this trajectory of maximum points, we may find a point $(\tilde{\mathbf{p}}, \tilde{\mathbf{q}}_p)$ such that π is a minimum among all π's on the trajectory. The point $(\tilde{\mathbf{p}}, \tilde{\mathbf{q}}_p)$ is a (mixed strategy) *minimax* point in **p**,**q**-space. A hostile, intelligent P will choose the mixed strategy $\tilde{\mathbf{p}}$ of one of the minimax points.

Similarly, for every $\mathbf{q}$ in **p**,**q**-space, there exists a mixed strategy $\mathbf{p}_q$ such that π is a minimum. Imagine plotting $(\mathbf{p}_q, \mathbf{q})$ as a trajectory in **p**,**q**-space. There may exist a point $(\tilde{\mathbf{p}}_q, \tilde{\mathbf{q}})$ on this trajectory such that π is a maximum among all π's on the trajectory. This is a (mixed strategy) *maximin* point in **p**,**q**-space. A hostile, intelligent Q will choose the mixed strategy $\tilde{\mathbf{q}}$ of one of the maximin points.

Von Neumann's *fundamental theorem of the theory of games* (also known as the *minimax theorem*) [11] states that in every finite game—i.e., every game in which $s < \infty$ and $t < \infty$—a maximin and a minimax point coincide, thereby constituting a *saddle point*. (This is illustrated in Figure 2.7 by the contour map of the expected payoff, treating the vectors $\mathbf{p}$, $\mathbf{q}$ as if they are scalars.) Consequently in every finite game there is a pair of optimal mixed strategies

$$(\hat{\mathbf{p}}, \hat{\mathbf{q}}) = (\tilde{\mathbf{p}}, \tilde{\mathbf{q}}_p) = (\tilde{\mathbf{p}}_q, \tilde{\mathbf{q}}) \tag{2.39}$$

such that if either player departs from his optimal strategy (while the other player adheres to his), his expected payoff will not increase. The expected payoff $\hat{\pi}$ for strategies $(\hat{\mathbf{p}}, \hat{\mathbf{q}})$ is called the *value* of the game. Note from Equation (2.38) that

$$\hat{\mathbf{q}}^T\mathbf{R}\hat{\mathbf{p}} = \hat{\pi}. \tag{2.40}$$

Let $\boldsymbol{\rho}_j$ denote the jth column of $\mathbf{R}$, so that $\hat{\mathbf{q}}^T\boldsymbol{\rho}_j$ is the expected payoff when Q uses the optimum mixed strategy $\hat{\mathbf{q}}$ and P uses a pure strategy P_j. By Von Neumann's Fundamental Theorem, it follows that $\hat{\mathbf{q}}^T\boldsymbol{\rho}_j \geqslant \hat{\pi}$ for every j, since the pure strategy P_j is not necessarily an optimum mixed strategy for P. Similarly, letting $\mathbf{r}_i^T$ denote the ith row of R, we have $\mathbf{r}_i^T\hat{\mathbf{p}} \leqslant \hat{\pi}$ for every i. Thus,

$$\mathbf{R}\hat{\mathbf{p}} \leqslant \hat{\pi}\mathbf{1} \leqslant \mathbf{R}^T\hat{\mathbf{q}}. \tag{2.41}$$

Hence

$$\hat{\pi} > 0 \quad \text{implies that} \quad \mathbf{R}^T\hat{\mathbf{q}} > 0 \tag{2.42}$$

and

$$\hat{\pi} \leqslant 0 \quad \text{implies that} \quad \mathbf{R}\hat{\mathbf{p}} \leqslant \mathbf{0}. \tag{2.43}$$

Furthermore

$$\hat{\pi} > 0 \quad \text{implies that} \quad \hat{\mathbf{q}}^T\mathbf{R}\hat{\mathbf{p}} = \hat{\pi} > 0, \tag{2.44}$$

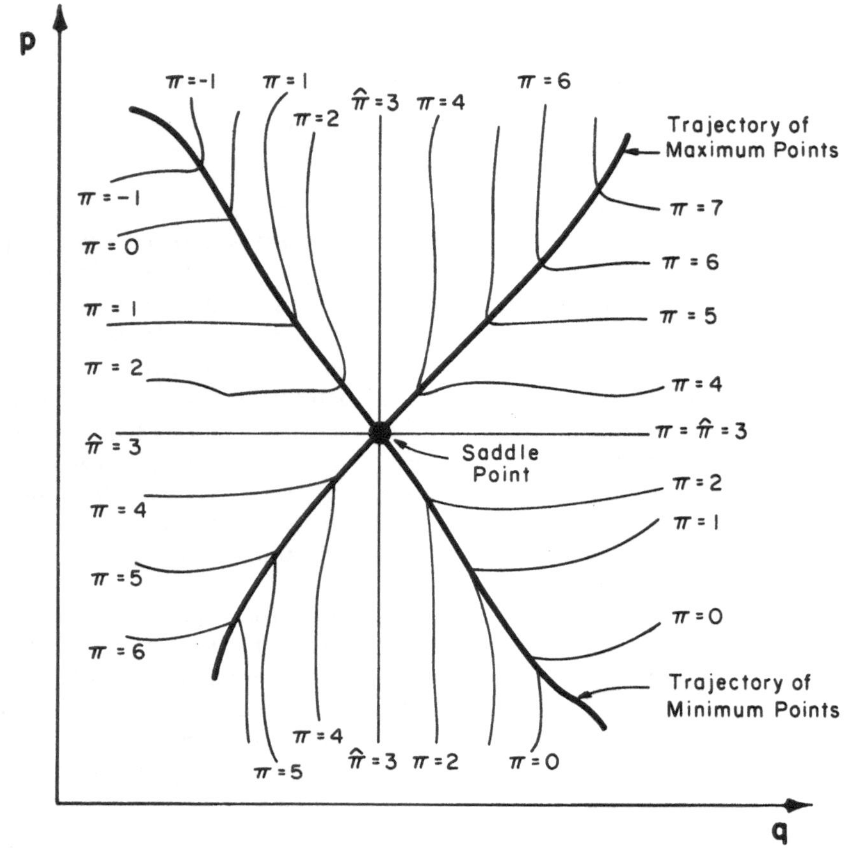

Figure 2.7. A saddle point in strategy space, with **p**, **q** treated, for illustrative purposes, as if they were scalars.

and hence $\mathbf{R}\hat{\mathbf{p}} \nleqslant \mathbf{0}$.

$$\hat{\pi} \leqslant 0 \quad \text{implies that} \quad \hat{\mathbf{q}}^T\mathbf{R}\mathbf{p} = \hat{\pi} \leqslant 0, \tag{2.45}$$

and hence $\mathbf{R}^T\hat{\mathbf{q}} \ngtr 0$. By Equations (2.42) and (2.43),

$$\hat{\pi} > 0 \quad \text{implies that} \quad \mathbf{R}^T\hat{\mathbf{q}} > \mathbf{0} \quad \text{and} \quad \mathbf{R}\hat{\mathbf{p}} \nleqslant \mathbf{0}, \tag{2.46}$$

$$\hat{\pi} \leqslant 0 \quad \text{implies that} \quad \mathbf{R}\hat{\mathbf{p}} \leqslant \mathbf{0} \quad \text{and} \quad \mathbf{R}^T\hat{\mathbf{q}} \ngtr \mathbf{0}. \tag{2.47}$$

Thus either $\mathbf{R}^T\hat{\mathbf{q}} > \mathbf{0}$ or $\mathbf{R}\hat{\mathbf{p}} \leqslant \mathbf{0}$, but not both.

If $\mathbf{R}^T\hat{\mathbf{q}} > \mathbf{0}$, then $\hat{\pi} > 0$. For if $\hat{\pi} \leqslant 0$ we would obtain a contradiction from Equation (2.47). Hence

$$\mathbf{R}^T\hat{\mathbf{q}} > \mathbf{0} \quad \text{implies} \quad \hat{\pi} > 0. \tag{2.48}$$

Similarly,

$$\mathbf{R}\hat{\mathbf{p}} \leqslant \mathbf{0} \quad \text{implies} \quad \hat{\pi} \leqslant 0. \tag{2.49}$$

By Equations (2.42), (2.43), (2.48), and (2.49),

$$\hat{\pi} > 0 \quad \text{if and only if} \quad \mathbf{R}^T\hat{\mathbf{q}} > 0, \tag{2.50}$$

$$\hat{\pi} \leqslant 0 \quad \text{if and only if} \quad \mathbf{R}\hat{\mathbf{p}} \leqslant 0. \tag{2.51}$$

Note that in all cases $\hat{\mathbf{q}} \geqslant \mathbf{0}$, $\hat{\mathbf{q}} \neq \mathbf{0}$, $\hat{\mathbf{p}} \geqslant \mathbf{0}$, $\hat{\mathbf{p}} \neq \mathbf{0}$. Hence, by Theorem 2.7,

$(\mathscr{X}_1, \mathscr{X}_2)$ is linearly separable if and only if $\hat{\pi} > 0$,

$(\mathscr{X}_1, \mathscr{X}_2)$ is not linearly separable if and only if $\hat{\pi} \leqslant 0$.

We say that a pure strategy is *worthwhile* if it appears with nonzero probability in an optimal mixed strategy. All other strategies are *not worthwhile*. Let P_i, Q_j denote pure strategies of players P, Q respectively. Using this terminology, if $\hat{\mathbf{q}}$ is an optimal mixed strategy and if $\hat{q}_i \neq 0$, then Q_i is worthwhile; and if $\hat{q}_i = 0$, then Q_i is not worthwhile.

Suppose $S(Q)$ is the set of subscripts of the worthwhile strategies of player Q, $\hat{\pi}$ is the value of the game, $\mathbf{r}_i$ is the ith row of $\mathbf{R}$, and $\hat{\mathbf{p}}$, $\hat{\mathbf{q}}$ are the optimal mixed strategies. Then one can show that

$$\mathbf{r}_i\hat{\mathbf{p}} = \hat{\pi} \quad \text{for every } i \in S(Q). \tag{2.52}$$

To prove this, note that the expected payoff π does not exceed $\hat{\pi}$ whenever P uses mixed strategy $\hat{\mathbf{p}}$ and Q uses any strategy $\mathbf{q}$. In particular if Q uses only the pure strategy Q_i, the expected payoff is

$$\pi = \mathbf{r}_i^T\hat{\mathbf{p}} \leqslant \hat{\pi}.$$

If it were true that $\mathbf{r}_i^T\hat{\mathbf{p}} < \hat{\pi}$ for some $i \in S(Q)$, then

$$\sum_{S(Q)} \hat{q}_i\mathbf{r}_i^T\hat{\mathbf{p}} < \hat{\pi}$$

since $\sum_{S(Q)} \hat{q}_i = 1$. But this contradicts Equation (2.40), namely,

$$\sum_{S(Q)} \hat{q}_i\mathbf{r}_i^T\hat{\mathbf{p}} = \hat{\pi}.$$

Hence $\mathbf{r}_i\hat{\mathbf{p}} = \hat{\pi}$ for every $i \in S(Q)$, thereby proving Equation (2.52).

Thus if Q_i is worthwhile, then $\mathbf{r}_i^T\hat{\mathbf{p}} = \hat{\pi}$; but—recalling Equation (2.41)—if Q_i is not worthwhile, then $\mathbf{r}_i^T\hat{\mathbf{p}} \leqslant \hat{\pi}$. Similarly, if P_i is worthwhile, then $\boldsymbol{\rho}_j^T\hat{\mathbf{q}} = \hat{\pi}$. But if P_i is not worthwhile, then $\boldsymbol{\rho}_j^T\hat{\mathbf{q}} \geqslant \hat{\pi}$.

Similarly one can show that

$$\hat{\mathbf{q}}^T\mathbf{R}\mathbf{p} = \hat{\pi}$$

for every $\mathbf{p}$ such that $p_i = 0$ whenever $\hat{p}_i = 0$.

The pair of optimal mixed strategies $(\hat{\mathbf{p}}, \hat{\mathbf{q}})$ is a saddle point of the expected payoff in $\mathbf{p}$,$\mathbf{q}$-space. This is illustrated in Figure 2.7.

Theorem 2.8. *If player P adheres to his optimal mixed strategy p, then the payoff π remains equal to the value of the game $\hat{\pi}$, regardless of Q's mixed strategy $\mathbf{q}$, provided Q adheres to worthwhile strategies—i.e., provided $q_i = 0$*

whenever $\hat{q}_i = 0$. *In other words:*

$$\mathbf{q}^T\mathbf{R}\hat{\mathbf{p}} = \hat{\pi} \quad \text{for every } \mathbf{q} \text{ such that } q_i^T = 0 \text{ whenever } \hat{q}_i = 0.$$

Proof. Let π denote the expected payoff when P uses an optimal mixed strategy $\hat{\mathbf{p}}$ and Q uses any mixed strategy $\mathbf{q}$ consisting of only worthwhile pure strategies. Then, by Equation (2.52),

$$\begin{aligned}\pi = \mathbf{q}^T\mathbf{R}\hat{\mathbf{p}} &= \sum_{i \in S(Q)} q_i \mathbf{r}_i \hat{\mathbf{p}} \\ &= \hat{\pi} \sum_{i \in S(Q)} q_i = \hat{\pi}\end{aligned} \tag{2.53}$$

provided $q_i = 0$ whenever $\hat{q}_i = 0$. □

Fictitious Play. The technique of fictitious play provides a means of determining whether or not the value of the game is positive, and hence whether or not $(\mathscr{X}_1, \mathscr{X}_2)$ is linearly separable. Imagine a sequence of fictitious games, in each of which P, Q alternately announce pure strategies. In each game, each of P, Q chooses his pure strategy as an optimum response to the empirical (i.e., apparent) mixed strategy of his opponent. Each of P and Q makes these choices in such a way as to minimize the expected payoff to his opponent. In this way all of the worthwhile elements of both $\mathbf{R}^T\mathbf{q}$ and $\mathbf{R}\mathbf{p}$ approach $\hat{\pi}$ as the fictitious play progresses. It can be shown that the expected payoffs in this sequence of games converge to $\hat{\pi}$ as the number of iterations in the fictitious play approaches infinity [12].

Let $\mathbf{p}(k)$, $\mathbf{q}(k)$ denote the empirical mixed strategy of players P, Q, respectively at stage k, and let $\partial\mathbf{p}(k)$, $\partial\mathbf{q}(k)$ denote the probability distributions of the pure strategies of players P, Q at stage k. (A stage consists of an announcement of P_i followed by an announcement of Q_j.) The play starts at $k = 0$, with P assuming that

$$\mathbf{q}(0) = \frac{1}{t}[1, 1, \ldots, 1]^T. \tag{2.54}$$

Since the linear separability of $(\mathscr{X}_1, \mathscr{X}_2)$ and the effectiveness of the solution vector $\mathbf{v}^*$, if it exists, are unaffected by a multiplying factor on $\mathbf{q}(0)$, we hereafter ignore the requirement that $\mathbf{q}(k)^T\mathbf{1} = \mathbf{p}(k)^T\mathbf{1}$. If we wish we may introduce this requirement at any time by a normalization of the form

$$\mathbf{q}'(k) = \mathbf{q}(k)/\mathbf{q}(k)^T\mathbf{1}.$$

Thus we replace Equation (2.54) by

$$\mathbf{q}(0) = [1, 1, \ldots, 1]^T. \tag{2.55}$$

The payoff vector in response to $\mathbf{q}(0)$ is

$$\boldsymbol{\pi}_q(0) = \mathbf{R}^T\mathbf{q}(0).$$

The ith component of $\boldsymbol{\pi}_q(0)$ is the expected payoff when P chooses pure strategy P_i. In the fictitious play, P_i is choosen so that the ith component of $\mathbf{R}^T\mathbf{q}$ is not greater than any other component of $\mathbf{R}^T\mathbf{q}$. Suppose P chooses pure strategy P_i at stage 0. Then

$$\partial\mathbf{p}(0) = (\delta_{i1}, \delta_{i2}, \ldots, \delta_{is})^T$$

where

$$\delta_{ij} = \begin{cases} 0, & \text{for } i \neq j \\ 1, & \text{for } i = j. \end{cases}$$

Thus $\mathbf{p}(0) = \partial\mathbf{p}(0)$.

The payoff vector in response to $\mathbf{p}(0)$ is

$$\boldsymbol{\pi}_p(0) = \mathbf{Rp}(0).$$

The jth component of $\boldsymbol{\pi}_p(0)$ is the expected payoff when Q chooses pure strategy Q_j. In fictitious play, Q chooses Q_j so that the jth component of $\pi_p(0)$ is not less than any other component of $\mathbf{Rp}(0)$. Let Q_j denote this locally optimum pure strategy. Then

$$\partial\mathbf{q}(1) = \{\delta_{jl} \mid l = 1, \ldots, t\}$$

and

$$\mathbf{q}(1) = \mathbf{q}(0) + \partial\mathbf{q}(1)$$

(Here it is assumed that $\mathbf{q}(0)$ is the empirical mixed strategy of m pure strategies, where $m = \mathbf{q}^T(0)\mathbf{1}$.) Similarly,

$$\mathbf{p}(1) = \mathbf{p}(0) + \partial\mathbf{p}(1).$$

In this way we obtain

$$\mathbf{p}(k) = \mathbf{p}(k-1) + \partial\mathbf{p}(k)$$
$$\mathbf{q}(k) = \mathbf{q}(k-1) + \partial\mathbf{q}(k).$$

The proof of the Robinson–Brown theorem [12] shows that

$$\mathbf{r}_i^T\mathbf{p}(k) \xrightarrow[k\to\infty]{} \hat{\pi} \quad \text{for } i \in S(Q),$$

where $S(Q) = \text{set } \{i \mid q_i > 0\}$.

$$\boldsymbol{\rho}_j^T\mathbf{q}(k) \xrightarrow[K\to\infty]{} \hat{\pi} \quad \text{for } j \in S(P).$$

Suppose $i \notin S(Q)$. Then

$$\pi_p(\infty) = \lim_{k\to\infty} \mathbf{r}_i^T\hat{\mathbf{p}}(k) \equiv \mathbf{r}_i^T\hat{\mathbf{p}} \leqslant \hat{\pi}.$$

If $j \notin S(P)$, then

$$\lim_{k\to\infty} \boldsymbol{\rho}_j^T\mathbf{q}(k) \geqslant \hat{\pi}.$$

Since the Robinson–Brown theorem states that $\pi(k)$ approaches $\hat{\pi}$ as k approaches ∞, it follows that $\mathbf{Rp}(k) \leqslant \hat{\pi}\mathbf{1} \leqslant \mathbf{R}^T\mathbf{q}(k)$ for some $k > 0$. Thus either $\mathbf{Rp}(k) \leqslant \mathbf{0}$ or $\mathbf{0} < \mathbf{R}^T\mathbf{q}(k)$ for some k. (Recall that it is impossible to find stochastic vectors $\mathbf{p}$ and $\mathbf{q}$ such that both $\mathbf{Rp} \leqslant \mathbf{0}$ and $\mathbf{R}^T\mathbf{q} > \mathbf{0}$.) It follows that if $\hat{\pi} \neq 0$, either $\mathbf{R}^T\mathbf{q}(k) > \mathbf{0}$ or $\mathbf{Rp}(k) \leqslant \mathbf{0}$ will be achieved by a finite k. If $\hat{\pi} = 0$, then convergence but nontermination to $\mathbf{0}$ by $\mathbf{Rp}(k)$ is possible as $k \to \infty$.

The fictitious play continues until either $\mathbf{R}^T\mathbf{q}(k) > \mathbf{0}$ or $\mathbf{Rp}(k) \leqslant \mathbf{0}$. If $\mathbf{R}^T\mathbf{q} > \mathbf{0}$, one uses $\mathbf{q}$ to find $\mathbf{v}$. If $\mathbf{Rp} \leqslant \mathbf{0}$, one knows that $(\mathscr{X}_1, \mathscr{X}_2)$ is not linearly separable.

This completes our explicit description of the fictitious play procedure. In Section 2.9, we illustrate this procedure by a numerical example.

2.8 Simplifying Techniques

In some cases one may reduce $\mathbf{R}$ to a smaller matrix before beginning fictitious play, without affecting the classification performance of the derived value of $\mathbf{v}$. In other cases one may find that $\mathscr{X}_1, \mathscr{X}_2$ are not linearly separable without carrying out fictitious play. In still other cases one may detect linear separability and find a solution vector without applying fictitious play to $\mathbf{R}$.

When the possibility of such simplifications exist, they may often be achieved by one or more of the five techniques described below. The first two of these techniques are reduction techniques, the next two are nonfictitious play detectors of the fact that $\mathscr{X}_1, \mathscr{X}_2$ are not linearly separable, and the last one detects linear separability and finds $\mathbf{v}$ without fictitious play.

Simplifying Technique 1 (Reduction). *Let* $\mathbf{r}_j^T$ *denote the* ith *row of* $\mathbf{R}$. *If*

$$\mathbf{r}_i \leqslant \sum_{j \neq i} c_j \mathbf{r}_j, \quad \textit{for some } \mathbf{c} \geqslant \mathbf{0},$$

then $\mathbf{r}_i^T$ *may be deleted from* $\mathbf{R}$ *in testing to see whether there exists a stochastic vector* $\mathbf{q}$ *such that* $\mathbf{R}^T\mathbf{q} > \mathbf{0}$.

Proof. (a) Suppose $\mathscr{X}_1$, $\mathscr{X}_2$ are linearly separable. Then there exists a stochastic vector $\mathbf{q}$ such that $\mathbf{R}^T\mathbf{q} > \mathbf{0}$. Consequently $\sum_k q_k \mathbf{r}_k > \mathbf{0}$. Since

$$\mathbf{r}_i \leqslant \sum_{j \neq i} c_j \mathbf{r}_j, \quad \text{with } \mathbf{c} \geqslant \mathbf{0},$$

it follows that

$$0 < \sum_k q_k \mathbf{r}_k \leqslant \sum_{k \neq i} (q_k + q_i c_k)\mathbf{r}_k = \sum_{k \neq i} q'_k \mathbf{r}_k,$$

where $q'_k = q_k + q_i c_k$. Note that $\mathbf{q}' > \mathbf{0}$. Let $\hat{\mathbf{q}} = \mathbf{q}'/|\mathbf{q}'|$. Note that $\hat{\mathbf{q}}$ is a stochastic vector. Hence if there exists a stochastic vector $\mathbf{q}$ such that $\mathbf{R}^T\mathbf{q} > \mathbf{0}$ and if $\mathbf{r}_i \leqslant \sum_{j \neq i} c_j \mathbf{r}_j$ with $c \geqslant \mathbf{0}$, then there exists a stochastic vector $\hat{\mathbf{q}}$ such that $(\mathbf{R}')^T\mathbf{q} > 0$, where $\mathbf{R}'$ is defined as the matrix obtained by removing $\mathbf{r}_i^T$ from $\mathbf{R}$.

(b) Suppose $\mathscr{X}_1, \mathscr{X}_2$ are not linearly separable. Then there exists a stochastic vector $\mathbf{p}$ such that $\mathbf{Rp} \leqslant \mathbf{0}$. Hence $\mathbf{r}_i^T\mathbf{p} \leqslant 0$ for every i. Since $\mathbf{r}_i \leqslant \sum_{j \neq i} c_j \mathbf{r}_j$, it follows that

$$\mathbf{r}_i^T\mathbf{p} \leqslant \sum_{j \neq i} c_j \mathbf{r}_j^T\mathbf{p} \leqslant 0.$$

Hence $\mathbf{r}_i^T$ may be removed from $\mathbf{R}$ without affecting the validity of the test for linear separability if $\mathscr{X}_1, \mathscr{X}_2$ are not linearly separable. □

Simplifying Technique 2 (Reduction). *Let $\{\boldsymbol{\rho}_j\}$ denote the columns of* $\mathbf{R}$. *If*

$$\boldsymbol{\rho}_k \leqslant \sum_{j \neq k} \lambda_j \boldsymbol{\rho}_j \quad \textit{with } \boldsymbol{\lambda} \leqslant \mathbf{0}, \boldsymbol{\lambda} \neq \mathbf{0},$$

then $\boldsymbol{\rho}_k$ may be eliminated from $\mathbf{R}$ *in the process of simultaneously testing for linear separability and finding a separating weight vector. The vector* $\mathbf{q}$ *obtained in this way can be used in the equation* $\mathbf{v} = \mathbf{F}^T\mathbf{q}'$ *to find the separating weight vector* $\mathbf{v}$ *as described in Equation* (2.33).

The proof of this technique is analogous to the proof of Simplifying Technique 1. To see this, interchange the roles of P and Q, replace $\mathbf{R}$ by $-\mathbf{R}$ (thereby changing the signs of the payoffs) and replace $\boldsymbol{\lambda}$ by $\mathbf{c}$.

Simplifying Techniques 1 and 2 may be used alternately in reducing $\mathbf{R}$. If only columns of $\mathbf{R}$ are removed in the reduction process, one may eliminate corresponding columns in $\mathbf{A}^T\mathbf{P}$, thereby reducing the size of the training set.

In the two reduction techniques just described, one may wish to save the eliminated rows and columns in the event one will want to find a solution vector $\mathbf{v}$. To do this one may shift the eliminated rows and columns to the bottom portion and the extreme right-hand portions, respectively, of the original matrix $\mathbf{R}$, absorbing the shifts in $\mathbf{F}$ and $\mathbf{P}$. In this way, if the reduced $\mathbf{R}$ indicates the existence of a separating hyperplane the full $\mathbf{R}$ (including the shifted rows and columns) may be used to find $\mathbf{q}$. This vector $\mathbf{q}$ may then be used to find the solution vector $\mathbf{v}$ by means of Equation (2.33).

Simplifying Technique 3 (Detecting That $\mathscr{X}_1$, $\mathscr{X}_2$ Are Not Linearly Separable). *If any column of* $\mathbf{R}$ *has no positive element, then $\mathscr{X}_1$, $\mathscr{X}_2$ are not linearly separable.*

PROOF. If any column of $\mathbf{R}$ has no positive element, then there is no stochastic vector $\mathbf{q}$ such that $\mathbf{R}^T\mathbf{q} > \mathbf{0}$. Hence $\mathscr{X}_1$, $\mathscr{X}_2$ are not linearly separable. □

Simplifying Technique 4 (Detecting That $\mathscr{X}_1$, $\mathscr{X}_2$ Are Not Linearly Separable). *Let $\{\boldsymbol{\rho}_j\}$ denote the columns of* $\mathbf{R}$. *Let $\mathscr{S}$ denote any nonempty subset of the indices among the $\boldsymbol{\rho}_j$. If $\sum_{j\in\mathscr{S}} \boldsymbol{\rho}_j \leqslant \mathbf{0}$, then $\mathscr{X}_1$, $\mathscr{X}_2$ are not linearly separable.*

PROOF. Let

$$b_i = \begin{cases} 1, & \text{for } i \in \mathscr{S}, \\ 0, & \text{for } i \notin \mathscr{S}. \end{cases}$$

Let $\mathbf{b} = \{b_i\}$, $\mathbf{p} = \mathbf{b}/|\mathbf{b}|$. Then

$$\mathbf{Rb} = \frac{1}{|b|} \sum_{j\in\mathscr{S}} \boldsymbol{\rho}_j \leqslant \mathbf{0}.$$

Hence $\mathbf{Rp} \leqslant \mathbf{0}$, which implies that $\mathscr{X}_1$, $\mathscr{X}_2$ are not linearly separable. □

Simplifying Technique 5 (Detecting Linear Separability and Finding a Solution $\mathbf{v}$). *Let* $\mathbf{F}$ *denote a nonsingular row operation matrix and let* $\mathbf{P}$ *denote* *a permutation matrix in a transformation of* $\mathbf{A}$ *given by* $\mathbf{F}\mathbf{A}^T\mathbf{P} = \mathbf{S}$. *Let* $\mathbf{s}_i$ *denote the* ith *column of* $\mathbf{S}$. *If there exist* $\mathbf{F}$, $\mathbf{P}$ *such that* $\mathbf{s}_i \geqslant \mathbf{0}$, $\mathbf{s}_i \neq \mathbf{0}$ *for every i, then* $\mathscr{X}_1$, $\mathscr{X}_2$ *are linearly separable. Furthermore, the weight vector* $\mathbf{v}$ *of a separating hyperplane may be obtained from*

$$\mathbf{v} = \mathbf{F}^T\mathbf{q},$$

where $\mathbf{q}$ *is any positive t-vector* (i.e., $\mathbf{q} > \mathbf{0}$ *and the number of elements in* $\mathbf{q}$ *is t*).

PROOF. Let $\mathbf{q}$ denote any positive t-vector. Then $\mathbf{s}_i^T\mathbf{q} > 0$ for every i. Hence $\mathbf{S}^T\mathbf{q} > \mathbf{0}$, i.e., $\mathbf{P}^T\mathbf{A}\mathbf{F}^T\mathbf{q} > \mathbf{0}$.

Let $\mathbf{v} = \mathbf{F}^T\mathbf{q}$. Then $\mathbf{P}^T\mathbf{A}\mathbf{v} > 0$. Since $\mathbf{P}$ is a permutation matrix, it follows that $\mathbf{A}\mathbf{v} > 0$. □

Another simplifying technique, describe by Gaston [13], tests whether or not a finite, two-class training set is linearly separable. This technique, however, would be a superfluous addition to the set of five already given here.

2.9 Illustrative Example [10]

This numerical example illustrates the fictitious play game-theoretic technique. Let $\hat{\mathbf{R}}$ denote either $\mathbf{R}$ or the reduced version of $\mathbf{R}$, i.e., the matrix resulting from the application of reduction techniques (Simplifying Techniques 1 and 2) to $\mathbf{R}$. Suppose

$$\hat{\mathbf{R}} = \begin{bmatrix} 2 & -2 & -2 \\ -1 & 1 & 1 \\ 1 & -1 & 0 \\ 1 & 0 & -1 \\ 0 & -1 & -1 \end{bmatrix}, \qquad \mathbf{F} = \begin{bmatrix} -1 & -1 & -1 & -1 & -1 \\ 1 & 0 & 0 & 0 & 0 \\ 0 & -1 & 0 & 0 & 0 \\ 0 & 0 & -1 & 0 & 0 \\ 0 & 0 & 0 & -1 & 0 \end{bmatrix}.$$

For convenience we omit the cap (^) in the remainder of this example.

Start the fictitious play with $\mathbf{q}(0) = \mathbf{1}$, ignoring the requirement that $\mathbf{q}(k)^T\mathbf{1} = 1$. (The sum-to-unity normalization would have no effect on the outcome, while adding unnecessary computations to the fictitious play.) Starting with $\mathbf{q}(0) = \mathbf{1}$ is equivalent to the assumption by player P that all the pure strategies of Q are equally likely. The steps in the fictitious play are shown in Table 2.2, where the 3×5 matrix in the upper left is $\mathbf{R}^T$.

In this table the columns to the right of $\mathbf{R}^T$ are nonnormalized instances of the vector $\mathbf{R}^T\mathbf{q}(k)$. Each element of $\mathbf{R}^T\mathbf{q}(k)$, after normalization of $\mathbf{q}(k)$, is the expected payoff for the corresponding pure strategy of player P.

The rows below $\mathbf{R}^T$ in Table 2.2 are nonnormalized instances of $\mathbf{R}\mathbf{p}(k)$. Each element of $\mathbf{R}\mathbf{p}(k)$, after normalization of $\mathbf{p}(k)$, is the expected payoff for the corresponding pure strategy of player Q.

Table 2.2. Steps in fictitious play for the numerical example.

2	−1	1	1	0	3	2	1	0	−1°	−2°	−3°	−2°	−1°	0°	1
−2	1	−1	0	−1	−3°	−2	−1°	0	1	2	3	2	2	1	1
−2	1	0	−1	−1	−3	−2°	−1	0°	1	2	3	3	2	2	1
−2	1*	−1	0	−1											
−4	2*	−1	−1	−2											
−6	3*	−2	−1	−3											
−8	4*	−2	−2	−4											
−6	3*	−1	−1	−4											
−4	2*	0	0	−4											
−2	1	1*	1	−4											
0	0	2	2*	−4											
2	−1	3*	3	−4											
4	−2	4	4*	−4											

The first step in the fictitious play in this example is to add the five columns of $\mathbf{R}^T$ to obtain the first column on the right. This column is proportional to the expected payoff for the three pure strategies associated with the rows of $\mathbf{R}^T$. In this column the smallest expected payoff occurs in the second and third rows. Either of these rows are acceptable for subsequent fictitious play. We arbitrarily chose the second row, indicating our choice by the small circle (°). This row is entered on the first line below $\mathbf{R}^T$ in Table 2.2. Each element of this row is proportional to the payoff when $\mathbf{p}(0) = [0,1,0]^T$ (a pure strategy). The maximum value of this row occurs in the second column, marked by an asterisk (*). Hence a pure strategy of $q(1) = [0,1,0,0,0]^T$ is played next, causing an empirical mixed strategy of $\mathbf{q}(1) = \mathbf{1} + [0,1,0,0,0]^T = [1,2,1,1,1]^T$, ignoring the normalizing factors. This results in our expected payoff on the second column of $\mathbf{R}^T$ plus the previously computed column at the right of $\mathbf{R}^T$. The result is written in Table 2.2 as the second column to the right of $\mathbf{R}^T$. The row of $\mathbf{R}^T$ corresponding to the minimum of this column is then added to the first row below $\mathbf{R}^T$ to form the second row below $\mathbf{R}^T$. In general, this process is continued until either the column on the extreme right is positive (i.e., every element of the column is positive), in which case $\mathscr{X}_1$, $\mathscr{X}_2$ are linearly separable, or all the elements of a row below $\mathbf{R}^T$ are nonpositive, in which case $\mathscr{X}_1$, $\mathscr{X}_2$ are not linearly separable. In the first case, with $\mathscr{X}_1$, $\mathscr{X}_2$ linearly separable, the vector $\mathbf{q}$ is obtained by computing the number of times each column of $\mathbf{R}^T$ has been added to the columns to the right of $\mathbf{R}^T$ to form the column on the extreme right. This computation may be done by counting the asterisks in the rows below $\mathbf{R}^T$.

In the present example the rightmost column of Table 2.2 is 1. Hence $(\mathscr{X}_1, \mathscr{X}_2)$ is linearly separable. By counting the asterisks and recalling that the initial value of $\mathbf{q}$ in the fictitious play was $\mathbf{1}$, we find that the final value of $\mathbf{q}$ is

$$\mathbf{q} = [1,1,1,1,1]^T + [0,6,2,2,0]^T = [1,7,3,3,1]^T.$$

To find the solution vector $\mathbf{v}$, we use Equation (2.35):

$$\mathbf{v} = \mathbf{F}^T\mathbf{q} = \begin{bmatrix} 6 \\ -4 \\ -4 \\ -2 \\ -1 \end{bmatrix}.$$

In these computations, we have ignored the requirements $\mathbf{q}^T\mathbf{1} = 1$ and $\mathbf{p}^T\mathbf{1} = 1$, because this does not affect the classification performance of the weight vector $\mathbf{v}$.

Although a pair of class regions may be linearly separable (LS), the possibility that it is non-LS usually must be dealt with.

The game-theoretic technique detects both the LS and non-LS conditions in a finite time, and—if the pair of classes is LS—finds a solution vector in a finite time. But if the classes are non-LS, the game-theoretic technique does not necessarily compute a near-optimal weight vector.

In Section 2.13 we discuss a somewhat similar technique—the Ho–Kashyap procedure. Like the game-theoretic method described above, it yields a solution vector in a finite time if the classes are LS, but it does not necessarily find a near-optimal weight vector in a finite time if the classes are are non-LS.

The Ho–Kashyap procedure is based on the concepts of gradient descent and least-mean-square-error solutions. These concepts are discussed below, before our discussion of the Ho–Kashyap procedure. In Chapters 4 and 5, we discuss other training procedures derivable from the concept of gradient descent.

2.10 Gradient Descent

Visualize a man standing on a hillside in a very dense fog. He wishes to find his home, which is located at the point of lowest altitude within a very large crater. If he begins his search within the crater, if the terrain is smooth, and if there are no dips or smaller craters along his path, he will find his home by always traveling against the direction of maximum increasing slope. In doing this, he would be applying the gradient descent technique.

Conceptually, then, the gradient descent technique is simple. A loss function $J(\mathbf{v})$ is first determined, where $\mathbf{v}$ is a controllable parameter vector. (In the example of the man on the hillside, $\mathbf{v}$ is a position vector.) One seeks the value or values of $\mathbf{v}$ where $J(\mathbf{v})$ is a minimum. In the example, the loss function $J(\mathbf{v})$ is the altitude of the terrain as a function of $\mathbf{v}$. Let $\mathbf{v}(t)$ denote $\mathbf{v}$ as a function of time t. An initial vector $\mathbf{v}(0)$ is arbitrarily selected. Motion of the parameter vector commences in parameter space from this point. The motion is always in a direction *which is exactly opposite* to the gradient of the loss function (i.e., the direction of $-\nabla J(\mathbf{v})$ which is the negative of the direction of maximum increasing $J(\mathbf{v})$ at location $\mathbf{v}$). If a loss function has been chosen so that the descent motion does not become entrapped at a local minimum and if a finite global minimum exists, the gradient descent procedure will find a parameter vector $\mathbf{v}$ such that $J(\mathbf{v})$ is minimized.

For recursive techniques some modification of this technique is made since discrete corrections to $\mathbf{v}$ require that the descent motion be an incremental approximation to the continuous gradient descent path. In these modified procedures, we converge to a minimum of a function $J(\mathbf{v})$ by making an initial guess $\mathbf{v}(0)$, finding the gradient (multidimensional derivative) of $J(\mathbf{v})$ at $\mathbf{v} = \mathbf{v}(0)$, and making a second guess $\mathbf{v}(1)$ by adding to $\mathbf{v}(0)$ a vector $\boldsymbol{\delta}$ having the direction of the negative of the gradient. Subsequent $\mathbf{v}(k)$'s are computed in a similar manner. Under proper constraints on $J(\mathbf{v})$ and $\mathbf{v}(0)$, $\{\mathbf{v}(k)\}$ will converge to a vector $\mathbf{v}^*$ where $J(\mathbf{v}^*)$ is a local minimum.

To understand the details of this process, it is helpful to consider the one-dimensional case first. Here the function $J(\theta)$ is defined over the scalar θ. Let $J'(\theta)$ denote the derivative of $J(\theta)$ with respect to θ, $J''(\theta)$ the second derivative, etc. We find an approximation to $J'(\theta + \delta)$ via a Taylor's series:

$$\begin{aligned} J'(\theta + \delta) &= J'(\theta) + \delta J''(\theta) + \frac{\delta^2}{2!} J'''(\theta) + \cdots \\ &= \left(1 + \delta p + \frac{(\delta p)^2}{2!} + \frac{(\delta p)^3}{3!} + \cdots\right) J'(\theta) \qquad (2.56) \\ &= e^{\delta p} J'(\theta), \end{aligned}$$

where p is the operator $d/d\theta$, and where $e^{\delta p}$ is the operator defined by

$$e^{\delta p} = 1 + \delta p + \frac{(\delta p)^2}{2!} + \frac{(\delta p)^3}{3!} + \cdots.$$

Make a first-order approximation of $J'(\theta + \delta)$ as follows:

$$J'(\theta + \delta) \cong J'(\theta) + \delta J''(\theta).$$

If $J'(\theta + \delta) = 0$, then

$$\delta \cong -J'(\theta)/J''(\theta).$$

If this approximation is applied iteratively, then the kth iteration of δ may be written as follows:

$$\delta(k) = \theta(k + 1) - \theta(k) = -J'[\theta(k)]/J''[\theta(k)],$$

which may be written in the following recursive form, known as *Newton's algorithm:*

$$\theta(k + 1) = \theta(k) - \frac{J'[\theta(k)]}{J''[\theta(k)]}.$$

Successive applications of this equation, under the constraints that (a) all derivatives of $J(\theta)$ exist in a neighborhood $\mathcal{N}(\theta^*)$ containing θ^* and (b) $\theta(0) \in \mathcal{N}(\theta^*)$, will yield a sequence $\{\theta(k)\}$ that converges to θ^* as $k \to \infty$.

Figure 2.8 illustrates Newton's algorithm in geometric terms. This algorithm is perhaps one of the oldest training procedures in formal mathematics.

An analogous procedure exists for functions of a vector variable $\mathbf{v}$. In $(d + 1)$-dimensional space, Equation (2.56) becomes

$$J(\mathbf{v} + \boldsymbol{\delta}) = e^{\boldsymbol{\delta}^T \nabla} J(\mathbf{v}),$$

where

$$\begin{aligned} \mathbf{v} &= [v_0, \ldots, v_d]^T, \\ \boldsymbol{\delta}^T &= [\boldsymbol{\delta}_0, \ldots, \boldsymbol{\delta}_d], \\ \nabla &= \left[\frac{\partial}{\partial v_0}, \ldots, \frac{\partial}{\partial v_d}\right]^T. \end{aligned}$$

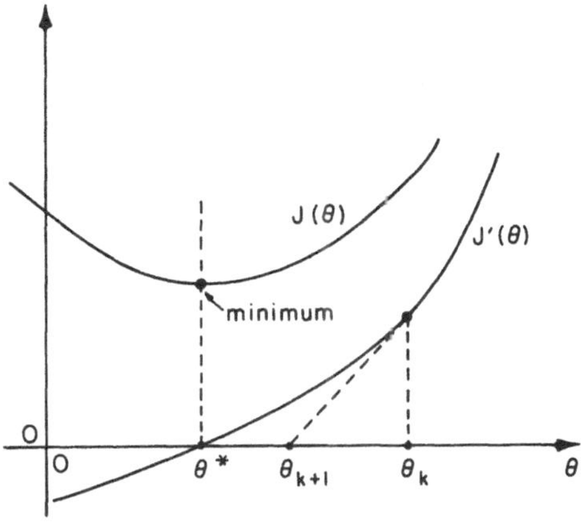

Figure 2.8. Newton's algorithm.

Thus

$$J(\mathbf{v} + \boldsymbol{\delta}) = \left[1 + \boldsymbol{\delta}^T \nabla + \frac{(\boldsymbol{\delta}^T \nabla)^2}{2!} + \cdots \right] J(\mathbf{v})$$

$$= J(\mathbf{v}) + \boldsymbol{\delta}^T \nabla J(\mathbf{v}) + \frac{1}{2!} \boldsymbol{\delta}^T \nabla \nabla^T J(\mathbf{v}) \boldsymbol{\delta} + \cdots$$

$$= J(\mathbf{v}) + \nabla^T J(\mathbf{v}) \boldsymbol{\delta} + \frac{1}{2!} \boldsymbol{\delta}^T \nabla \nabla^T J(\mathbf{v}) \boldsymbol{\delta} + \cdots$$

and

$$\nabla J(\mathbf{v} + \boldsymbol{\delta}) = \nabla J(\mathbf{v}) + \nabla \nabla^T J(\mathbf{v}) \boldsymbol{\delta} + \cdots.$$

Setting $\nabla J(\mathbf{v} + \boldsymbol{\delta}) = \mathbf{0}$ yields the approximation

$$\boldsymbol{\delta} \cong -[\nabla \nabla^T J(\mathbf{v})]^{-1} \nabla J(\mathbf{v}). \tag{2.57}$$

If this approximation is applied iteratively, then Equation (2.57) may be written in the following recursive form:

$$\mathbf{v}(k + 1) = \mathbf{v}(k) - [\nabla \nabla^T J(\mathbf{v}(k))]^{-1} \nabla J(\mathbf{v}(k)), \tag{2.58}$$

where k is the iteration number, and $\boldsymbol{\delta}(k)$ has been replaced by $\mathbf{v}(k + 1) - \mathbf{v}(k)$. This is the vector form of Newton's algorithm.

Gradient descent procedures often use the following modification of Equation (2.58):

$$\mathbf{v}(k + 1) = \mathbf{v}(k) + \rho_k[-\nabla J(\mathbf{v}(k)], \tag{2.59}$$

where ρ_k is a positive number that may be constant or may depend on k.

Provided the conditions presented in the following section are met, Equation (2.58) will yield a sequence $\{\mathbf{v}(k)\}$ that converges to $\mathbf{v}^*$ as $k \to \infty$.

In pattern recognition training procedures, a complete description of the class probability densities is generally unavailable. Therefore, the actual gradient at a point cannot be calculated; instead, a statistical estimate of the gradient using a finite sample must be used. When this estimate is used, the training procedure becomes *stochastic approximation* rather than gradient descent. Further discussion of gradient descent appears in Chapter 4, where the theory of stochastic approximation is also presented.

2.11 Conditions for Ensuring Desired Convergence

Let $\mathbf{v} = \mathbf{v}^*$ denote any vector in some set of interest and let this set be denoted by $\mathscr{V}^*$. The necessary and sufficient conditions for ensuring that $\mathscr{V}^*$ represents the set of augmented weight vectors associated with a *local* minimum of $J(\mathbf{v})$ are given below.

Condition 1. The set $\mathscr{V}^*$ is connected and

$$\nabla J(\mathbf{v})\big|_{\mathbf{v} \in \mathscr{V}^*} = \mathbf{0}. \tag{2.60}$$

Condition 2.

$$\left.\frac{\partial^2 J(\mathbf{v})}{\partial \mathbf{v}^2}\right|_{\mathbf{v} \in \partial\mathscr{V}^*} > 0, \tag{2.61}$$

where $\partial\mathscr{V}^*$ denotes the boundary of $\mathscr{V}^*$. The left member of Inequality (2.61) is a positive definite (i.e., all eigenvalues positive) matrix whose elements are $\partial^2 J(\mathbf{v})/\partial v_i \, \partial v_j$.

Condition 1 states that $J(\mathbf{v})$ must be constant over the connected region $\mathscr{V}^*$, and Condition 2 states that it must be increasing at the outer edges of the boundary of $\mathscr{V}^*$ as one proceeds outward from $\mathscr{V}^*$.

If a region $\mathscr{N} \equiv \mathscr{N}(\mathbf{v}^*)$ of augmented weight space contains just one such connected set $\mathscr{V}^*$, then $J(\mathbf{v})$ has precisely one minimum value in $\mathscr{N}$. If $\mathbf{v}(0)$ is chosen within $\mathscr{N}$ and all the derivatives of $J(\mathbf{v})$ exist in $\mathscr{N}$, convergence to $\mathscr{V}^*$ is ensured when a gradient descent procedure is used. This, however, will not be guaranteed when a stochastic approximation to a gradient descent procedure is used.

2.12 Gradient Descent for Designing Classifiers

In the process of designing classifiers, one may define a *loss function* that numerically summarizes the defects in the classifier's performance. We usually denote this loss function by $J(\mathbf{v})$, where $\mathbf{v}$ is an augmented weight vector. We seek a value, say $\mathbf{v}^*$, where $J(\mathbf{v})$ is a minimum—i.e.,

$$J(\mathbf{v}^*) \leqslant J(\mathbf{v}) \quad \text{for all } \mathbf{v}.$$

If the conditions of the preceding section are met, then Equation (2.59) will generate a sequence $\{\mathbf{v}(k)\}$ that converges to $\mathbf{v}^*$ as $k \to \infty$ whenever $\mathbf{v}(0)$ lies within a neighborhood of $\mathbf{v}^*$, say $\mathcal{N}'(\mathbf{v}^*)$, that is a subset of $\mathcal{N}(\mathbf{v}^*)$.

One of the simplest nontrivial forms of $J(\mathbf{v})$ that has such differentiability almost everywhere in $\mathbf{v}$-space is

$$J(\mathbf{v}) = \sum_{n=0}^{N-1} [\mathbf{v}^T\boldsymbol{\eta}(n) - b_n]^2, \qquad b_n > 0,$$

where N denotes the number of feature vectors in the training set. Let $\mathbf{A}$ denote the $N \times (d+1)$ matrix

$$\mathbf{A} = \begin{bmatrix} \boldsymbol{\eta}^T(0) \\ \vdots \\ \boldsymbol{\eta}^T(N-1) \end{bmatrix},$$

and let $\mathbf{b}$ denote the N-dimensional vector

$$\mathbf{b} = \begin{bmatrix} b_0 \\ \vdots \\ b_{N-1} \end{bmatrix}.$$

Then

$$\begin{aligned} J(\mathbf{v}) &= \|\mathbf{A}\mathbf{v} - \mathbf{b}\|^2 \\ &= (\mathbf{A}\mathbf{v} - \mathbf{b})^T(\mathbf{A}\mathbf{v} - \mathbf{b}). \end{aligned} \tag{2.62}$$

This loss function is the sum of the squared differences between $\mathbf{v}^T\boldsymbol{\eta}(n)$ and a desired positive number b_n, for all n.

The loss function defined in Equation (2.62) has a useful geometric interpretation. The magnitude of $\mathbf{v}^T\boldsymbol{\eta}(n)$ is $\|\mathbf{w}\|$ times the distance of $\mathbf{x}(n)$ from the hyperplance $\mathbf{v}^T\mathbf{y}(n) = 0$ in $\mathbf{x}$-space. The sign of $\mathbf{v}^T\boldsymbol{\eta}(n)$ is positive if $\mathbf{x}(n)$ falls on the side of the hyperplane which yields a correct classification and is negative on the other side. The quantity b_n is $\|\mathbf{w}\|$ times the desired distance of $\mathbf{x}$ from this hyperplane. For correct classification, this desired distance is positive, thereby requiring that b_n be positive. The difference between the desired distance and the actual distance is $\|\mathbf{w}\|^{-1}|\mathbf{v}^T\boldsymbol{\eta}(n) - b_n|$. This difference is $\|\mathbf{w}\|^{-1}$ times the error $|\mathbf{v}^T\boldsymbol{\eta}(n) - b_n|$, as illustrated in Figure 2.9. Thus, $\|\mathbf{A}\mathbf{v} - b\|^2$ is the sum of the squared errors when the weight vector is $\mathbf{v}$. Consequently, we refer to the loss function defined in Equation (2.62) as the *sum-square error*.

For every choice of $\mathbf{b}$, the value of $\mathbf{v}$ that minimizes the sum-square error $J(\mathbf{v})$ is determined as follows.

$$\begin{aligned} \nabla J(\mathbf{v}) &= \nabla[(\mathbf{A}\mathbf{v} - \mathbf{b})^T(\mathbf{A}\mathbf{v} - \mathbf{b})] \\ &= 2\mathbf{A}^T(\mathbf{A}\mathbf{v} - \mathbf{b}) = 0. \end{aligned} \tag{2.63}$$

Hence

$$\mathbf{A}^T\mathbf{A}\mathbf{v} = \mathbf{A}^T\mathbf{b}.$$

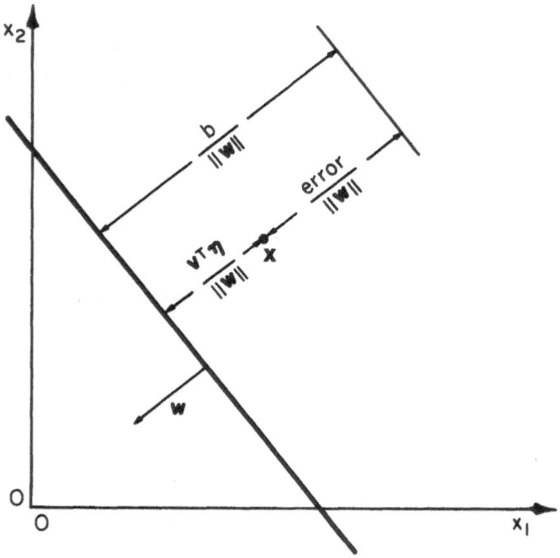

Figure 2.9. Illustration of the error $\mathbf{v}^T\boldsymbol{\eta} - b$.

If $\mathbf{A}^T\mathbf{A}$ is nonsingular, a solution to Equation (2.63) is given by

$$\begin{aligned} \mathbf{v} &= (\mathbf{A}^T\mathbf{A})^{-1}\mathbf{A}^T\mathbf{b} \\ &= \mathbf{A}^{\#}\mathbf{b}, \end{aligned} \tag{2.64}$$

where $\mathbf{A}^{\#}$ is defined as

$$\mathbf{A}^{\#} = (\mathbf{A}^T\mathbf{A})^{-1}\mathbf{A}^T. \tag{2.65}$$

This matrix is called the *pseudoinverse* of $\mathbf{A}$. It is equivalent to the inverse of $\mathbf{A}$ when $\mathbf{A}$ is square and nonsingular. One of the problems in using Equation (2.64) is that $\mathbf{A}^T\mathbf{A}$ may be singular or almost singular, leading to impossible computational requests or to large round-off errors in the computation of $\mathbf{A}^{\#}$.

One may overcome this problem by using a gradient descent on $J(\mathbf{v})$, thereby eliminating the direct inversion of $\mathbf{A}^T\mathbf{A}$. This descent may be formulated by applying the gradient operator to Equation (2.62):

$$\nabla J(\mathbf{v}) = 2\mathbf{A}^T(\mathbf{A}\mathbf{v} - \mathbf{b}). \tag{2.66}$$

Substituting this result in Equation (2.59) yields the descent equation

$$\mathbf{v}(k+1) = \mathbf{v}(k) - \rho_k\mathbf{A}^T(\mathbf{A}\mathbf{v}(k) - \mathbf{b}). \tag{2.67}$$

If $\rho_k = c/(k+1)$ (c = positive constant; $k = 0, 1, 2, 3, \ldots$), then Equation (2.67) generates a sequence $\{\mathbf{v}(k)\}$ that converges to a vector $\mathbf{v}^*$ for which $\nabla J(\mathbf{v}^*) = \mathbf{0}$. (See Exercise 2.28).

A difficulty in applying the above descent procedure lies in choosing $\mathbf{b}$. For the case

$$\mathbf{b} = \mathbf{1},$$

this descent procedure becomes the *minimum mean-square-error training procedure*. This procedure, as well as the case $\mathbf{b} = [b, b, \ldots, b]^T$, $b > 0$, is discussed in Subsection 4.5.2 and Section 5.4.

2.13 The Ho–Kashyap Procedure

Although Equation (2.67) minimizes $J(\mathbf{v})$ with respect to $\mathbf{v}$, it does not minimize $J(\mathbf{v})$ with respect to $\mathbf{b}$. Thus, the $J(\mathbf{v}^*)$ achieved by Equation (2.67) is a function of $\mathbf{b}$. The Ho–Kashyap procedure improves on this situation by an iterative adjustment of both $\mathbf{b}$ and $\mathbf{v}$.

Equation (2.67) converges regardless of whether the classes are LS or non-LS. If the classes are LS, however, it is desirable that the equation converge to a solution vector $\mathbf{v}^*$. To ensure this, it is necessary that $\mathbf{b} > \mathbf{0}$, i.e., that $b_n > 0$ for every n. The Ho–Kashyap procedure implements the constraint $\mathbf{b} > \mathbf{0}$ using an initial value of $\mathbf{b}$, namely $\mathbf{b}(0)$, all of whose components are positive, and by prohibiting adjustments that decrease b_n. (If the classes are non-LS, the Ho–Kashyap procedure does not ensure that $\mathbf{v}(n)$ will converge to a vector $\mathbf{v}^*$ that minimizes the sum-square error. To converge to such a $\mathbf{v}^*$ in the non-LS case, we suggest the use of stochastic approximation training procedures such as those discussed in Chapters 4 and 5.)

The Ho–Kashyap procedure achieves these results by using a loss function of both $\mathbf{v}$ and $\mathbf{b}$, and by adjustments on $\mathbf{b}$ that are always nonnegative. The loss function is

$$J(\mathbf{v}, \mathbf{b}) = \|\mathbf{A}\mathbf{v} - \mathbf{b}\|^2 \equiv (\mathbf{A}\mathbf{v} - \mathbf{b})^T(\mathbf{A}\mathbf{v} - \mathbf{b}) \tag{2.68}$$

subject to the constraint $\mathbf{b} > \mathbf{0}$. Two gradients are brought to zero:

$$\nabla_{\mathbf{v}} J(\mathbf{v}, \mathbf{b}) \quad \text{and} \quad \nabla_{\mathbf{b}} J(\mathbf{v}, \mathbf{b}),$$

i.e., the gradient with respect to $\mathbf{v}$ and the gradient with respect to $\mathbf{b}$. For each newly computed value of $\mathbf{b}$ at each stage of iteration, the gradient $\nabla_{\mathbf{v}} J(\mathbf{v}, \mathbf{b})$ is brought to zero in one step by Equation (2.64). The gradient $\nabla_{\mathbf{b}} J(\mathbf{v}, \mathbf{b})$ is brought to zero by a modification of Equation (2.59):

$$\mathbf{b}(k+1) = \mathbf{b}(k) - \frac{\rho}{2}\left[\nabla_{\mathbf{b}} J(\mathbf{v}, \mathbf{b}) - |\nabla_{\mathbf{b}} J(\mathbf{v}, \mathbf{b})|\right],$$

where $|\mathbf{a}|$ is defined as the vector $[|a_0|, |a_1|, \ldots, |a_d|]^T$. The ith component of the vector quantity

$$\tfrac{1}{2}\left[\nabla_{\mathbf{b}} J(\mathbf{v}, \mathbf{b}) - |\nabla_{\mathbf{b}} J(\mathbf{v}, \mathbf{b})|\right]$$

equals either the ith component of $\nabla_{\mathbf{b}} J(\mathbf{v}, \mathbf{b})$ or zero, depending on whether that component is respectively negative or positive.

In this manner the components of $\mathbf{b}(k)$ are never reduced. Since $\mathbf{b}(0)$, the initial value of $\mathbf{b}(k)$, is a positive vector, all the $\mathbf{b}(k)$'s will be positive vectors.

From Equation (2.68) one obtains the following expressions for the gradient of J with respect to $\mathbf{v}$ and $\mathbf{b}$:

$$\nabla_{\mathbf{v}} J(\mathbf{v}, \mathbf{b}) = 2\mathbf{A}^T(\mathbf{A}\mathbf{v} - \mathbf{b}) \tag{2.69}$$

$$\nabla_{\mathbf{b}} J(\mathbf{v}, \mathbf{b}) = -2(\mathbf{A}\mathbf{v} - \mathbf{b}). \tag{2.70}$$

Thus the Ho–Kashyap algorithm is as follows:

$$\begin{aligned} &\mathbf{b}(0) > \mathbf{0}, \\ &\mathbf{b}(k+1) = \mathbf{b}(k) + \rho[\mathbf{A}\mathbf{v}(k) - \mathbf{b}(k) + |\mathbf{A}\mathbf{v}(k) - \mathbf{b}(k)|], \\ &\mathbf{v}(k+1) = \mathbf{A}^{\#}\mathbf{b}(k+1) \equiv \mathbf{v}(k) + \rho\mathbf{A}^{\#}|(\mathbf{A}\mathbf{v}(k) - \mathbf{b}(k)|. \end{aligned} \tag{2.71}$$

The last expression in Equations (2.71) follows from Equation (2.63), since

$$\mathbf{A}^{\#}[\mathbf{A}\mathbf{v}(k) - \mathbf{b}(k)] = (\mathbf{A}^T\mathbf{A})^{-1}\mathbf{A}^T[\mathbf{A}\mathbf{v}(k) - \mathbf{b}(k)] = 0.$$

The quantities ρ and $\mathbf{b}(0)$ are based on the amount of knowledge one has about the asymptotic value of $\mathbf{b}(k)$, namely $\mathbf{b}(\infty)$. To ensure convergence of Equation (2.71), choose ρ so that $0 < \rho \leqslant 1$.

If for some value of k, say $k = K$, we obtain

$$\mathbf{A}\mathbf{v}(K) - \mathbf{b}(K) = 0, \tag{2.72}$$

then the pair of class regions is LS, and $\mathbf{v}(K)$ is a solution vector. This solution is reached in a finite time. If the class regions are not LS, then either k reaches a value K for which

$$\mathbf{A}\mathbf{v}(K) - \mathbf{b}(K) \leqslant \mathbf{0} \quad \text{and} \quad \neq \mathbf{0}$$

or the sequence $\{\mathbf{A}\mathbf{v}(k) - \mathbf{b}(k)\}$ converges to a nonpositive nonzero vector as $k \to \infty$. Thus, LS is detected in a finite time, but non-LS is detected in either a finite time or an infinite time.

If the classes are not LS, the Ho–Kashyap algorithm does not necessarily yield a least sum-squared-error choice of $\mathbf{b}$. (By sum-squared error we mean $\|\mathbf{A}\mathbf{v} - \mathbf{b}\|^2$.)

A sufficient, but not necessary, condition for convergence of the $\{\mathbf{v}_k\}$ and $\{\mathbf{b}(k)\}$ generated by Equation (2.71) is that $0 < \rho \leqslant 1$. Proofs of this convergence appear in References [14] and [15].

Although Equation (2.71) requires the inversion of $\mathbf{A}^T\mathbf{A}$ only once, this inversion is time consuming and is impossible or difficult to carry out if $\mathbf{A}^T\mathbf{A}$ is singular or nearly singular. The inversion may be replaced by a recursive process in which a new feature vector is entered into the algorithm at each iteration. In this case the kth iteration occurs when the nth feature vector is selected so that $k = n$. The process proceeds as follows [16]. Let

$$\mathbf{A}_n = \begin{bmatrix} \boldsymbol{\eta}^T(0) \\ \vdots \\ \boldsymbol{\eta}^T(n) \end{bmatrix},$$

where $\boldsymbol{\eta}(n)$ denotes the signed augmented feature vector entered into the algorithm at trial n. Then

$$\mathbf{A}_{n+1} = \left[\begin{array}{c} \mathbf{A}_n \\ \hline \boldsymbol{\eta}^T(n+1) \end{array}\right]$$

and

$$\begin{aligned} \mathbf{A}_{n+1}^T\mathbf{A}_{n+1} &= [\mathbf{A}_n^T \mid \boldsymbol{\eta}(n+1)]\left[\begin{array}{c} \mathbf{A}_n \\ \hline \boldsymbol{\eta}^T(n+1) \end{array}\right] \\ &= \mathbf{A}_n^T\mathbf{A}_n + \boldsymbol{\eta}(n+1)\boldsymbol{\eta}^T(n+1). \end{aligned} \tag{2.73}$$

Let $\mathbf{B}_n = (\mathbf{A}_n^T\mathbf{A}_n)^{-1}$. We shall assume $\mathbf{B}_{n+1} \cong \mathbf{B}_n$. By Equation (2.73),

$$\begin{aligned} \mathbf{B}_{n+1} &= [\mathbf{B}_n^{-1} + \boldsymbol{\eta}(n+1)\boldsymbol{\eta}^T(n+1)]^{-1} \\ &= \{[\mathbf{I} + \boldsymbol{\eta}(n+1)\boldsymbol{\eta}^T(n+1)\mathbf{B}_n]\mathbf{B}_n^{-1}\}^{-1} \\ &= \mathbf{B}_n[\mathbf{I} + \boldsymbol{\eta}(n+1)\boldsymbol{\eta}^T(n+1)\mathbf{B}_n]^{-1} \\ &= \mathbf{B}_n\{\mathbf{I} - \boldsymbol{\eta}(n+1)\boldsymbol{\eta}^T(n+1)\mathbf{B}_n + [\boldsymbol{\eta}(n+1)\boldsymbol{\eta}^T(n+1)\mathbf{B}_n]^2 - \cdots\}. \end{aligned}$$

Thus when $\mathbf{B}_{n+1} \cong \mathbf{B}_n$

$$\mathbf{B}_{n+1} \cong \mathbf{B}_n[\mathbf{I} - \boldsymbol{\eta}(n+1)\boldsymbol{\eta}^T(n+1)\mathbf{B}_n]. \tag{2.74}$$

By Equations (2.65) and (2.74),

$$\begin{aligned} \mathbf{A}_{n+1}^{\#} &= \mathbf{B}_{n+1}\mathbf{A}_{n+1}^T \\ &= (\mathbf{A}_n^T\mathbf{A}_n)^{-1}\{\mathbf{I} - \boldsymbol{\eta}(n+1)\boldsymbol{\eta}^T(n+1)[\mathbf{A}_n^T\mathbf{A}_n]^{-1}\}[\mathbf{A}_n^T \mid \boldsymbol{\eta}(n+1)]. \end{aligned} \tag{2.75}$$

Thus Equation (2.71) is replaced by

$$\begin{aligned} \mathbf{b}(0) &> \mathbf{0} \\ \mathbf{b}(n+1) &= \mathbf{b}(n) + \rho[\mathbf{A}_n\mathbf{v}(n) - \mathbf{b}(n) + |\mathbf{A}_n\mathbf{v}(n) - \mathbf{b}(n)|] \\ \mathbf{v}(n+1) &= \mathbf{A}_{n+1}^{\#}\mathbf{b}(n+1), \end{aligned} \tag{2.76}$$

where $\mathbf{A}_{n+1}^{\#}$ is computed by Equation (2.75).

To exploit the recursive process in Equations (2.75) and (2.76), one needs to add a new component of $\mathbf{b}$ at each new trial (i.e., for each new value of n). This additional component is initially set to an arbitrary positive value, say unity.

Another way of eliminating the direct inversion of $\mathbf{A}^T\mathbf{A}$ is to modify Equation (2.71) as follows. Set $\rho = 1$, express $\mathbf{A}^{\#}$ as $(\mathbf{A}^T\mathbf{A})^{-1}\mathbf{A}^T$, and replace $(\mathbf{A}^T\mathbf{A})^{-1}$ by γ, where $0 < \gamma \leqslant |\mathbf{A}^T\mathbf{A}|^{-1}$, $|\mathbf{A}^T\mathbf{A}| \equiv$ determinant of $\mathbf{A}^T\mathbf{A}$. This yields

$$\begin{aligned} \mathbf{b}(0) &> \mathbf{0} \\ \mathbf{b}(k+1) &= \mathbf{b}(k) + \mathbf{A}\mathbf{v}(k) - \mathbf{b}(k) + |\mathbf{A}\mathbf{v}(k) - \mathbf{b}(k)| \\ \mathbf{v}(k+1) &= \mathbf{v}(k) + \gamma\mathbf{A}^T|\mathbf{A}\mathbf{v}(k) - \mathbf{b}(k)|, \end{aligned}$$

where γ must satisfy $0 < \gamma < |\mathbf{A}^T\mathbf{A}|^{-1}$. Other forms of this algorithm are described in References [14] and [17].

Exercises

2.1. Show that $\mathscr{H}$ and $\mathscr{V}$ yield a 3-summable function U, regardless of the way in which the x_i are assigned to the 3×3 grid. (See Example 2.4.)

2.2. Show that if $\mathscr{X}_1$ and $\mathscr{X}_2$ are linearly separable then there exist a $(d+1)$-vector $\mathbf{v}$ and an N-vector $\mathbf{t}$ such that

$$\mathbf{Av} \geqslant \mathbf{t} > \mathbf{0},$$

where $\mathbf{0} = [0, \ldots, 0]^T$. (The matrix $\mathbf{A}$ is defined in Section 2.3.)

2.3. Show that the intersection of two convex sets is convex.

2.4. Show that the convex hull of any set of vectors is convex.

2.5. Show that the decision regions of a linear machine are convex. (A linear machine is a pattern classifier consisting of a set of linear discriminant functions followed by a maximum selector.)

2.6. Show that if the classes of patterns $\mathscr{X}_1, \ldots, \mathscr{X}_p$ are linearly separable (i.e., there exists a linear machine that separates the $\mathscr{X}_i$ perfectly), then $(\mathscr{X}_i, \mathscr{X}_j)$ is linearly separable for every $i, j, (i \neq j)$.

2.7. (a) Using the definitions of x_1, x_2 shown in Figure 2.10(a) find $\mathbf{x} = [x_1, x_2]^T$ for each of the examples of A and U illustrated in Figure 2.10(b). (a,b,c,d, and e are measured distances.) Let class ω_1 be the set of A's and class ω_2 be the set of U's.

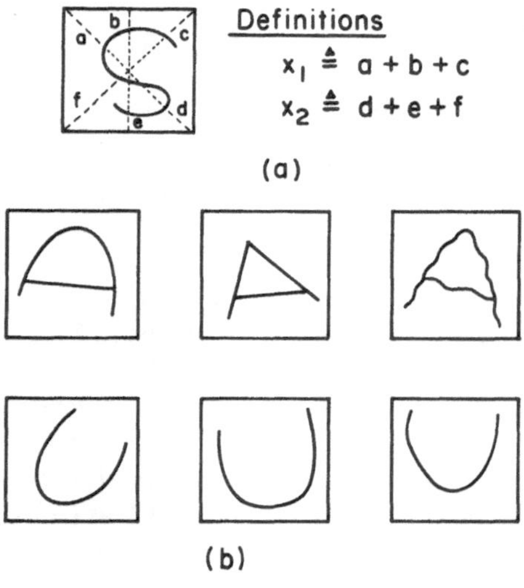

Figure 2.10. Examples for Exercise 2.7.

(b) Plot $\mathbf{x}$ for each example in the x_1,x_2-plane. Label each point by either A or U. Plot a separating hyperplane (i.e., straight line) if it exists.

(c) Let $\xi_i = x_i/\|\mathbf{x}\|$ and plot $\boldsymbol{\xi} = [\xi_1, \xi_2]^T$ for each example in the ξ_1, ξ_2-plane. Plot a separating hyperplane if it exists and find the augmented weight vector $\mathbf{v}$ which describes this decision hyperplane.

(d) Use the test of summability to verify your conclusions on the linear separability of the $\boldsymbol{\xi}$'s of part (c).

2.8. Repeat Exercise 2.7 using three unique A's and three unique U's drawn by yourself.

2.9. Consider a one-dimensional feature space. Suppose $\mathscr{X}_1$ and $\mathscr{X}_2$ are nonempty, binary-valued ($x(n) = 0$ or 1), and disjoint.
(a) Find the matrix **A**.
(b) Show that there is no vector **b** such that $\mathbf{A}^T\mathbf{b} = \mathbf{0}, \mathbf{b} \geqslant \mathbf{0}, \mathbf{b} \neq \mathbf{0}$.

2.10. Show that if a switching function U is k-summable, then $k \geqslant 2$.

2.11. Let

$$\mathscr{C}\mathscr{X} = \left\{\mathbf{x} \,\middle|\, \sum_{n=0}^{k} b_n\mathbf{x}(n), \mathbf{x}(n) \in \mathscr{X}, b_n \geqslant 0, \Sigma b_n = 1; k = 0, 1, 2, \ldots, n-1.\right\}$$

Let

$$\mathscr{Q}^{(1)} = \{\mathbf{x} \,|\, \mathbf{x} = \alpha\mathbf{x}(i) + (1-\alpha)\mathbf{x}(j), \mathbf{x}(i) \in \mathscr{X}, \mathbf{x}(j) \in \mathscr{X}, 0 \leqslant \alpha \leqslant 1\}$$
$$\mathscr{Q}^{(2)} = \{\mathbf{x} \,|\, \mathbf{x} = \alpha\mathbf{x}(i) + (1-\alpha)\mathbf{x}(j), \mathbf{x}(i) \in \mathscr{Q}^{(1)}, \mathbf{x}(j) \in \mathscr{Q}^{(1)}, 0 \leqslant \alpha \leqslant 1\}$$
$$\vdots$$
$$\mathscr{Q}^{(r)} = \{\mathbf{x} \,|\, \mathbf{x} = \alpha\mathbf{x}(i) + (1-\alpha)\mathbf{x}(j), \mathbf{x}(i) \in \mathscr{Q}^{(r-1)}, \mathbf{x}(j) \in \mathscr{Q}^{(r-1)}, 0 \leqslant \alpha \leqslant 1\}$$
$$\vdots$$

Show that

$$\mathscr{C}\mathscr{X} = \lim_{r \to \infty} \mathscr{Q}^{(r)}.$$

2.12. Consider the following training procedure in a p-class linear classifier in which every member of $\bigcup_i \mathscr{X}_i$ occurs at least once in every M successive trials. Suppose $\mathbf{x}(n) \in \mathscr{X}_i$ and $\max_l (\mathbf{v}_l^T\mathbf{y}(n)) = \mathbf{v}_j^T\mathbf{y}(n) > \mathbf{v}_i^T\mathbf{y}(n), j \neq i$. Then

$$\mathbf{v}_i(n+1) = \mathbf{v}_i(n) + \rho\mathbf{y}(n)$$
$$\mathbf{v}_j(n+1) = \mathbf{v}_j(n) - \rho\mathbf{y}(n)$$
$$\mathbf{v}_k(n+1) = \mathbf{v}_k(n), \qquad \text{if } k \neq i, \quad k \neq j.$$

Otherwise $\mathbf{v}_i(n+1) = \mathbf{v}_i(n)$ for all i.
Show that this training procedure terminates at a set of solution vectors $\{\mathbf{v}_i^*\}$ in a finite number of trials, provided at least one set of solution vectors exists.

2.13. Suppose **BP** is a permutation of the columns of **B** for all **B** having a number of columns equal to the number of rows of **P**. Show that **P** is the same permutation of the columns of **I** as the permutation of the columns of **B** obtained by **BP**.

2.14. Assume a two-feature problem, and the use of the proportional increment training procedure with $\rho = \frac{1}{2}$. At the fifty-first trial during the training phase, $\mathbf{v}(51) = [-2, 2, 1]^T$. Furthermore, $\mathbf{x}(51) = [2, 2]^T \in \omega_2$, and $\mathbf{x}(52) = [2, 0]^T \in \omega_1$.
(a) Find the augmented weight vector $\mathbf{v}(53)$.
(b) Plot in **x**-space the hyperplanes specified by $\mathbf{v}(51)$ and $\mathbf{v}(53)$.

2.15. Repeat exercise 2.14 using the fixed fraction training procedure with:
(a) $\lambda = \frac{1}{2}$.
(b) $\lambda = 1$.
(c) $\lambda = \frac{3}{2}$.

2.16. Show that if any column of **R** has no positive element, then $(\mathscr{X}_1, \mathscr{X}_2)$ is not linearly separable.

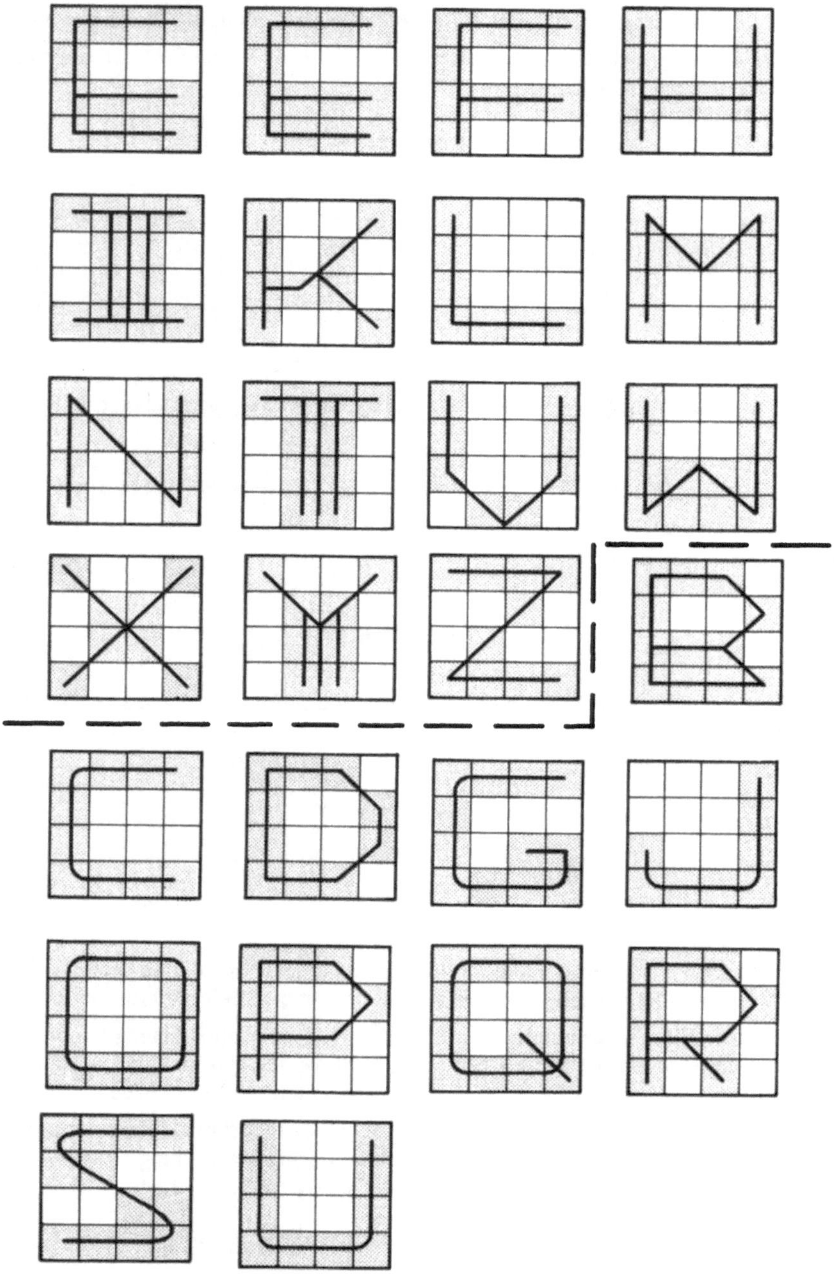

Figure 2.11. Mosaics for Exercise 2.17.

2.17. Suppose the capital letters A, B, . . . , Z are sensed on a 4×4 mosaic, as shown in Figure 2.11 (designed by J. R. Fridell). Let ω_1 denote the class of characters containing no curved arcs, and let ω_2 denote the class of characters containing curved arcs. Let $\mathscr{X}_1$ denote the set of characters above the dashed line in the figure, and let $\mathscr{X}_2$ denote the set of characters below the dashed line. Assume that

$$\mathscr{X}_1 \subseteq \omega_1, \mathscr{X}_2 \equiv \omega_2.$$

Show that $(\mathscr{X}_1, \mathscr{X}_2)$ is linearly separable.

2.18. (a) Find the matrix **A** for Exercise 2.7(c) and use fictitious play to determine whether the classes of ξ's are linearly separable.

(b) Using the fictitious play technique, find an augmented weight vector **v** which separates the classes.

2.19. Given training sets $\mathscr{X}_1$ and $\mathscr{X}_2$ shown in Table 2.3:

(a) Find the matrix **A** and use fictitious play to determine whether or not $\mathscr{X}_1$ and $\mathscr{X}_2$ are linearly separable.

(b) Find an augmented weight vector **v** which separates the classes if such a **v** exists. Use the results of fictitious play.

Table 2.3. Training sets for Exercise 2.19.

x's in $\mathscr{X}_1$	x's in $\mathscr{X}_2$
$\begin{bmatrix}0\\1\\1\\0\end{bmatrix}, \begin{bmatrix}1\\1\\0\\0\end{bmatrix}, \begin{bmatrix}0\\1\\0\\1\end{bmatrix}$	$\begin{bmatrix}1\\0\\0\\1\end{bmatrix}, \begin{bmatrix}0\\0\\1\\0\end{bmatrix}, \begin{bmatrix}1\\0\\1\\1\end{bmatrix}, \begin{bmatrix}0\\1\\1\\0\end{bmatrix}$

2.20. Given the two classes of characters on the mosaics of Figure 2.12(b), define a vector $\mathbf{x} = [x_1, x_2, x_3, x_4, x_5, x_6]^T$ for each character so that $x_i = 1$ or 0 according

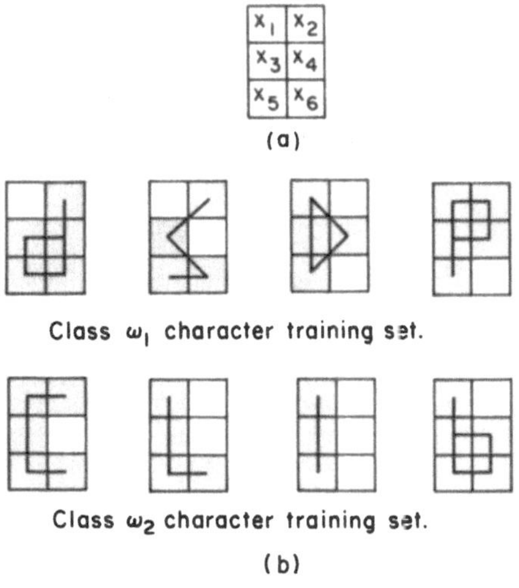

Figure 2.12. Character mosiacs for Exercise 2.20.

to whether or not the corresponding square is respectively dark or light. The x_i-to-square correspondence is given in Figure 2.12.(a)

(a) Find the matrix **A** and use fictitious play to determine whether or not $\mathscr{X}_1$ and $\mathscr{X}_2$ are linearly separable.

(b) Find an augmented weight vector **v** which separates the classes if such a **v** exists.

2.21. Given the loss function

$$J(\mathbf{v}) = \mathbf{v}^T E(\mathbf{Y}),$$

where E is the expectation operator, find

$$\nabla J(\mathbf{v}) = \partial J(\mathbf{v})/\partial \mathbf{v}.$$

2.22. Given the scalar function of a vector

$$J(\mathbf{v}) = \|\mathbf{v}\|^2, \quad \text{where} \quad \mathbf{v} = \begin{bmatrix} v_0 \\ v_1 \\ v_2 \end{bmatrix},$$

find

$$\nabla J(\mathbf{v}) = \partial J(\mathbf{v})/\partial \mathbf{v}.$$

2.23. Find the matrix **A** for the training sets of Exercise 2.7(c) and find the **b**(1) and **v**(1) vectors after one trial when using the Ho-Kashyap procedure. Choose $\mathbf{b}(0) = \mathbf{1}$, $\mathbf{v}(0) = [0, 1, -1]^T$ and $\rho = \frac{1}{2}$. Is **v**(1) a solution vector?

2.24. Find the matrix **A** for the training set of Exercise 2.19 and find the **b**(1) and **v**(1) vectors after one trial when using the Ho–Kashyap procedure. Choose $\mathbf{b}(0) = \mathbf{1}$, $\rho = \frac{1}{2}$, and a **v**(0) of your choice. Is your **v**(1) a solution vector?

2.25. Given training sets $\mathscr{X}_1$ and $\mathscr{X}_2$ in Table 2.4, find the matrix **A** and find the **b**(1) and **v**(1) vectors after one trial using the Ho–Kashyap procedure. Choose $\mathbf{b}(0) = \mathbf{1}$, $\rho = \frac{1}{2}$, and $\mathbf{v}(0) = [0, 0, 1, 0]^T$. Is **v**(1) a solution vector?

Table 2.4. Training sets for Exercise 2.25.

x's in $\mathscr{X}_1$	x's in $\mathscr{X}_2$
$\begin{bmatrix} 1.2 \\ 0.2 \\ 1 \end{bmatrix}, \begin{bmatrix} 0.2 \\ 0 \\ 0.9 \end{bmatrix}, \begin{bmatrix} 0.8 \\ 0 \\ 0 \end{bmatrix}$	$\begin{bmatrix} 0 \\ 0 \\ 0 \end{bmatrix}, \begin{bmatrix} 0 \\ 0.5 \\ 0 \end{bmatrix}, \begin{bmatrix} 1 \\ 0.8 \\ 0.2 \end{bmatrix}$

2.26. Find the matrix **A** for the training sets of Exercise 2.20 and find the **b**(1) and **v**(1) vectors after one trial using the Ho–Kashyap procedure. Choose $\mathbf{b}(0) = \mathbf{1}$, $\rho = \frac{1}{2}$, and a **v**(0) of your choice. Is your **v**(1) a solution vector?

2.27. Suppose $\mathscr{X}_1$ consists of the position vectors for the positions marked by the crosses in Figure 2.13, and suppose $\mathscr{X}_2$ consists of the position vectors for the positions marked by the circles in Figure 2.13.

(a) What is the matrix **A** in this case?

(b) Use the Ho–Kashyap procedure to find the equation of a straight line that separates the crosses from the circles. Assume $\mathbf{b} = [1, \ldots, 1]^T$, and $\rho = 0.5$.

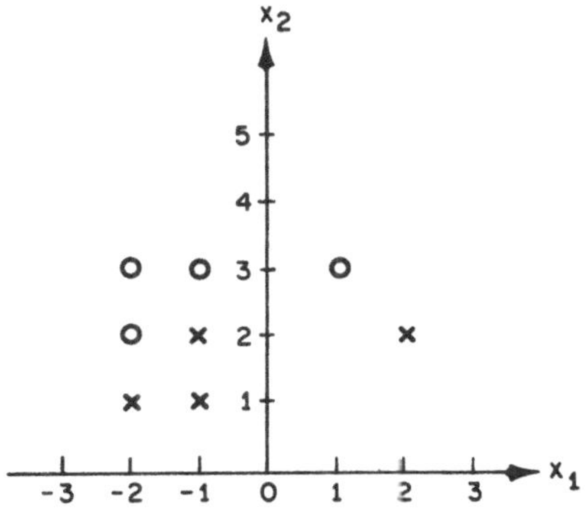

Figure 2.13. Position vectors for Exercise 2.27.

2.28. (a) Apply the gradient descent procedure to find a straight line that separates $\mathscr{X}_1$ from $\mathscr{X}_2$ in Exercise 2.27. (*Suggestion*: Use Equation (2.67), with $\rho_k = 2/(k+1)$.)

(b) Show that if $\rho_k = c/(k+1)$, where c is any constant, then

1. Equation (2.67) generates a sequence $\{\mathbf{v}_k\}$ that approaches a limit $\mathbf{v}^*$ as $k \to \infty$, and
2. $\mathbf{v}^*$ satisfies $\nabla J(\mathbf{v}^*) = \mathbf{0}$.

References

1. P. J. Kelly and M. L. Weiss, *Geometry and Convexity*. John Wiley & Sons, New York, 1979.
2. F. B. Hildebrand, *Methods of Applied Mathematics*. Prentice-Hall, New York, 1952.
3. R. O. Winder, Enumeration of seven-argument threshold functions. *IEEE Transactions on Electronic Computers*, **EC-14** (3): 315–325 (1965).
4. S. Yajima and T. Ibaraki, A theory of completely monotonic functions and its applications to threshold logic. *IEEE Transactions on Computers*, **C-17**. (3) 214–229 (1968).
5. H. D. Block and S. A. Levin, On the boundedness of an iterative procedure for solving a system of linear inequalities. *Proceedings of the American Mathematical Society*, **26**. (2) 229–235 (1970).
6. W. C. Ridgway, An adaptive logic system with generalizing properties. Stanford Electronics Laboratories Technical Report 1556-1, Stanford University, Stanford, California, April 1962.
7. N. J. Nilsson, *Learning Machines*. McGraw-Hill, New York, 1965.
8. J. McKinsey, *Introduction to the Theory of Games*. McGraw-Hill, New York, 1952.
9. E. S. Venttsel, *An Introduction to the Theory of Games*. D. C. Heath, Boston, 1963.
10. R. C. Singleton, A test for linear separability as applied to self-organizing machines. In: *Self-Organizing Systems 1962*, M. C. Yovits et al. (eds.), Spartan Books, Washington, D.C., 1962, pp. 503–524. Reprinted in *Pattern Recognition*, J. Sklansky (ed.), Dowden, Hutchinson & Ross, Stroudsburg, PA, 1973, pp. 55–76.

11. J. Von Neumann, Oskar Morgenstern, *Theory of Games and Economic Behavior*, 3rd ed. Princeton University Press, Princeton, NJ, 1953.
12. Julia Robinson, An iterative method of solving a game. *Annals of Mathematics*, **54**; 296–301 (1951).
13. C. A. Gaston, A simple test for linear separability. *IEEE Transactions on Electronic Computers*, **EC-12** (2): 134–135 (1963).
14. R. O. Duda, P. E. Hart, *Pattern Classification and Scene Analysis*. John Wiley & Sons, New York, 1973.
15. Y.-C. Ho and R. L. Kashyap, An algorithm for linear inequalities and its applications. *IEEE Transactions on Electronic Computers*, **EC-14** (5): 683–688 (1965).
16. E. Bodewig, *Matrix Calculus*. Interscience, New York, 1956, p. 30.
17. Y.-C. Ho and R. L. Kashyap, *J. SIAM*, **4** (1): 112–115 (1966).

CHAPTER 3

Nonlinear Classifiers

3.1 Introduction

The classifier design techniques we have discussed so far have been, for the most part, restricted to linear decision surfaces (hyperplanes). These techniques are useful only in cases where the error probability obtained by the best linear decision surface—a hyperplane—is acceptably small. In many practical situations, no such hyperplane exists, so that a nonlinear decision surface is necessary. An example of such a situation is illustrated in Figure 3.1 for a two-dimensional feature space. Here the smallest possible error probability obtainable by any hyperplane (straight line in this illustration) is 0.25. On the other hand, the *pair* of hyperplanes shown in the figure achieves zero error.

Another example is the case where the class densities are normal, i.e.,

$$p(\mathbf{x} \mid \omega_j) = (2\pi)^{-d/2} |\Sigma_j|^{-1/2} \exp\left[-\tfrac{1}{2}(\mathbf{x} - \boldsymbol{\mu}_j)^T \Sigma_j^{-1} (\mathbf{x} - \boldsymbol{\mu}_j)\right],$$

and the number of classes is two. The notation $p(\mathbf{x} \mid \omega_j) = N(\boldsymbol{\mu}_j, \Sigma_j)$ is often used to denote this density form. For this case, the optimum decision surface is of the form

$$b - \tfrac{1}{2}(\mathbf{x} - \boldsymbol{\mu}_1)^T \Sigma_1^{-1} (\mathbf{x} - \boldsymbol{\mu}_1) + \tfrac{1}{2}(\mathbf{x} - \boldsymbol{\mu}_2)^T \Sigma_2^{-1} (\mathbf{x} - \boldsymbol{\mu}_2) = 0.$$

The constant b depends on the a priori probabilities and the covariance matrices Σ_1, Σ_2. This decision surface is quadratic whenever $\Sigma_1 \neq \Sigma_2$ and linear otherwise.

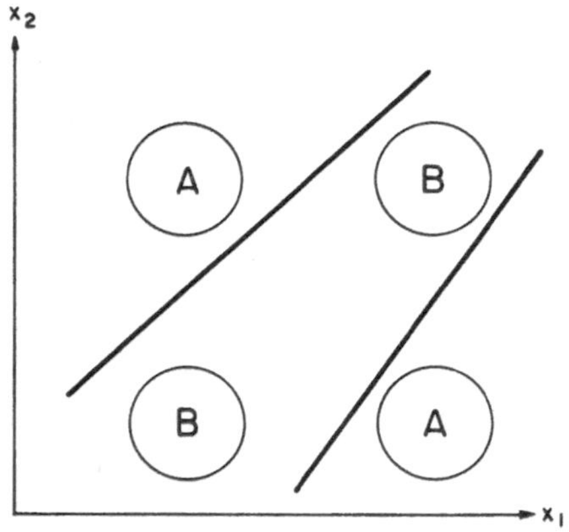

Figure 3.1. Class A has two clusters of equal size and equal a priori probability; so has Class B.

In this chapter we describe design techniques—most of which involve training processes—that can compute optimal or near optimal nonlinear decision surfaces. These techniques are applicable to both separable and nonseparable classes.

3.2 Φ-Classifiers

In Section 1.6 we described how the derivation of linear decision surfaces in an expanded feature space can be a means of deriving nonlinear decision surfaces in the original feature space. Such derivations are suitable in cases where the training sets are linearly separable in the expanded feature space but not linearly separable in the original feature space. This approach can be extended to cases where the optimum decision surfaces yield nonzero error probabilities.

Consider the following example. Suppose the statistically optimum decision surface in **x**-space is given by

$$w_0 + w_1x_1 + w_2x_2 + w_3x_1^2 + w_4x_1x_2 + w_5x_2^2 = 0.$$

This is a quadric surface (an ellipse, hyperbola, parabola, or circle). Now let $\xi_1 = x_1$, $\xi_2 = x_2$, $\xi_3 = x_1^2$, $\xi_4 = x_1x_2$, $\xi_5 = x_2^2$. Then, in $\boldsymbol{\xi}$ space, the above decision surface is given by

$$\mathbf{v}^T\mathbf{y} = 0,$$

where $\mathbf{v} = [w_0, \ldots, w_5]^T$, $\mathbf{y} = [1, \xi_1, \ldots, \xi_5]^T$. This surface is a hyperplane in $\boldsymbol{\xi}$-space.

To find or to converge to this hyperplane by a training procedure, one may use, for example, the window training procedure described in Chapter 5:

$$\mathbf{v}(n+1) = \mathbf{v}(n) + \rho(n)\mathbf{Z}(n)$$

where

$$\mathbf{Z}(n) = \begin{cases} \|\mathbf{w}\|\mathbf{y}(n), & \text{if } |\mathbf{v}(n)^T\mathbf{y}(n)| \leqslant c(n)\|\mathbf{w}(n)\|, \ \omega(n) = \omega_2 \\ -\|\mathbf{w}\|\mathbf{y}(n), & \text{if } |\mathbf{v}(n)^T\mathbf{y}(n)| \leqslant c(n)\|\mathbf{w}(n)\|, \ \omega(n) = \omega_1 \\ \mathbf{0}, & \text{otherwise,} \end{cases}$$

$c(n) \geqslant 0$, and $\rho(n)$ and $c(n)$ satisfy certain constraints presented in Section 5.3.2. Since the optimum decision surface is a hyperplane in ξ-space, the window training procedure in ξ-space will converge to the optimum decision surface, provided the required constraints on the constituent densities are satisfied.

In the above example, each element of ξ is a function of one or two elements of $\mathbf{x}$. More generally, one may view ξ as a transformation of $\mathbf{x}$ of the following form:

$$\xi_i = \Phi_i(\mathbf{x}), \quad i = 1, \ldots, r$$

or

$$\xi = \boldsymbol{\Phi}(\mathbf{x}).$$

Our comments made in Chapter 1 on the advantages and disadvantages of the use of these Φ-functions are applicable here as well.

3.3 Bayes Estimation: Parametric Training

To estimate nonlinear decision surfaces, one may use training techniques that estimate the class-conditional densities. Then, if the a priori probabilities $\{P(\omega_j)\}$ are known, one obtains estimates of the constituent densities. The constituent densities may then be used as discriminant functions for the classifier. In this and the next several sections we describe a few trainable estimators of the class-conditional densities $p(\mathbf{x}|\omega_j)$. In these estimators it is assumed that $\{P(\omega_j)\}$ are known, and that all of the observed feature vectors $\{\mathbf{x}(i)\}$ belong to the same class.

In the present section we describe the Bayes procedure for estimating class-conditional densities. In this procedure one assumes that the class-conditional densities are completely determined by a finite set of parameters $\{\theta_i\}$, and that an a priori probability of these parameters is known. This procedure is thus a form of *parametric training*.

In the Bayes procedure one assumes that each class density is of some chosen form

$$p(\mathbf{x}|\boldsymbol{\theta}, \omega_j)$$

where $\boldsymbol{\theta}$ is an M-dimensional vector of parameters that completely determines all of the class-conditional densities of $\mathbf{x}$. Thus, one assumes that $p(\mathbf{x}|\omega_j, \boldsymbol{\theta})$ is a known function of $\boldsymbol{\theta}$ and j. One assumes that when training begins $\boldsymbol{\theta}$ is not known, but that the a priori density $p(\boldsymbol{\theta})$ is known. (In practice the assumption that $p(\boldsymbol{\theta})$ is known is often generous, since only a crude estimate is commonly assumed.) One also assumes that

$$\mathscr{X}_n = \{\mathbf{x}(1), \ldots, \mathbf{x}(n)\}$$

is a statistically independent sequence of n feature vectors drawn from the density $p(\mathbf{x}|\boldsymbol{\theta},\omega_j)$ for some $\boldsymbol{\theta}$ and some j, and that $\{P(\omega_j)\}$ are all known. (This notation for $\mathscr{X}_n$ should be distinguished from the $\mathscr{X}_j$ introduced in Section 1.6.) Let $\mathscr{X}_0$ denote the empty training set.

Under these assumptions one may compute the densities $p(\mathbf{x}|\omega_j,\mathscr{X}_n)$ as a function of $\mathscr{X}_n$. These densities are the best estimates of the class densities for the observation $\mathscr{X}_n$. Using these densities, one obtains the following set of discriminant functions in terms of the training set $\mathscr{X}_n$.

$$g_j(\mathbf{x}|\mathscr{X}_n) = P(\omega_j)p(\mathbf{x}|\omega_j,\mathscr{X}_n), \quad j = 1, 2, \ldots. \tag{3.1}$$

Since the $P(\omega_j)$ are known, $p(\mathbf{x}|\omega_j,\mathscr{X}_n)$ is the portion of $g_j(\mathbf{x})$ that must be computed by the training algorithm.

We now derive expressions for $p(\mathbf{x}|\omega_j,\mathscr{X}_n)$. Note that

$$p(\mathbf{x}|\omega_j,\mathscr{X}_n) = \int d\boldsymbol{\theta}\, p(\mathbf{x}|\omega_j,\boldsymbol{\theta},\mathscr{X}_n)p(\boldsymbol{\theta}|\omega_j,\mathscr{X}_n) \tag{3.2}$$

where $\int d\boldsymbol{\theta}$ denotes $\int_{-\infty}^{\infty}\cdots\int_{-\infty}^{\infty} d\theta_1 \cdots d\theta_M$.

Since $\boldsymbol{\theta}$ and ω_j provide full information about $p(\mathbf{x}|\omega_j,\boldsymbol{\theta},\mathscr{X}_n)$, it follows that $p(\mathbf{x}|\omega_j,\boldsymbol{\theta},\mathscr{X}_n) = p(\mathbf{x}|\omega_j,\boldsymbol{\theta})$. Thus Equation (3.2) simplifies to

$$p(\mathbf{x}|\omega_j,\mathscr{X}_n) = \int d\boldsymbol{\theta}\, p(\mathbf{x}|\omega_j,\boldsymbol{\theta})p(\boldsymbol{\theta}|\omega_j,\mathscr{X}_n). \tag{3.3}$$

We shall assume that the components of $\boldsymbol{\theta}$ may be partitioned into statistically independent subvectors associated with each class, i.e.,

$$\boldsymbol{\theta} = (\boldsymbol{\theta}_1^T, \ldots, \boldsymbol{\theta}_c^T)^T, \tag{3.4}$$

where $\boldsymbol{\theta}_j$ = subvector of $\boldsymbol{\theta}$ associated with ω_j and the $\boldsymbol{\theta}_j$ are statistically independent. In other words, we assume that knowledge of $\boldsymbol{\theta}_i$ provides no information about the probability density of $\boldsymbol{\theta}_j$ when $j \neq i$. Examples of possible components of $\boldsymbol{\theta}_j$ are $E(\mathbf{x}|\omega_j)$, $E[(\mathbf{x} - E(\mathbf{x}|\omega_j))^2|\omega_j]$, etc.

The a priori probability $P(\omega_j)$ may also be a component of $\boldsymbol{\theta}_j$. This is a reasonable component of $\boldsymbol{\theta}_j$ whenever the correct classifications of the members of $\mathscr{X}_n$ are uncertain (imperfect training) and $P(\omega_j)$ is unknown. As indicated earlier, we assume here that $P(\omega_j)$ is known, and hence that $P(\omega_j)$ is not a component of $\boldsymbol{\theta}_j$.

Under assumption (3.4),

$$p(\boldsymbol{\theta}|\omega_j,\mathscr{X}_n) = p(\boldsymbol{\theta}_1, \ldots, \boldsymbol{\theta}_c|\omega_j,\mathscr{X}_n) = \prod_{i=1}^{c} p(\boldsymbol{\theta}_i|\omega_j,\mathscr{X}_n). \tag{3.5}$$

Combining Equations (3.3) and (3.5), we obtain

$$\begin{aligned} p(\mathbf{x}|\omega_j,\mathscr{X}_n) &= \int d\boldsymbol{\theta}_1 \cdots d\boldsymbol{\theta}_c\, p(\mathbf{x}|\omega_j,\boldsymbol{\theta}_1, \ldots, \boldsymbol{\theta}_c) \prod_{i=1}^{c} p(\boldsymbol{\theta}_i|\omega_j,\mathscr{X}_n) \\ &= \int d\boldsymbol{\theta}_j\, p(\mathbf{x}|\omega_j,\boldsymbol{\theta}_j)p(\boldsymbol{\theta}_j|\omega_j,\mathscr{X}_n) \prod_{\substack{i=1 \\ i\neq j}}^{c} \left[\int d\boldsymbol{\theta}_i\, p(\boldsymbol{\theta}_i|\omega_j,\mathscr{X}_n)\right]. \end{aligned}$$

This reduces to

$$p(\mathbf{x}|\omega_j, \mathscr{X}_n) = \int d\boldsymbol{\theta}_j\, p(\mathbf{x}|\omega_j, \boldsymbol{\theta}_j) p(\boldsymbol{\theta}_j|\omega_j, \mathscr{X}_n). \tag{3.6}$$

This equation tells us that the best estimate of $p(\mathbf{x}|\omega_j)$, namely $p(\mathbf{x}|\omega_j, \mathscr{X}_n)$, is the average of $p(\mathbf{x}|\omega_j, \boldsymbol{\theta}_j)$ weighted by the density $p(\boldsymbol{\theta}_j|\omega_j, \mathscr{X}_n)$. Since $p(\mathbf{x}|\omega_j, \boldsymbol{\theta}_j)$ is assumed to be known, we need only find $p(\boldsymbol{\theta}_j|\omega_j, \mathscr{X}_n)$ in order to estimate $p(\mathbf{x}|\omega_j)$.

In our subsequent discussion, we simplify the notation by omitting the explicit condition on ω_j. Thus

$$p(\boldsymbol{\theta}_j|\omega_j, \mathscr{X}_n) \equiv p(\boldsymbol{\theta}|\mathscr{X}_n)$$

and

$$p(\mathbf{x}|\omega_j) \equiv p(\mathbf{x}),$$

where it is understood that $\boldsymbol{\theta} = \boldsymbol{\theta}_j$, and that the densities are conditioned on class ω_j. For each class, we must find a distinct density $p(\boldsymbol{\theta}|\mathscr{X}_n)$ in order to estimate a distinct $p(\mathbf{x})$.

By exploiting the independence of successive observations $\{\mathbf{x}(i)\}$ one obtains a recursive equation for updating $p(\boldsymbol{\theta}|\mathscr{X}_n)$ for successive values of n:

$$p(\boldsymbol{\theta}|\mathscr{X}_n) = \frac{p(\mathscr{X}_n|\boldsymbol{\theta})p(\boldsymbol{\theta})}{p(\mathscr{X}_n)}. \tag{3.7}$$

Since the $\{\mathbf{x}(i)\}$ are statistically independent,

$$p(\mathscr{X}_n|\boldsymbol{\theta}) = \prod_{k=1}^{n} p(\mathbf{x}(k)|\boldsymbol{\theta}) \tag{3.8}$$

and

$$p(\mathscr{X}_n|\boldsymbol{\theta}) = p(\mathbf{x}(n)|\boldsymbol{\theta})p(\mathscr{X}_{n-1}|\boldsymbol{\theta}). \tag{3.9}$$

Thus Equation (3.7) becomes

$$p(\boldsymbol{\theta}|\mathscr{X}_n) = \frac{p(\mathbf{x}(n)|\boldsymbol{\theta})p(\mathscr{X}_{n-1}|\boldsymbol{\theta})p(\boldsymbol{\theta})}{p(\mathscr{X}_n)}. \tag{3.10}$$

Note that

$$p(\mathscr{X}_{n-1}|\boldsymbol{\theta})p(\boldsymbol{\theta}) = p(\mathscr{X}_{n-1})p(\boldsymbol{\theta}|\mathscr{X}_{n-1}). \tag{3.11}$$

Hence

$$p(\boldsymbol{\theta}|\mathscr{X}_n) = p(\mathbf{x}(n)|\boldsymbol{\theta})p(\boldsymbol{\theta}|\mathscr{X}_{n-1})\frac{p(\mathscr{X}_{n-1})}{p(\mathscr{X}_n)}. \tag{3.12}$$

Since $p(\mathscr{X}_{n-1})/p(\mathscr{X}_n)$ is not a function of $\boldsymbol{\theta}$, we may compute it by integrating both sides of Equation (3.12) with respect to $d\boldsymbol{\theta}$ over the entire $\boldsymbol{\theta}$-space, and noting that the result is equal to unity. Thus,

$$\frac{p(\mathscr{X}_{n-1})}{p(\mathscr{X}_n)} = \left[\int d\boldsymbol{\theta}\, p(\mathbf{x}(n)|\boldsymbol{\theta})p(\boldsymbol{\theta}|\mathscr{X}_{n-1})\right]^{-1}. \tag{3.13}$$

Combining Equations (3.12) and (3.13), we obtain the following recursive formula for $p(\boldsymbol{\theta}|\mathscr{X}_n)$:

$$p(\boldsymbol{\theta}|\mathscr{X}_n) = \frac{p(\mathbf{x}(n)|\boldsymbol{\theta})p(\boldsymbol{\theta}|\mathscr{X}_{n-1})}{\int d\boldsymbol{\theta}\, p(\mathbf{x}(n)|\boldsymbol{\theta})p(\boldsymbol{\theta}|\mathscr{X}_{n-1})}. \tag{3.14}$$

Alternatively, substituting Equation (3.8) in Equation (3.7) yields

$$p(\boldsymbol{\theta}|\mathscr{X}_n) = \frac{p(\boldsymbol{\theta}) \prod_{i=1}^{n} p(\mathbf{x}(i)|\boldsymbol{\theta})}{\int d\boldsymbol{\theta}\, p(\boldsymbol{\theta}) \prod_{i=1}^{n} p(x(i)|\boldsymbol{\theta})}. \tag{3.15}$$

EXAMPLE 3.1. The application of Bayes estimation is demonstrated by a single-feature example. We assume that for a particular class the feature probability density is uniformly distributed between zero and some unknown positive value b; i.e.,

$$p(x|b) = \begin{cases} 1/b, & \text{if } 0 \leqslant x \leqslant b, \\ 0, & \text{otherwise.} \end{cases}$$

In this example the unknown parameter $\theta \equiv b$. We are also confident that b is at least equal to one but not greater than three, so we assume an initial $p(b)$ as follows:

$$p(b) \equiv p(b|\mathscr{X}_0) = \begin{cases} \frac{1}{2}, & \text{if } 1 \leqslant b \leqslant 3, \\ 0, & \text{otherwise.} \end{cases} \tag{3.16}$$

This density is illustrated in Figure 3.2(a). The available training set for this class contains 25 values of x ranging in magnitude from 0.2 to 2.5. The first two members of the set are $x(1) = 2$ and $x(2) = 1.3$. We will use the Bayes estimate method to find $p(b|\mathscr{X}_{25})$ and $p(\mathbf{x}|\mathscr{X}_{25})$.

First we note that

$$p(x(1)|b) = \begin{cases} 1/b, & \text{if } 2 \leqslant b, \\ 0, & \text{otherwise,} \end{cases}$$

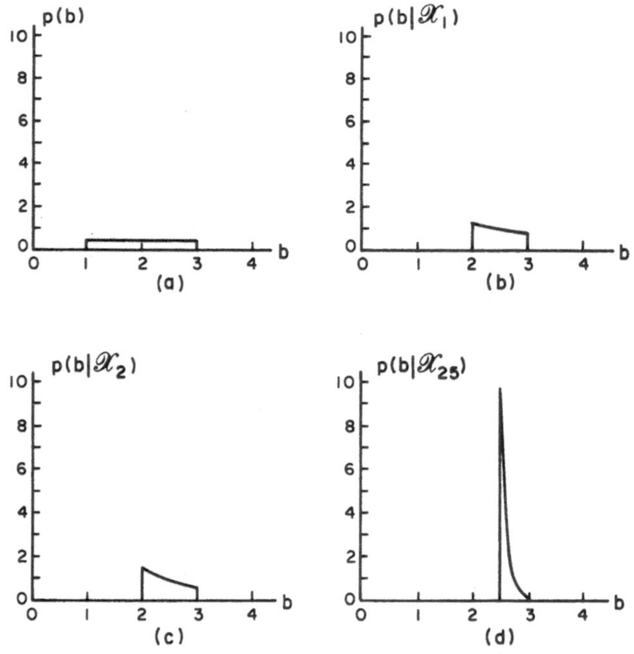

Figure 3.2. Successive estimates of $p(b)$ for Example 3.1.

so that

$$p(x(1)|b)p(b) = \begin{cases} 1/(2b), & \text{if } 2 \leqslant b \leqslant 3, \\ 0, & \text{otherwise.} \end{cases}$$

Since

$$\int_{-\infty}^{\infty} p(x(1)|b)p(b)\, db = \int_{2}^{3} (1/(2b))\, db = \tfrac{1}{2} \ln\left(\tfrac{3}{2}\right).$$

Equation (3.14) gives

$$p(b|\mathscr{X}_1) = \begin{cases} \dfrac{1}{b \ln\left(\frac{3}{2}\right)}, & \text{if } 2 \leqslant b \leqslant 3, \\ 0, & \text{otherwise.} \end{cases}$$

This first improved estimate of the distribution of b is illustrated in Figure 3.2(b). Similarly we apply $x(2)$ by first finding

$$p(x(2)|b)P(b|\mathscr{X}_1) = \begin{cases} \dfrac{1}{b^2 \ln\left(\frac{3}{2}\right)}, & \text{if } 2 \leqslant b \leqslant 3, \\ 0, & \text{otherwise,} \end{cases}$$

so that Equation (3.14) yields

$$p(b|\mathscr{X}_2) = \begin{cases} \dfrac{6}{b^2}, & \text{if } 2 \leqslant b \leqslant 3, \\ 0, & \text{otherwise.} \end{cases}$$

This second improved estimate is illustrated on Figure 3.2.(c). Note that for this example, successive applications of **x**'s from the training set will always give a density estimate of the form

$$p(b|\mathscr{X}_n) = \begin{cases} k/b^n, & \text{if } \max\limits_{i \leqslant n} x(i) \leqslant b \leqslant 3, \\ 0, & \text{otherwise,} \end{cases} \tag{3.17}$$

where k is chosen so that the probability density area of $p(b|\mathscr{X}_n)$ equals one. For $n = 25$, $k \approx 8.64 \times 10^{10}$ and for the given training set, $\max_i x(i) = 2.5$. Figure 3.2(d) illustrates the solution, $p(b|\mathscr{X}_{25})$. Note that each time n increases, the variance of $p(b|\mathscr{X}_n)$ decreases. This indicates an increased confidence that a good choice of b is one which is near the peak value of $p(b|\mathscr{X}_n)$.

To find $p(x|\mathscr{X}_{25})$ we apply Equation (3.6) while restricting ourselves to a single class. First we note that

$$p(x|b)p(b|\mathscr{X}_{25}) = \begin{cases} \dfrac{8.64 \times 10^{10}}{b^{26}}, & \text{for } 0 \leqslant x \leqslant 2.5,\ 2.5 \leqslant b \leqslant 3; \text{ or } 2.5 \leqslant x \leqslant 3,\ x \leqslant b \leqslant 3, \\ 0, & \text{otherwise,} \end{cases}$$

so that Equation (3.6) yields the solution

$$p(x|\mathscr{X}_{25}) = \int_{-\infty}^{\infty} p(x|b)p(b|\mathscr{X}_{25})\, db = \begin{cases} 0.390, & \text{if } 0 \leqslant x \leqslant 2.5, \\ \dfrac{3.5 \times 10^9}{x^{25}} - 0.004, & \text{if } 2.5 \leqslant x \leqslant 3, \\ 0, & \text{otherwise.} \end{cases} \tag{3.18}$$

Equation (3.18) is plotted in Figure 3.3.

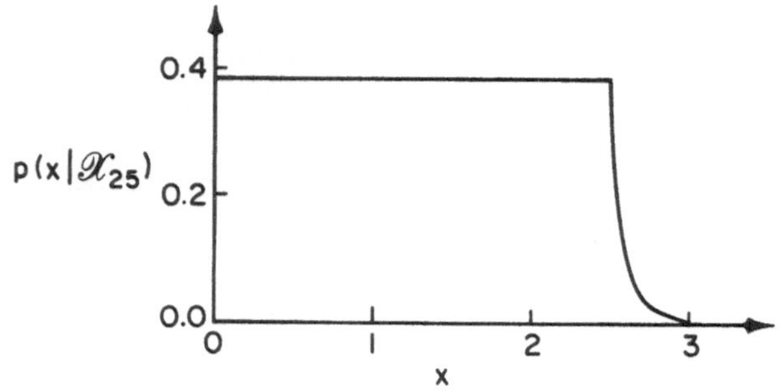

Figure 3.3. $p(x|\mathscr{X}_{25})$ for Example 3.1.

3.3.1 Bayesian Learning and Sufficient Statistics

Another effective approach to computing $p(\boldsymbol{\theta}|\mathscr{X}_n)$ is through the use of *sufficient statistics.* Suppose there exists a simple vector-valued function of $\mathscr{X}_n$, denoted by $\boldsymbol{\Phi}(\mathscr{X}_n) = \{\Phi_i(\mathscr{X}_n) | i = 1, \ldots, m\}$ for some m, such that $p(\boldsymbol{\theta}|\mathscr{X}_n)$ depends on only $\boldsymbol{\Phi}(\mathscr{X}_n)$ and $\boldsymbol{\theta}$; i.e.,

$$p(\boldsymbol{\theta}|\mathscr{X}_n) = f[\boldsymbol{\Phi}(\mathscr{X}_n), \boldsymbol{\theta}] \tag{3.19}$$

for some function f. A function $\boldsymbol{\Phi}(\mathscr{X}_n)$ having this property is called a *sufficient statistic** for $\boldsymbol{\theta}$, because knowledge of the value of $\boldsymbol{\Phi}(\mathscr{X}_n)$ is sufficient for the computation of $p(\boldsymbol{\theta}|\mathscr{X}_n)$ with no further information about $\mathscr{X}_n$ or n. Note that the value of n is not needed for computing this density after $\boldsymbol{\Phi}(\mathscr{X}_n)$ is known.

To gain an insight into the concept of sufficient statistics, consider the outcome of n tosses of a coin, with θ = probability of a head; hence $0 \leqslant \theta \leqslant 1$. Suppose θ is unknown. (Thus here the vector of unknown parameters $\boldsymbol{\theta}$ is just the scalar θ.) Let $x(i) = 0$ when the outcome of the ith toss is a tail, and $x(i) = 1$ when the outcome of the ith toss is a head. The quantities $\{x(1), \ldots, x(n)\}$ are independent, identically distributed random variables whose statistical behavior is fully specified by the parameter θ. To estimate θ from $\{x(1), \ldots, x(n)\}$, it is sufficient to know just the number of heads in the sequence. The order of occurrence of heads and tails cannot improve our estimate of θ, once we know the number of heads. Let

$$\mathscr{X}_n = \{x(1), \ldots, x(n)\}, \qquad \Phi(\mathscr{X}_n) = \sum_{i=1}^{n} x(i).$$

The conditional probability of occurrence of $\mathscr{X}_n$, given $\Phi(\mathscr{X}_n) = j$, is the same for every $\mathscr{X}_n$ that has precisely j ones. This probability is equal to

* An alternative, but equivalent, definition of a sufficient statistic for $\boldsymbol{\theta}$ is a vector-valued function $\boldsymbol{\Phi}(\mathscr{X}_n)$ for which $p(\mathscr{X}_n|\boldsymbol{\Phi}, \boldsymbol{\theta}) = p(\mathscr{X}_n|\boldsymbol{\Phi})$. For further discussion of this matter, see References [1] and [2].

$\binom{n}{j}^{-1}$, which is independent of θ. Thus the number of heads carries all the information in $\mathscr{X}_n$ about the unknown parameter θ. Consequently we say that $\Phi(\mathscr{X}_n)$ is a sufficient statistic for θ.

Formally a sufficient statistic for a vector parameter $\boldsymbol{\theta}$ is defined as follows. A statistic $\boldsymbol{\Phi}(\mathscr{X}_n)$ is sufficient for $\boldsymbol{\theta}$ if and only if the conditional distribution of $\mathscr{X}_n$, given $\boldsymbol{\Phi}(\mathscr{X}_n)$, does not involve $\boldsymbol{\theta}$. This definition is of little practical use in finding simple forms of $\boldsymbol{\Phi}$, since the use of the definition requires the derivation of an expression for the conditional distribution, which is often difficult.

An important computational advantage of sufficient statistics is that the functional form of $p(\boldsymbol{\theta}|\mathscr{X}_n)$ is invariant with respect to n; i.e., the density of $\boldsymbol{\theta}$ for any $\mathscr{X}_n$ is of the form $f(\boldsymbol{\Phi}, \boldsymbol{\theta})$. This is the so-called *reproducing property* of the probability density $p(\boldsymbol{\theta}|\mathscr{X}_n)$.

The main problem in using sufficient statistics for computing $p(\boldsymbol{\theta}|\mathscr{X}_n)$ is to find simple forms of the functions $\boldsymbol{\Phi}$ and f. In particular one often seeks a $\boldsymbol{\Phi}(\mathscr{X}_n)$ that has a simple recursive form—i.e., a simple function of $\mathbf{x}(n)$ and $\boldsymbol{\Phi}(\mathscr{X}_{n-1})$.

To find an appropriate $\boldsymbol{\Phi}$ and f, one may exploit the factorization theorem stated below. In this theorem, one assumes that the number of components in the vector $\boldsymbol{\Phi}(\mathscr{X}_n)$ is finite.

Factorization Theorem. *A necessary and sufficient condition for a finite-dimensional $\boldsymbol{\Phi}(\mathscr{X}_n)$ to be sufficient for $\boldsymbol{\theta}$ is that there exists the following form of a factorization of the likelihood of $\boldsymbol{\theta}$ with respect to $\mathscr{X}_n$:*

$$p(\mathscr{X}_n|\boldsymbol{\theta}) = g(\boldsymbol{\Phi}, \boldsymbol{\theta})h(\mathscr{X}_n), \tag{3.20}$$

where the first factor may depend on $\boldsymbol{\theta}$ but depends on $\mathscr{X}_n$ only through $\boldsymbol{\Phi}(\mathscr{X}_n)$, while the second factor is independent of $\boldsymbol{\theta}$.

In using this theorem one tries to express $p(\mathscr{X}_n|\boldsymbol{\theta})$ as a product of a function of $\boldsymbol{\Phi}$ and $\boldsymbol{\theta}$ and a function of $\mathscr{X}_n$, as indicated in Equation (3.20).

PROOF. To demonstrate the validity of this theorem, write Equation (3.7) in the form

$$p(\boldsymbol{\theta}|\mathscr{X}_n) = p(\mathscr{X}_n|\boldsymbol{\theta})p(\boldsymbol{\theta}) \Big/ \int d\boldsymbol{\theta}\, p(\mathscr{X}_n|\boldsymbol{\theta})p(\boldsymbol{\theta}) \tag{3.21}$$

and substitute Equation (3.20) in Equation (3.21). This yields

$$p(\boldsymbol{\theta}|\mathscr{X}_n) = g(\boldsymbol{\Phi}, \boldsymbol{\theta})p(\boldsymbol{\theta}) \Big/ \int d\boldsymbol{\theta}\, g(\boldsymbol{\Phi}, \boldsymbol{\theta})p(\boldsymbol{\theta}), \tag{3.22}$$

which depends only on $\boldsymbol{\Phi}$ and $\boldsymbol{\theta}$. Thus the existence of g and h satisfying Equation (3.20) implies the existence of f satisfying Equation (3.19), which demonstrates that $\boldsymbol{\Phi}$ is a sufficient statistic for $\boldsymbol{\theta}$. Conversely, if $\boldsymbol{\Phi}$ is a sufficient statistic for $\boldsymbol{\theta}$, then by Equations (3.19) and (3.21)

$$p(\mathscr{X}_n|\boldsymbol{\theta}) = \frac{f(\boldsymbol{\Phi}, \boldsymbol{\theta})}{p(\boldsymbol{\theta})} \int d\boldsymbol{\theta}\, p(\mathscr{X}_n|\boldsymbol{\theta})p(\boldsymbol{\theta}). \tag{3.23}$$

The integral in the Equation (3.23) is a function of $\mathscr{X}_n$ only, while $f(\boldsymbol{\Phi}, \boldsymbol{\theta})/p(\boldsymbol{\theta})$ depends on only $\boldsymbol{\Phi}$ and $\boldsymbol{\theta}$. This proves the factorization theorem, namely, that $\boldsymbol{\Phi}$ is a sufficient statistic for $\boldsymbol{\theta}$ if and only if $p(\mathscr{X}_n|\boldsymbol{\theta}) = g(\boldsymbol{\Phi}, \boldsymbol{\theta})h(\mathscr{X}_n)$ for some g and h. □

The computational advantage of a sufficient statistic depends on the ease with which one may compute $\boldsymbol{\Phi}(\mathscr{X}_n)$ and f. Thus it is important that $\boldsymbol{\Phi}(\mathscr{X}_n)$ be simple or that $\boldsymbol{\Phi}(\mathscr{X}_{n+1})$ be easily computed from $\boldsymbol{\Phi}(\mathscr{X}_n)$ and $\mathbf{x}(n+1)$. The trivial sufficient statistic

$$\boldsymbol{\Phi}(\mathscr{X}_n) = [\mathbf{x}(1), \ldots, \mathbf{x}(n), \mathbf{0}, \mathbf{0}, \ldots, \mathbf{0}] \quad \text{for } n \leqslant m$$

clearly yields no computational advantage.

The following is an example of a useful sufficient statistic:

$$\boldsymbol{\Phi}(\mathscr{X}_n) = \hat{\boldsymbol{\mu}}(n) \equiv \frac{1}{n} \sum_{k=1}^{n} \mathbf{x}(k).$$

In this case

$$p(\boldsymbol{\theta}|\mathscr{X}_{n+1}) = f(\hat{\boldsymbol{\mu}}(n+1), \boldsymbol{\theta}) = f\left[\frac{n\hat{\boldsymbol{\mu}}(n) + \mathbf{x}(n+1)}{n+1}, \boldsymbol{\theta}\right].$$

This equation displays both the recursive computation of the sufficient statistic $\hat{\boldsymbol{\mu}}(n)$ and the reproducing property of $p(\boldsymbol{\theta}|\mathscr{X}_n)$.

EXAMPLE 3.2.

$$\boldsymbol{\theta} = \mathbf{M} = E[\mathbf{X}(n)|\mathbf{M}], \tag{3.24}$$

$$\mathbf{X}(n) = \mathbf{Z}(n) + \mathbf{M}, \tag{3.25}$$

where $\mathbf{X}(n) \in \omega_j$ for all n, $\mathbf{Z}(n)$ and $\mathbf{M}$ are statistically independent of each other, the members of the sequence $\{\mathbf{Z}(n)\}$ are statistically independent, and the densities of $\mathbf{Z}(n)$ and $\mathbf{M}$ are normal densities given by

$$p_{\mathbf{Z}(n)}(\mathbf{z}) = N(\mathbf{0}, \boldsymbol{\Sigma}), \qquad \text{for all } n \tag{3.26}$$

$$p_{\mathbf{M}}(\boldsymbol{\mu}) \equiv p_{\mathbf{M}}(\boldsymbol{\mu}|\mathscr{X}_0) = N(\boldsymbol{\mu}(0), \mathbf{K}(0)). \tag{3.27}$$

It follows from Equations (3.25)–(3.27) that

$$p_{\mathbf{X}}(\mathbf{x}|\mathscr{X}_0) = N(\boldsymbol{\mu}(0), \boldsymbol{\Sigma} + \mathbf{K}(0)). \tag{3.28}$$

We now proceed to apply the theory presented above to the derivation of $p_{\mathbf{M}}(\boldsymbol{\mu}|\mathscr{X}_n)$ and $p_{\mathbf{X}}(\mathbf{x}|\mathscr{X}_n)$. Hereafter we omit the subscripts $\mathbf{M}$ and $\mathbf{X}$ wherever the omission seems to cause no confusion. We shall derive $p(\boldsymbol{\mu}|\mathscr{X}_n)$ by two methods—one based on Equation (3.14) and the other based on sufficient statistics.

First Method of Deriving $p(\boldsymbol{\mu}|\mathscr{X}_n)$. By Equation (3.14),

$$p(\boldsymbol{\mu}|\mathscr{X}_n) = \frac{p(\mathbf{x}(n)|\boldsymbol{\mu})p(\boldsymbol{\mu}|\mathscr{X}_{n-1})}{\int d\boldsymbol{\mu}\, p(\mathbf{x}(n)|\boldsymbol{\mu})p(\boldsymbol{\mu}|\mathscr{X}_{n-1})}. \tag{3.29}$$

We will show that $p(\boldsymbol{\mu}|\mathscr{X}_n)$ is Gaussian for all n, so that $p(\boldsymbol{\mu}|\mathscr{X}_n)$ has the reproducing property. To demonstrate this property, we use mathematic induction. First note that $p(\boldsymbol{\mu}|\mathscr{X}_0)$ is Gaussian, from Equation (3.27). Now suppose that $p(\boldsymbol{\mu}|\mathscr{X}_{n-1}) = N(\boldsymbol{\mu}(n-1), \mathbf{K}(n-1))$ = Gaussian density. Note that $p(\mathbf{x}(n)|\mathbf{M} = \boldsymbol{\mu}) = N(\boldsymbol{\mu}, \boldsymbol{\Sigma})$ by Equations (3.25) and 3.26). Then, by Equation (3.29),

$$\begin{aligned} p(\boldsymbol{\mu}|\mathscr{X}_n) &= C_n' \exp[-\tfrac{1}{2}(\mathbf{x}(n)-\boldsymbol{\mu})^T\boldsymbol{\Sigma}^{-1}(\mathbf{x}(n)-\boldsymbol{\mu})]\, C_n'' \\ &\quad \times \exp[-\tfrac{1}{2}(\boldsymbol{\mu}-\boldsymbol{\mu}(n-1))^T\mathbf{K}^{-1}(n-1)(\boldsymbol{\mu}-\boldsymbol{\mu}(n-1))] \\ &= C_n \exp\{-\tfrac{1}{2}[(\mathbf{x}(n)-\boldsymbol{\mu})^T\boldsymbol{\Sigma}^{-1}(\mathbf{x}(n)-\boldsymbol{\mu}) \\ &\quad + (\boldsymbol{\mu}-\boldsymbol{\mu}(n-1))^T\mathbf{K}^{-1}(n-1)(\boldsymbol{\mu}-\boldsymbol{\mu}(n-1))]\}, \end{aligned} \tag{3.30}$$

where C_n', C_n'', and C_n are constants with respect to $\boldsymbol{\mu}$, and in particular C_n is a constant which normalizes $p(\boldsymbol{\mu}|\mathscr{X}_n)$ to have a hypervolume of 1. Since the exponent in Equation (3.30) is a negative definite quadratic function of $\boldsymbol{\mu}$, it follows that $p(\boldsymbol{\mu}|\mathscr{X}_n)$ is Gaussian for all $n \geqslant 0$. We now proceed to derive expressions for $\boldsymbol{\mu}(n)$ and $\mathbf{K}(n)$, the mean and covariance matrix of $p(\boldsymbol{\mu}|\mathscr{X}_n)$. By taking the logarithm of each side of Equation (3.30) and simplifying, we obtain

$$\begin{aligned} &(\boldsymbol{\mu}-\boldsymbol{\mu}(n))^T\mathbf{K}^{-1}(n)(\boldsymbol{\mu}-\boldsymbol{\mu}(n)) + b_1 \\ &\quad = (\mathbf{x}(n)-\boldsymbol{\mu})^T\boldsymbol{\Sigma}^{-1}(\mathbf{x}(n)-\boldsymbol{\mu}) + (\boldsymbol{\mu}-\boldsymbol{\mu}(n-1))^T\mathbf{K}^{-1}(n-1)(\boldsymbol{\mu}-\boldsymbol{\mu}(n-1)) \\ &\quad = \boldsymbol{\mu}^T[\boldsymbol{\Sigma}^{-1} + \mathbf{K}^{-1}(n-1)]\boldsymbol{\mu} - 2\boldsymbol{\mu}^T[\boldsymbol{\Sigma}^{-1}\mathbf{x}(n) + \mathbf{K}^{-1}(n-1)\boldsymbol{\mu}(n-1)] + b_2, \end{aligned}$$

where the b_i are terms that are independent of $\boldsymbol{\mu}$. Hence

$$\mathbf{K}^{-1}(n) = \boldsymbol{\Sigma}^{-1} + \mathbf{K}^{-1}(n-1) \tag{3.31}$$

$$\mathbf{K}^{-1}(n)\boldsymbol{\mu}(n) = \boldsymbol{\Sigma}^{-1}\mathbf{x}(n) + \mathbf{K}^{-1}(n-1)\boldsymbol{\mu}(n-1). \tag{3.32}$$

By Equations (3.31) and (3.32),

$$\begin{aligned} \boldsymbol{\mu}(n) &= (\boldsymbol{\Sigma}^{-1} + \mathbf{K}^{-1}(n-1))^{-1}[\boldsymbol{\Sigma}^{-1}\mathbf{x}(n) + \mathbf{K}^{-1}(n-1)\boldsymbol{\mu}(n-1)] \\ &\equiv \mathbf{K}(n-1)[\boldsymbol{\Sigma} + \mathbf{K}(n-1)]^{-1}\mathbf{x}(n) + \boldsymbol{\Sigma}[\boldsymbol{\Sigma} + \mathbf{K}(n-1)]^{-1}\boldsymbol{\mu}(n-1). \end{aligned} \tag{3.33}$$

It follows from Equations (3.31) and (3.33) that

$$\boldsymbol{\mu}(n) = \frac{1}{n}\boldsymbol{\Sigma}\left(\mathbf{K}(0) + \frac{1}{n}\boldsymbol{\Sigma}\right)^{-1}\boldsymbol{\mu}(0) + \mathbf{K}(0)\left(\mathbf{K}(0) + \frac{1}{n}\boldsymbol{\Sigma}\right)^{-1}\hat{\boldsymbol{\mu}}(n) \tag{3.34}$$

$$\mathbf{K}(n) = \mathbf{K}(0)\left(\mathbf{K}(0) + \frac{1}{n}\boldsymbol{\Sigma}\right)^{-1}\frac{1}{n}\boldsymbol{\Sigma}, \tag{3.35}$$

where $\hat{\boldsymbol{\mu}}(n) = (1/n)[\mathbf{x}(1) + \cdots + \mathbf{x}(n)]$ = sample mean.

Second Method of Deriving $p(\boldsymbol{\mu}|\mathscr{X}_n)$. The method of sufficient statistics for deriving $p(\boldsymbol{\mu}|\mathscr{X}_n)$ starts with finding an expression for $p(\mathscr{X}_n|\boldsymbol{\mu})$:

$$\begin{aligned} p(\mathscr{X}_n|\boldsymbol{\mu}) &= \prod_{i=1}^{n} p(\mathbf{x}(i)|\boldsymbol{\mu}) \\ &= C \exp\left\{-\tfrac{1}{2}\sum_{i=1}^{n} [(\mathbf{x}(i)-\boldsymbol{\mu})^T\boldsymbol{\Sigma}^{-1}(\mathbf{x}(i)-\boldsymbol{\mu})]\right\}, \end{aligned} \tag{3.36}$$

where C is a constant. Note that $\sum_{i=1}^{n} [(\mathbf{x}(i)-\boldsymbol{\mu})^T\boldsymbol{\Sigma}^{-1}(\mathbf{x}(i)-\boldsymbol{\mu})]$ is a sum of positive definite quadratic forms in $\boldsymbol{\mu}$, and hence is itself a positive definite quadratic form in $\boldsymbol{\mu}$.

Since $p(\boldsymbol{\mu})$ is Gaussian by Equation (3.27), it follows that $p(\mathscr{X}_n|\boldsymbol{\mu})p(\boldsymbol{\mu})$ is a Gaussian density in $\boldsymbol{\mu}$ except for a normalizing factor. Hence $p(\boldsymbol{\mu}|\mathscr{X}_n)$ is Gaussian, by Equation (3.7), in which $\boldsymbol{\theta}$ is replaced by $\boldsymbol{\mu}$.

Expanding the exponent in the right-hand side of Equation (3.36), we get

$$p(\mathscr{X}_n|\boldsymbol{\mu}) = h(\mathscr{X}_n)\exp\left\{-\tfrac{1}{2}\sum_{i=1}^{n}[\boldsymbol{\mu}^T\boldsymbol{\Sigma}^{-1}\boldsymbol{\mu} - 2\boldsymbol{\mu}^T\boldsymbol{\Sigma}^{-1}\mathbf{x}(i)]\right\}$$

$$= h(\mathscr{X}_n)\exp\left\{-\frac{n\boldsymbol{\mu}^T\boldsymbol{\Sigma}^{-1}}{2}(\boldsymbol{\mu} - 2\hat{\boldsymbol{\mu}}(n))\right\}, \tag{3.37}$$

where

$$\hat{\boldsymbol{\mu}}(n) = \frac{1}{n}\sum_{i=1}^{n}\mathbf{x}(i). \tag{3.38}$$

Hence

$$g(\boldsymbol{\Phi}, \boldsymbol{\mu}) = \exp\left\{-\frac{n\boldsymbol{\mu}^T\boldsymbol{\Sigma}^{-1}}{2}[\boldsymbol{\mu} - 2\hat{\boldsymbol{\mu}}(n)]\right\}, \tag{3.39}$$

where $\boldsymbol{\Phi} = \hat{\boldsymbol{\mu}}(n)$ = sufficient statistic for $\boldsymbol{\mu}$. By Equation (3.22),

$$p(\boldsymbol{\mu}|\mathscr{X}_n) = Cg(\hat{\boldsymbol{\mu}}(n), \boldsymbol{\mu})p(\boldsymbol{\mu})$$

$$= C'\exp\left\{-\frac{n\boldsymbol{\mu}^T\boldsymbol{\Sigma}^{-1}}{2}(\boldsymbol{\mu} - 2\hat{\boldsymbol{\mu}}(n))\right\}\exp[-\tfrac{1}{2}(\boldsymbol{\mu} - \boldsymbol{\mu}(0))^T\mathbf{K}^{-1}(0)(\boldsymbol{\mu} - \boldsymbol{\mu}(0))]. \tag{3.40}$$

Since $p(\boldsymbol{\mu}|\mathscr{X}_n)$ is Gaussian, we write

$$p(\boldsymbol{\mu}|\mathscr{X}_n) = C''\exp[-\tfrac{1}{2}(\boldsymbol{\mu} - \boldsymbol{\mu}(n))^T\mathbf{K}^{-1}(n)(\boldsymbol{\mu} - \boldsymbol{\mu}(n))]. \tag{3.41}$$

Equating corresponding coefficients in the exponents of Equations (3.40) and (3.41), we get the following formulas for $\mathbf{K}(n)$ and $\boldsymbol{\mu}(n)$:

$$\mathbf{K}^{-1}(n) = n\boldsymbol{\Sigma}^{-1} + \mathbf{K}^{-1}(0) \tag{3.42}$$

$$\boldsymbol{\mu}(n) = (n\boldsymbol{\Sigma}^{-1} + \mathbf{K}^{-1}(0))^{-1}(\mathbf{K}^{-1}(0)\boldsymbol{\mu}(0) + n\boldsymbol{\Sigma}^{-1}\hat{\boldsymbol{\mu}}(n)). \tag{3.43}$$

These equations are equivalent to Equations (3.34) and (3.35). (Prove this as an exercise.)

From Equations (3.34) and (3.35) as well as Equation (3.38), note that

$$\lim_{n\to\infty}\boldsymbol{\mu}(n) = \lim_{n\to\infty}\hat{\boldsymbol{\mu}}(n) = \boldsymbol{\mu} = \text{“true” mean} \tag{3.44}$$

and

$$\lim_{n\to\infty}\mathbf{K}(n) = \mathbf{0}. \tag{3.45}$$

Thus we see that $p(\boldsymbol{\mu}|\mathscr{X}_n)$ approaches a Dirac delta function $\delta(\boldsymbol{\mu} - \boldsymbol{\mu}(n))$ as the covariance matrix $\mathbf{K}(n) \to 0$ as $n \to \infty$. (For a discussion of Dirac delta functions in multidimensional space, see Reference [3].)

If $\mathbf{K}(0) = (1/\alpha)\boldsymbol{\Sigma}$, then Equations (3.31) and (3.32) become

$$\boldsymbol{\mu}(n) = \frac{\alpha}{n+\alpha}\boldsymbol{\mu}(0) + \frac{n}{n+\alpha}\hat{\boldsymbol{\mu}}(n)$$

$$\mathbf{K}(n) = \frac{1}{\alpha+n}\boldsymbol{\Sigma}.$$

The parameter α is often called the *dogmatism*, because it is a measure of the a priori confidence in $\boldsymbol{\mu}(0)$ as an estimate of the mean of $\mathbf{X}$ [1, 4, 5].

We are now in a position to find the Bayes estimate of $p(\mathbf{x}|\omega_j)$, namely $p(\mathbf{x}|\omega_j, \mathscr{X}_n)$. Again, omitting ω_j and the subscript j, we have from Equation (3.6):

$$p(\mathbf{x}|\mathscr{X}_n) = \int d\boldsymbol{\mu}\, p(\mathbf{x}|\boldsymbol{\mu})p(\boldsymbol{\mu}|\mathscr{X}_n)$$

$$= C\int d\boldsymbol{\mu} \exp[-\tfrac{1}{2}(\mathbf{x}-\boldsymbol{\mu})^T\boldsymbol{\Sigma}^{-1}(\mathbf{x}-\boldsymbol{\mu})]\exp[-\tfrac{1}{2}(\boldsymbol{\mu}-\boldsymbol{\mu}(n))^T\mathbf{K}^{-1}(n)(\boldsymbol{\mu}-\boldsymbol{\mu}(n))]$$

$$= N[\boldsymbol{\mu}(n), \boldsymbol{\Sigma} + \mathbf{K}(n)]. \tag{3.46}$$

(Prove this as an exercise.) In this particular case, a simpler way of deriving Equation (3.46) is available to us. Recall that $\mathbf{X} = \mathbf{Z} + \mathbf{M}$ and that $\mathbf{Z}$ and $\mathbf{M}$ are independent.

$$p_{\mathbf{M}}(\boldsymbol{\mu}|\mathscr{X}_n) = N(\boldsymbol{\mu}(n), \mathbf{K}(n)),$$
$$p_{\mathbf{Z}}(\mathbf{z}|\mathscr{X}_n) = p(\mathbf{z}) = N(0, \boldsymbol{\Sigma}).$$

Hence

$$p_{\mathbf{X}}(\mathbf{x}|\mathscr{X}_n) = p_{\mathbf{Z}+\mathbf{M}}(\mathbf{x}|\mathscr{X}_n) = N[\boldsymbol{\mu}(n), \boldsymbol{\Sigma} + \mathbf{K}(n)].$$

3.3.2 An Intuitive Interpretation of Example 3.2

The concepts underlying Example 3.2 may perhaps be clarified by the following interpretation. Suppose $\mathbf{X}(n)$ is the feature vector of the nth observed printed character on a page of a book. Suppose all of the printed characters in the book belong to the same font f_i. Suppose the book is selected at random from a library each of whose books is restricted to a single font, but that the fonts may vary from book to book.

Because of variations in the way in which ink adheres to the paper during the printing process, the feature vectors of different instances of the same character ω_j in a given book with font f_i have variations $\mathbf{Z}(n)$ about a mean value $\mathbf{M} = \boldsymbol{\mu}_i$ with a covariance matrix $\boldsymbol{\Sigma}$. Assume that the density of variations are described by $P_{\mathbf{X}}(\mathbf{x}|\mathbf{M} = \boldsymbol{\mu}_i)$. $\mathbf{M}$ varies randomly from book to book when books are selected at random, assuming that different books have different fonts. Although the value of $\mathbf{M}$ for a particular character and font is unknown, successive observations of feature vectors from ω_j, namely $\{\mathbf{X}(n)|\mathbf{X}(n) \in \omega_j\}$, contribute to estimating this $\mathbf{M}$, since all of the $\mathbf{X}(n)$'s in a given training phase belong to the same book and hence to the same density $P_{\mathbf{X}}(\mathbf{x}|\mathbf{M})$.

In any particular training process both $\boldsymbol{\mu}_i$ and the covariance matrix $\boldsymbol{\Sigma}$ are fixed. Assume that $\boldsymbol{\mu}_i$ is unknown, but that $p_{\mathbf{M}}(\boldsymbol{\mu})$, the initially assumed (i.e., a priori) density of $\mathbf{M}$ for character ω_j over an ensemble of books, is known. Assume that each training process is restricted to a single book. The degree of ignorance of $\mathbf{M}$, given $\mathscr{X}_n$, may be expressed by the covarianc matrix of $\mathbf{M}$, given $\mathscr{X}_n$. Let $\mathbf{K}(n)$ denote this covariance matrix. Thus:

$$\mathbf{K}(n) = E\{[\mathbf{M} - E(\mathbf{M})][\mathbf{M} - E(\mathbf{M})]^T|\mathscr{X}_n\},$$

where the expectation $E(\mathbf{M})$ is taken over the ensemble of books. Thus, $E(\mathbf{M})$ represents the a priori ensemble mean of $\mathbf{M}$. To simplify matters, assume

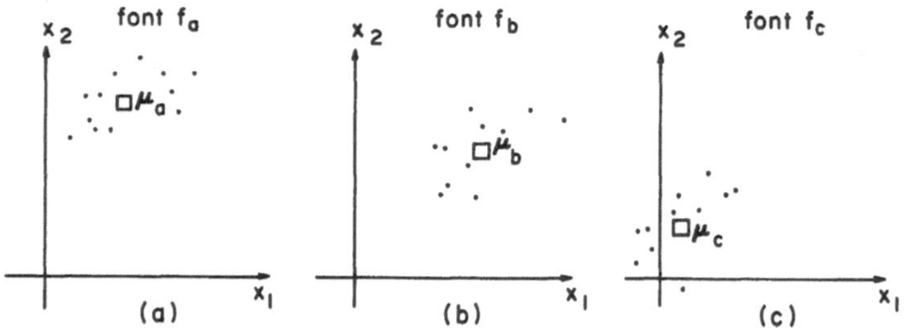

Figure 3.4. Illustration of distributions of **x** for three different fonts of a single character.

that $p_{\mathbf{M}}(\boldsymbol{\mu})$ and $p_{\mathbf{X}}(\mathbf{x}\,|\,\boldsymbol{\mu})$ are both Gaussian—i.e.,

$$P_{\mathbf{M}}(\boldsymbol{\mu}\,|\,\mathscr{X}_n) = N(\boldsymbol{\mu}, \mathbf{K}(n))$$
$$P_{\mathbf{X}}(\mathbf{x}\,|\,\boldsymbol{\mu}) = N(\boldsymbol{\mu}, \boldsymbol{\Sigma}).$$

Also assume that $\mathbf{X}(n) = \mathbf{Z}(n) + \mathbf{M}$. The vector $\mathbf{M}$ represents the average feature vector of a character for a particular font. $\mathbf{Z}(n)$ denotes the deviations of the feature vector $\mathbf{X}(n)$ from $\mathbf{M}$ caused by fluctuations in the adhesion of the ink.

This situation is illustrated in Figures 3.4 and 3.5. Each of parts (a), (b), and (c) of Figure 3.4 illustrates a member of the ensemble of training processes, all three members for the same class, say ω_j. The black dots in each part of Figure 3.4 represent instances of $\{\mathbf{X}(n)\,|\,\mathbf{X}(n) \in \omega_j\}$ with $\mathbf{M}$ fixed. The small white squares denote various values of $\mathbf{M}$. Note that the dots in each training process cluster around their means. The white squares in Figure 3.5 illustrate various values of $\mathbf{M}$ over several members of the ensemble of training processes. The black square in Figure 3.5 represents the mean value of $\mathbf{M}$ at $n = 0$, namely $\boldsymbol{\mu}(0)$. (The vector $\boldsymbol{\mu}(0)$ may be viewed as the feature vector of ω_j in the "average font" of the library.) The spread of white squares clustering around $\boldsymbol{\mu}(0)$ indicates the size of $|\mathbf{K}(0)|^{1/2}$, the square root of the determinant of the a priori covariance matrix of $\mathbf{M}$.

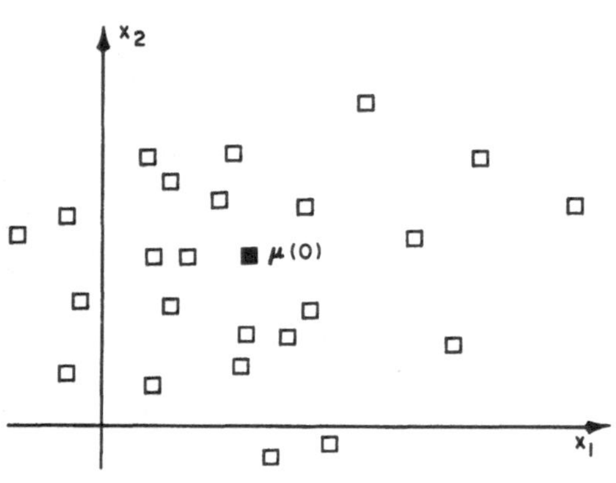

Figure 3.5. Illustration of a distribution of **M** for a particular character.

With the Bayes training procedure, one sees how the elements of the covariance matrix $\mathbf{K}(n)$ become smaller as n increases, thereby displaying the reduction of uncertainty of $\mathbf{M}$ with increasing n. This training procedure also finds the optimal discriminant function as a function of n, using Equation (3.1). In this case, the optimal discriminant function for class ω_j in the selected book is

$$g_j(\mathbf{x}) = P(\omega_j)N(\boldsymbol{\mu}(n), \boldsymbol{\Sigma} + \mathbf{K}(n)),$$

where $\boldsymbol{\mu}(n)$ and $\mathbf{K}(n)$ are given by Equations (3.34) and (3.35).

3.3.3 Conjugate Prior Probability Densities

In Example 3.2 we saw that if $p(\boldsymbol{\mu})$ is normal with mean $\boldsymbol{\mu}(0)$ and covariance matrix $\mathbf{K}(0)$, then $p(\boldsymbol{\mu}|\mathscr{X}_n)$ is a normal density $N[\boldsymbol{\mu}(n), \mathbf{K}(n)]$, where $\boldsymbol{\mu}(n)$ and $\mathbf{K}(n)$ may be computed recursively from $\boldsymbol{\mu}(0)$ and $\mathbf{K}(0)$ by Equations (3.42) and (3.43). Thus in this example $p(\boldsymbol{\mu}|\mathscr{X}_n)$ has a reproducing property if the prior density $p(\boldsymbol{\mu})$ is normal; but if $p(\boldsymbol{\mu})$ is not normal $p(\boldsymbol{\mu}|\mathscr{X}_n)$ may not have this property. Thus in this example a normal prior density $p(\boldsymbol{\mu})$ yields not only a $p(\boldsymbol{\mu}|\mathscr{X}_n)$ having the reproducing property, but one which is normal—i.e., a density $p(\boldsymbol{\mu}|\mathscr{X}_n)$ that has the same form as the prior density $p(\boldsymbol{\mu})$. Such a prior density is called a *conjugate prior* probability density.

In applying the factorization theorem (Equation (3.22)) one must assume a form of $p(\boldsymbol{\theta})$ that yields a density $p(\boldsymbol{\theta}|\mathscr{X}_n)$ having the reproducing property. We refer to such a $p(\boldsymbol{\theta})$ as a *reproducing density*. Usually, because of the resulting computational simplicity, one is tempted to use the same form of $p(\boldsymbol{\theta})$ as the form of the function of $\boldsymbol{\theta}$ obtained in $g(\boldsymbol{\Phi}, \boldsymbol{\theta})$ when $\boldsymbol{\Phi}$ is constant; i.e., the conjugate prior density, if it exists, yields an attractive procedure for computing $p(\boldsymbol{\theta}|\mathscr{X}_n)$. For this purpose the following relation between sufficient statistics and conjugate prior densities is useful: if a sufficient statistic exists (i.e., if the factorization theorem can be applied in a nontrivial manner) then a conjugate prior density exists. More generally, one can show that if a sufficient statistic exists, then $p(\boldsymbol{\theta})$ is a reproducing density if and only if it is of the form

$$p(\boldsymbol{\theta}) = \frac{g(\boldsymbol{\Phi},\boldsymbol{\theta})s(\boldsymbol{\theta})}{\int d\boldsymbol{\theta}\, g(\boldsymbol{\Phi},\boldsymbol{\theta})s(\boldsymbol{\theta})},$$

where $s(\boldsymbol{\theta})$ is any nonnegative function of $\boldsymbol{\theta}$ such that the denominator of the right-hand side of this equation exists [6]. This equation yields a class of reproducing densities for each admissible $s(\boldsymbol{\theta})$. If $s(\boldsymbol{\theta})$ is a constant, then $p(\boldsymbol{\theta})$ is a conjugate prior density.

This result may be used for the derivation of conjugate prior densities. Suppose, for example, that we are not given that $\mathbf{M}$ is normally distributed in the example and suppose Equation (3.27) is deleted from the statement of the problem. This change in the problem statement does not affect the derivation of Equation (3.39). By Equation (3.39) and the equation in the last paragraph, with $s(\boldsymbol{\theta}) = 1$, it follows that $p_{\mathbf{M}}(\boldsymbol{\mu})$ (or, equivalently, $p(\boldsymbol{\mu})$) is a normal conjugate prior density with mean $\boldsymbol{\mu}(0)$ and covariance matrix $\mathbf{K}(0)$. In this situation, since $\boldsymbol{\mu}(0)$ may not be given explicitly, we may have to estimate $\boldsymbol{\mu}(0)$ and $\mathbf{K}(0)$ as best we can from other a priori information available to us about the population from which $\mathscr{X}_n$ is drawn.

The reproducing property has been shown to exist, and has been tabulated, for a broad family of exponential densities, among which the normal density is a special case. On the other hand for certain other densities no sufficient statistic other than the trivial case of the sample itself exists, and hence no reproducing density exists [6].

A warning: sometimes the true prior density is quite different from the conjugate prior density. In such a case the estimates obtained by using the conjugate prior densities may not be satisfactory for small values of n [7].

3.4 Smoothing Techniques: Nonparametric Training

If the class-conditional densities are not easily described in terms of a small set of parameters, and if the optimum decision surface is known to be highly nonlinear, one should consider nonparametric training techniques for estimating these densities.

Nonparametric training techniques exploit the fact that for many practical situations the class densities are smooth functions over the sample space. (By a *smooth function* we mean that changes in values of the function are relatively small in any small neighborhood of a point in the function's domain and, if the domain is a continuum, that these changes become infinitesimal as the size of the neighborhood approaches zero.) The assumption of smoothness in the class densities leads to the replacement of each observed feature vector by a positive, single-peaked, piecewise continuous function that contributes linearly to an estimate of the class density. We refer to this replacement as *smoothing*. The smoothing techniques we will discuss in the sections to follow are: (a) bar graphs, (b) Parzen windows (potential functions), and (c) prototypes and clusters.

3.5 Bar Graphs

The essential concept of the smoothing of a sample set is illustrated by the bar graph in part (a) of Figure 3.6. The bar graph is constructed from a training set $\mathscr{X}_n = \{\mathbf{X}(1), \ldots, \mathbf{X}(n)\}$. This set consists of n observed d-dimensional feature vectors, all drawn statistically independently from a population whose probability density is $p(\mathbf{x})$. In the figure, $n = 10$ and $d = 1$. The tic marks on the x axis indicate the elements of $\mathscr{X}_n$. Let I_k denote the kth interval, $(kV, (k+1)V)$. Each bar in the figure has a width V and height $\beta_{nk}H$, where β_{nk} is the number of members of $\mathscr{X}_n$ lying in I_k. If the feature space is d-dimensional, it is partitioned into a regular array of d-cubes of equal size, each having a hypervolume of V. Let I_k denote the kth d-cube, and let

$$u(\mathbf{x}, I_k) = \begin{cases} 1, & \text{for } \mathbf{x} \in I_k \\ 0, & \text{otherwise.} \end{cases} \tag{3.47}$$

The case where $\mathbf{x}$ is a scalar—i.e., the function $u(x, I_k)$—is illustrated in part (b) of Figure 3.6. Each observation $X(i)$ contributes a rectangular pulse $Hu(x, I_k)$ of height H and width V to one of the bars.

We define the *window function*

$$\Psi(x - X(i), V) = u(x, I_k)/V, \quad \text{for } X(i) \in I_k$$
$$= \begin{cases} \frac{1}{V}, & \text{if both } x \text{ and } X(i) \in I_k \text{ for some } k, \\ 0, & \text{otherwise.} \end{cases} \tag{3.48}$$

Let $p_n(x)$ denote the bar graph approximation of $p(x)$, where n denotes the number of observations.

By the above definitions and an appropriate choice of H,

$$p_n(x) = \sum_{\mathscr{X}_n \cap I_k} Hu(x, I_k)$$
$$= \sum_{i=1}^{n} HV\Psi(x - X(i), V).$$

In d-dimensional feature space, let

$$\Psi(\mathbf{x} - \mathbf{X}(i), V) = \begin{cases} \frac{1}{V}, & \text{if both } \mathbf{x} \text{ and } \mathbf{X}(i) \text{ belong to } I_k \\ 0, & \text{otherwise.} \end{cases} \tag{3.49}$$

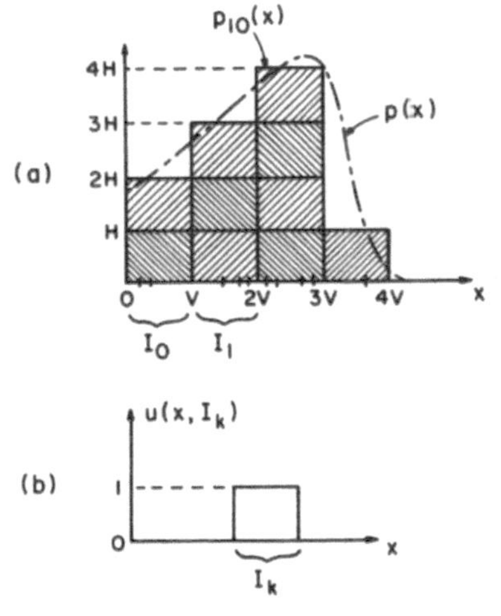

Figure 3.6. One-dimensional bar graph.

Let $p_n(\mathbf{x})$ denote the extension of $p_n(x)$ to d-dimensional feature space, so that $p_n(\mathbf{x})$ is an approximation of $p(\mathbf{x})$. Then

$$p_n(\mathbf{x}) = \sum_{i=1}^{n} HV\Psi(\mathbf{x} - \mathbf{X}(i), V). \tag{3.50}$$

The kth bar in $p_n(\mathbf{x})$ is of the form

$$b_{nk}(\mathbf{x}) = \sum_{\mathscr{X}_n \cap I_k} HV\Psi(\mathbf{x} - \mathbf{X}(i), V). \tag{3.51}$$

Thus

$$p_n(\mathbf{x}) = \sum_{k} b_{nk}(\mathbf{x}). \tag{3.52}$$

In order that $p_n(\mathbf{x})$ be sufficiently accurate, we must choose appropriate values of n, V, and H. A proper choice of n depends on $p(\mathbf{x})$ as well as V and H. Since $p(\mathbf{x})$ is initially unknown, it seems reasonable to adjust both V and H as a function of n until $p_n(\mathbf{x})$ undergoes little change even under large increases in n. Hence hereafter we view V, H, I_k, and $\Psi(\mathbf{x} - \mathbf{X}(i), V)$ as functions of n, denoted by V_n, H_n, I_{nk}, and $\Psi_n(\mathbf{x} - \mathbf{X}(i), V_n)$, respectively.

Note from Equation (3.50) that $p_n(\mathbf{x})$ is a random variable, since $\{\mathbf{X}(i)\}$ are random variables. Hence we must seek a strategy of choosing n, V_n, and H_n so that $p_n(\mathbf{x})$ converges in a stochastic sense to $p(\mathbf{x})$. We now derive such a strategy.

Since $p_n(\mathbf{x})$ is an estimate of a probability density, we require that

$$\int p_n(\mathbf{x})\, d\mathbf{x} = 1 \quad \text{for all } n. \tag{3.53}$$

By Equation (3.50), this becomes

$$\sum_{i=1}^{n} \int H_n V_n \Psi_n(\mathbf{x} - \mathbf{X}(i), V_n)\, d\mathbf{x} = 1, \tag{3.54}$$

which, by Equation (3.49) leads to

$$nV_n H_n = 1. \tag{3.55}$$

Now suppose that n is increased while V_n and H_n are constrained so that Equation (3.55) is always satisfied, and that on the average the number of examples in each I_{nk} increases. Under this condition the statistical variability of the height of each bar, $b_{nk}(\mathbf{x})$, decreases. As n increases one must reduce V_n in order to make negligible the error in $p_n(\mathbf{x})$ caused by the quantization of $\mathbf{x}$-space into $\{I_{nk}\}$. But since the average number of samples in each I_{nk} is proportional to nV_n, it follows that if V_n decreases more rapidly than n increases, then the average number of samples in each I_{nk} decreases. This decrease would lead to an increase of the statistical variability of $p_n(\mathbf{x})$.

Thus it seems that we must require both $V_n \to 0$ and $nV_n \to \infty$ as $n \to \infty$ in order to achieve a convergence of $p_n(\mathbf{x})$ to $p(\mathbf{x})$ as $n \to \infty$. We will prove this conjecture below. Specifically we will show that if $V_n \to 0$ and $nV_n \to \infty$ as $n \to \infty$, then, for any $\varepsilon > 0$, $[|p_n(\mathbf{x}) - p(\mathbf{x})| > \varepsilon] \to 0$ as $n \to \infty$, i.e., $p_n(\mathbf{x})$ converges to $p(\mathbf{x})$ in probability. We will present two proofs. The first proof is simpler. The second proof is a modification of the first proof which provides it with increased generality, in the sense that it is easily adapted to prove the convergence of Parzen window estimators (Section 3.6).

Theorem 3.1. *If $\{X(i)\}$ are statistically independent samples drawn from $p(\mathbf{x})$, $p(\mathbf{x}) \leqslant M$ for some positive number M,*

$$p_n(\mathbf{x}) = \frac{1}{n} \sum_{i=1}^{n} \Psi_n(\mathbf{x} - \mathbf{X}(i), V_n), \tag{3.56}$$

$V_n \to 0$ as $n \to \infty$, and $nV_n \to \infty$ as $n \to \infty$, then $p_n(\mathbf{x})$ converges to $p(\mathbf{x})$ in probability.

First Proof. We will prove this theorem by first showing that (a) $E(p_n(\mathbf{x})) \to p(\mathbf{x})$ as $n \to \infty$ and (b) $E([p_n(\mathbf{x}) - E(p_n(\mathbf{x}))]^2) \to 0$ as $n \to \infty$, i.e., $\text{Var}[p_n(\mathbf{x})] \to 0$ as $n \to \infty$. Then we will show that (a) and (b) imply the theorem. In statistical terminology, part (a) states that $p_n(\mathbf{x})$ is an *asymptotically unbiased* estimator of $p(\mathbf{x})$. Part (b) states that $p_n(\mathbf{x})$ is a *consistent* estimator of $p(\mathbf{x})$.

(a) Let P_{nk} denote the probability that $\mathbf{X}(i)$ falls in I_{nk}; i.e.,

$$P_{nk} = \int_{I_{nk}} p(\mathbf{x})\, d\mathbf{x}. \tag{3.57}$$

Let $Y_{nk}(\mathbf{X}(i))$ denote the binary-valued random variable defined by

$$Y_{nk}(\mathbf{X}(i)) = \begin{cases} 1, & \text{if } \mathbf{X}(i) \in I_{nk} \\ 0, & \text{if } \mathbf{X}(i) \in I_{nk} \end{cases} \tag{3.58}$$

Thus $Y_{nk}(X(i))$ is an indicator of the event $\mathbf{X}(i) \in I_{nk}$, and

$$E[Y_{nk}(\mathbf{X}(i))] = P_{nk}. \tag{3.59}$$

Since the $\{\mathbf{X}(i)\}$ are statistically independent, so are the $\{Y_{nk}(X(i))\}$.

Note that

$$b_{nk}(\mathbf{x}) = H_n u(\mathbf{x}, I_{nk}) \sum_{i=1}^{n} Y_{nk}(\mathbf{X}(i)).$$

Hence

$$\begin{aligned} E[b_{nk}(\mathbf{x})] &= H_n u(\mathbf{x}, I_{nk}) E\left[\sum_{i=1}^{n} Y_{nk}(\mathbf{X}(i))\right] \\ &= H_n u(\mathbf{x}, I_{nk}) n E[Y_{nk}(\mathbf{X}(i))] \\ &= H_n u(\mathbf{x}, I_{nk}) n P_{nk}, \end{aligned} \tag{3.60}$$

by Equation (3.59). Hence

$$E[b_{nk}(\mathbf{x})] = \frac{P_{nk}}{V_n} u(\mathbf{x}, I_{nk}) \tag{3.61}$$

by Equation (3.55). By Equation (3.57), the hypothesis $p(\mathbf{x}) \leqslant M$, and the mean value theorem of integral calculus, there exists $\mathbf{x}_{nk}^* \in I_{nk}$ such that

$$P_{nk} = V_n p(\mathbf{x}_{nk}^*) \tag{3.62}$$

Combining Equations (3.61) and (3.62), we get

$$E[b_{nk}(\mathbf{x})] = p(\mathbf{x}_{nk}^*) u(\mathbf{x}, I_{nk}) \quad \text{for } \mathbf{x} \in I_{nk}.$$

Since $V_n \to 0$ as $n \to \infty$, it follows that

$$E[b_{nk}(\mathbf{x})] \xrightarrow[n\to\infty]{} p(\mathbf{x}) \quad \text{for } \mathbf{x} \in I_{nk}. \tag{3.63}$$

Hence, by Equation (3.52),

$$\bar{p}_n(\mathbf{x}) \equiv E(p_n(\mathbf{x})) \xrightarrow[n\to\infty]{} p(\mathbf{x}) \quad \text{for all } \mathbf{x}. \tag{3.64}$$

(b) Also,

$$\begin{aligned} \operatorname{Var}[b_{nk}(\mathbf{x})] &= \operatorname{Var}\left[H_n u(\mathbf{x}, I_{nk}) \sum_{i=1}^{n} Y_{nk}(\mathbf{X}(i))\right] \\ &= H_n^2 u(\mathbf{x}, I_{nk}) \operatorname{Var}\left[\sum_{i=1}^{n} Y_{nk}(\mathbf{X}(i))\right]. \end{aligned} \tag{3.65}$$

Since $\{Y_{nk}(\mathbf{X}(i))\}$ for any k are statistically independent and identically distributed,

$$\operatorname{Var}\left[\sum_{i=1}^{n} Y_{nk}(\mathbf{X}(i))\right] = \sum_{i=1}^{n} \operatorname{Var}[Y_{nk}(\mathbf{X}(i))] = n \operatorname{Var}[Y_{nk}(\mathbf{X}(i))]. \tag{3.66}$$

Using Equation (3.59) and the identity $Y_N^2(\mathbf{X}(i)) \equiv Y_N(\mathbf{X}(i))$,

$$\begin{aligned} \operatorname{Var}[Y_{nk}(\mathbf{X}(i))] &= E[(Y_{nk}(\mathbf{X}(i)) - P_{nk})^2] \\ &= E[Y_{nk}(\mathbf{X}(i))] - P_{nk}^2 \\ &= P_{nk}(1 - P_{nk}). \end{aligned} \tag{3.67}$$

Combining Equations (3.65), (3.66), and (3.67), we get

$$\operatorname{Var}[b_{nk}(\mathbf{x})] = H_n^2 u(\mathbf{x}, I_{nk}) n P_{nk}(1 - P_{nk}). \tag{3.68}$$

Using Equations (3.55) and (3.62), we obtain

$$\operatorname{Var}[b_{nk}(\mathbf{x})] = \frac{p(\mathbf{x}_{nk}^*)}{nV_n}[1 - p(\mathbf{x}_{nk}^*)V_n] \tag{3.69}$$

for some $\mathbf{x}_{nk}^* \in I_{nk}$ and all $\mathbf{x} \in I_{nk}$.

Since $V_n \to 0$ and $nV_n \to \infty$ as $n \to \infty$, Equation (3.69) becomes

$$\operatorname{Var}[b_{nk}(\mathbf{x})] \to 0 \quad \text{as } n \to \infty \text{ for } \mathbf{x} \in I_{nk}. \tag{3.70}$$

Hence

$$\operatorname{Var} p_n(\mathbf{x}) \to 0 \quad \text{as } n \to \infty. \tag{3.71}$$

Recapitulating: we have shown (a) that $\bar{p}_n(\mathbf{x}) \to p(\mathbf{x})$ and (b) that

$$E\{[p_n(\mathbf{x}) - \bar{p}_n(\mathbf{x})]^2\} \to 0 \text{ as } n \to \infty.$$

(c) To show that $p_n(\mathbf{x})$ approaches $p(\mathbf{x})$ in probability, note that

$$\begin{aligned} E\{(p_n(\mathbf{x}) - p(\mathbf{x}))^2\} &\equiv E\{[(p_n(\mathbf{x}) - \bar{p}_n(\mathbf{x})) - (p(\mathbf{x}) - \bar{p}_n(\mathbf{x}))]^2\} \\ &= E[(p_n(\mathbf{x}) - \bar{p}_n(\mathbf{x}))^2] + E[(p(\mathbf{x}) - \bar{p}_n(\mathbf{x}))^2] \\ &= E[(p_n(\mathbf{x}) - \bar{p}_n(\mathbf{x}))^2] + (p(\mathbf{x}) - \bar{p}_n(\mathbf{x}))^2 \xrightarrow[n\to\infty]{} 0, \end{aligned} \tag{3.72}$$

by Equations (3.64) and (3.71).

Chebychev's Inequality [8] states that, for any $\varepsilon > 0$ and any random variable Z,

$$P(|Z| \geqslant \varepsilon) \leqslant \frac{E(Z^2)}{\varepsilon^2}.$$

Let $Z = p_n(\mathbf{x}) - p(\mathbf{x})$. Then, for any $\varepsilon > 0$,

$$P[|p_n(\mathbf{x}) - p(\mathbf{x})| \geqslant \varepsilon] \leqslant \frac{E[(p_n(\mathbf{x}) - p(\mathbf{x}))^2]}{\varepsilon^2} \xrightarrow[n\to\infty]{} 0,$$

by Chebychev's Inequality and Equation (3.72). □

In the second proof the role of the indicators $\{Y_{nk}(\mathbf{X}(i))\}$ is transferred to the functions $\Psi_n(\mathbf{x} - \mathbf{X}(i), V_n)$ via the identity

$$\Psi_n(\mathbf{x} - \mathbf{X}(i), V_n) = \frac{1}{V_n} \sum_k Y_{nk}(\mathbf{X}(i)) u(\mathbf{x}, I_{nk}).$$

The resulting proof has the advantage that it is easily extended to the problem of proving the convergence of Parzen window estimators.

Second Proof. In this proof, too, we show that (a) $E(p_n(\mathbf{x})) \to p(\mathbf{x})$ and (b) $\operatorname{Var}[p_n(\mathbf{x})] \to 0$ as $n \to \infty$. The remainder of the proof is identical to part (c) of the first proof.

(a) Let

$$\bar{p}_n(\mathbf{x}) = E(p_n(\mathbf{x})). \tag{3.73}$$

By Equation (3.56)

$$\begin{aligned} \bar{p}_n(\mathbf{x}) &= \sum_{i=1}^{n} E\left[\frac{1}{n} \Psi_n(\mathbf{x} - \mathbf{X}(i), V_n)\right] \\ &= E[\Psi_n(\mathbf{x} - \mathbf{X}(i), V_n)] \end{aligned} \tag{3.74}$$

since $E[(1/n)\Psi_n(\mathbf{x} - \mathbf{X}(i), V_n)]$ is independent of i. In Equation (3.49), we note that $\Psi(\mathbf{x} - \mathbf{X}(i), V)$ is symmetric with respect to $\mathbf{x}$ and $\mathbf{X}(i)$, i.e., $\Psi(\mathbf{x} - \mathbf{X}(i), V) \equiv \Psi(\mathbf{X}(i) - \mathbf{x}, V)$. Hence

$$\int \Psi_n(\mathbf{x} - \mathbf{x}(i), V_n)\, d\mathbf{x}(i) = nV_n H_n = 1. \tag{3.75}$$

Let $\delta(\mathbf{x})$ denote the Dirac delta function—a function that satisfies

$$\delta(\mathbf{x}) = 0 \text{ for } \mathbf{x} \neq 0 \quad \text{and} \quad \int \delta(\mathbf{x})\, d\mathbf{x} = 1.$$

Note that $V_n \to 0$ as $n \to \infty$ and that $\Psi_n(\mathbf{x} - \mathbf{X}(i), V_n) > 0$ only when both $\mathbf{x}$ and $\mathbf{X}(i)$ belong to the same d-cube I_k. It follows from these observations and Equation (3.75) that

$$\Psi_n(\mathbf{x} - \mathbf{X}(i), V_n) \to \delta(\mathbf{x} - \mathbf{X}(i)) \quad \text{as } n \to \infty. \tag{3.76}$$

Hence

$$\begin{aligned} E[\Psi_n(\mathbf{x} - \mathbf{X}(i), V_n)] &= \int \Psi_n(\mathbf{x} - \mathbf{x}(i), V_n)\, p(\mathbf{x}(i))\, d\mathbf{x}(i) \\ &\xrightarrow[n\to\infty]{} \int \delta(\mathbf{x} - \mathbf{x}(i))\, p(\mathbf{x}(i))\, d\mathbf{x}(i) = p(\mathbf{x}). \end{aligned} \tag{3.77}$$

(b) Note that

$$\begin{aligned} \operatorname{Var}[p_n(\mathbf{x})] &= E\{[p_n(\mathbf{x}) - \bar{p}_n(\mathbf{x})]^2\} \\ &= E\left[\left\{\sum_{i=1}^{n} \left[\frac{1}{n} \Psi_n(\mathbf{x} - \mathbf{X}(i), V_n) - \frac{1}{n} \bar{p}_n(\mathbf{x})\right]\right\}^2\right]. \end{aligned} \tag{3.78}$$

By Equation (3.74), the quantity $\Psi_n(\mathbf{x} - \mathbf{X}(i), V_n) - \bar{p}_n(\mathbf{x})$ is a random variable having a mean of zero. Furthermore it is statistically independent with respect to i, since the

$\mathbf{X}(i)$ are statistically independent. Hence the expectation of the square of the sums of this random variable is the sum of the expectation of the squares of the variable. Thus Equation (3.78) becomes

$$\begin{aligned}
\operatorname{Var}[p_n(\mathbf{x})] &= \sum_{i=1}^{n} E\left\{\left[\frac{1}{n}\Psi_n(\mathbf{x}-\mathbf{X}(i), V_n) - \frac{1}{n}\bar{p}_n(\mathbf{x})\right]^2\right\} \\
&= \frac{1}{n^2}\sum_{i=1}^{n} E[\Psi_n^2(\mathbf{x}-\mathbf{X}(i), V_n) - \bar{p}_n^2(\mathbf{x})] \\
&\leqslant \frac{1}{n} E[\Psi_n^2(\mathbf{x}-\mathbf{X}(i), V_n)] = \frac{1}{n}\int \Psi_n^2(\mathbf{x}-\mathbf{x}(i), V_n)p(\mathbf{x}(i))\,d\mathbf{x}(i) \\
&\leqslant \frac{M}{n}\int \Psi_n^2(\mathbf{x}-\mathbf{x}(i), V_n)\,d\mathbf{x}(i), \quad \text{by the hypothesis in Theorem 3.1} \\
&= M/(nV_n)
\end{aligned}$$

by Equation (3.55). Since

$$nV_n \to \infty \quad \text{as } n \to \infty,$$

it follows that

$$\operatorname{Var}[p_n(\mathbf{x})] \to 0 \quad \text{as } n \to \infty. \tag{3.79}$$

Using part (c) of the first proof, this second proof is completed. □

3.6 Parzen Windows and Potential Functions

3.6.1 Generalization of Bar Graph Technique

The bar graph technique may be generalized in a natural way as follows. Replace the rectangular pulses by functions $\{\Psi_n(\mathbf{x}-\mathbf{X}(i), V_n)\}$ satisfying the following constraints:

$$\Psi_n(\mathbf{x}-\mathbf{z}, V_n) \geqslant 0, \quad \text{for all } \mathbf{x} \text{ and } \mathbf{z} \tag{3.80}$$

$$\int \Psi_n(\mathbf{x}-\mathbf{z}, V_n)\,d\mathbf{x} = 1 \tag{3.81}$$

$$\frac{1}{n}\int \Psi_n^2(\mathbf{x}-\mathbf{z}, V_n)\,d\mathbf{x} \xrightarrow[n\to\infty]{} 0 \tag{3.82}$$

$$\Psi_n(\mathbf{x}-\mathbf{z}, V_n) \xrightarrow[n\to\infty]{} \delta(\mathbf{x}-\mathbf{z}). \tag{3.83}$$

Under this scheme the pulses need not be rectangular; they may even have infinite skirts. Furthermore, coincidence of pulses in the sum

$$\sum_{i=1}^{n} \Psi_n(\mathbf{x}-\mathbf{z}, V_n)$$

need not occur unless $\mathbf{X}(i) = \mathbf{X}(j)$ for $i \neq j$. These generalized pulses are referred to as *Parzen windows*.

Often each pulse is symmetric with respect to $\mathbf{X}(i)$—i.e.,

$$\Psi_n(\mathbf{x} - \mathbf{X}(i), V_n) = \Psi_n(|\mathbf{x} - \mathbf{X}(i)|, V_n)$$

where $|\mathbf{a}| = \{|a_i|\}$ = vector of absolute values. Under this condition the estimate of $p(\mathbf{x})$ is formed by a sum of symmetric functions, each centered at an example $\mathbf{X}(i)$. This is illustrated for the one-dimensional case in Figure 3.7. In this figure $p_n(x)$ is formed from a set of symmetric Parzen windows of the form

$$\Psi_n(|x - X(i)|, V_n) = \begin{cases} \dfrac{2}{V_n}\cos^2\left[\dfrac{\pi}{V_n}(x - X(i))\right], & \text{for } |x - X(i)| \leqslant V_n/2, \\ 0, & \text{for } |x - X(i)| > V_n/2, \end{cases}$$

drawn as dashed curves.

Other frequently used Parzen windows are rectangular pulses centered at $\mathbf{X}(i)$ (recall that in the bar graph the pulses were typically not centered at $\mathbf{X}(i)$) and Gaussian pulses centered at $\mathbf{X}(i)$. Parzen windows tend to produce smoother estimates of $p(\mathbf{x})$ than that achieved by a bar graph for the same value of n.

Parzen windows are similar in many respects to *potential functions*. The term "potential function" arises from visualizing each example $\mathbf{X}(i)$ as an electrically charged particle in feature space, and imagining $p_n(\mathbf{x})$ as the electric potential at point $\mathbf{x}$. From this point of view each $\mathbf{X}(i)$ contributes linearly to $p_n(\mathbf{x})$ an amount of potential which is a decreasing function of $|\mathbf{x} - \mathbf{X}(i)|$. Thus

$$p_n(\mathbf{x}) = \frac{1}{n}\sum_{i=1}^{n} \Psi_n(|\mathbf{x} - \mathbf{X}(i)|, V_n).$$

Usually the potential function Ψ_n is so chosen that Ψ_n is positive for $|\mathbf{x} - \mathbf{X}(i)| = \mathbf{0}$ and Ψ_n is a nonincreasing function of $|\mathbf{x} - \mathbf{X}(i)|$.

The smoothing process obtained by Parzen windows and potential functions is similar to the blurring process performed by the human eye when it transforms a half-tone reproduction of a photograph consisting of great numbers of black and white dots into a subjective impression of a continuous-tone photograph.

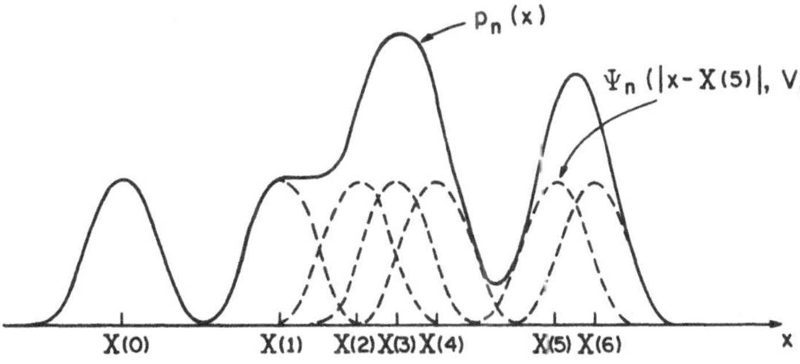

Figure 3.7. Formation of $p_n(x)$ by Parzen windows.

The theoretical support for bar graphs extends easily to Parzen windows and potential functions. In particular, Theorem 3.1 can be extended to Parzen windows and potential functions by introducing Equations (3.80)–(3.83) as constraints. The proof is a straightforward extension of the second proof of Theorem 3.1 (Exercise 3.7). Thus we have the following theorem.

Theorem 3.2. *If*

$$p_n(\mathbf{x}) = \frac{1}{n} \sum_{i=1}^{n} \Psi_n(\mathbf{x} - \mathbf{X}(i), V_n),$$

and $\Psi_n(\mathbf{x} - \mathbf{z}, V_n)$ *satisfies Equations* (3.80)–(3.83), *then* $p_n(\mathbf{x})$ *converges to* $p(\mathbf{x})$ *in probability.*

3.6.2 Designing a Parzen Window Training Procedure

The design of a Parzen window training procedure depends heavily on (a) the initial window, $\Psi_1(\mathbf{x} - \mathbf{z}, V_1)$, and (b) the rate of decrease of V_n with respect to n. We suggest the following guidelines.

1. Choose Ψ_1 so that it is a nonincreasing function of $|\mathbf{x} - \mathbf{z}|$, and so that if $\mathbf{z} = \mathbf{X}(1)$ is at the centroid (i.e., barycenter) of $p(\mathbf{x})$, then Ψ_1 will be an overall approximation of $p(\mathbf{x})$. Even if practically nothing is known about $p(\mathbf{x})$, one almost always knows a region $\mathscr{R}$ of feature space where occurrences of $\mathbf{X}(i)$ are extremely unlikely. The complement of this region, $\bar{\mathscr{R}}$, is usually bounded. Then one chooses Ψ_1 so that the significant portions of Ψ_1 will cover almost all of $\bar{\mathscr{R}}$ when $\mathbf{z}$ is chosen to approximate the centroid of $\bar{\mathscr{R}}$.

This is illustrated in Figure 3.8 for a one-dimensional feature space. The solid curve represents a best guess of the shape and size of $p(x)$. The dashed line is the designer's choice of Ψ_1.

2. Choose $V_n = (1/\sqrt{n})V_1$.

This choice of V_n is a midway compromise between the constraints

$$V_n \xrightarrow[n\to\infty]{} 0$$

and

$$nV_n \xrightarrow[n\to\infty]{} \infty$$

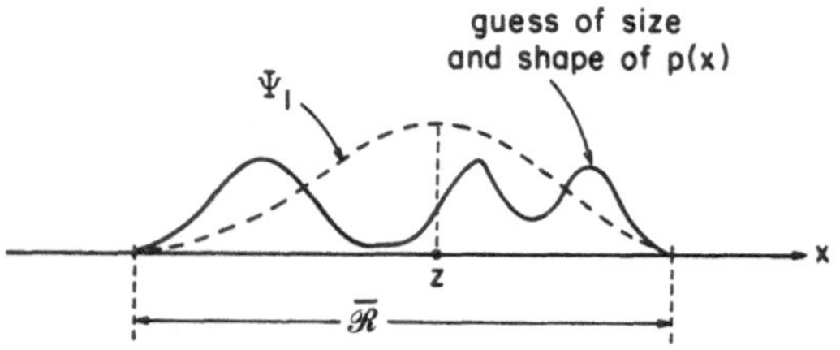

Figure 3.8. Matching Ψ_1 to a guess of $p(x)$.

by the following reasoning. Suppose V_n is of the form $V_n = V_1 n^{-\gamma}$. Then

$$V_n \xrightarrow[n\to\infty]{} 0$$

implies that $\gamma > 0$, and

$$nV \xrightarrow[n\to\infty]{} \infty$$

implies that $\gamma < 1$. Hence a midway compromise is $\gamma = \frac{1}{2}$.

Once Ψ_n has been chosen, then $p_n(\mathbf{x})$ is observed as a function of n. Training is stopped when $p_n(\mathbf{x})$ is smooth and undergoes little change with large changes in n.

3.7 Storage Economies

In the practical implementation of estimates of $p(\mathbf{x})$, requirements for memory storage are often just as important as the accuracy of the estimates. Usually any reduction in storage requirements must be paid for by a corresponding reduction in the accuracy of the estimator. In this section we describe various ways in which significant reductions in storage requirements may be obtained at relatively low cost in accuracy reduction.

3.7.1 Bar Graphs Versus Parzen Windows

A simple approach to reducing the storage requirements of a Parzen window or potential function estimator is to replace the window functions by bars. We see this by examining the expressions for the Parzen window and the bar graph estimators. The Parzen window estimator is

$$p_n^P(\mathbf{x}) = \frac{1}{n}\sum_{i=1}^{n} \Psi_n(|\mathbf{x} - \mathbf{X}(i)|, V_n), \tag{3.84}$$

and the bar graph estimator is

$$p_n^B(\mathbf{x}) = \sum_k b_{nk}(\mathbf{x}), \tag{3.85}$$

where

$$b_{nk}(\mathbf{x}) = \frac{1}{n}\sum_{\mathscr{X}_n \cap I_k} \Psi_n(\mathbf{x} - \mathbf{X}(i), V_n)$$

and

$$\Psi_n(\mathbf{x} - \mathbf{X}(i), V_n) = \begin{cases} 1/V_n, & \text{for } \mathbf{x}, \mathbf{X}(i) \in I_k, \\ 0, & \text{otherwise.} \end{cases}$$

In Equations (3.84) and (3.85), both $p_n^P(\mathbf{x})$ and $p_n^B(\mathbf{x})$ are expressed as sums of components or building blocks. The building blocks of $p_n^P(\mathbf{x})$ are the window functions or potential functions $\{\Psi_n\}$, while the building blocks of $p_n^B(\mathbf{x})$ are the bars $\{b_{nk}(\mathbf{x})\}$. In general, for any given $p(\mathbf{x})$ and $\mathscr{X}_n$ and over any given large finite region of $\mathbf{x}$-space, the number of bars $\{b_{nk}(\mathbf{x})\}$ will be much fewer than the number of window functions $\{\Psi_n\}$, since $\{b_{nk}(\mathbf{x})\}$ is a denumerable set while $\{\Psi_n\}$ is a nondenumerable set—i.e., all of the bars $\{b_{nk}(\mathbf{x})\}$ can be placed in a sequence and counted, whereas the window functions $\{\Psi_n\}$ cannot be placed in such a sequence. Thus over any large but finite region of **x**-space, $p_n^P(\mathbf{x})$ is constructed from a much larger number of building blocks than $p_n^B(\mathbf{x})$, and hence, for any given $p(\mathbf{x})$ and $\mathscr{X}_n$, $p_n^P(\mathbf{x})$ will usually be smoother and somewhat more accurate than $p_n^B(\mathbf{x})$ as an estimator of $p(\mathbf{x})$.

On the other hand, the sparseness of $\{b_{nk}(\mathbf{x})\}$ in comparison to $\{\Psi_n\}$ gives $p_n^B(\mathbf{x})$ an important advantage over $p_n^P(\mathbf{x})$. The storage space required for $p_n^B(\mathbf{x})$ in a computer memory is approximately proportional to the number of nonzero-valued bars, while the storage space required for $p_n^P(\mathbf{x})$ is approximately proportional to the size of the data set when the smoothing effects of the roundoff in **x** may be ignored. Thus the storage space needed for $p^B(\mathbf{x})$ in a computer memory is generally much less than the storage space needed for $p_n^P(\mathbf{x})$, for a given $p(\mathbf{x})$ and $\mathscr{X}_n$.

Yet if $p(\mathbf{x})$ is relatively smooth the reduction in accuracy incurred by replacing the rectangular pulses in the bar graph estimate by the Parzen windows is much smaller than the reduction in accuracy incurred by reducing the value of n. This is illustrated in Exercise 3.12. Thus the bar graph offers an attractive way of trading some of the accuracy of a Parzen window estimator for a reduction in storage requirements.

Other approaches to achieving storage economies are (a) to replace the rectangular bars by overlapping curvilinear interpolation functions such as Gaussian functions, (b) to use a training procedure to minimize the number of component densities, and (c) to eliminate component densities that contribute little to the decision surface.

3.7.2 Interpolation Functions

Bar graphs are examples of a technique of approximating $p(\mathbf{x})$ by a set of interpolation functions. More specifically, recall that the bar graph estimate is

$$p_n^B(\mathbf{x}) = \sum_k b_{nk}(\mathbf{x}) \cong p(\mathbf{x}). \tag{3.86}$$

Let B_{nk} denote the height of $b_{nk}(\mathbf{x})$. Then, recalling Equations (3.48) and (3.51),

$$b_{nk}(\mathbf{x}) = B_{nk}u(\mathbf{x}, I_{nk}) \tag{3.87}$$

and

$$p_n^B(\mathbf{x}) = \sum_k B_{nk}u(\mathbf{x}, I_{nk}). \tag{3.88}$$

Now since $p(\mathbf{x})$ is a smooth function of **x**, $p(\mathbf{x})$ may also be approximated as a sum of bars, the kth bar centered at $\boldsymbol{\mu}_{nk}$, where $\boldsymbol{\mu}_{nk}$ is the centroid of I_{nk}, the bar's base. The height of each bar equals the value of $p(\mathbf{x})$ at the center of the bar's base. Let $p_n^*(\mathbf{x})$ denote this sum of bars. Then

$$p(\mathbf{x}) \cong p_n^*(\mathbf{x}) = \sum_k p(\boldsymbol{\mu}_{nk})u(\mathbf{x}, I_{nk}). \tag{3.89}$$

Note that

$$p_n^*(\boldsymbol{\mu}_{nk}) \equiv p(\boldsymbol{\mu}_{nk}). \tag{3.90}$$

The $\{p(\boldsymbol{\mu}_{nk})\}$ are *sampled values* or *samples* of $p(\mathbf{x})$, and each of $\{u(\mathbf{x}, I_{nk})\}$ is a form of *interpolation function.* Here *sample* is used in a completely different sense from that in probability theory. This form of sampling is a quantization of $\mathbf{x}$-space, and results in the *sampled-data error*

$$H_S(\mathbf{x}) = |p(\mathbf{x}) - p_n^*(\mathbf{x})|. \tag{3.91}$$

Note, by Equation (3.90), that $H_S(\boldsymbol{\mu}_{nk}) \equiv 0$ for all n and k.

The $\{u(\mathbf{x}, I_{nk})\}$ are interpolation functions in the sense that the approximation in Equation (3.89) becomes an equality at the sample points, where $\mathbf{x} = \boldsymbol{\mu}_{nk}$, and the $\{u(\mathbf{x}, I_{nk})\}$ serve to estimate all the remaining points on the basis of the assumed smoothness of $p(\mathbf{x})$.

Other forms of interpolation functions are possible, so that Equation (3.89) may be generalized to

$$p_n^*(\mathbf{x}) = \sum_k p(\boldsymbol{\mu}_{nk}) t_n(\mathbf{x} - \boldsymbol{\mu}_{nk}),$$

where $t_n(\mathbf{y})$ is a generalized interpolation function satisfying

$$t_n(\mathbf{0}) = 1$$
$$t_n(\mathbf{y}) > 0, \quad \text{for all } \mathbf{y} \neq \mathbf{0}$$
$$\int t_n(\mathbf{y})\, d\mathbf{y} = V_n.$$

It follows from Theorem 3.1 that

$$\bar{p}_n^B(\mathbf{x}) \equiv E(p_n^B(\mathbf{x})) \xrightarrow[n\to\infty]{} p(\mathbf{x}) \quad \text{for all } \mathbf{x} \tag{3.92}$$

(see Equation (3.64)) and

$$\operatorname{Var}(B_{nk}) \xrightarrow[n\to\infty]{} 0. \tag{3.93}$$

Hence

$$B_{nk} \cong p(\boldsymbol{\mu}_{nk}) \tag{3.94}$$

in the sense described in Equations (3.92) and (3.93).

The *truncation error* caused by a finite n is

$$H_T(\mathbf{x}) = |p(\boldsymbol{\mu}_{nk}) - B_{nk}|\, u(\mathbf{x}, I_{nk}).$$

By Equations (3.89) and (3.94),

$$p(\mathbf{x}) \cong p_n^*(x) = \sum_k p(\boldsymbol{\mu}_{nk}) u(\mathbf{x}, I_{nk}) \cong \sum_k B_{nk} u(\mathbf{x}, I_{nk}) = p_n^B(\mathbf{x}).$$

Thus the error $|p_n^B(\mathbf{x}) - p(\mathbf{x})|$ may be viewed as consisting of two principal components: (a) the sampled-data error, $H_S(\mathbf{x})$, caused by the quantization of $\mathbf{x}$-space into $\{I_{nk}\}$, and (b) the truncation error, $H_T(\mathbf{x})$, as exhibited in the statistical variation of B_{nk}, which is caused by the finite value of n.

To reduce $H_T(\mathbf{x})$ we adjust n until $p_n(\mathbf{x})$ is sufficiently close to, say, $p_{2n}(\mathbf{x})$. To reduce $H_S(\mathbf{x})$ we must use effective interpolation functions. If the interpolation functions are chosen effectively, we may achieve a satisfactory approximation of $p(\mathbf{x})$ by a relatively small n and hence by a relatively large V_n and low storage requirement (i.e., few values of $\{\boldsymbol{\mu}_{nk}\}$ need to be stored).

3.8 Fixed-Base Bar Graphs

Sometimes it is possible to estimate a probability density as a sum of bars over a regularly arranged array of d-dimensional rectangular parallelepipeds. We refer to each such parallelepiped as a *d-box.*

We assume that the d-boxes are arranged in a nonoverlapping Cartesian array that fills the entire feature space—i.e., we assume that each of the edges of the d-boxes is parallel to one of the axes of feature space, and that every d-box shares a distinct $(d-1)$-dimensional face with each of $2d$ neighboring d-boxes. We assume that the centroid of one of the d-boxes is at the origin of feature space, and that the principal diagonal of each d-box is a vector $\mathbf{c} > \mathbf{0}$. (By $\mathbf{c} > \mathbf{0}$ we mean $c_i > 0$ for every i.) We identify each d-box by $I(\boldsymbol{\mu})$, where $\boldsymbol{\mu}$ is the centroid of that d-box. Thus the endpoints of the principal diagonal of $I(\boldsymbol{\mu})$ are $\boldsymbol{\mu} - \frac{1}{2}\mathbf{c}$ and $\boldsymbol{\mu} + \frac{1}{2}\mathbf{c}$. The set of points in $I(\boldsymbol{\mu})$ is defined by

$$\mathbf{x} \in I(\boldsymbol{\mu}) \quad \text{if and only if} \quad \boldsymbol{\mu} - \tfrac{1}{2}\mathbf{c} \leqslant \mathbf{x} \leqslant \boldsymbol{\mu} + \tfrac{1}{2}\mathbf{c}.$$

It is now convenient to define a restriction of $I(\boldsymbol{\mu})$, denoted by $\hat{I}(\boldsymbol{\mu})$, such that

$$\mathbf{x} \in \hat{I}(\boldsymbol{\mu}) \quad \text{if and only if} \quad \boldsymbol{\mu} - \tfrac{1}{2}\mathbf{c} \leqslant \mathbf{x} < \boldsymbol{\mu} + \tfrac{1}{2}\mathbf{c}.$$

Thus $\hat{I}(\boldsymbol{\mu})$ and $I(\boldsymbol{\mu})$ are identical except for half of the faces of $I(\boldsymbol{\mu})$, which are not included in $\hat{I}(\boldsymbol{\mu})$. Thus we refer to $\hat{I}(\boldsymbol{\mu})$ as a *semi-open d-box.*

Under these conditions every $\mathbf{X}(n)$ is contained in precisely one semi-open d-box. One may easily identify this semi-open d-box by finding the largest integer portion of

$$\frac{1}{c_i}\left(X_i(n) - \frac{c_i}{2}\right)$$

for $i = 1, \ldots, d$; namely,

$$m_i = \left[\frac{1}{c_i}\left(X_i(n) - \frac{c_i}{2}\right)\right]$$

where $[y]$ denotes "greatest integer not exceeding y." Thus

$$m_i c_i + \frac{c_i}{2} = \mu_i - \frac{c_i}{2},$$

from which we deduce

$$\mu_i = (m_i + 1)c_i$$

and

$$m_i = \frac{\mu_i}{c_i} - 1.$$

Note that it is not necessary to find the distance of $\mathbf{x}$ from all possible $\boldsymbol{\mu}$'s in order to find the $\hat{I}(\boldsymbol{\mu})$ that contains $\mathbf{x}$.

Let $\mathbf{m}$ denote the vector $\{m_i\}$. Note that for every $\boldsymbol{\mu}$ of a d-box there is a unique $\mathbf{m}$, and vice versa. The set of the $\mathbf{m}$'s (and hence that of the $\boldsymbol{\mu}$'s) of the array of d-boxes is denumerable; that is, they can be counted by first counting the origin of $\mathbf{m}$-space, then counting all the $\mathbf{m}$'s at a unit Hamming distance* from the origin of $\mathbf{m}$-space, then all $\mathbf{m}$'s at a Hamming distance of 2 from the origin, etc.

Let $\{\boldsymbol{\mu}_k | k = 1, 2, \ldots\}$ denote an enumeration of the $\boldsymbol{\mu}$'s. The bar graph estimate of each class-conditional density can then be expressed in the form

$$p(\mathbf{x}) \cong p^B(\mathbf{x}) = \sum_k b_k(\mathbf{x}),$$

where

$$b_k(\mathbf{x}) = \begin{cases} B_k, & \text{for } \mathbf{x} \in \hat{I}(\boldsymbol{\mu}_k), \\ 0, & \text{otherwise.} \end{cases}$$

Under this condition both the training procedure and the working phase are greatly simplified with respect to the bar-graph and window estimation techniques described in preceding sections. The simplification comes about for three reasons:

(a) Whenever a member of the sample set or an unknown feature vector $\mathbf{X}$ enters the training procedure or the working-phase decision procedure, the problem of finding the nearest prototype is solved by merely finding the integer-valued point $\mathbf{m}$ in feature space that is nearest to $\mathbf{X}$. To do this there is no need to measure the distance of $\mathbf{X}$ to all of the prototypes.

(b) There is no need to store the feature-space coordinates of each prototype—the prototypes may merely be stored in a list following a natural sequence: for example, left to right along each row and top to bottom for successive rows.

(c) Let c denote the number of classes. The use of bars as interpolation functions means that each classification decision during the working phase depends only on the B_k's at each $I(\boldsymbol{\mu})$—i.e., the relative values of the c prototypes at one particular d-box. Thus, with respect to other forms of interpolation functions, the bar graph yields one of the simplest forms of classification decision processes.

In the fixed-base bar graph technique the main problem is to find an appropriate value of $\mathbf{c}$. We suggest an approach based on the theory of

* In an integer-valued space, such as $\mathbf{m}$-space, the *Hamming distance* between two points $\mathbf{m}$ and $\mathbf{n}$ is defined as

$$d_H(\mathbf{m}, \mathbf{n}) = \sum_{i=1}^{d} |m_i - n_i|.$$

sampled-data signals [9]. We assume that the observed feature vectors, denoted by $\{\mathbf{X}'(n)\}$, have approximately bandwidth-limited class-conditional densities—i.e., that the Fourier transform of each density is negligibly small outside a d-box of edge dimensions $(u_1, \ldots, u_d)$. By a scaling transformation of the form

$$\mathbf{x} = \left\{\frac{x_i'}{u_i}\right\}$$

each d-box in $\mathbf{x}$-space is transformed to a d-cube of unit volume—i.e., $V = 1$—since all of the edges of the transformed d-box have unit length. Then one may place all of the $\boldsymbol{\mu}_k$'s at the integer-valued centers of the d-boxes.

We assume that each class-conditional density $p(\mathbf{x})$ may be fully represented by the Fourier transform

$$P(\mathbf{s}) = \mathscr{F}[p(\mathbf{x})] = \int d\mathbf{x}\, p(\mathbf{x}) \exp[-i2\pi \mathbf{s}^T \mathbf{x}].$$

$P(\mathbf{s})$ is the so-called characteristic function of $p(\mathbf{x})$. We assume that $P(\mathbf{s})$ is bandwidth limited in the sense that $|P(\mathbf{s})|$ is negligibly small when $\mathbf{s}$ lies outside a d-dimensional rectangular parallelepiped whose edges have lengths $(S_1, \ldots, S_d)$. We express this condition as follows: for some $\varepsilon > 0$,

$$|P(\mathbf{s})| < \varepsilon \quad \text{whenever} \quad \max_i \left(\frac{s_i}{S_i}\right) \geqslant 1.$$

Under these conditions, an appropriate value of $\mathbf{c}$ is

$$\mathbf{c} = \left[\frac{2}{S_1}, \ldots, \frac{2}{S_d}\right]^T.$$

This yields a spacing of prototypes at approximately π times the Nyquist sampling rate.

The above technique is useful when the computing time for estimating the class density must be made small, and there is available a sufficiently large data base to yield a statistically significant population—say 25—in every d-box near or on the Bayes-optimum decision boundary. When a data base sufficiently large for this purpose is not available, we must resort to computationally expensive design algorithms that make efficient use of the data. The cluster formers, prototype formers, and sample set formers described below, are examples of such algorithms.

3.9 Sample Sets and Prototypes

When the amount of data is relatively small and limited, one may increase the accuracy of the estimates of the class densities by generalizing the rectangular d-box to a *sample set* [10], *prototype region* [11], or *cluster* [12].

The prototype regions are usually not rectangular, they are usually not uniformly distributed, and they often vary in size. Each cluster is usually represented by a prototype—a feature vector approximately equal to the position vector of the centroid of the cluster.

These prototypes form the basis for estimating the class densities and the decision surfaces. Following Sebestyen's concept of *sample set* [10], one may express each class density as a sum of estimates of Gaussian densities:

$$g(\mathbf{x}|\omega_i) = K \sum_{i=1}^{N_i} N_i \exp -[\tfrac{1}{2}(\mathbf{x} - \boldsymbol{\mu}_i)^T \boldsymbol{\Sigma}_i^{-1}(\mathbf{x} - \boldsymbol{\mu}_i)]$$

where N_i is the cardinality of the ith cluster of the training set, $\boldsymbol{\mu}_i$ is the sample mean of the ith cluster, and $\boldsymbol{\Sigma}_i$ is the sample covariance matrix of the ith cluster. The sample mean and the sample covariance matrix may be computed as follows [1]:

$$\boldsymbol{\mu}_i = \frac{1}{N_i} \sum_{j=1}^{N_i} \mathbf{x}_i(j)$$

$$\boldsymbol{\Sigma}_i = \frac{1}{N_i - 1} \sum_{j=1}^{N_i} (\mathbf{x}_i(j) - \boldsymbol{\mu}_i)(\mathbf{x}_i(j) - \boldsymbol{\mu}_i)^T,$$

where $\{\mathbf{x}_i(j)\}$ is the set of feature vectors in the ith cluster.

A simpler approach is to ignore the problem of estimating the class densities, and concentrate on the problem of estimating the decision surfaces. To this end we may express the discriminant function for class ω_i as

$$g_i(\mathbf{x}) = \sum_j g_{ij}(\mathbf{x}),$$

where $g_{ij}(\mathbf{x})$ is a *cluster discriminant function* associated with the jth cluster of class ω_i. Typical forms of these cluster discriminant functions are:

(a) the negative of the square of the Euclidean distance of $\mathbf{x}$ from the prototype:

$$\begin{aligned} g_{ij}(\mathbf{x}) &= -\|\mathbf{x} - \boldsymbol{\mu}_{ij}\|^2 \\ &= -\|\mathbf{x}\|^2 + 2\boldsymbol{\mu}_{ij}^T\mathbf{x} - \|\boldsymbol{\mu}_{ij}\|^2, \end{aligned} \tag{3.95}$$

where $\boldsymbol{\mu}_{ij}$ is the prototype of the jth cluster of class ω_i; and

(b) the negative of the Mahalanobis distance of $\mathbf{x}$ from the prototype:

$$g_{ij}(\mathbf{x}) = -(\mathbf{x} - \boldsymbol{\mu}_{ij})^T \boldsymbol{\Sigma}_{ij}^{-1}(\mathbf{x} - \boldsymbol{\mu}_{ij}), \tag{3.96}$$

where $\boldsymbol{\Sigma}_{ij}$ is the sample covariance matrix of the jth cluster of class ω_i.

The first of these two forms of cluster discriminant functions reduces to the linear form

$$g_{ij}(\mathbf{x}) = 2\boldsymbol{\mu}_{ij}^T\mathbf{x} - \|\boldsymbol{\mu}_{ij}\|^2, \tag{3.97}$$

since the term $-\mathbf{x}^T\mathbf{x}$ in the expansion of Equation (3.96) is invariant over i and j. Thus the Euclidean-distance discriminant function leads to a *piecewise linear* decision surface (i.e., a decision surface consisting only of segments of hyperplanes) determined by the following rule: $\mathbf{x}$ is assigned to class ω_i if

$$\max_j (2\boldsymbol{\mu}_{ij}^T\mathbf{x} - \|\boldsymbol{\mu}_{ij}\|^2) \geqslant \max_j (2\boldsymbol{\mu}_{lj}^T\mathbf{x} - \|\boldsymbol{\mu}_{lj}\|^2) \quad \text{for all } l. \tag{3.98}$$

We refer to the classifier produced by this decision rule as the *nearest-prototype classifier*. This rule is sometimes extended by assigning $\mathbf{x}$ to class ω_i if a plurality of the k nearest prototypes belong to class ω_i, thereby producing the *k-nearest-prototype classifier*. The quantity k is an empirically chosen parameter. If the number of feature vectors forming each prototype is just one, the nearest-prototype and k-nearest-prototype classifiers become the *nearest-neighbor* and *k-nearest-neighbor* classifiers [1]. Prototypes formed from single feature vectors may be especially useful when the size of the design set is small or when a priori knowledge of the shapes of the class densities may be exploited in constructing economical searches of the k nearest neighbors.

When the number of clusters in each class is just one, and when there are just two classes, the decision surface determined by Inequality (3.98) reduces to the hyperplane that bisects the two prototypes. This is known as the *minimum distance* classifier, since every feature vector $\mathbf{x}$ entering the classifier is assigned to the class associated with the prototype closest to $\mathbf{x}$.

Below we describe two procedures for forming prototypes and clusters: (a) diameter-invariant cluster grower, and (b) the isodata procedure.

3.9.1 Diameter-Invariant Cluster Grower

In this procedure all of the clusters are hyperspherical, and all have radius R. It is similar in concept to Sebestyen's adaptive sample set construction [10].

One chooses R from empirical evidence and practical experience. Let $\mathscr{X}$ denote the training set. (Usually $\mathscr{X}$ contains members of just a single class.) Choose any feature vector $\mathbf{x}(j)$ belonging to $\mathscr{X}$. Let $\{\boldsymbol{\mu}_j\}$ denote the prototypes formed so far (at the beginning of the algorithm's execution, there are no $\boldsymbol{\mu}_j$'s.) Search through all of the $\boldsymbol{\mu}_i$'s until $\boldsymbol{\mu}_l$ is found such that

$$\|\boldsymbol{\mu}_l - \mathbf{x}(j)\| \leqslant \|\boldsymbol{\mu}_i - \mathbf{x}(j)\| \quad \text{for all } i.$$

Suppose, so far, there are $k\boldsymbol{\mu}_i$'s, specifically $\{\boldsymbol{\mu}_i\} = \{\boldsymbol{\mu}_1, \ldots, \boldsymbol{\mu}_k\}$. If $\|\mathbf{x}(j) - \boldsymbol{\mu}_l\| \geqslant R$, let $\mathbf{x}(j)$ become the $(k+1)$th prototype: $\mathbf{x}(j) = \boldsymbol{\mu}_{k+1}$. If $\|\mathbf{x}(j) - \boldsymbol{\mu}_l\| < R$, form a revision of $\boldsymbol{\mu}_l$ as follows:

$$\boldsymbol{\mu}_l \leftarrow \frac{n_l\boldsymbol{\mu}_l + \mathbf{x}(j)}{n_l + 1}$$

$$n_l \leftarrow n_l + 1,$$

where the arrow denotes "is replaced by," and where n_l is the number of feature vectors entering into the formation of $\boldsymbol{\mu}_l$. Then delete $\mathbf{x}(j)$ from the training set $\mathscr{X}$.

If the resulting $\mathscr{X}$ is not empty, choose another member of $\mathscr{X}$ and repeat the above steps. If the resulting $\mathscr{X}$ is empty, check whether the set of prototypes $\{\boldsymbol{\mu}_j\}$ obtained at the present iteration (iteration S) is close to or identical to the set of prototypes obtained at the previous iteration (iteration $S - 1$). Alternatively, check whether the number of prototypes at iteration S is equal to the number of prototypes at iteration $S - 1$. (This alternative procedure is faster, but possibly somewhat less accurate.) If it is, the algorithm stops. If not, $\mathscr{X}$ is replaced by the original training set, the present set $\{\boldsymbol{\mu}_i\}$ is retained as the initial set for iteration $S + 1$, and iteration $S + 1$ is entered. In iteration $S + 1$, the steps of the preceding iteration are repeated.

The set of prototype regions obtained in this manner are all the same size. Depending on the sequence in which the $\mathbf{x}(j)$'s are drawn from $\mathscr{X}$, the regions are likely to drift [11] so that each $\boldsymbol{\mu}_i$ is not at the centroid of the set of $\mathbf{x}_j$'s that entered into the computation of that $\boldsymbol{\mu}_i$. On the other hand, the number of prototypes is determined by the data.

In some applications it is important that each prototype be the mean of the $\mathbf{x}(i)$'s in the associated prototype region or cluster. The isodata procedure, described below, yields such a set of prototypes and clusters.

3.9.2 The Isodata Procedure

Isodata procedures are a class of procedures that compute the shapes, sizes, and locations of clusters for a given training set, while the number of clusters is specified by the user of the procedure. The isodata procedure described below was devised by Forgy (see page 157 of [12]). In this procedure each cluster's prototype is the centroid of that cluster. This property may be effectively exploited in the construction of piecewise linear classifiers in which each segment of a decision surface is a minimum distance classifier for a pair of clusters in distinct classes.

Let M denote the number of clusters. The user chooses M from empirical evidence, often involving a series of computational experiments. Let $\mathscr{X}$ denote the training set—usually restricted to a single class. Choose any M members of $\mathscr{X}$. Let these be the set of prototypes at iteration 1. Form a set of M clusters by assigning each member of $\mathscr{X}$ to the cluster containing the closest prototype.

After all the members of $\mathscr{X}$ have been assigned to clusters in iteration S, compute a new set of M prototyes as the centroids of the subset of $\mathscr{X}$ in each cluster. In iteration $S + 1$, repeat the process of assigning members of $\mathscr{X}$ to clusters on the basis of the nearest prototypes. Repeat the iterations until the set of prototypes in two successive iterations is unchanged.

The use of the Euclidean distance as a distance function is a sufficient condition for convergence of Forgy's algorithm after a finite number of iterations.

To determine an appropriate value of M, the user of the isodata procedure usually must experiment until he finds a value of M that yields best results. This number should be as small as possible, while large enough so that the set of prototypes serves adequately as a skeleton for the data.

3.10 Close Opposed Pairs of Prototypes

In Section 3.9, we showed how the use of prototypes and Euclidean-distance discriminant functions leads to piecewise linear decision surfaces. The classifiers associated with these decision surfaces are known as *piecewise linear classifiers.* In this section we indicate how the number of prototypes in a linear classifier may be reduced—often substantially—with little or no reduction in performance. The technique for doing this is based on the observation that the Bayes surface often passes through regions of feature space where two or more classes overlap or where the data from these classes are very close to one another. Bayes theory suggests that the construction of the optimum decision boundary depends principally on the data within these regions.

The use of *close opposed pairs of prototypes* for the design of piecewise linear classifiers is based on the following property of Bayes-optimum decision surfaces: the Bayes surface often passes through regions of feature space where the hulls of subsets of feature vectors from different classes overlap or where the data from these classes are very close to one another. We refer to such regions as *encounter zones.* The *close-opposed-pair* technique finds pairs of prototypes that lie in these encounter zones and that belong to opposite or different classes [13].

Let $\mathcal{M}_i$ denote the set of prototypes formed from feature vectors in $\omega_i (i = 1, 2)$. A pair of prototypes $(\boldsymbol{\mu}_i, \boldsymbol{\mu}_j)$ is *opposed* if and only if $\boldsymbol{\mu}_i \in \mathcal{M}_1$ and $\boldsymbol{\mu}_j \in \mathcal{M}_2$. A pair of prototypes is *close opposed* if and only if the distance $d(\boldsymbol{\mu}_i, \boldsymbol{\mu}_j)$ between $\boldsymbol{\mu}_i$ and $\boldsymbol{\mu}_j$ satisfies

$$d(\boldsymbol{\mu}_i, \boldsymbol{\mu}_j) = \min_{\boldsymbol{\mu}_k \in \mathcal{M}_2} d(\boldsymbol{\mu}_i, \boldsymbol{\mu}_k) = \min_{\boldsymbol{\mu}_k \in \mathcal{M}_1} d(\boldsymbol{\mu}_k, \boldsymbol{\mu}_j).$$

Let π denote the set of close opposed pairs. This set contains at least one member, namely the opposed pair $(\boldsymbol{\mu}_i, \boldsymbol{\mu}_j)$ for which $d(\boldsymbol{\mu}_i, \boldsymbol{\mu}_j)$ is a minimum. The procedure for finding π consists of the following steps:

1. For each $\boldsymbol{\mu}_i \in \mathcal{M}_1$ find the closest prototype $\boldsymbol{\mu}_j(\boldsymbol{\mu}_i) \in \mathcal{M}_2$. Save the link set
$$\mathcal{L}_{12} = \{(\boldsymbol{\mu}_i, \boldsymbol{\mu}_j(\boldsymbol{\mu}_i)) \mid \boldsymbol{\mu}_i \in \mathcal{M}_1\}.$$
2. For each $\boldsymbol{\mu}_j \in \mathcal{M}_2$, find the closest prototype $\boldsymbol{\mu}_i(\boldsymbol{\mu}_j) \in \mathcal{M}_1$. Save the link set
$$\mathcal{L}_{21} = \{(\boldsymbol{\mu}_i(\boldsymbol{\mu}_j), \boldsymbol{\mu}_j) \mid \boldsymbol{\mu}_j \in \mathcal{M}_2\}.$$
3. Find the intersection of $\mathcal{L}_{12}$ and $\mathcal{L}_{21}$. This is π. Thus $\pi = \mathcal{L}_{12} \cap \mathcal{L}_{21}$.

The set π represents the gap or the region of overlap between the two classes of data. Sometimes it is useful to enlarge this set by extending the concept of close opposed pairs to that of *k-close opposed* pairs, $\pi^{(k)}$.

The algorithm for finding $\pi^{(k)}$ is like that for finding π, except that the instruction "find the closest prototype" in steps 1 and 2 is replaced by "find the k closest prototypes." This means, for example, that $(\boldsymbol{\mu}_i, \boldsymbol{\mu}_j) \in \mathscr{L}_{12}$ if and only if no more than $k - 1$ prototypes in $\mathscr{M}_2$ are closer to $\boldsymbol{\mu}_i$ than is $\boldsymbol{\mu}_j$. Clearly,

$$\pi^{(k)} \subseteq \pi^{(k+1)}$$

and

$$\pi \equiv \pi^{(1)} = \bigcap_k \pi^{(k)}.$$

The best choice of k depends on empirical evidence. As a first guess we suggest $k = d =$ the number of dimensions in feature space.

3.11 Locally Trained Piecewise Linear Classifiers

The set of close opposed pairs of prototypes provides a basis for a locally trained piecewise linear classifier [13]. This localized training seems to provide a substantial increase in versatility and performance over other piecewise linear classifiers—such as the perceptron—that are trained on the entire training set.

In this technique the hyperplanes are found sequentially: First find the hyperplane that separates the closest among the k-close opposed pairs of prototypes. Let $(\boldsymbol{\mu}_I, \boldsymbol{\mu}_J)$ denote this pair. For simplicity, choose the perpendicular bisector of $(\boldsymbol{\mu}_I, \boldsymbol{\mu}_J)$, satisfying the equation

$$[\mathbf{x} - \tfrac{1}{2}(\boldsymbol{\mu}_I + \boldsymbol{\mu}_J)]^T(\boldsymbol{\mu}_I - \boldsymbol{\mu}_J) = 0.$$

Call this hyperplane $\hat{H}_1$.

Next find those pairs of prototypes that are correctly classified by $\hat{H}_1$. The feature vectors in the corresponding clusters are combined to form a local training set for training $\hat{H}_1$. Many of the linear-classifier training algorithms described in this book will be useful for this purpose.

Let H_1^* denote the hyperplane obtained by the training process using $\hat{H}_1$ as the initial hyperplane. H_1^* may or may not separate the same set of pairs of k-close opposed prototypes as $\hat{H}_1$. If it does not, then repeat the training process, treating H_1^* as the initial hyperplane. Repeat the training until the training does not change the set of correctly classified close opposed pairs of prototypes. Let H_1 denote the final hyperplane.

Now remove the prototypes separated by H_1 from the set of k-close opposed pairs.

Next remove from the set of k-close opposed pairs those pairs that are correctly classified by H_1. Compute the closest among the remaining k-close opposed pairs. Now find H_2 for these remaining pairs in a manner similar to that in which H_1 was found.

Continue finding the H_i's in this way until the set of k-close pairs is empty.

The set of H_i's are now combined to form a switching function, as follows. For each H_i, the classifier determines whether $\mathbf{x}$ lies on the "negative" or "positive" side of H_i. In particular, if the equation of H_i is

$$\mathbf{v}_i^T\mathbf{y} = 0,$$

where $\mathbf{y}$ is the augmented vector $[1, x_1, \ldots, x_d]^T$, then the classifier determines whether $\mathbf{v}_i^T\mathbf{y}$ is negative or positive. If $\mathbf{v}_i^T\mathbf{y}$ is negative, a 0 is generated. If $\mathbf{v}_i^T\mathbf{y}$ is positive, a 1 is generated. The 0's and 1's of these $\mathbf{v}_i$'s (which may be computed sequentially to suppress the computation of redundant $(\mathbf{v}_i^T\mathbf{y})$'s) are combined in a logical network or switching function to produce one of three classifications of $\mathbf{x}$: ω_1, ω_2, or δ, where δ stands for "undecided." As soon as either ω_1 or ω_2 is first produced by this logical network the sequence of computations of $\{\mathbf{v}_i^T\mathbf{y}\}$ is terminated.

To design the logical network, let $z_i(\mathbf{x})$ denote a binary-valued function of $\mathbf{x}$ such that

$$z_i(\mathbf{x}) = \begin{cases} 0, & \text{if and only if} \quad \mathbf{v}_i^T\mathbf{y} < 0 \\ 1, & \text{if and only if} \quad \mathbf{v}_i^T\mathbf{y} > 0. \end{cases}$$

Let $\mathbf{z}(\mathbf{x})$ = column vector = $[z_1(\mathbf{x}), \ldots, z_m(\mathbf{x})]^T$, where m is the number of hyperplanes.

Construct a population table for the $\mathbf{x}$'s in the training set $\mathscr{X}$. For each $\mathbf{x} \in \mathscr{X}$, find $\mathbf{z}(\mathbf{x})$. For each possible $\mathbf{z}$, count the number of $\mathbf{x}$'s in ω_1 for which $\mathbf{z}(\mathbf{x}) = \mathbf{z}$. Call this the population $N_1(\mathbf{z})$. Similarly, count the number of $\mathbf{x}$'s in ω_2 for which $\mathbf{z}(\mathbf{x}) = \mathbf{z}$. Call this $N_2(\mathbf{z})$. If $N_1(\mathbf{z}) + N_2(\mathbf{z})$ is small, let $\Omega(\mathbf{z}) = \delta$, representing "don't care". (The number separating small from not small must be determined empirically.) If $N_1(\mathbf{z}) + N_2(\mathbf{z})$ is not small, let $\Omega(\mathbf{z}) = 0$ or 1 depending on whether $N_1(\mathbf{z})/[N_1(\mathbf{z}) + N_2(\mathbf{z})]$ is greater than or less than $\frac{1}{2}$, respectively. For each binary m-vector $\mathbf{z}$ not included among the $\mathbf{z}$'s in the population table, let $\Omega(\mathbf{z}) = \delta$. $\Omega(\mathbf{z})$ is a switching function that can be simplified by various techniques, such as Karnaugh maps [14], iterated consensus [15], or computer programs [16].

To illustrate the role of $\Omega(\mathbf{z})$ in the design of the classifier, suppose the set of prototypes yields three hyperplanes, H_1, H_2, H_3, and suppose the population table for the $\mathbf{z}$'s yields the function $\Omega(\mathbf{z})$ specified by Table 3.1. Using a Karnaugh map, one may obtain the following Boolean expression for $\Omega(\mathbf{z})$:

$$\Omega(\mathbf{z}) = z_3 \wedge (z_1 \vee \bar{z}_2),$$

where $\wedge$ denotes AND, $\vee$ denotes OR and $\bar{z}_2$ denotes the complement of z_2.

The equation in the last paragraph suggests the following sequence of operations in the classifier for each incoming feature $\mathbf{x}$. First the classifier finds z_3, corresponding to hyperplane H_3. If $z_3 = 0$, then $\Omega(\mathbf{z}) = 0$; hence the classifier assigns $\mathbf{x}$ to ω_1. If $z_3 = 1$, then the classifier must find either z_1 or z_2. If $z_1 = 1$, then $\Omega(\mathbf{z}) = 1$; hence the classifier assigns $\mathbf{x}$ to ω_2. If $z_2 = 0$, then $\Omega(\mathbf{z}) = 1$; hence the classifier assigns $\mathbf{x}$ to ω_2. If both $z_1 = 0$ and $z_2 = 1$, then $\Omega(\mathbf{z}) = 0$; hence the classifier assigns $\mathbf{x}$ to ω_1.

Table 3.1.

$\mathbf{z}$	$\Omega(\mathbf{z})$
$[0,0,0]^T$	δ
$[0,0,1]^T$	1
$[0,1,0]^T$	δ
$[0,1,1]^T$	0
$[1,0,0]^T$	0
$[1,0,1]^T$	δ
$[1,1,0]^T$	δ
$[1,1,1]^T$	1

The following numerical example, devised by L. Michelotti, illustrates the power of the locally trained piecewise linear classifier using close opposed pairs and a window training procedure (Chapter 5). Two classes of data form a tic-tac-toe figure, as indicated in Figure 3.9. The members of one class are represented by O's, and the other class by +'s. Each class has a mean of 0 and a variance of 1. Thus the covariance matrix of each class is the identity matrix. Consequently second-order statistics contain no information leading to an effective quadratic classifier. Thus this example represents a challenge for most automated procedures for designing nonlinear classifiers.

The decision regions produced by the close-opposed-pair algorithm and a single iteration of a localized window training procedure yielded the decision regions shown in Figure 3.10. The region for +'s is shown heavily shaded, and the region for O's is shown lightly shaded.

Although this example is in two-dimensional feature space, this local training procedure classifier has achieved similar successes in three-dimensional and higher-dimensional spaces.

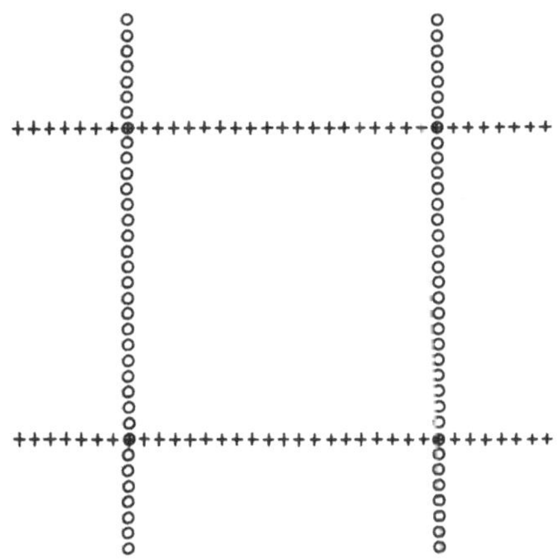

Figure 3.9. Training set for example of a locally trained piecewise linear classifier.

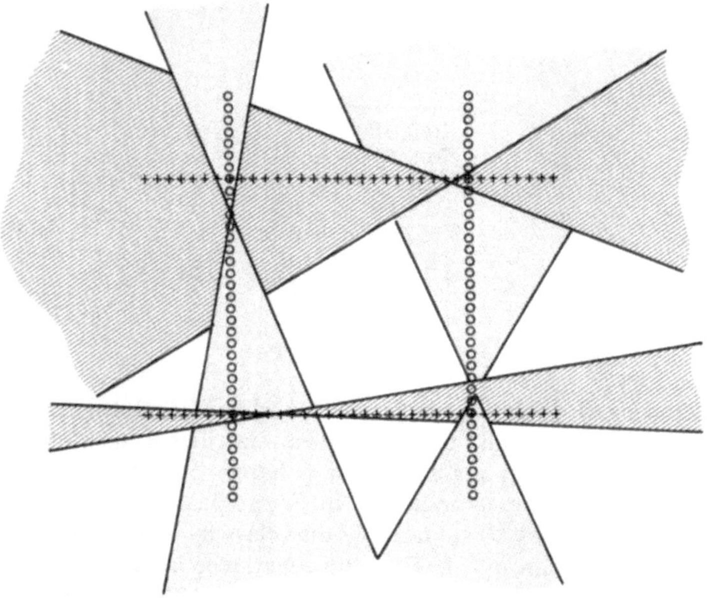

Figure 3.10. Decision regions of a locally trained piecewise linear classifier.

EXERCISES

3.1. Consider a single-feature classifier and assume the probability density for a particular class to be of the form

$$p(x|\mu) = \frac{1}{\sqrt{2\pi}} \exp[-\tfrac{1}{2}(x-u)^2].$$

Use Bayes estimate to find (a) $p(\mu|\mathscr{X}_2)$ and (b) $p(x|\mathscr{X}_2)$ if we assume initially that

$$p(\mu) = \frac{1}{\sqrt{2\pi}} \exp[-\tfrac{1}{2}(\mu-1)^2]$$

for this class. The first two samples of the training set for this class are given by $\mathscr{X}_2 = \{1,1\}$.

3.2. Suppose $p(\mathbf{x}|\omega_j) = N(\boldsymbol{\mu}, \boldsymbol{\Sigma})$, $\boldsymbol{\mu}$ is known, $\boldsymbol{\Sigma}$ is unknown, the a priori density of $\boldsymbol{\Sigma}$ is $p(\boldsymbol{\Sigma})$. Show that

$$p(\boldsymbol{\Sigma})|\mathscr{X}_n) = C|\boldsymbol{\Sigma}^{-1}|^{n/2} \exp\left[-\frac{n}{2}\operatorname{tr}(\hat{\boldsymbol{\Sigma}}^{-1}\boldsymbol{\Sigma}_n)\right]p(\boldsymbol{\Sigma})$$

where the sample convariance matrix is

$$\hat{\boldsymbol{\Sigma}}_n = \frac{1}{n}\sum_{i=1}^{n} [(\mathbf{x}_i - \boldsymbol{\mu})(\mathbf{x}_i - \boldsymbol{\mu})^T]$$

so that $\hat{\boldsymbol{\Sigma}}$ is a sufficient statistic for $\boldsymbol{\Sigma}$.
Hints.

1. Show that $\mathbf{x}^T\mathbf{A}\mathbf{x} = \operatorname{tr}(\mathbf{A}\mathbf{x}\mathbf{x}^T)$ for any square matrix $\mathbf{A}$.
2. Show that

$$\sum_{i=1}^{n} (\mathbf{x}_i - \boldsymbol{\mu})^T\boldsymbol{\Sigma}^{-1}(\mathbf{x}_i - \boldsymbol{\mu}) = n \operatorname{tr}(\boldsymbol{\Sigma}^{-1}\hat{\boldsymbol{\Sigma}}_n).$$

3.3. Demonstrate whether or not $p(b|\mathscr{X}_n)$ of Example 3.1 has the reproducing property of sufficient statistics. If it does, find a sufficient statistic $\Phi(\mathscr{X}_n)$ and a function $g(\Phi, b)$ defined in Equation (3.20).

3.4. Suppose $p(\mathbf{x}|\omega_j) = N(\boldsymbol{\mu}, \boldsymbol{\Sigma})$, both $\boldsymbol{\mu}$ and $\boldsymbol{\Sigma}$ are unknown, and the priori density of $\boldsymbol{\mu}$ and $\boldsymbol{\Sigma}$ is $p(\boldsymbol{\mu}, \boldsymbol{\Sigma})$.

(a) Show that

$$p(\mathscr{X}_1|\boldsymbol{\mu},\boldsymbol{\Sigma}) = (2\pi)^{-nd/2}|\boldsymbol{\Sigma}|^{-n/2}\exp\left[-\tfrac{1}{2}\sum_{k=1}^{n} (\mathbf{x}(i) - \boldsymbol{\mu})^T\boldsymbol{\Sigma}^{-1}(\mathbf{x}(i) - \boldsymbol{\mu})\right].$$

(b) Using the above equation, show that

$$p(\boldsymbol{\mu}, \boldsymbol{\Sigma}|\mathscr{X}_n) = C|\boldsymbol{\Sigma}|^{-n/2}\exp[-\tfrac{1}{2}(n-1)\operatorname{tr}(\boldsymbol{\Sigma}^{-1}\hat{\boldsymbol{\Sigma}}_n) + n\operatorname{tr}[\boldsymbol{\Sigma}^{-1}(\hat{\boldsymbol{\mu}}_n - \boldsymbol{\mu})(\boldsymbol{\mu}_n - \boldsymbol{\mu})^T]p(\boldsymbol{\mu}, \boldsymbol{\Sigma})$$

where

$$\hat{\boldsymbol{\mu}}_n = \frac{1}{n}\sum_{k=1}^{n} \mathbf{x}(k)$$

$$\hat{\boldsymbol{\Sigma}}_n = \frac{1}{n-1}\sum_{k=1}^{n} (\mathbf{x}(k) - \hat{\boldsymbol{\mu}}_n)(\mathbf{x}(k) - \hat{\boldsymbol{\mu}}_n)^T,$$

so that $\hat{\boldsymbol{\mu}}_n$ and $\hat{\boldsymbol{\Sigma}}_n$ are sufficient statistics for $\boldsymbol{\mu}$ and $\boldsymbol{\Sigma}$.

3.5. Suppose

$$p(x|\theta) = \begin{cases} \theta e^{-\theta x}, & \text{for } x \geqslant 0, \theta > 0, \\ 0, & \text{otherwise.} \end{cases}$$

(a) Show that

$$\Phi(\mathscr{X}_n) = \sum_{i=1}^{n} x(i)$$

is a sufficient statistic for θ.

(b) Find a formula for $p(x|\mathscr{X}_n)$.

3.6. Let $\Gamma(u)$ denote the Γ function $\Gamma(u) = \int_0^\infty e^{-t}t^{u-1}\,dt$. Suppose

$$p(x|\boldsymbol{\theta}) = \begin{cases} \dfrac{\Gamma(\theta_1 + \theta_2 + 2)}{\Gamma(\theta_1 + 1)\Gamma(\theta_2 + 1)} x^{\theta_1}(1-x)^{\theta_2}, & \text{for } 0 \leqslant x \leqslant 1, \theta_1 > -1, \theta_2 > -1, \\ 0, & \text{otherwise.} \end{cases}$$

(a) Show that

$$\boldsymbol{\Phi} = \begin{bmatrix} \Phi_1 \\ \Phi_2 \end{bmatrix} = \begin{bmatrix} \prod_{i=1}^{n} x(i) \\ \prod_{i=1}^{n} (1 - x(i)) \end{bmatrix}$$

is a sufficient statistic for $\boldsymbol{\theta}$.

(b) Find a formula for $p(x|\mathscr{X}_n)$.

3.7. Prove Theorem 3.2.

3.8. Prove that Equations (3.42) and (3.43) are equivalent to Equations (3.34) and (3.35).

3.9. Prove Equation (3.46).

3.10. (a) Plot and label a fixed-base bar graph estimate of the constituent densities $f_1(x)$ and $f_2(x)$ from the data given in Table 3.2. Assume that each of the bars in these estimates has a width of 1, and begins and ends at integer values of x. Let $f_i^B(x) (i = 1, 2)$ denote these bar graph estimates. Assume that the relative sizes of the data sets for ω_1 and ω_2 indicate the class probabilities.
(b) Repeat part (a), replacing the bars by rectangular Parzen windows of width 1. Let $f_i^P(x)$ denote these Parzen window estimates.
(c) Plot $|f_i^P(x) - f_i^B(x)|$ for $i = 1, 2$.

Table 3.2. Data for Exercise 3.10.

$x \in \omega_1$	$x \in \omega_2$	
−0.80	1.20	4.30
−0.20	2.70	4.60
0.20	3.10	4.70
0.25	3.20	4.85
0.30	3.60	5.05
0.90	3.90	5.40
1.10	4.15	5.60
1.80	4.20	5.80

3.11. (a) Repeat part (a) of Exercise 3.10, using bars of width 0.5 located so that

$$x = 0, \pm 0.5, \pm 1.0, \pm 1.5, \ldots$$

are endpoints of the bars. Let $f_i^B(x)$ denote these bar graph estimates.
(b) Repeat part (b) of Exercise 3.10 using Parzen windows of width 0.5. Let $f_i^P(x)$ denote these Parzen window estimates.
(c) Plot $|f_i^P(x) - f_i^B(x)|$ for $i = 1, 2$.

3.12. (a) Plot and label a rectangular Parzen window estimate of $f_2(x)$. Use every other example in the data set of Exercise 3.10, as shown in Table 3.3. Use a window width of $\sqrt{2}$. Let $f_2^P(x)$ denote the Parzen window estimate.
(b) Plot $|f_2^P(x) - f_2^B(x)|$. Compare the area under this curve to the area under the curve of $|f_2^P(x) - f_2^B(x)|$ obtained in part (c) of Exercise 3.10. Compare the storage requirements for representing $f_2(x)$ by $f_2^B(x)$ and by $f_2^P(x)$.

Table 3.3. Reduced set for Exercise 3.12.

$x \in \omega_1$	$x \in \omega_2$	
−0.80	1.20	4.30
0.20	3.10	4.70
0.30	3.60	5.05
1.10	4.15	5.60

3.13. Refer to the training set given in Exercise 3.10. Note that the sample size for class ω_1 is 8, i.e., $N_1 = 8$, and for class ω_2, $N_2 = 16$. Using the bar graph technique of Section 3.5 with $V_n = 2/\sqrt{n}$, find the following *for each class* utilizing the entire training set for each class:

(a) $\Psi_n(x - X(i), V_n)$ for interval I_2.
(b) $b_{n2}(x)$.
(c) $p_n(x)$.
(d) the bar graph estimate of the constituent densities $f_1(x)$ and $f_2(x)$.

3.14. Using the training set given in Exercise 3.10, find $p_n(x)$ for each class using the Parzen window technique and the following window function (use the entire training set for each class):

$$\Psi_n(x - z, V_n) = \begin{cases} 1/(2\sqrt{n}), & \text{for } |x - z| \leq 1/\sqrt{n}, \\ 0, & \text{otherwise.} \end{cases}$$

Sketch the graph of the solutions.

3.15. Repeat Exercise 3.13 using the following window function:

$$\Psi_n(x - z, V_n) = (n/2\pi)^{1/2} \exp\left[-\frac{n}{2}(x - z)^2 \right].$$

Do not attempt to obtain exact plots of the solutions, but demonstrate the technique of finding sketches of these solutions.

3.16. Assume a two-feature problem where we wish to use the fixed-base bar graph technique of Section 3.8. Further assume that, for the class of interest, the features are restricted to the region $-2 < x_1 < 3$, $-1 < x_2 < 4$. Let the principal diagonal vector be $\mathbf{c} = [2, 4]^T$.

(a) Find the applicable set of vectors $\{\mathbf{m}_j\}$.
(b) Find the associated set of box centroid vectors $\{\boldsymbol{\mu}_j\}$.
(c) Sketch the boxes and their centroids in feature space.

3.17. Suppose the number of classes is two, and that the two class densities $p(\mathbf{x}|\omega_1)$ and $p(\mathbf{x}|\omega_2)$ are Gaussian:

$$p(\mathbf{x}|\omega_1) = N(\boldsymbol{\mu}_1, \boldsymbol{\Sigma}_1)$$
$$p(\mathbf{x}|\omega_2) = N(\boldsymbol{\mu}_2, \boldsymbol{\Sigma}_2).$$

Let P_1, P_2 denote the a priori probabilities of ω_1, ω_2, respectively.

(a) Find the equation of the decision surface determined by Equation (3.95) for this pair of classes.
(b) Show that if $\boldsymbol{\Sigma}_1 = \boldsymbol{\Sigma}_2 = K\mathbf{I}$, where K is a positive constant and $\mathbf{I}$ is an identity matrix, then the decision surface of part (a) is the perpendicular bisector of the line segment joining the points $\boldsymbol{\mu}_1$ and $\boldsymbol{\mu}_2$ in feature space.

3.18. In Figure 3.11, each × denotes a prototype of the training set of class ω_1, and and each ○ denotes a prototype of the training set of class ω_2.

(a) Find the set of close opposed pairs of prototypes.
(b) Find the set of 2-close opposed pairs of prototypes.
(c) Find the set of hyperplanes associated with the set of close opposed pairs of part (a).
(d) Assume that the training set of feature vectors is, in this problem, identical to the set of prototypes. Find the vector $\mathbf{z}(\mathbf{x})$ corresponding to the hyperplanes obtained in part (c).

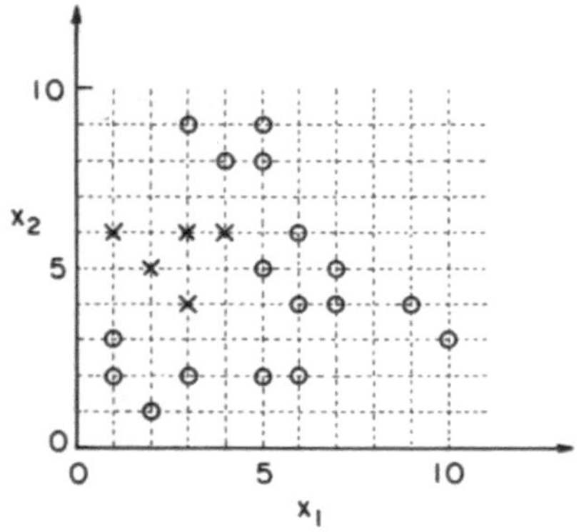

Figure 3.11. Illustration for Exercise 3.18.

References

1. R. O. Duda and P. E. Hart, *Pattern Classification and Scene Analysis*. John Wiley & Sons, New York, 1973.
2. E. L. Lehman, *Testing Statistical Hypotheses*. John Wiley & Sons, New York, 1959.
3. J. Arsac, *Fourier Transforms and the Theory of Distributions*. Prentice-Hall, Englewood Cliffs, NJ, 1966.
4. N. Abramson and D. Braverman, Learning to Recognize Patterns in a Random Environment. *Trans. IRE on Information Theory*, **IT-8** (5): 558–563 (1962).
5. N. J. Nilsson, *Learning Machines*. McGraw-Hill, New York, 1965.
6. J. Spragins, A note on the iterative application of the Bayes rule. *IEEE Trans. on Information Theory*, **IT-11** (4): 544–549 (1965).
7. H. Rubin, Robust Bayesian estimation. In: *Statistical Decision Theory and Related Topics*, Vol. 11, S. S. Gupta and D. S. Moore (eds.), Academic Press, New York, 1977.
8. R. B. Ash, *Basic Probability Theory*. John Wiley & Sons, New York, 1970.
9. E. I. Jury, *Theory and Application of the Z-transform Method*. John Wiley & Sons, New York, 1964.
10. G. S. Sebestyen, Pattern recognition by an adaptive process of sample set construction. *IRE Trans. on Information Theory*, **IT-8** (5): 582–591 (1962). *Pattern Recognition—Introduction and Foundations*, J. Sklansky (ed.), Dowden, Hutchinson & Ross, Stroudsburg, PA, 1973, pp. 117–126.
11. J. Sklansky, G. A. Davison, Recognizing three-dimensional objects by their silhouettes. *J. of the Society of Photo-Optical Instrumentation Engineers*, **10** (1): 10–17 (1971).
12. M. R. Anderberg, *Cluster Analysis for Applications*. Academic Press, New York, 1973.

13. J. Sklansky and L. Michelotti, Locally trained piecewise linear classifiers. *IEEE Trans. on Pattern Analysis and Machine Intelligence*, **PAMI-2** (2): 101–111 (1980).

14. Z. Kohavi, *Switching and Finite Automata Theory*. McGraw-Hill, New York, 1970.

15. E. J. McCluskey, *Introduction to the Theory of Switching Circuits*. McGraw-Hill, New York, 1965.

16. A. Svoboda and D. E. White, *Advanced Logical Circuit Design Techniques*. Garland Publishing Company, New York, 1978.

CHAPTER 4

Loss Functions and Stochastic Approximation

4.1 Introduction

Gradient descent as a technique for finding the minimum of a loss function $J(\mathbf{v})$ was introduced in Section 2.10. Recall that the technique consists of finding the gradient $\nabla J(\mathbf{v})$ and then adjusting the parameter vector $\mathbf{v}$ so that the change in $\mathbf{v}$ is in the direction of the negative of the gradient.

The conditions for ensuring convergence when applying a gradient descent technique were discussed in Section 2.11. These conditions require that $J(\mathbf{v})$ be constant over the solution region $\mathscr{V}^*$, that it must be increasing at the outer edges of the boundary of $\mathscr{V}^*$ as one proceeds outward from $\mathscr{V}^*$, that there be precisely one connected region $\mathscr{V}^*$ in a neighborhood $\mathscr{N}(\mathbf{v}^*)$ which includes the range of $\mathbf{v}$, and that all the derivatives of $J(\mathbf{v})$ exist in $\mathscr{N}(\mathbf{v}^*)$.

The pattern recognition training procedures which are derived from the loss functions of this chapter are approximation procedures, since they use a finite sample to estimate the gradient. They are stochastic approximation procedures when $\mathbf{v}$ is a random variable $\mathbf{V}$ and the loss functions are expectations of functions of $\mathbf{V}$.

The training procedures in this chapter often use very small samples. (In fact they often use samples containing only one element.) Thus they are extreme approximations to gradient descent methods. Conversely when an approximation to the gradient based on a large sample is made, the method is generally referred to as an approximate gradient descent rather than a stochastic approximation technique. The term *approximate gradient descent* overlooks the probabilistic or stochastic nature of the approximation.

In this chapter we shall derive a generalized gradient approximation based upon a finite discrete sample, present several basic loss functions associated with certain important pattern recognition training procedures, derive their gradients, and describe the stochastic approximation descent technique. This material leads to the stochastic training procedures described in the next chapter and will help us evaluate the effectiveness of these procedures for various applications.

4.2 A Loss Function for the Proportional Increment Procedure

Recall the proportional increment training procedure (Section 2.4) which may be described by the following recursive equation:

$$\mathbf{V}(n+1) = \mathbf{V}(n) + \rho(n)\mathbf{Z}(n) \tag{4.1}$$

where

$$\mathbf{Z}(n) = \begin{cases} \mathbf{Y}(n), & \text{if } \mathbf{V}(n)^T\mathbf{Y}(n) < 0 \text{ and } \omega(n) = \omega_2, \\ -\mathbf{Y}(n), & \text{if } \mathbf{V}(n)^T\mathbf{Y}(n) > 0 \text{ and } \omega(n) = \omega_1, \\ \mathbf{0}, & \text{otherwise.} \end{cases} \tag{4.2}$$

If the classes are known to be linearly separable, $\rho(n)$ is usually chosen to be a positive constant; in more general cases, $\rho(n)$ is usually chosen to be a positive decreasing function of n. In general $\mathbf{Z}(n)$ is a vector random variable whose value depends on the classification $\omega(n)$ of the augmented feature vector $\mathbf{Y}(n)$ and the position of $\mathbf{Y}(n)$ relative to the hyperplane determined by $\mathbf{V}(n)$. In the techniques of stochastic approximation to be discussed later in this chapter, this procedure is viewed as a method for seeking the minimum in $\mathbf{v}$-space of some loss function whose gradient at any point $\mathbf{v}$ is the negative of the average $\mathbf{Z}$ for that $\mathbf{v}$, i.e., $\nabla J(\mathbf{v}) = -E(\mathbf{Z}|\mathbf{v})$. A simple form of such a loss function for the proportional increment training procedure is

$$J(\mathbf{v}) = E(-\mathbf{v}^T\mathbf{Z}|\mathbf{V} = \mathbf{v}) \tag{4.3}$$

where $\mathbf{Z}$ is given by Equation (4.2). The demonstration that $-E(\mathbf{Z}|\mathbf{v})$ is the gradient of $-E(\mathbf{v}^T\mathbf{Z}|\mathbf{v})$ is nontrivial since $\mathbf{Z}$ as given by Equation (4.2) is a function of $\mathbf{v}$. Some of the techniques to be presented in Section 4.5.1 can be used to demonstrate its validity. This demonstration is left as an exercise.

In order to exemplify the gradient descent technique and to display some of the properties of the multifeature proportional increment training procedure, it is instructive to analyze this loss function.

Referring to Figure 1.7, Equation (4.3) can be written in terms of the distance from the decision hyperplane by simply multiplying the entire expression by $\|\mathbf{w}\|$ while dividing the expectation's operand by $\|\mathbf{w}\|$:

$$J(\mathbf{v}) = \|\mathbf{w}\| E\left(\frac{-\mathbf{v}^T\mathbf{Z}}{\|\mathbf{w}\|}\bigg|\mathbf{v}\right). \tag{4.4}$$

Applying Equation (4.2), we obtain

$$J(\mathbf{v}) = \|\mathbf{w}\| \left[\int_{\mathscr{R}_1} \frac{(-\mathbf{v}^T\mathbf{y})}{\|\mathbf{w}\|} P(\omega_2)p(\mathbf{x}|\omega_2)\,d\mathbf{x} + \int_{\mathscr{R}_2} \frac{-\mathbf{v}^T(-\mathbf{y})}{\|\mathbf{w}\|} P(\omega_1)p(\mathbf{x}|\omega_1)\,d\mathbf{x} \right] \tag{4.5}$$

where $d\mathbf{x}$ denotes the hypervolume differential element $dx_1\,dx_2 \cdots dx_d$ and where $\int_{\mathscr{R}_j} d\mathbf{x}$, $j = 1, 2$, denotes integration over the half-space decision region, delineated by the decision hyperplane $\mathbf{v}^T\mathbf{y} = 0$, where class ω_j decisions are made. Specifically $\mathscr{R}_1 = \{\mathbf{x}|\mathbf{v}^T\mathbf{y} < 0\}$ and $\mathscr{R}_2 = \{\mathbf{x}|\mathbf{v}^T\mathbf{y} > 0\}$. Therefore note that $\mathbf{v}^T\mathbf{y}/\|\mathbf{w}\| < 0$ in the first integral of Equation (4.5) and $\mathbf{v}^T\mathbf{y}/\|\mathbf{w}\| > 0$ in the second. Equation (4.5) can also be written in terms of the constituent densities $f_1(\mathbf{x})$ and $f_2(\mathbf{x})$:

$$J(\mathbf{v}) = \|\mathbf{w}\| \left[\int_{\mathscr{R}_1} \frac{(-\mathbf{v}^T\mathbf{y})}{\|\mathbf{w}\|} f_2(\mathbf{x})\,d\mathbf{x} + \int_{\mathscr{R}_2} \frac{\mathbf{v}^T\mathbf{y}}{\|\mathbf{w}\|} f_1(\mathbf{x})\,d\mathbf{x} \right]. \tag{4.6}$$

This can be developed further by noting that the probability density $p(\mathbf{x}|\omega_2, \mathbf{y} \in \mathscr{R}_1)$ is equal to $f_2(\mathbf{x})/P(\omega_2, \mathbf{y} \in \mathscr{R}_1)$ in region $\mathscr{R}_1$ and is zero otherwise. Similarly $p(\mathbf{x}|\omega_1, \mathbf{y} \in \mathscr{R}_2) = f_1(\mathbf{x})/P(\omega_1, \mathbf{y} \in \mathscr{R}_2)$ in region $\mathscr{R}_2$. Therefore

$$\begin{aligned} J(\mathbf{v}) = \|\mathbf{w}\| \Bigg[& P(\omega_2, \mathbf{y} \in \mathscr{R}_1) E\left(\frac{-\mathbf{v}^T\mathbf{y}}{\|\mathbf{w}\|} \middle| \mathbf{v}, \omega_2, \mathbf{y} \in \mathscr{R}_1 \right) \\ & + P(\omega_1, \mathbf{y} \in \mathscr{R}_2) E\left(\frac{\mathbf{v}^T\mathbf{y}}{\|\mathbf{w}\|} \middle| \mathbf{v}, \omega_1, \mathbf{y} \in \mathscr{R}_2 \right) \Bigg]. \end{aligned} \tag{4.7}$$

But $P(\omega_1, \mathbf{y} \in \mathscr{R}_2)$ and $P(\omega_2, \mathbf{y} \in \mathscr{R}_1)$ are just the hypervolumes of the error tails of $f_1(\mathbf{x})$ and $f_2(\mathbf{x})$, respectively. Therefore, this proportional increment loss function can be interpreted in the following way:

$$J(\mathbf{v}) = \|\mathbf{w}\| [M_1(\mathbf{v}) + M_2(\mathbf{v})], \tag{4.8}$$

where $M_i(\mathbf{v})$ denotes the *magnitude* of the first moment of the $f_i(\mathbf{x})$ error tail about the decision hyperplane specified by the vector $\mathbf{v}$.

Another interpretation can be found by further developing Equation (4.7). Let N denote the total number of feature vectors $\mathbf{Y}_i$ in the sample and let N_1 and N_2 denote the number of these feature vectors misclassified in regions $\mathscr{R}_1$ and $\mathscr{R}_2$, respectively. Note that $N_1 + N_2 \leqslant N$. Then, from Equation (4.7),

$$J(\mathbf{v}) = \|\mathbf{w}\| \lim_{\substack{N_1 \to \infty \\ N_2 \to \infty}} \left\{ \left[\frac{N_1}{N}\right] \left[\frac{1}{N_1} \sum_{i=1}^{N_1} \frac{(-\mathbf{v}^T\mathbf{Y}_i)}{\|\mathbf{w}\|} \right]_{\substack{\omega = \omega_2 \\ Y_i \in R_1}} + \left[\frac{N_2}{N}\right] \left[\frac{1}{N_2} \sum_{i=1}^{N_2} \frac{\mathbf{v}^T\mathbf{Y}_i}{\|\mathbf{w}\|} \right]_{\substack{\omega = \omega_1 \\ Y_i \in R_2}} \right\} \tag{4.9}$$

or

$$J(\mathbf{v}) = \|\mathbf{w}\| \lim_{\substack{N_1 \to \infty \\ N_2 \to \infty}} \frac{N_1 + N_2}{N} \left\{ \frac{1}{N_1 + N_2} \left[\sum_{i=1}^{N_1} \frac{(-\mathbf{v}^T \mathbf{Y}_i)}{\|\mathbf{w}\|} \bigg|_{\substack{\omega = \omega_2 \\ Y_i \in R_1}} + \sum_{i=1}^{N_2} \frac{\mathbf{v}^T \mathbf{Y}_i}{\|\mathbf{w}\|} \bigg|_{\substack{\omega = \omega_1 \\ Y_i \in R_2}} \right] \right\}, \tag{4.10}$$

or, since

$$\left[\sum_{i=1}^{N_1} \frac{(-\mathbf{v}^T \mathbf{Y}_i)}{\|\mathbf{w}\|} \bigg|_{\substack{\omega = \omega_2 \\ Y_i \in R_1}} + \sum_{i=1}^{N_2} \frac{\mathbf{v}^T \mathbf{Y}_i}{\|\mathbf{w}\|} \bigg|_{\substack{\omega = \omega_1 \\ Y_i \in R_2}} \right]$$

represents the sum of the distances of the misclassified features from the decision hyperplane,

$$J(\mathbf{v}) = \|\mathbf{w}\| P(\text{error} \mid \mathbf{v}) E(\delta \mid \mathbf{v}) \tag{4.11}$$

where $E(\delta \mid \mathbf{v})$ denotes the average distance of misclassified features from the decision hyperplane specified by the vector $\mathbf{v}$.

In order to better visualize the magnitude of this proportional increment loss function for different positions in weight space, a single-feature example will be considered. In this example we assume that the a priori probabilities $P(\omega_1)$ and $P(\omega_2)$ are equal. We also assume that the class densities $p(x|\omega_1)$ and $p(x|\omega_2)$ are Gaussian with variances of unity. Figure 4.1 illustrates the constituent densities for this example plotted in augmented feature space, i.e., y-space. This space may also be viewed as augmented weight space, i.e., v-space, when the weight vector is of interest. This figure illustrates an arbitrary weight vector $\mathbf{v}(n)$ together with the decision threshold $\theta = -w_0/w_1$ which $\mathbf{v}(n)$ determines. Using Equation (4.8) or Equation (4.11), we compute $J(\mathbf{v})$ as a function of $\mathbf{v}$. The resulting surface is depicted in Figure 4.2. The

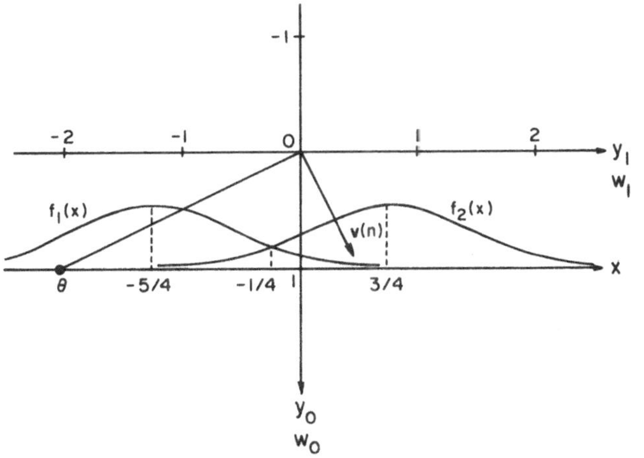

Figure 4.1. A Gaussian one-dimensional feature example superimposed on augmented feature space and weight space.

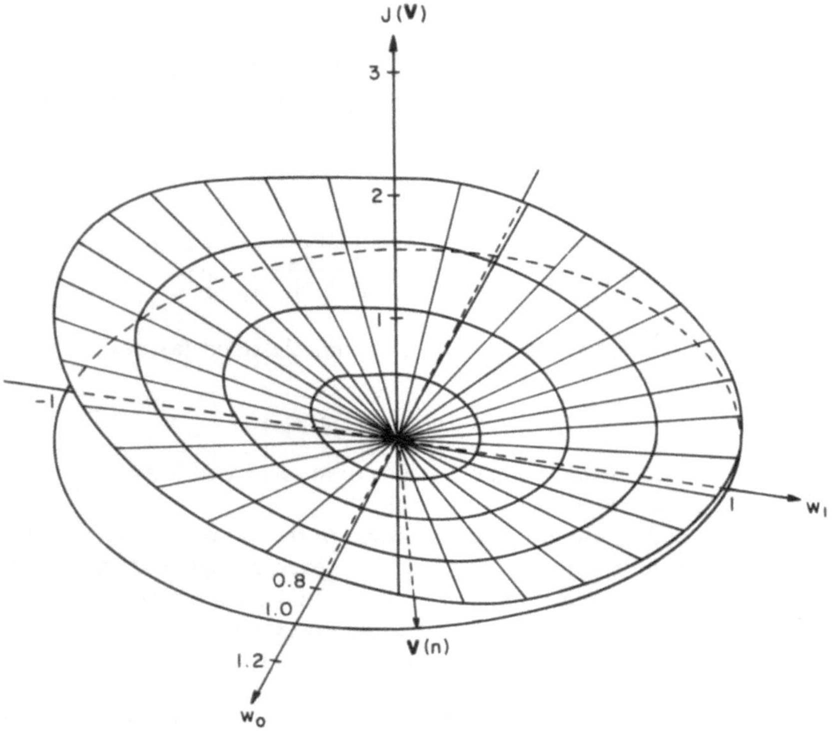

Figure 4.2. Proportional increment loss function for the example of Figure 4.1.

particular vector $\mathbf{v}(n)$ displayed in Figure 4.1 is identified in Figure 4.2 for reference. Note in Figure 4.2 that $J(\mathbf{v})$ has a single minimum at the origin. However, a weight vector with all components equal to zero has no direction and is useless. This then demonstrates one of the flaws in the proportional increment loss function, and its training procedure. If a constraint $\|\mathbf{v}\| = 1$ is made, then the possible values of $J(\mathbf{v})$ are described by the outer edge of the surface shown in Figure 4.2 and the difficulty at the origin is excluded. There is another more fundamental problem with this training procedure when it is applied to overlapping feature cases. However, it is not important to our present use and will therefore be discussed later in Section 4.5.3.

When a gradient descent technique is used, the initial vector $\mathbf{v}(0)$ can be chosen anywhere within some restricted region. (In the above example the restricted region might be the half space where $w_1 > 0$.) A restriction of this kind on $\mathbf{v}(0)$ is sometimes needed in order to guarantee or improve convergence and/or stability properties when some a priori knowledge of the class densities is available. Often no such restriction is necessary. In other cases such a restriction may be needed but cannot be specified precisely.

The direction and rate of motion along the path which seeks a minimum loss point must be determined for gradient techniques solely from the rate of change of loss at the point $\mathbf{v}(t)$. In a gradient descent procedure, the velocity vector $d\mathbf{v}/dt$ points in a direction exactly opposite to the gradient of the loss $\nabla J(\mathbf{v})$, i.e., opposite to the direction of *maximum increasing* loss. The magnitude of $d\mathbf{v}/dt$ is proportional to the magnitude of the gradient. Thus

$$d\mathbf{v}/dt = -\rho \nabla J(\mathbf{v}) \tag{4.12}$$

where ρ is a constant. For the proportional increment training procedure,

$$\nabla J(\mathbf{v}) = \begin{bmatrix} \partial J(\mathbf{v})/\partial w_0 \\ \hline \partial J(\mathbf{v})/\partial \mathbf{w} \end{bmatrix} = -E(\mathbf{Z} \mid \mathbf{v}) \tag{4.13}$$

as found from Equation (4.3). Therefore, combining Equations (4.12) and (4.13),

$$d\mathbf{v}/dt = \rho E(\mathbf{Z} \mid \mathbf{v}), \tag{4.14}$$

where the restrictions on $\mathbf{v}$ have been removed temporarily to facilitate our explanation.

Figure 4.3 illustrates two paths of $\mathbf{v}(t)$ under a gradient descent procedure applied to the proportional increment loss function. The two paths are associated with two initial vectors: $\mathbf{v}(0) = \mathbf{v}_a$ and $\mathbf{v}(0) = \mathbf{v}_b$. This example demonstrates a case for which the gradient descent procedure will not

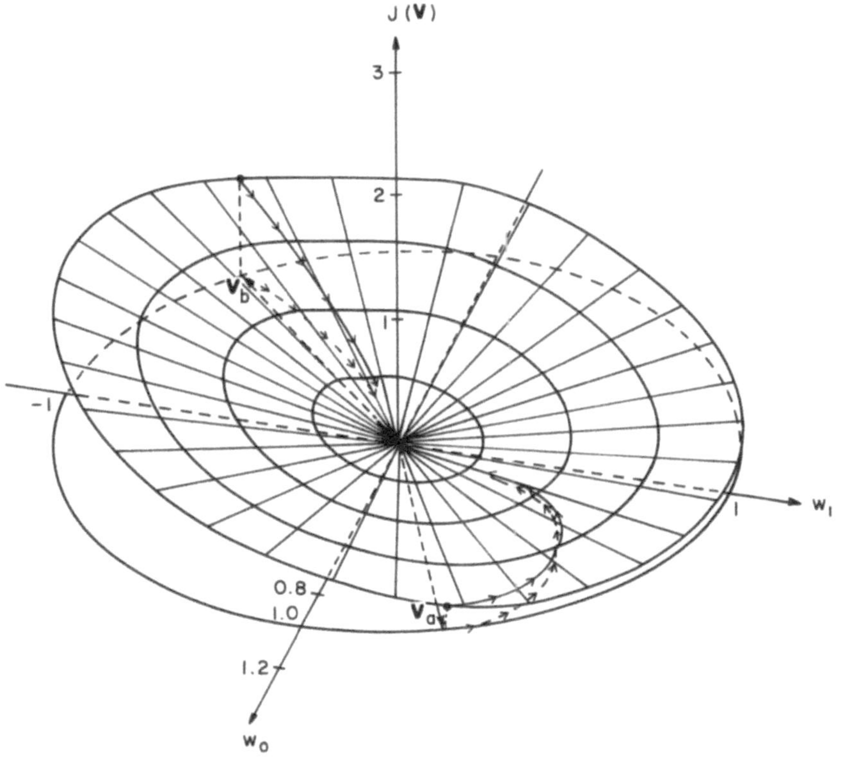

Figure 4.3. Possible gradient descent paths.

always approach a desired vector for the loss function defined. If conditions which ensure convergence do not exist, it is sometimes possible to choose a new loss function which will ensure convergence. Alternatively, a restriction on $\mathbf{v}(0)$ will ensure the convergence to a desirable $\mathbf{v}(\infty)$.

Although continuous calculation of the gradient might be desirable, it is not feasible in practical applications. A piecewise approximation to the gradient path must be made by calculating the gradient at discrete points. This gradient descent training procedure, analogous to Equation (4.12), is given by

$$\mathbf{v}(n+1) = \mathbf{v}(n) - \rho(n)\nabla J(\mathbf{v}(n)) \tag{4.15}$$

or, as discussed in Section 2.10,

$$\mathbf{v}(n+1) = \mathbf{v}(n) - \rho(n)[\nabla\nabla^T J(\mathbf{v}(n))]^{-1}\nabla J(\mathbf{v}(n)) \tag{4.16}$$

where $\rho(n)$ must approach zero as $n \to \infty$ to ensure convergence in non-separable cases.

4.3 The Sample Gradient

For pattern recognition training procedures in general, a priori knowledge of the statistical distribution of feature vectors is not available. Normally, a sample of classified feature vectors is the only available information. From this we wish in some probabilistic sense to approximate a desired function. Our present concern is to find the gradient of $J(\mathbf{v})$, or some reasonable estimate to it, at any point $\mathbf{v}$.

The sample gradient technique will necessarily require the calculation of sample values of the loss function at various values of $\mathbf{v}$. We will denote this sample value of $J(\mathbf{v})$ by $\hat{J}(\mathbf{v})$. Fortunately, this calculation is normally quite easy for well defined loss functions. As an example for the case where $J(\mathbf{v})$ is the probability of misclassification of a feature vector, it is given by

$$\hat{J}(\mathbf{v}) = \left.\frac{N_1 + N_2}{N}\right|_{\mathbf{v}}, \tag{4.17}$$

where $N_1 + N_2$ denotes the number of feature vectors incorrectly classified by $\mathbf{v}$ and, as before, N denotes the total number of feature vectors in the available training set $(\mathscr{X}_1, \mathscr{X}_2)$. Note that

$$\hat{J}(\mathbf{v}) \to J(\mathbf{v}) \quad \text{as} \quad N \to \infty. \tag{4.18}$$

The gradient of $J(\mathbf{v})$ can be written as

$$\begin{bmatrix} \dfrac{\partial J(\mathbf{v})}{\partial w_0} \\ \vdots \\ \dfrac{\partial J(\mathbf{v})}{\partial w_d} \end{bmatrix} = \begin{bmatrix} \lim\limits_{\Delta w_0 \to 0} \dfrac{J(\mathbf{v} + [\Delta w_0, 0, \cdots, 0]^T) - J(\mathbf{v})}{\Delta w_0} \\ \vdots \\ \lim\limits_{\Delta w_d \to 0} \dfrac{J(\mathbf{v} + [0, \cdots, 0, \Delta w_d]^T) - J(\mathbf{v})}{\Delta w_d} \end{bmatrix}. \tag{4.19}$$

Let $\mathbf{v}_k$ denote a randomly selected location within an ε-hypercube neighborhood of $\mathbf{v}$, i.e., $|w_{ik} - w_i| < \varepsilon$, $i = 0, 1, \ldots, d$, in augmented weight space. Now also let $\Delta w_i = w_{ik} - w_i$. Then from Equation (4.19)

$$\nabla J(\mathbf{v}) = \begin{bmatrix} \lim\limits_{\varepsilon \to 0} \dfrac{J(w_{0k}, w_1, w_2, \ldots, w_d) - J(\mathbf{v})}{w_{0k} - w_0} \\ \vdots \\ \lim\limits_{\varepsilon \to 0} \dfrac{J(w_0, \ldots, w_{d-1}, w_{dk}) - J(\mathbf{v})}{w_{dk} - w_d} \end{bmatrix}, \tag{4.20}$$

with $|w_{ik} - w_i| < \varepsilon$, $i = 0, 1, \ldots, d$, whenever $w_{ik} \neq w_i$. Several $\mathbf{v}$'s are now chosen from the ε-hypercube neighborhood to form the sample set $\{\mathbf{v}_k\}$ with the number of $\mathbf{v}_k$'s in the sample denoted by m. The choice of these $\mathbf{v}_k$'s are theoretically arbitrary but would normally be chosen to be uniformly distributed in the ε-hypercube. An approximate sample gradient, denoted by $\widehat{\nabla J}$, can now be defined from Equation (4.20) by substituting the sample loss function $\hat{J}(\mathbf{v})$ for $J(\mathbf{v})$, maintaining ε constant, and averaging the components using the weight vector sample set $\{\mathbf{v}_k\}$. It is therefore given by

$$\widehat{\nabla J} = \begin{bmatrix} \dfrac{1}{m} \sum\limits_{k=1}^{m} \dfrac{\hat{J}(w_{0k}, w_1, w_2, \ldots, w_d) - \hat{J}(\mathbf{v})}{w_{0k} - w_0} \\ \vdots \\ \dfrac{1}{m} \sum\limits_{k=1}^{m} \dfrac{\hat{J}(w_0, w_1, \ldots, w_{d-1}, w_{dk}) - \hat{J}(\mathbf{v})}{w_{dk} - w_d} \end{bmatrix}, \tag{4.21}$$

with $|w_{ik} - w_i| < \varepsilon$, $i = 0, 1, \ldots, d$. Note that the calculation of this sample gradient requires the calculation of $\hat{J}(\mathbf{v})$ plus the $(d + 1)m$ calculations of $\hat{J}([w_0, \ldots, w_{i-1}, w_{ik}, w_{i+1}, \ldots, w_d]^T)$, $i = 1, \ldots, d$; $k = 1, \ldots, m$. Further, each of these calculations of $\hat{J}$ requires the classification of each of the N feature vectors in the training set in order to determine $N_1 + N_2$. Therefore the calculation of a reasonable approximation $\widehat{\nabla J}$ at any point $\mathbf{v}$ may be quite lengthy.

4.4 The Use of Prior Knowledge

Clearly, precise prior knowledge of the optimum decision surfaces and the best assignment of classes to decision regions is never available for the design of trainable classifiers. (Otherwise the training process would be unnecessary.) Nevertheless *some* prior knowledge is usually available. In particular, some a priori description of the spatial relations among the class regions in feature space is often available. For example, in single-feature two-class classifiers, one often knows a priori whether or not the centroid of class ω_2 exceeds the centroid of class ω_1. Under this condition the better choice of the two possible class decisions associated with any decision threshold θ is also most likely known (see Chapter 2). Specifically, if the centroid of ω_2 exceeds that of ω_1, it is usually assumed that all observations of the feature x exceeding θ are best classified as belonging to ω_2.

Similarly, in multiple-feature two-class classifiers, we may know a priori that the centroid of ω_2, say $\boldsymbol{\mu}_2$, is related to the centroid of ω_1, say $\boldsymbol{\mu}_1$, by the inequality

$$\mu_{1k} < \mu_{2k} \tag{4.22}$$

for some k, where μ_{ik} is the component of $\boldsymbol{\mu}_i$ along the kth coordinate of feature space. Since any $\mathbf{v}^*$ points to the half-space decision region $\mathscr{R}_2$, we may be reasonably confident that $v_k^* > 0$ when Inequality (4.22) holds. Alternatively, when the inequality is reversed, we would guess with reasonable confidence that $v_k^* < 0$.

These observations become significant when combined with the property that essentially every hyperplane in $\mathbf{y}$ space can be described by adjusting the components w_i, $i \neq k$, of $\mathbf{v}$ while holding w_k (i.e., v_k) equal to a nonzero constant. This is demonstrated by expressing the equation of the decision hyperplane $\mathbf{v}^T\mathbf{y} = 0$ as follows

$$\mathbf{v}^T\mathbf{y} = w_0 + \sum_{i=1}^{d} w_i x_i = 0 \tag{4.23}$$

and noting that, except for $w_k = 0$, a division by $\pm w_k$ does not reduce its generality. The only exception is for hyperplanes which are parallel to the x_k axis, since this requires that w_k be zero. However, by allowing the remaining components of $\mathbf{w}$ to become large—within the constraints of the word length in the digital computer—while holding w_k constant, a hyperplane which is nearly parallel to the x_k axis can be described. Of course, if one has accurate priori knowledge that the best linear classification is made for one of the two choices of $v_k^* \gtrless 0$, then the decision hyperplane can not be parallel to this axis and the restriction is of no consequence. Since we have no control over $\mathbf{y}$, a division of Equation (4.23) by some positive or negative constant has the effect of defining a new weight vector whose magnitude is different from that of $\mathbf{v}$ and whose direction is respectively either the same as or opposite to that of $\mathbf{v}$. Therefore there are two sets of augmented weight vectors which describe the same hyperplane. One contains all the vectors of varying length which have the exact same direction and the other set contains all the vectors of varying length which have directions exactly opposite to the vectors of the first set. Vectors from different sets make opposite assignments of the classes ω_1 and ω_2 to the half-spaces $\mathscr{R}_1$ and $\mathscr{R}_2$ delineated by the hyperplane.

It follows that an assignment of an initial weight vector can be made such that its kth component has a sign consistent with the a priori knowledge of the sign of v_k^*—for example,

$$v_k(0) = \begin{cases} +1, & \text{for } v_k^* > 0 \\ -1, & \text{for } v_k^* < 0. \end{cases} \tag{4.24}$$

A gradient descent from the initial $\mathbf{v}(0)$ can then be made based upon the gradient of a loss function calculated while w_k is constrained to be constant. This constrained gradient will be given by

$$\nabla J(\mathbf{v})\Big|_{w_k=\text{const}} = \widetilde{\frac{\partial J(\mathbf{v})}{\partial \mathbf{v}}} = \begin{bmatrix} \dfrac{\partial J(\mathbf{v})}{\partial w_0} \\ \vdots \\ \dfrac{\partial J(\mathbf{v})}{\partial w_{k-1}} \\ 0 \\ \dfrac{\partial J(\mathbf{v})}{\partial w_{k+1}} \\ \vdots \\ \dfrac{\partial J(\mathbf{v})}{\partial w_d} \end{bmatrix} \tag{4.25}$$

where the notation $\tilde{\mathbf{a}}$ denotes any vector $\mathbf{a}$ with a_k, $k \neq 0$, equal to zero. Note that when using this constrained gradient in a gradient descent procedure, changes are never made in the kth component of the weight vector, so that this component remains constant at the initially assigned value.

4.5 Loss Functions and Gradients of Some Important Training Procedures

The loss functions and continuous gradients utilized in the development of the window, the minimum (or least) mean square error, and equalized error training procedures of Chapter 5 will be presented in this section. The respective loss functions will be referred to as the probability of error, the mean square error, and the error tail moment loss functions. A method of modifying these loss functions to account for different costs of the types of misclassification is presented in Section 4.7.

4.5.1 Using the Probability of Error as a Loss Function

Let $\mathscr{S}$ denote a decision surface which is a member of a specified family of decision surfaces $\mathscr{T}$, i.e., $\mathscr{S} \in \mathscr{T}$. For the case where the cost of correct classification can be neglected and the costs of the types of misclassification are equal, an $\mathscr{S}$ which achieves the minimum average rate of misclassification possible for any member of $\mathscr{T}$ is usually viewed as an optimum surface for the specified family. Let $\mathscr{S}^*$ denote any such optimum $\mathscr{S}$. We shall refer to any misclassification as an error. Therefore the average rate of misclassification will be referred to as the probability of error.

If we expect to use a gradient descent or gradient descent–related technique to seek this desired $\mathscr{S}^*$, then we must select a loss function which is minimized at an argument where the probability of error is minimized. Several earlier proposed loss functions were mistakenly believed to have this desired minimum [1,2]. Such mistakes can be averted if we let the loss function be the probability of error and restrict its argument to the specification of members of $\mathscr{T}$. We restrict our attention to linear classifiers, so that $\mathscr{T}$ will hereafter denote the family of all linear decision surfaces (i.e., hyperplanes).

This loss function will be denoted by $P(\text{error}\,|\,\mathbf{v})$. The conditioning $\mathbf{v}$ implies a hyperplane decision surface specified by the augmented weight vector $\mathbf{v}$. Stated mathematically for the two-class problem, the probability-of-error loss function is

$$J(\mathbf{v}) = P(\text{error}\,|\,\mathbf{v}) = \int_{\mathscr{R}_1} f_2(\mathbf{x})\,d\mathbf{x} + \int_{\mathscr{R}_2} f_1(\mathbf{x})\,d\mathbf{x} \tag{4.26}$$

where, as before, $\mathscr{R}_1 = \{\mathbf{x}\,|\,\mathbf{v}^T\mathbf{y} < 0\}$ and $\mathscr{R}_2 = \{\mathbf{x}\,|\,\mathbf{v}^T\mathbf{y} > 0\}$. For the distributions illustrated in Figure 4.1, $J(\mathbf{v})$ is plotted in Figure 4.4.

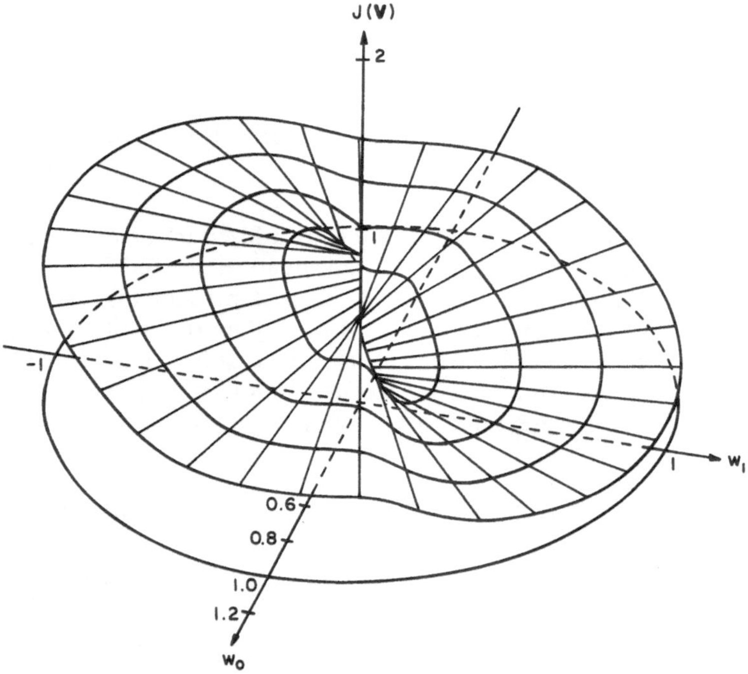

Figure 4.4. Probability of error loss function for the example of Figure 4.1.

To apply the gradient descent technique, it is necessary to find an expression for the gradient, i.e., partial derivative, of $J(\mathbf{v})$ in $\mathbf{v}$ space. Therefore Equation (4.26) is first expressed as follows.

$$J(\mathbf{v}) = \int_{-\infty}^{\infty} dx_1 \cdots \int_{-\infty}^{\infty} dx_{k-1} \int_{-\infty}^{\infty} dx_{k+1} \cdots \int_{-\infty}^{\infty} dx_d \int_{-\infty}^{x_k(\mathbf{v})} dx_k f_2(\mathbf{x}) + \int_{-\infty}^{\infty} dx_1 \cdots \int_{-\infty}^{\infty} dx_{k-1} \int_{-\infty}^{\infty} dx_{k-1} \cdots \int_{-\infty}^{\infty} dx_d \int_{x_k(\mathbf{v})}^{\infty} dx_k f_1(\mathbf{x}) \tag{4.27}$$

where

$$x_k(\mathbf{v}) = -\frac{1}{w_k}\left(w_0 + \sum_{\substack{i=1\\ i\neq k}}^{d} w_i x_i\right)$$

represents the values of x_k which fall on the hyperplane $\mathbf{v}^T\mathbf{y} = 0$ expressed in terms of the other variables. Equation (4.27) is presented for the case where $v_k > 0$ for some k. For the cases where $v_k < 0$, the limits of the integrations along the x_k axis should be reversed. From Equation (4.27) and the above comment we perform differentiation of an integral function which has variable limits, giving

$$\frac{\partial J(\mathbf{v})}{\partial \mathbf{v}} = -\int_{-\infty}^{\infty} dx_1 \cdots \int_{-\infty}^{\infty} dx_{k-1} \int_{-\infty}^{\infty} dx_{k+1} \cdots \int_{-\infty}^{\infty} \frac{dx_d}{|w_k|}\, \mathbf{y} \times [f_2([x_1, \ldots, x_{k-1}, x_k(\mathbf{v}), x_{k+1}, \ldots, x_d]^T) - f_1([x_1, \ldots, x_{k-1}, x_k(\mathbf{v}), x_{k+1}, \ldots, x_d]^T)] \tag{4.28}$$

In this integral the endpoint of the position vector

$$[x_1, x_2, \ldots, x_{k-1}, x_k(\mathbf{v}), x_{k+1}, \ldots, x_d]^T$$

is constrained to lie on the hyperplane $\{\mathbf{x} \mid \mathbf{v}^T\mathbf{y} = 0\}$. Let $\mathscr{S}$ denote this hyperplane, and let ds denote a differential hyperarea in $\mathscr{S}$. Let

$$d\mathbf{a}_k = dx_1 \cdots dx_{k-1}\, dx_{k+1} \cdots dx_d.$$

This quantity is a projections of ds onto a hyperplane normal to the x_k axis. In Figure 4.5, we see that

$$\frac{d\mathbf{a}_k}{ds} = \frac{|w_k|}{\|\mathbf{w}\|}.$$

Thus

$$d\mathbf{a}_k = \frac{|w_k|}{\|\mathbf{w}\|}\, ds.$$

Substituting this into Equation (4.28) yields

$$\nabla J(\mathbf{v}) \equiv \frac{\partial J(\mathbf{v})}{\partial \mathbf{v}} = -\frac{1}{\|\mathbf{w}\|} \int_{\mathscr{S}} \mathbf{y}[f_2(\mathbf{x}) - f_1(\mathbf{x})]\, ds, \tag{4.29}$$

where $\int_{\mathscr{S}} ds$ denotes an integration over the surface of the decision hyperplane in $\mathbf{x}$-space. Specifically, the hyperplane surface $\mathscr{S}$ is given by $\mathscr{S} = \{\mathbf{x} \mid \mathbf{v}^T\mathbf{y} = 0\}$.

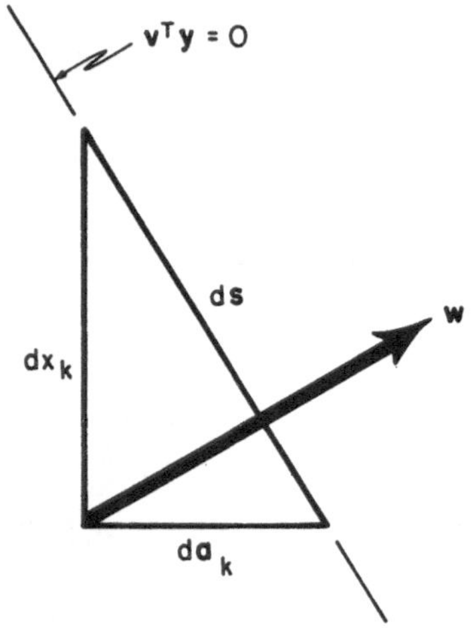

Figure 4.5. Relationship of $d\mathbf{a}_k$ to $d\mathbf{s}$.

Note that $\nabla J(\mathbf{v})$ is always normal to the vector $\mathbf{v}$, i.e., $\mathbf{v}^T[\nabla J(\mathbf{v})] = 0$. This follows because $\mathbf{v}^T\mathbf{y} = 0$ everywhere on the surface $\mathscr{S}$.

The necessary and sufficient conditions for ensuring that there exists precisely one connected region $\mathscr{V}^*$ are given by conditions 1 and 2 of Section 2.11. Condition 1 requires that the gradient be zero in $\mathscr{V}^*$ so that the first component of Equation (4.29) must satisfy

$$\int_{\mathscr{S}^*} f_1(\mathbf{x})\, ds = \int_{\mathscr{S}^*} f_2(\mathbf{x})\, ds \tag{4.30}$$

where $\int_{\mathscr{S}^*} ds$ denotes an integration over the surface of any minimum-probability-of-error decision hyperplane $\mathbf{v}^{*T}\mathbf{y} = 0$. This states that the magnitudes obtained by integrating $f_1(\mathbf{x})$ and $f_2(\mathbf{x})$ over an optimum decision hyperplane surface must be equal. For the two-feature case the integrations may be viewed as occurring over a planar slice of the constituent densities at the position of the hyperplane. The remaining components of Equation (4.29) satisfying condition 1 can be written

$$\int_{\mathscr{S}^*} \mathbf{x} f_1(\mathbf{x})\, ds = \int_{\mathscr{S}^*} \mathbf{x} f_2(\mathbf{x})\, ds, \tag{4.31}$$

which states that the first moments about each coordinate axis of the densities described by $f_1(\mathbf{x})$ and $f_2(\mathbf{x})$ with $\mathbf{x}$ restricted to the surface of an optimum decision hyperplane must be equal. When combined with the criterion from the first component, this simply states that the means of these density slices at the hyperplane must coincide.

If there is some a priori knowledge about any feature, say the feature x_k, of the better of the two possible assignments of classes ω_1 and ω_2 relative to larger or smaller values of x_k, then convergence properties can be improved by using the technique discussed in Section 4.4. By Equations (4.25) and

(4.27),

$$\nabla J(\mathbf{v})\big|_{w_k=\text{constant}} = -\frac{1}{\|\mathbf{w}\|}\int_{\mathscr{S}} \tilde{\mathbf{y}}[f_2(\mathbf{x}) - f_1(\mathbf{x})]\,ds \tag{4.32}$$

where $\tilde{\mathbf{y}}$ denotes the vector

$$\tilde{\mathbf{y}} = \begin{bmatrix} 1 \\ \hline \tilde{\mathbf{x}} \end{bmatrix} = \begin{bmatrix} 1 \\ x_1 \\ \vdots \\ x_{k-1} \\ 0 \\ x_{k+1} \\ \vdots \\ x_d \end{bmatrix}.$$

Note that $\tilde{\mathbf{y}}$ is equal to $\mathbf{y}$ except for the kth component x_k, which is zero in $\tilde{\mathbf{y}}$.

4.5.2 A Loss Function Based on the Mean Square Error

A mean square error loss function can be written as

$$J_m(\mathbf{v}) = \tfrac{1}{2}E(e^2 \mid \mathbf{v}) \tag{4.33}$$

where e denotes an error measure. In a linear classifier,

$$e = \mathbf{v}^T\mathbf{y} - g(\mathbf{x}), \tag{4.34}$$

where $g(\mathbf{x})$ denotes some function with a desired classification property. For the two-class problem, $g(\mathbf{x}) \lessgtr 0$ when $P(\omega_2 \mid \mathbf{x}) \lessgtr P(\omega_1 \mid \mathbf{x})$. A function $g(\mathbf{x})$ with this property is a two-class discriminant function. If $g(\mathbf{x})$ is an optimum Bayes discriminant function, it is given by

$$g(\mathbf{x}) = P(\omega_2 \mid \mathbf{x}) - P(\omega_1 \mid \mathbf{x}). \tag{4.35}$$

Since such a function is generally unknown a priori, another approach to minimizing Equation (4.33) is necessary.

Suppose we define $J(\mathbf{v})$ as

$$J(\mathbf{v}) = \tfrac{1}{2}E[(\mathbf{v}^T\mathbf{y} - \alpha)^2 \mid \mathbf{v}] \tag{4.36}$$

for a two-class classifier, where α is a classification random variable defined by

$$\alpha = \begin{cases} 1, & \text{for } \mathbf{x} \in \omega_2 \\ -1, & \text{for } \mathbf{x} \in \omega_1. \end{cases} \tag{4.37}$$

A vector $\mathbf{v}$ which minimizes the $J(\mathbf{v})$ given by Equations (4.36) and (4.37) will also minimize the $J_m(\mathbf{v})$ specified by Equations (4.33)–(4.35). This can be demonstrated by expanding $J(\mathbf{v})$ as follows:

$$\begin{aligned} J(\mathbf{v}) &= \tfrac{1}{2}E\{[(\mathbf{v}^T\mathbf{y} - g(\mathbf{x})) - (\alpha - g(\mathbf{x}))]^2 | \mathbf{v}\} \\ &= \tfrac{1}{2}E[(\mathbf{v}^T\mathbf{y} - g(\mathbf{x}))^2 | \mathbf{v}] - E[(\mathbf{v}^T\mathbf{y} - g(\mathbf{x}))(\alpha - g(\mathbf{x})) | \mathbf{v}] \\ &\quad + \tfrac{1}{2}E[(\alpha - g(\mathbf{x}))^2]. \end{aligned} \tag{4.38}$$

The middle term of the second line of this equation can be expanded as

$$\begin{aligned} E[(\mathbf{v}^T\mathbf{y} - g(\mathbf{x}))(\alpha - g(\mathbf{x})) | \mathbf{v}] &= E\{E[(\mathbf{v}^T\mathbf{y} - g(\mathbf{x}))(\alpha - g(\mathbf{x})) | \mathbf{x}, \mathbf{v}] | \mathbf{v}\} \\ &= E\{(\mathbf{v}^T\mathbf{y} - g(\mathbf{x}))(E(\alpha | \mathbf{x}, \mathbf{v}) - g(\mathbf{x})) | \mathbf{v}\}. \end{aligned} \tag{4.39}$$

Now from the definition of α,

$$E(\alpha | \mathbf{x}, \mathbf{v}) = P(\omega_2 | \mathbf{x}) - P(\omega_1 | \mathbf{x}) = g(\mathbf{x}).$$

Therefore the middle term of the second line of Equation (4.38) vanishes, leaving

$$J(\mathbf{v}) = \tfrac{1}{2}E[(\mathbf{v}^T\mathbf{y} - g(\mathbf{x}))^2 | \mathbf{v}] + \tfrac{1}{2}E[(\alpha - g(\mathbf{x}))^2]. \tag{4.40}$$

Since the second term on the right-hand side is not a function of $\mathbf{v}$, a $\mathbf{v}$ which minimizes $J(\mathbf{v})$ also minimizes $J_m(\mathbf{v})$.

The basic two-class loss function used for finding the minimum mean squared error can now be written in deterministic form. From Equation (4.36),

$$J(\mathbf{v}) = \frac{1}{2}\left\{\int (\mathbf{v}^T\mathbf{y} + 1)^2 f_1(\mathbf{x})\, d\mathbf{x} + \int (\mathbf{v}^T\mathbf{y} - 1)^2 f_2(\mathbf{x})\, d\mathbf{x}\right\} \tag{4.41}$$

where the integration is over the entire feature space. For the Gaussian constituent densities illustrated in Figure 4.1, the mean squared error loss function given by Equation (4.41) is plotted on Figure 4.6.

For the case where the classes are finite, and all the members of the training set are equally likely to occur, Equation (4.41) becomes

$$J(\mathbf{v}) = K(\mathbf{A}\mathbf{v} - \mathbf{1})^T(\mathbf{A}\mathbf{v} - \mathbf{1}),$$

where K is a positive constant with respect to $\mathbf{v}$, and where $\mathbf{1}$ is the N-vector

$$\mathbf{1} = \begin{bmatrix} 1 \\ \vdots \\ 1 \end{bmatrix},$$

as discussed in Section 2.12.

Several words of caution are advisable at this point. Although minimizing $J(\mathbf{v})$ minimizes the mean of the square of the difference between $\mathbf{v}^T\mathbf{y}$ and $g(\mathbf{x})$, it cannot make them equal. Indeed, the decision surface determined by $\mathbf{v}^T\mathbf{y}$ is a hyperplane whereas $g(\mathbf{x})$ will at the very least be an S-shaped hypersurface. Consider the case of a single-feature classifier. Here the discriminant function $g(\mathbf{x}) = P(\omega_2 | \mathbf{x}) - P(\omega_1 | \mathbf{x})$ is plotted in Figure 4.7, assuming the constituent densities are Gaussian as illustrated in Figure 4.1. (The relationship $P(\omega_j | \mathbf{x}) = f_j(\mathbf{x})/p(\mathbf{x})$, $p(\mathbf{x}) \neq 0$, is helpful in deriving this figure.) Not surprisingly, then, the $\mathbf{v}$ found from minimizing $J(\mathbf{v})$ will not in general give

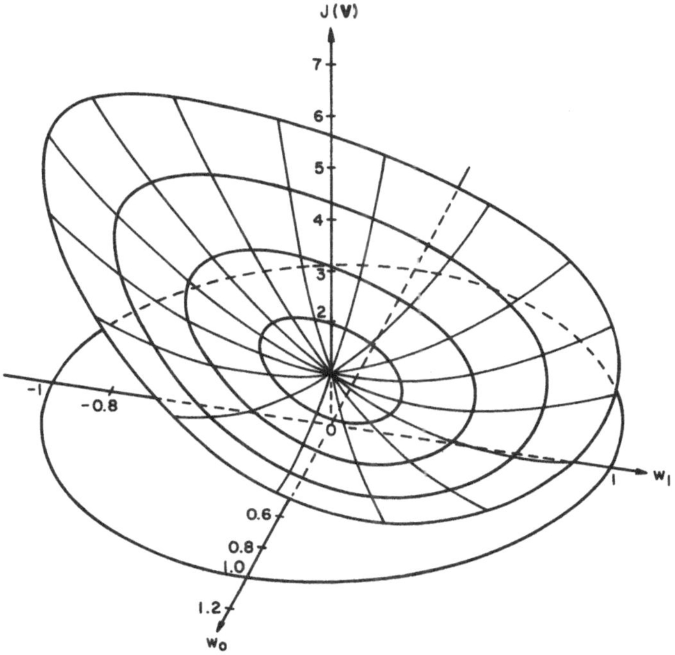

Figure 4.6. Mean square error loss function for the example of Figure 4.1.

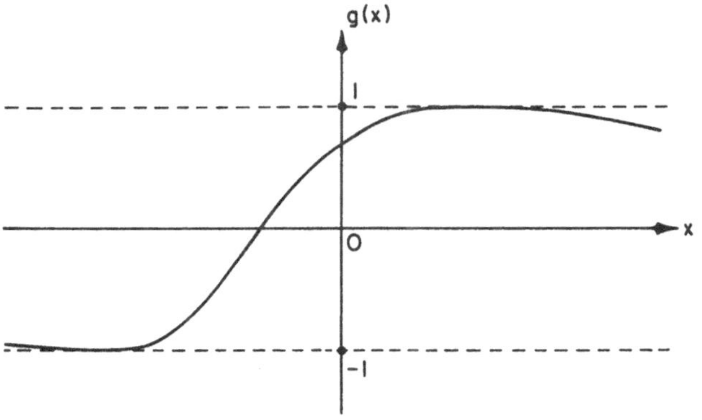

Figure 4.7. Plot of $g(x) = P(\omega_2|x) - P(\omega_1|x)$ for the example of Figure 4.1.

the minimum probability of error hyperplane. This is true since the best mean square error fit of the $\mathbf{v}^T\mathbf{y}$ hyperplane to $g(\mathbf{x})$ occurs in the regions where the probability density of the features is high, rather than in the preferred decision region where $g(\mathbf{x}) = 0$.

A commonly discussed approach to the improvement of this situation is to replace $\mathbf{y}$ by the $\boldsymbol{\Phi}(\mathbf{x})$ transformations discussed in Section 3.2 or by transformations to the normalized Hermite functions [3]. However, the classifier can then no longer be considered linear. Furthermore, the transformations normally greatly increase the dimensionality of the problem.

We now give a precise description of the result of minimizing $J(\mathbf{v})$ without these transformations. Let ξ_1 and ξ_2 be defined by the following expressions:

$$\begin{aligned} \xi_1 &= \frac{\mathbf{v}^T\mathbf{y}+1}{\|\mathbf{w}\|} \\ \xi_2 &= \frac{\mathbf{v}^T\mathbf{y}-1}{\|\mathbf{w}\|}. \end{aligned} \tag{4.42}$$

By reference to the relationships of Figure 1.7 it can be seen that ξ_1 and ξ_2 represent the signed distance of a feature vector $\mathbf{x}$ from the hyperplanes $\mathbf{v}^T\mathbf{y}+1=0$ and $\mathbf{v}^T\mathbf{y}-1=0$, respectively. These hyperplanes and their relation to the decision hyperplane $\mathbf{v}^T\mathbf{y}=0$ are illustrated in the two-dimensional feature space of Figure 4.8. Equation (4.41) can now be written in terms of the distances ξ_1 and ξ_2:

$$J(\mathbf{v}) = \tfrac{1}{2}\|\mathbf{w}\|^2\left[\int \xi_1^2 f_1(\mathbf{x})\,d\mathbf{x} + \int \xi_2^2 f_2(\mathbf{x})\,d\mathbf{x}\right]. \tag{4.43}$$

The integral $\int \xi_1^2 f_1(\mathbf{x})\,d\mathbf{x}$ is the second moment of $f_1(\mathbf{x})$ about the hyperplane $\mathbf{v}^T\mathbf{y}+1=0$, and $\int \xi_2^2 f_2(\mathbf{x})\,d\mathbf{x}$ is the second moment of $f_2(\mathbf{x})$ about $\mathbf{v}^T\mathbf{y}-$

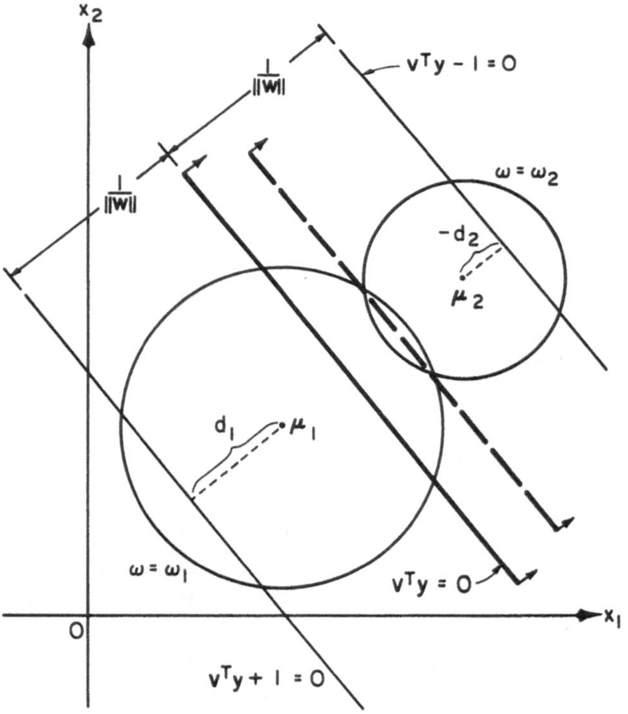

Figure 4.8. A minimum mean square error decision hyperplane; $P(\omega_2) = 2P(\omega_1)$.

$1 = 0$. But the second moment with respect to a hyperplane is at a minimum only if the hyperplane passes through the centroid of the constituent density. Therefore the portion of Equation (4.43) in square brackets can be a minimum only if (a) $\mathbf{v}^T\mathbf{y} + 1 = 0$ passes through the centroid of $f_1(\mathbf{x})$ and (b) $\mathbf{v}^T\mathbf{y} - 1 = 0$ passes through the centroid of $f_2(\mathbf{x})$. However, the bracketed terms are multiplied by $\|\mathbf{w}\|^2$. Therefore, in minimizing $J(\mathbf{v})$ as given by Equation (4.43) it is also desirable to make the $\|\mathbf{w}\|^2$ factor small and thereby make the separating distance of the hyperplanes large (see Figure 4.8). Making $\|\mathbf{w}\|$ small has two effects. First, the training process rotates the hyperplanes about their constituent density centroids until a compromise is achieved between an orientation giving a large separation between the hyperplanes and small second moments of the constituent densities with respect to their respective hyperplanes. Second, the hyperplanes are translated away from each other and from the centroids of $f_1(\mathbf{x})$ and $f_2(\mathbf{x})$ until the resulting increases in second moments dominate the decrease of $\|\mathbf{w}\|^2$ resulting from a larger separation of the hyperplanes.

Let d_1 or d_2 denote the signed distance of the hyperplane $\mathbf{v}^T\mathbf{y} + 1 = 0$ or $\mathbf{v}^T\mathbf{y} - 1 = 0$ from the centroids of $f_1(\mathbf{x})$ or $f_2(\mathbf{x})$, respectively. Specifically, let

$$d_1 = \frac{\mathbf{v}^T\begin{bmatrix} 1 \\ \hdashline \boldsymbol{\mu}_1 \end{bmatrix} + 1}{\|\mathbf{w}\|} \quad \text{and} \quad d_2 = \frac{\mathbf{v}^T\begin{bmatrix} 1 \\ \hdashline \boldsymbol{\mu}_2 \end{bmatrix} - 1}{\|\mathbf{w}\|} \tag{4.44}$$

Note that $d_j > 0$ when $\boldsymbol{\mu}_j$ lies in the half-space into which $\mathbf{w}$ is directed. Therefore $d_1 > 0$ and $d_2 < 0$ for the solution hyperplane. Now

$$\begin{aligned} \int \xi_j^2 f_j(\mathbf{x})\,d\mathbf{x} &= \int [(\xi_j - d_j) + d_j]^2 f_j(\mathbf{x})\,d\mathbf{x} \\ &= \int (\xi_j - d_j)^2 f_j(\mathbf{x})\,d\mathbf{x} + 2d_j \int (\xi_j - d_j) f_j(\mathbf{x})\,d\mathbf{x} \\ &\quad + d_j^2 \int f_j(\mathbf{x})\,d\mathbf{x}, \qquad j = 1, 2. \end{aligned} \tag{4.45}$$

The first term is the second moment of the feature vectors in ω_j with respect to a hyperplane parallel to the decision hyperplane and passing through their mean, i.e.,

$$P(\omega_j)\sigma_{\lambda_j}^2 = \int (\xi_j - d_j)^2 f_j(\mathbf{x})\,d\mathbf{x}, \tag{4.46}$$

where $\lambda_j = [\mathbf{w}^T\mathbf{x}/\|\mathbf{w}\| \,|\, \omega_j]$. It is left to the reader to prove the validity of Equation (4.46). Note that $\int(\xi_j - d_j) f_j(\mathbf{x})\,d\mathbf{x} = 0$, since it represents the first moment of the constituent density of ω_j with respect to a hyperplane passing through the centroid of this density. Furthermore $\int f_j(\mathbf{x})d\mathbf{x} = P(\omega_j)$. Therefore Equation (4.43) can be written

$$J(\mathbf{v}) = \frac{\|\mathbf{w}\|^2}{2}\left[\sigma_{\lambda_1}^2 P(\omega_1) + \sigma_{\lambda_2}^2 P(\omega_2) + d_1^2 P(\omega_1) + d_2^2 P(\omega_2)\right].$$

The increase in second moments as the hyperplanes are each translated by the distance d_j away from the centroid is indicated by the $d_j^2 P(\omega_j)$ terms.

From Equation (4.41), the gradient of the minimum square error loss function is

$$\nabla J(\mathbf{v}) = \int (\mathbf{v}^T\mathbf{y} + 1)\mathbf{y} f_1(\mathbf{x})\, d\mathbf{x} + \int (\mathbf{v}^T\mathbf{y} - 1)\mathbf{y} f_2(\mathbf{x})\, d\mathbf{x}. \tag{4.48}$$

From condition 1 of Section 2.11, $\nabla J(\mathbf{v}) = 0$ for $\mathbf{v} \in \mathscr{V}^*$. Therefore, for $\mathbf{v} = \mathbf{v}^*$,

$$\int (\mathbf{v}^{*T}\mathbf{y} + 1)\mathbf{y} f_1(\mathbf{x})\, d\mathbf{x} = -\int (\mathbf{v}^{*T}\mathbf{y} - 1)\mathbf{y} f_2(\mathbf{x})\, d\mathbf{x}. \tag{4.49}$$

Premultiplying each side of Equation (4.49) by $\mathbf{v}^{*T}$ yields

$$\int (\mathbf{v}^{*T}\mathbf{y} + 1)\mathbf{v}^{*T}\mathbf{y} f_1(\mathbf{x})\, d\mathbf{x} = -\int (\mathbf{v}^{*T}\mathbf{y} - 1)\mathbf{v}^{*T}\mathbf{y} f_2(\mathbf{x})\, d\mathbf{x}.$$

Replacing $\mathbf{v}^{*T}\mathbf{y}$ by $(\mathbf{v}^{*T}\mathbf{y} + 1 - 1)$, expanding, and substituting Equation (4.42) yields

$$\int (\|\mathbf{w}^*\|\xi_1^2 - \xi_1) f_1(\mathbf{x})\, d\mathbf{x} = -\int (\|\mathbf{w}^*\|\xi_2^2 + \xi_2) f_2(\mathbf{x})\, d\mathbf{x}. \tag{4.50}$$

Now replacing ξ_j with $(\xi_j - d_j + d_j)$ and expanding in a manner similar to that used in developing Equation (4.47), we obtain

$$\|\mathbf{w}^*\|(\sigma_{\lambda_1^*}^2 + d_1^2)P(\omega_1) - d_1 P(\omega_1) = -\|\mathbf{w}^*\|(\sigma_{\lambda_2^*}^2 + d_2^2)P(\omega_2) - d_2 P(\omega_2),$$

where $\lambda_j^* = [\mathbf{w}^{*T}\mathbf{x}/\|\mathbf{w}^*\| \,|\, \omega_j]$. Solving for $\|\mathbf{w}^*\|$ we find

$$\|\mathbf{w}^*\| = \frac{d_1 P(\omega_1) - d_2 P(\omega_2)}{(\sigma_{\lambda_1^*}^2 + d_1^2)P(\omega_1) + (\sigma_{\lambda_2^*}^2 + d_2^2)P(\omega_2)}. \tag{4.51}$$

The first component of Equation (4.49), with Equation (4.42) substituted, can be expressed as

$$\int \xi_1 f_1(\mathbf{x})\, d\mathbf{x} = -\int \xi_2 f_2(\mathbf{x})\, d\mathbf{x}.$$

Again substituting $\xi_j = (\xi_j - d_j) + d_j$ and simplifying as before, we find that the relationship between the optimum d_j's is given by

$$d_1 P(\omega_1) = -d_2 P(\omega_2) \tag{4.52}$$

The relationship among $\|\mathbf{w}\|$ and the d_j's is

$$\frac{2}{\|\mathbf{w}\|} = \frac{\mathbf{w}^T}{\|\mathbf{w}\|}(\boldsymbol{\mu}_2 - \boldsymbol{\mu}_1) + d_1 - d_2, \tag{4.53}$$

which can be derived from the relationships illustrated in Figure 4.8. By substituting Equations (4.52) and (4.53) into Equation (4.51), d_1 and d_2 can be found for a given optimum direction $\mathbf{w}^*$, i.e.,

$$d_1 P(\omega_1) = -d_2 P(\omega_2) = \frac{\sigma_{\lambda_1}^2 P(\omega_1) + \sigma_{\lambda_2}^2 P(\omega_2)}{\mathbf{w}^{*T}(\boldsymbol{\mu}_2 - \boldsymbol{\mu}_1)/\|\mathbf{w}^*\|}. \tag{4.54}$$

Figure 4.8 also illustrates the decision hyperplane $\mathbf{v}^T\mathbf{y} = 0$ which gives the minimum of the mean square error loss function for the two constituent densities whose contours are shown and whose a priori probabilities are related by $P(\omega_2) = 2P(\omega_1)$. The dashed line represents the minimum probability of error decision hyperplane. The forms of the constituent densities were chosen to illustrate the difference between the minimum probability of error and the minimum square error decision hyperplanes.

Recall the loss function defined in Equation 4.36. This function becomes a square error loss function when $\alpha = \pm 1$, in accordance with Equation 4.37. Other loss functions are obtained when $\alpha \neq \pm 1$ [1]. When this choice of α is based on some a priori knowledge, the performance of the training process can be improved. In particular, when α is defined as

$$\alpha = \begin{cases} \dfrac{1}{\hat{P}(\omega_2)}, & \text{for } \mathbf{x} \in \omega_2, \\ -\dfrac{1}{\hat{P}(\omega_2)}, & \text{for } \mathbf{x} \in \omega_1, \end{cases} \tag{4.55}$$

where $\hat{P}(\omega_1)$ and $\hat{P}(\omega_2)$ denote the sample class probabilities, the $\mathbf{v}$ that minimizes $J(\mathbf{v})$ will often give a smaller probability of misclassification. Note, however, that under these conditions we are no longer finding the $\mathbf{v}$ which minimizes the mean square difference between $\mathbf{v}^T\mathbf{y}$ and $g(\mathbf{x})$.

The Ho–Kashyap training procedure discussed in Section 2.13 is also related to the mean square error training procedures. It finds a vector $\mathbf{b}$, similar in its role to that of α, which determines a vector $\mathbf{v}$ in the solution set $\mathscr{V}^*$ in cases where the class densities are linearly separable. Unfortunately, the Ho–Kashyap procedure does not always converge in a finite number of iterations for nonseparable classes and, furthermore, requires both matrix multiplication and calculation of the generalized inverse of a large matrix. This matrix is large—often very large—because it contains *all* the feature vectors of the training set.

4.5.3 Error Tail Moments

As discussed in Section 4.2 and exemplified in Equations (4.8) and (4.11), the proportional increment (perceptron) training procedure loss function includes a multiplying $\|\mathbf{w}\|$ factor. Recall that an unfortunate result of including this factor is the resulting minimum of $J(\mathbf{v})$ at the origin of $\mathbf{v}$-space. This was illustrated in Figure 4.2. Note that for the case of overlapping class regions this zero will be the only minimum since the terms of Equations (4.8) and (4.11), other than $\|\mathbf{w}\|$, must always be greater than zero. As mentioned earlier, another problem exists with the proportional increment training procedure when it is applied to pairs of class regions that overlap. Since its form is based on a single-feature error correction training procedure, which in two-class problems finds an equal error solution (i.e., the probability of a feature vector being misclassified in category ω_1 is equal to that of being misclassified in category ω_2), a user might expect it also to seek an equal error solution. This, however, is not the case, regardless of any constraint on the magnitude of the vector $\mathbf{v}$. For example, the minimum $J(\mathbf{v})$ on the perimeter of the surface depicted in Figure 4.2 does not define the $\mathbf{v}$ which gives an equal error solution. This can be demonstrated in general by first noting that the component of the gradient parallel to $\mathbf{v}$ is given by

$$-\frac{E(\mathbf{v}^T\mathbf{Z}\,|\,\mathbf{v})}{\|\mathbf{v}\|^2}\,\mathbf{v}. \tag{4.56}$$

Therefore, corrections normal to $\mathbf{v}$ based on the gradient descent approach will cease when

$$E(\mathbf{Z}\,|\,\mathbf{v}) - \frac{E(\mathbf{v}^T\mathbf{Z}\,|\,\mathbf{v})}{\|\mathbf{v}\|^2}\,\mathbf{v} = \mathbf{0}, \tag{4.57}$$

which can be written as

$$E\left(\mathbf{Z} - \frac{\mathbf{v}^T\mathbf{Z}}{\|\mathbf{v}\|^2}\,\mathbf{v}\,\middle|\,\mathbf{v}\right) = \mathbf{0}. \tag{4.58}$$

This equation therefore describes the properties of a solution sought by the proportional increment procedure whether or not the magnitude $\|\mathbf{v}\|$ is kept constant. By reference to the techniques leading to Equation (4.7), Equation (4.58) can be expanded to give

$$P(\omega_2, \mathbf{Y} \in \mathscr{R}_1)E\left(\mathbf{Y} - \frac{\mathbf{v}^T\mathbf{Y}}{\|\mathbf{v}\|^2}\,\mathbf{v}\,\middle|\,\omega_2, \mathbf{Y} \in \mathscr{R}_1\right) = P(\omega_1\,|\,\mathbf{Y} \in \mathscr{R}_2)E\left(\mathbf{Y} - \frac{\mathbf{v}^T\mathbf{Y}}{\|\mathbf{v}\|^2}\,\mathbf{v}\,\middle|\,\omega_1, \mathbf{Y} \in \mathscr{R}_2\right). \tag{4.59}$$

The zeroth component of this vector equation expresses the conditions on the error tail areas. Recall that $Y_0 = 1$ for all $\mathbf{Y}$. The zeroth component of (4.59) can therefore be written as

$$P(\omega_2, \mathbf{Y} \in \mathscr{R}_1)\left[1 - E\left(\frac{\mathbf{v}^T\mathbf{Y}}{\|\mathbf{v}\|}\,\middle|\,\omega_2, \mathbf{Y} \in \mathscr{R}_1\right)\frac{w_0}{\|\mathbf{v}\|}\right] = P(\omega_1, \mathbf{Y} \in \mathscr{R}_2)\left[1 - E\left(\frac{\mathbf{v}^T\mathbf{Y}}{\|\mathbf{v}\|}\,\middle|\,\omega_1, \mathbf{Y} \in \mathscr{R}_2\right)\frac{w_0}{\|\mathbf{v}\|}\right]. \tag{4.60}$$

Note that $P(\omega_2, \mathbf{Y} \in \mathscr{R}_1) = P(\omega_1, \mathbf{Y} \in \mathscr{R}_2)$ (the equal error condition) only in either the very special case where $w_0 = 0$ or in the linearly separable case where

$$E\left(\frac{\mathbf{v}^T\mathbf{Y}}{\|\mathbf{v}\|}\,\middle|\,\omega_2, \mathbf{Y} \in \mathscr{R}_1\right) = E\left(\frac{\mathbf{v}^T\mathbf{Y}}{\|\mathbf{v}\|}\,\middle|\,\omega_1, \mathbf{Y} \in \mathscr{R}_2\right) = 0. \tag{4.61}$$

Therefore, this training procedure does not in general seek an equal error solution. Furthermore, the difference from an equal error solution becomes large for large values of the w_0 component of the solution vector $\mathbf{v}$ compared to $\|\mathbf{v}\|$ (i.e., optimum decision hyperplanes located far from the origin).

A loss function for two-class problems which preserves the intent of the proportional increment training procedure, including the equal error solution, while eliminating the problem of the $\|\mathbf{w}\|$ factor is, from (4.8),

$$J(\mathbf{v}) = M_1(\mathbf{v}) + M_2(\mathbf{v}), \tag{4.62}$$

where $M_i(\mathbf{v})$, $i = 1, 2$, denotes the *magnitude* of the first moment of the $f_i(\mathbf{x})$ error tail about the decision hyperplane. Equivalently, from Equation (4.11), this can be written

$$J(\mathbf{v}) = P(\text{error}\,|\,\mathbf{v})E(\delta\,|\,\mathbf{v}) \tag{4.63}$$

where δ denotes the distance of a misclassified feature vector from the decision hyperplane described by $\mathbf{v}$. Analytically, as determined from (4.2), this is expressed as

$$\begin{aligned} J(\mathbf{v}) &= E\left(\frac{-\mathbf{v}^T\mathbf{Z}}{\|\mathbf{w}\|}\,\middle|\,\mathbf{v}\right) = \frac{1}{\|\mathbf{w}\|}E(-\mathbf{v}^T\mathbf{Z}\,|\,\mathbf{v}) \\ &= \frac{1}{\|\mathbf{w}\|}\left[\int_{\mathscr{R}_2}(-\mathbf{v}^T\mathbf{y})f_1(\mathbf{x})\,d\mathbf{x} + \int_{\mathscr{R}_1}\mathbf{v}^T\mathbf{y}f_2(\mathbf{x})\,d\mathbf{x}\right], \end{aligned} \tag{4.64}$$

where $\mathbf{Z}$ is given by Equation (4.2). For the Gaussian distributions illustrated in Figure 4.1, $J(\mathbf{v}) = M_1(\mathbf{v}) + M_2(\mathbf{v}) = P(\text{error}\,|\,\mathbf{v})E(\delta\,|\,\mathbf{v})$ is plotted in Figure 4.9. Note that the minimum no longer occurs at $\|\mathbf{v}\| = 0$. Also note that $J(\mathbf{v})$ increases rapidly as the direction of $\mathbf{v}$ moves away from that of $\mathbf{v}^*$. This is true not only in this example but also in the general case. Indeed $J(\mathbf{v}) \to \infty$

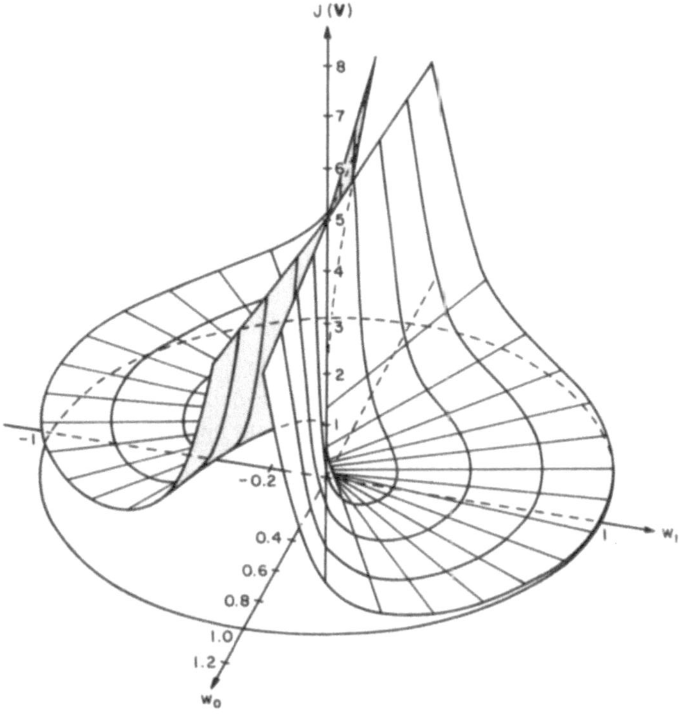

Figure 4.9. The loss function $[M_1(\mathbf{v}) + M_2(\mathbf{v})]$ for the example of Figure 4.1.

as $w_0/\|\mathbf{w}\| \to \infty$. This is simply due to the error moments about the decision hyperplane becoming large as the hyperplane translates away from its optimum position. In the extreme, the moment arm approaches an infinite length as $w_0/\|\mathbf{w}\| \to \infty$, i.e., the hyperplane is translated an infinite distance. This has the advantage of giving a large gradient for positions of the hyperplane far from its optimum. However, an initial vector $\mathbf{v}(0)$ chosen in the wrong half-space will never allow the optimum solution to be found when a true gradient descent is used.

To find the expression for the gradient of $J(\mathbf{v})$, note that

$$\begin{aligned} \nabla J(\mathbf{v}) &= \nabla E\left(-\frac{\mathbf{v}^T\mathbf{Z}}{\|\mathbf{w}\|}\middle|\mathbf{v}\right) \\ &= \frac{1}{\|\mathbf{w}\|}\nabla E(-\mathbf{v}^T\mathbf{Z}|\mathbf{v}) + E(-\mathbf{v}^T\mathbf{Z}|\mathbf{v})\nabla\left(\frac{1}{\|\mathbf{w}\|}\right). \end{aligned} \tag{4.65}$$

Recalling the discussion of Equation 4.3, we note that

$$\nabla E(-\mathbf{v}^T\mathbf{Z}|\mathbf{v}) = -E(\mathbf{Z}|\mathbf{v}).$$

Also,

$$\nabla\left(\frac{1}{\|\mathbf{w}\|}\right) = -\frac{1}{\|\mathbf{w}\|^3}\begin{bmatrix} 0 \\ \hdashline \mathbf{w} \end{bmatrix}$$

Hence Equation (4.65) reduces to

$$\nabla J(\mathbf{v}) = -\frac{1}{\|\mathbf{w}\|} E\left[\mathbf{Z} - \frac{\mathbf{v}^T\mathbf{Z}}{\|\mathbf{w}\|^2}\begin{bmatrix} 0 \\ \hdashline \mathbf{w} \end{bmatrix}\middle|\mathbf{v}\right]. \tag{4.66}$$

By reference to the definition of $\mathbf{Z}$ given by Equation (4.2), Equation (4.66) may be expanded as follows:

$$\begin{aligned} \nabla J(\mathbf{v}) = -\frac{1}{\|\mathbf{w}\|}\Bigg\{&\int_{\mathscr{R}_1}\left(\mathbf{y} - \frac{\mathbf{v}^T\mathbf{y}}{\|\mathbf{w}\|^2}\begin{bmatrix} 0 \\ \hdashline \mathbf{w} \end{bmatrix}\right) f_2(\mathbf{x})\,d\mathbf{x} \\ &-\int_{\mathscr{R}_2}\left(\mathbf{y} - \frac{\mathbf{v}^T\mathbf{y}}{\|\mathbf{w}\|^2}\begin{bmatrix} 0 \\ \hdashline \mathbf{w} \end{bmatrix}\right) f_1(\mathbf{x})\,d\mathbf{x}\Bigg\}. \end{aligned} \tag{4.67}$$

Recall by reference to Figure 1.7 that $\mathbf{v}^T\mathbf{y}/\|\mathbf{w}\|$ is the signed distance in nonaugmented feature space from the decision hyperplane to the feature vector $\mathbf{x}$. This distance is positive for $\mathbf{x} \in \mathscr{R}_2$ and negative for $\mathbf{x} \in \mathscr{R}_1$. Therefore the vector

$$\frac{\mathbf{v}^T\mathbf{y}}{\|\mathbf{w}\|^2}\begin{pmatrix} 0 \\ \hdashline \mathbf{w} \end{pmatrix}$$

represents a vector which lies in nonaugmented feature space, begins at and is orthogonal to the decision hyperplane in nonaugmented feature space, and terminates at the feature vector $\mathbf{x}$.

$$\mathbf{y} - \frac{\mathbf{v}^T\mathbf{y}}{\|\mathbf{w}\|^2}\begin{bmatrix} 0 \\ \hdashline \mathbf{w} \end{bmatrix}$$

is therefore an orthogonal projection in $\mathbf{x}$-space of any augmented feature vector $\mathbf{y}$ onto the hyperplane $\mathbf{v}^T\mathbf{y} = 0$. Although the projection is orthogonal to the decision hyperplane in nonaugmented feature space, it is not orthogonal in augmented feature space (i.e., $\mathbf{y}$ space) since the zeroth component of $\mathbf{y}$, namely y_0, is maintained at a constant value of 1. Note that $\nabla J(\mathbf{v})$ is always normal to $\mathbf{v}$, since

$$\mathbf{y} - \frac{\mathbf{v}^T\mathbf{y}}{\|\mathbf{w}\|^2}\begin{bmatrix} 0 \\ \cdots \\ \mathbf{w} \end{bmatrix}$$

lies on the hyperplane $\mathbf{v}^T\mathbf{y} = 0$.

Recall that conditions 1 and 2 of Section 2.11 are necessary and sufficient conditions for the existence of a region in $\mathbf{v}$-space within which there is precisely one connected region $\mathscr{V}^*$ where $J(\mathbf{v})$ has a local minimum. Condition 1 requires the gradient to be zero in $\mathscr{V}^*$. Therefore the first component of Equation (4.67) must satisfy

$$\int_{\mathscr{R}_1} f_2(\mathbf{x})\,d\mathbf{x} = \int_{\mathscr{R}_2} f_1(\mathbf{x})\,d\mathbf{x} \tag{4.68}$$

when $\mathbf{v} \in \mathscr{V}^*$. This states that the $f_1(\mathbf{x})$ and $f_2(\mathbf{x})$ error tails must be equal when $J(\mathbf{v})$ is a minimum. The remaining components of Equation (4.67), while satisfying condition 1, can be written

$$\int_{\mathscr{R}_1}\left[\mathbf{x} - \frac{\mathbf{v}^T\mathbf{y}}{\|\mathbf{w}\|^2}\mathbf{w}\right] f_2(\mathbf{x})\,d\mathbf{x} = \int_{\mathscr{R}_2}\left[\mathbf{x} - \frac{\mathbf{v}^T\mathbf{y}}{\|\mathbf{w}\|^2}\mathbf{w}\right] f_1(\mathbf{x})\,d\mathbf{x} \tag{4.69}$$

when $\mathbf{v} \in \mathscr{V}^*$. This states that the moments about each coordinate axis of the normal projection onto the optimum decision hyperplane in nonaugmented feature space of the error portion of the constituent densities must be equal. These requirements are similar to those described by the discussion following expressions (4.30) and (4.31) for the probability of error loss function.

An illustration of contours of constituent densities in two-dimensional feature space is given in Figure 4.10, where the decision hyperplanes labeled A, B, and C all satisfy the equal classification error criterion described by Equation (4.68). The small arrows on the ends of each hyperplane are directed into the half-space in which the choice of category ω_2 is made. It is, however, only hyperplane C which also satisfies the equal moment criteria of Equation (4.69). Note that since both (4.68) and (4.69) must be satisfied simultaneously for a desired $\mathbf{v} = \mathbf{v}^*$, the projections onto the hyperplane of the centroids of the error tails of $f_1(\mathbf{x})$ and $f_2(\mathbf{x})$ must coincide.

From our earlier discussion, we see that it is important that the initial vector $\mathbf{v}(0)$ be chosen in the correct half-space. Generally a reasonably good guess can be made, based on some best understood feature, say x_k, of whether the best assignment of class ω_2 by a linear classifier is made for feature samples which occur in the direction of increasing x_k or decreasing x_k from the ω_1 feature region. As discussed in Section 4.4, if the guess is correct, $v_k^* > 0$ for the increasing x_k case or $v_k^* < 0$ for the decreasing x_k

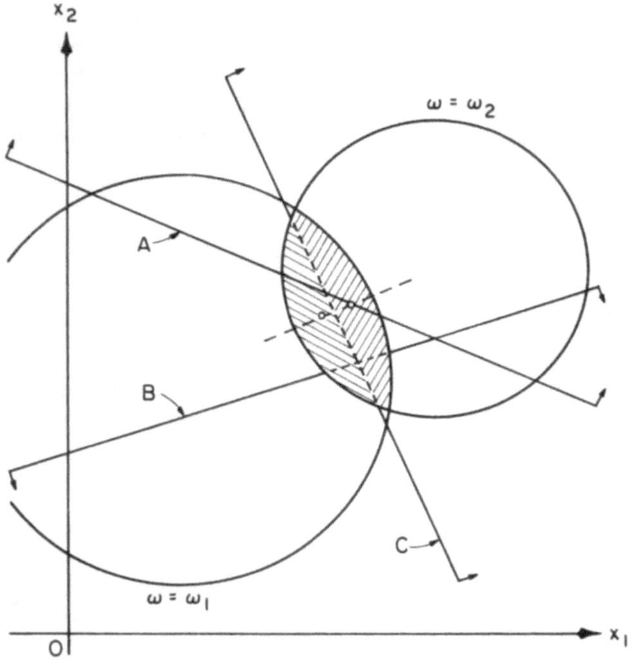

Figure 4.10. Example of hyperplanes which satisfy the equal error criterion.

case. As demonstrated in that section, we can minimize $J(\mathbf{v})$ by holding some $v_k = \pm$const., based on our guess that $v_k^* \gtrless 0$, while using the constrained gradient $\nabla J(\mathbf{v})|_{v_k=\text{const.}}$. Fortunately, if our confidence in the guess is not high, the gradient procedure can be repeated with the opposite sign of v_k chosen. Based on the training feature vectors, the performance of the two resulting linear classifiers can then be compared and the one which gives the least error rate chosen. From Equation (4.64), the constrained gradient is found to be

$$\nabla J(\mathbf{v})|_{v_k=\text{const.}} = -\frac{1}{\|\mathbf{w}\|} E\left(\tilde{\mathbf{Z}} - \frac{\mathbf{v}^T\mathbf{Z}}{\|\mathbf{w}\|^2}\begin{bmatrix} 0 \\ \cdots \\ \tilde{\mathbf{w}} \end{bmatrix}\right) \tag{4.70}$$

where

$$\tilde{\mathbf{Z}} = \begin{bmatrix} Z_0 \\ Z_1 \\ \vdots \\ Z_{k-1} \\ 0 \\ Z_{k+1} \\ \vdots \\ Z_d \end{bmatrix}, \qquad k \neq 0, \tag{4.71}$$

i.e., the vector $\mathbf{Z}$ given by Equation (4.2) with Z_k equal to zero. The vector $\tilde{\mathbf{w}}$ is similarly defined as $\mathbf{w}$ with w_k equal to zero.

This gradient will be utilized in the equalized error training procedure discussed in Chapter 5.

4.6 Loss Functions Compared

We shall now compare the decision hyperplanes which are specified by the minima of the loss functions. This comparison will be made for two examples. The constituent densities for both of these examples will be bivariate Gaussian in form.

Figure 4.11 depicts the contours of the constituent densities $f_1(\mathbf{x})$ and $f_2(\mathbf{x})$ for the first example. The contours illustrated are for values of $\mathbf{x}$ where the magnitudes of the densities $f_1(\mathbf{x})$ and $f_2(\mathbf{x})$ are equal and where the boundary of the union of these contours represents an approximate 80 percent probability boundary. The a priori probabilities are $P(\omega_1) = \frac{1}{4}$ and $P(\omega_2) = \frac{3}{4}$. The means are located at $\boldsymbol{\mu}_1 = [-0.5/\sqrt{2}, -0.5/\sqrt{2}]^T$ and $\boldsymbol{\mu}_2 = [1.5/\sqrt{2}, 1.5/\sqrt{2}]^T$. (Thus $\|\boldsymbol{\mu}_2 - \boldsymbol{\mu}_1\| = 2$). The covariance matrix of both

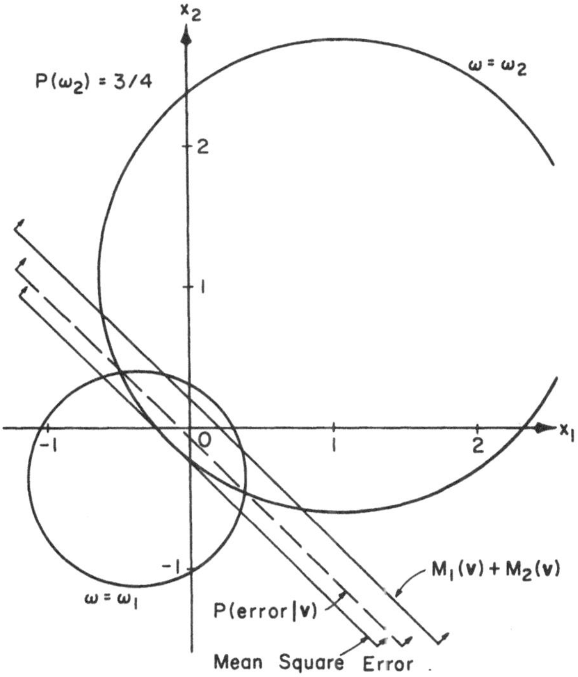

Figure 4.11. Decision hyperplanes corresponding to the minimum of the labeled loss functions for an example where the class probabilities differ but the variances are the same.

class densities is

$$\Sigma_1 = \Sigma_2 = \begin{bmatrix} 1 & 0 \\ 0 & 1 \end{bmatrix}.$$

The decision hyperplanes illustrated in Figure 4.11 are specified by a $\mathbf{v}^*$ corresponding to the minimum of each of the three loss functions discussed in Section 4.5. Each hyperplane has been labeled by its associated loss function. The probability of error associated with each of these loss functions is: for minimum $P(\text{error}|\mathbf{v})$, $P(\text{error}) = 0.128$; for minimum mean square error, $P(\text{error}) = 0.129$; for minimum error tail moments, $P(\text{error}) = 0.131$. Note that each of these hyperplanes provide a reasonably good decision surface.

Figure 4.12 depicts contours of the constituent densities for the second example. Again the densities are Gaussian. The a priori class probabilities are $P(\omega_1) = \frac{1}{3}$ and $P(\omega_2) = \frac{2}{3}$. The mean of the density of class ω_1 is at the origin while the mean of the density of class ω_2 is $\boldsymbol{\mu}_2 = [6/\sqrt{2}, 6/\sqrt{2}]^T$. The covariance matrices of the class densities are

$$\Sigma_1 = \begin{bmatrix} 4 & 0 \\ 0 & 4 \end{bmatrix} \quad \text{and} \quad \Sigma_2 = \begin{bmatrix} 1 & 0 \\ 0 & 1 \end{bmatrix}.$$

Decision hyperplanes corresponding to respective loss function minima are also illustrated in Figure 4.12. The probability of error associated with each

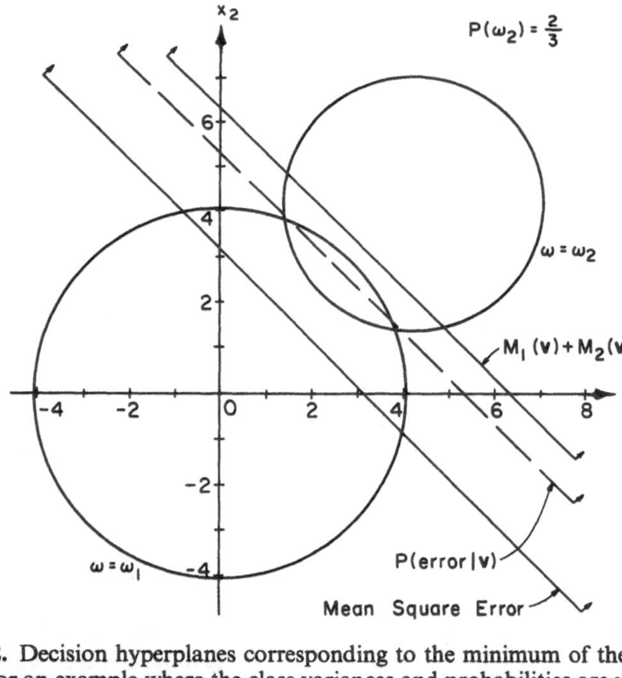

Figure 4.12. Decision hyperplanes corresponding to the minimum of the labeled loss functions for an example where the class variances and probabilities are unequal.

hyperplane is: for minimum $P(\text{error}\,|\,\mathbf{v})$, $P(\text{error}) = 0.066$; for minimum error tail moments, $P(\text{error}) = 0.088$; for minimum mean square error, $P(\text{error}) = 0.096$.

4.7 Unequal Costs of Category Decisions

Often certain types of misclassification are more critical than some others. Therefore it is desirable to establish within the loss functions the costs of the types of misclassification. We will assume that the cost of correct classification is zero. For the misclassified cases, let c_{ij} denote the cost of choosing class ω_i when the correct class is ω_j. Then, for two-class problems, analogously to Equation (4.26), the expected cost associated with a decision hyperplane described by $\mathbf{v}$ is

$$E(\text{cost}\,|\,\mathbf{v}) = c_{12} \int_{\mathscr{R}_1} f_2(\mathbf{x})\, d\mathbf{x} + c_{21} \int_{\mathscr{R}_2} f_1(\mathbf{x})\, d\mathbf{x}. \tag{4.72}$$

Since $f_j(\mathbf{x}) = P(\omega_j)p(\mathbf{x}\,|\,\omega_j)$, Equation (4.72) can be written as

$$E(\text{cost}\,|\,\mathbf{v}) = \int_{\mathscr{R}_1} c_{12}P(\omega_2)p(\mathbf{x}\,|\,\omega_2)\, d\mathbf{x} + \int_{\mathscr{R}_2} c_{21}P(\omega_1)p(\mathbf{x}\,|\,\omega_1)\, d\mathbf{x}. \tag{4.73}$$

It is possible to obtain an expression proportional to the right-hand side of Equation (4.73) by replacing the $c_{ij}P(\omega_j)$ terms by terms having the properties of a priori probabilities. To do this we find a positive constant c such that

$$\frac{c_{12}P(\omega_2)}{c} + \frac{c_{21}P(\omega_1)}{c} = 1.$$

This yields $c = c_{12}P(\omega_2) + c_{21}P(\omega_1)$. (Note that c denotes the expected cost of worst possible misclassification—i.e., the cost for the case where every feature vector is misclassified.)

Define

$$P_{\text{eq}}(\omega_j) = \frac{c_{ij}}{c_{12}P(\omega_2) + c_{21}P(\omega_1)} P(\omega_j), \qquad i, j = 1, 2;\ i \neq j, \tag{4.74}$$

Replacing $c_{ij}P(\omega_j)$ in Equation (4.73) by $cP_{\text{eq}}(\omega_j)$ yields

$$\frac{E(\text{cost}\,|\,\mathbf{v})}{c} = \int_{\mathscr{R}_1} P_{eq}(\omega_2)p(\mathbf{x}|\omega_2)\, d\mathbf{x} + \int_{\mathscr{R}_2} P_{eq}(\omega_1)p(\mathbf{x}|\omega_1) d\mathbf{x} \tag{4.75}$$

This result can also be obtained by replacing $f_j(\mathbf{x})$ in the expression for $P(\text{error}|\mathbf{v})$—see Equation (4.26)—by $P_{\text{eq}}(\omega_j)p(\mathbf{x}|\omega_j)$. Therefore the relative costs can be accounted for by scaling up one constituent density and correspondingly scaling the other one down. Note that the constituent density whose class is associated with the greater cost of error is the one which is scaled up.

Even though only one of the loss functions discussed in Section 4.5 has as its minimum the minimum probability of error for all problems, the goal in establishing each of the loss functions was to have their minima yield error probabilities that are equal to or close to the minimum probability of error for a large class of problems. Therefore, if instead the goal is determining loss functions whose minima yield error costs that are close to the least expected cost of misclassification, the constituent densities $f_j(\mathbf{x})$ in each of the loss functions of Section 4.5 should be replaced by $P_{\text{eq}}(\omega_j)p(\mathbf{x}|\omega_j)$. This replacement must then also be made in the gradient expressions which are utilized for the gradient descent technique.

Note that for the case of equal costs, i.e., $c_{12} = c_{21}$, Equation (4.74) gives

$$P_{\text{eq}}(\omega_j) = \frac{c_{ij}}{c_{ij}[P(\omega_1) + P(\omega_2)]} P(\omega_j) = P(\omega_j).$$

Therefore, in the equal cost case, the gradient descent techniques which tend to seek the least cost decision hyperplane reduce to those which tend to seek the minimum probability of error hyperplane.

4.8 Stochastic Approximation

4.8.1 Stochastic Optimization Techniques

Let us reexamine the sample gradient technique of Section 4.3. N denotes the number of feature vectors in the training set. Since N is finite, $\hat{J}(\mathbf{v})$ in Equation (4.17) and the approximation $\widehat{\nabla J}(\mathbf{v})$ in Equation (4.21) are random variables. The training procedures presented in Chapter 5, which are based on the loss functions described in the preceding sections, often determine the motion from point $\mathbf{v}$ on the basis of a single feature vector in the training set. Therefore practical procedures are not truly gradient descent procedures, but only probabilistic approximations, where the approximation improves as N increases. Interestingly, the specific procedures presented in the next chapter place a restriction on the magnitude of the steps. These procedures, then, together with certain constraints, cause the weight vector to migrate, on the average, in the gradient descent direction. The technique which formally describes and places constraints on these procedures is known as the *stochastic approximation* method. The following is a brief description of how stochastic approximation may be used for the design of weight vectors for linear classifiers.

Let $\mathbf{G}(\mathbf{v})$ be a $(d + 1)$-dimensional vector valued function which is expressable in the form $\mathbf{G}(\mathbf{v}) = E(\mathbf{Z}(n)|\mathbf{V}(n) = \mathbf{v})$, where $\mathbf{Z}(n)$ is a function of a $(d + 1)$-dimensional parameter vector $\mathbf{V}(n)$ and a random vector which we denote by $\mathbf{X}(n)$. If $\mathbf{G}(\mathbf{v})$ can be described in this way, it is called a *vector regression function.* As an example, $\mathbf{G}(\mathbf{v})$ might be equal to the negative of the gradient of a loss function, i.e., $\mathbf{G}(\mathbf{v}) = -\nabla J(\mathbf{v})$. In linear classifiers, the random variable $\mathbf{Z}(n)$ is a function of the random feature vector $\mathbf{X}(n)$, the class $\omega(n)$ to which $\mathbf{X}(n)$ belongs, and the value of the augmented weight vector $\mathbf{V}(n)$ describing the decision hyperplane at the nth trial. Suppose that all components of $\mathbf{G}(\mathbf{v})$ are zero (i.e., $\mathbf{G}(\mathbf{v}) = \mathbf{0} = [0, 0, \ldots, 0]^T$) only for v's which are elements of the solution set $\mathscr{V}^*$. Then, under some reasonable constraints on $\mathbf{G}(\mathbf{v})$, and some fairly weak constraints on $\mathbf{Z}$ and $\mathbf{V}(n)$, an algorithm (i.e., a training procedure) can be found such that the sequence of vectors $\{\mathbf{V}(n)\}$ converges on the average to a member in the "solution" set $\mathscr{V}^*$.

Figure 4.13 illustrates a surface $G_i(\mathbf{v})$ which represents an example of the ith component of a vector regression function $\mathbf{G}(\mathbf{v})$ in two of the coordinates of augmented weight space. The region where the ith component of $\mathbf{G}(\mathbf{v})$ is zero is shown hatched. The set intersection of this shaded region with similarly defined regions for the remaining components of $\mathbf{G}(\mathbf{v})$, i.e., the region of $\mathbf{v}$ where $\mathbf{G}(\mathbf{v}) = \mathbf{0}$, determines the region $\mathscr{V}^*$. The average value of Z_i for any $\mathbf{v}$ is equal to $G_i(\mathbf{v})$, i.e., $G_i(\mathbf{v}) = E\{Z_i|\mathbf{v}\}$ is a regression function. If $\mathbf{G}(\mathbf{v})$ represents the negative of the gradient of a loss function, then a randomly selected $\mathbf{Z}$ at a point $\mathbf{v}$ can be thought of as a noisy approximation to the negative of the gradient. Suppose $\mathbf{G}(\mathbf{v}) = -\nabla J(\mathbf{v})$ could be utilized in a gradient descent procedure which seeks the optimum region $\mathscr{V}^*$. Then if $\mathbf{G}(\mathbf{v})$ is replaced by its noisy or stochastic approximation $\mathbf{Z}$, and if the the required constraints are met, the procedure will still seek the region $\mathscr{V}^*$, although along a more erratic path. The constraints on $G(\mathbf{v})$, as in the gradient descent procedures, must ensure that the training procedure will always make corrections to $\mathbf{V}(n)$ so that on the average the new value $\mathbf{V}(n+1)$ will move a sufficient distance toward $\mathscr{V}^*$ to ensure stochastic convergence to $\mathscr{V}^*$ as $n \to \infty$. The constraints on $\mathbf{Z}$ not only require that its average value equal $\mathbf{G}(\mathbf{v})$ but also that its variance is sufficiently bounded so that random fluctuations will not hinder convergence.

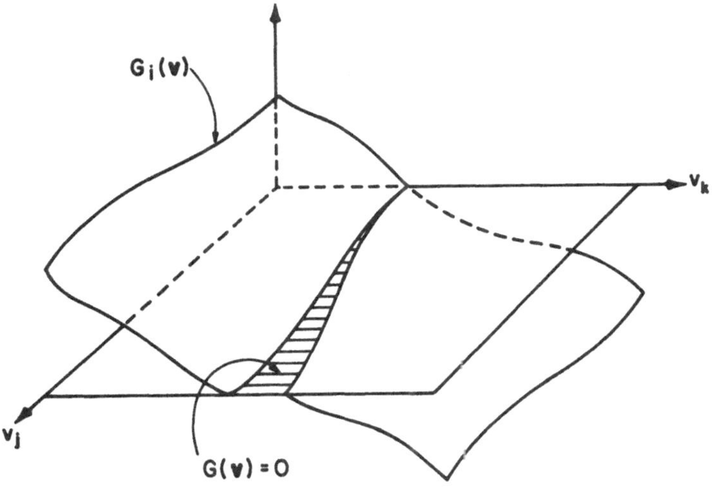

Figure 4.13. An example of the ith component of a vector regression function $\mathbf{G}(\mathbf{v})$ in two-dimensional weight space.

EXAMPLE 4.1. As an illustrative example of an application of stochastic approximation, we consider the classical problem of fitting a curve, in a least mean square (LMS) error sense, to a set of sample points. (Note that this is *not* a classification problem, so that we are neither concerned with classification regions $\mathscr{R}_j$ nor with classes ω_j.) In the classical solution, all the sample points are considered together in a nonrecursive manner from which some best fitting curve is established. (For example, see Papoulis, Section 7.4 [5].) In the stochastic approximation solution, we shall assume an initial curve and then recursively change the curve on the basis of each selected sample point. This latter method will approach the best curve stochastically.

For simplicity, we restrict the form of the curve to a straight line passing through the origin. The generalized equation of this line is

$$x_2 = vx_1.$$

Such a straight line specified by some arbitrarily chosen $v(0)$ is illustrated in Figure 4.14. Note that the parameter vector $\mathbf{v}$ is simply the scalar v for this example. The set of sample points $\{\mathbf{X}(i)\}$ to which we wish to fit such a line is also illustrated. (We represent each point by its position vector.)

The error ε_i for sample point $\mathbf{X}(i)$ is defined as

$$\varepsilon_i = |X_2(i) - vX_1(i)|$$

where $\mathbf{X}(i) = [X_1(i), X_2(i)]^T$. For least mean square curve fitting, we wish to find the v which minimizes the average of the square of the ε_i errors. Ideally we would like to minimize a function $J(v)$ which is proportional to the average of the square errors over the entire probability distribution of $\mathbf{X}$. Such a $J(v)$ is given by

$$J(v) = \tfrac{1}{2}E[(X_2 - VX_1)^2 \,|\, V = v]$$

where $\mathbf{X} = [X_1, X_2]^T$ is a vector random variable. We now define a function $G(v)$ by

$$G(v) = -\nabla J(v)$$

so that

$$G(v) = E[(X_2 - VX_1)X_1 \,|\, V = v].$$

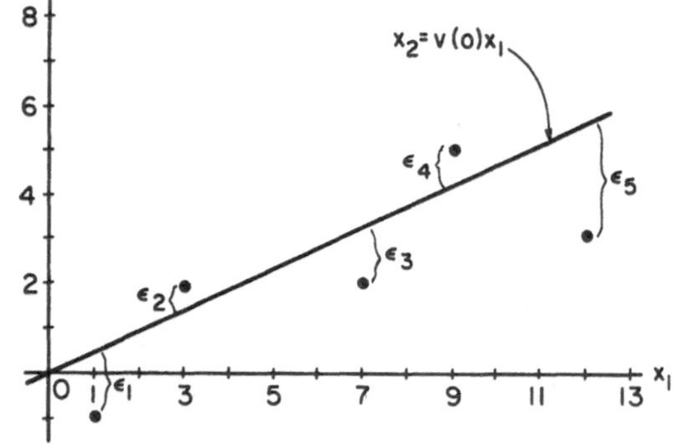

Figure 4.14. Set of sample points $\{\mathbf{x}(i)\}$ and an arbitrary initial line described by $V(0)$ for Example 4.1.

Let us now define a random variable Z by

$$Z = (X_2 - VX_1)X_1$$

and note that $G(v)$ can now be expressed as

$$G(v) = E(Z \mid V = v).$$

Therefore $G(v)$ is a regression function. (In this example $G(v)$ is a scalar rather than a vector regression function.)

The set of sample points illustrated in Figure 4.14 determines a different set of values of Z for each different value of v. Five such sets are illustrated in Figure 4.15 for $v = 0$, 0.2, 0.4, 0.6, and 0.8. $G(v)$ is also illustrated, where it has been assumed that the average of the given sample of Z's for a given v is equal to $E(Z \mid V = v)$. Notice that $G(v) = 0$ at $v = 0.352$. Since $G(v) = -\nabla J(v)$, $v = 0.352$ is the solution point v^*. (Note the analogy between Figures 4.15 and 4.13.)

Since $G(v)$ is a regression function, its noisy approximation Z can be used in a stochastic approximation algorithm such as

$$V(n + 1) = V(n) + \rho(n)Z(n).$$

As each successive $\mathbf{X}(n)$ is tried its associated $Z(n) = [X_2(n) - V(n)X_1(n)]X_1(n)$ is calculated and applied to the algorithm. If the form of the step size $\rho(n)$ is properly chosen, $V(n)$ will converge probabilistically to v^*. Following the next section in which the

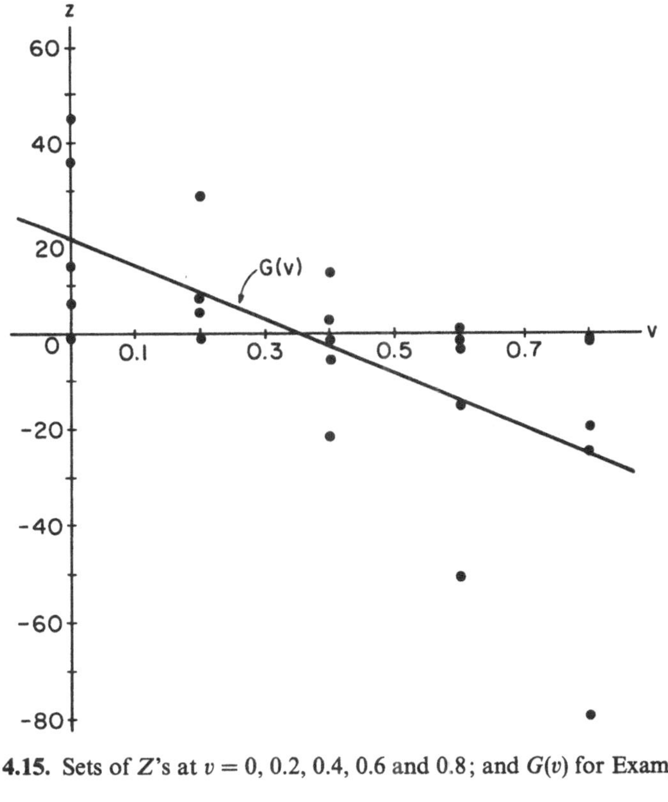

Figure 4.15. Sets of Z's at $v = 0$, 0.2, 0.4, 0.6 and 0.8; and $G(v)$ for Example 4.1.

stochastic approximation theorem and its constraints are formally presented, this example will be continued and an appropriate $\rho(n)$ will be established.

4.8.2 The Stochastic Approximation Theorem

The stochastic approximation technique was devised and described by Robbins and Monro in 1951 [6], extended to minimizing a regression function by Kiefer and Wolfowitz [7], extended to multidimensions by Blum [8], and further generalized in a major paper by Dvoretzky [9].

The Robbins–Monro technique as extended by Blum finds the root, or roots, of a vector regression function $\mathbf{G}(\mathbf{v}) = E(\mathbf{Z}|\mathbf{v})$ by the use of the procedure

$$\mathbf{v}(n+1) = \mathbf{v}(n) + \rho(n)\mathbf{Z}(n). \tag{4.76}$$

The multidimensional extension of Dvoretzky's generalization of Equation (4.76) is

$$\mathbf{v}(n+1) = \mathbf{t}[\mathbf{v}(0), \ldots, \mathbf{v}(n)] + \mathbf{R}(n), \tag{4.77}$$

where $\mathbf{t}[\mathbf{v}(0), \ldots, \mathbf{v}(n)]$ is a deterministic transformation and $\mathbf{R}(n)$ is an unbiased random variable, i.e., $E[\mathbf{R}(n)|\mathbf{v}(n)] = \mathbf{0}$. As an example of the application of this generalization, it might be desirable to modify the recursion procedure given by (4.76) so that the magnitude of $\mathbf{v}$ is maintained at unity. The following is an example of such a modified recursion procedure:

$$\mathbf{v}(n+1) = \frac{\mathbf{v}(n) + \rho(n)\mathbf{Z}(n)}{\|\mathbf{v}(n) + \rho(n)\mathbf{Z}(n)\|}. \tag{4.78}$$

The Robbins–Monro theory is not applicable to the analysis of this process, but Dvoretzky's theory is. For this example, Dvoretzky's transformation $\mathbf{t}$ is given by

$$\mathbf{t}[\mathbf{v}(0), \ldots, \mathbf{v}(n)] \equiv \mathbf{t}[\mathbf{v}(n)] \equiv E\left\{\frac{\mathbf{v}(n) + \rho(n)\mathbf{Z}(n)}{\|\mathbf{v}(n) + \rho(n)\mathbf{Z}(n)\|}\,\middle|\,\mathbf{v}(n)\right\}$$

and

$$\mathbf{R}(n) \equiv \frac{\mathbf{v}(n) + \rho(n)\mathbf{Z}(n)}{\|\mathbf{v}(n) + \rho(n)\mathbf{Z}(n)\|} - \mathbf{t}[\mathbf{v}(n)].$$

Dvoretzky's Theorem (in multidimensional form). *Let* α_n, β_n, *and* γ_n, $n = 0, 1, 2, \ldots$, *be nonnegative real numbers satisfying:*

1. $$\alpha_n \to 0$$

2. $$\sum_{n=0}^{\infty} \beta_n < \infty$$

3. $$\sum_{n=0}^{\infty} \gamma_n = \infty.$$

Let $\mathbf{v}$ *be a real vector and let* $\mathbf{t}_n$ *denote measurable transformations satisfying*

4. $\|\mathbf{t}_n(\mathbf{v}(0), \ldots, \mathbf{v}(n)) - \mathbf{v}^*\| \leqslant \max\{\alpha_n, \quad (1 + \beta_n)\|\mathbf{v}(n) - \mathbf{v}^*\| - \gamma_n\}$

for all real $\mathbf{v}(0), \ldots, \mathbf{v}(n)$. *Let* $\mathbf{R}_n$ *be a vector random variable, and let*

5. $$\mathbf{V}(n+1) = \mathbf{t}_n[\mathbf{V}(0), \ldots, \mathbf{V}(n)] + \mathbf{R}_n, \quad \text{for } n > 0.$$

Then the conditions $E\|\mathbf{V}(0)\|^2 < \infty$,

6. $$\sum_{n=0}^{\infty} E\|\mathbf{R}_n\|^2 < \infty,$$

and

7. $$E(\mathbf{R}_n) = \mathbf{0}$$

with probability 1 *for all n, imply convergence in mean square:*

$$\lim_{n\to\infty} E\|\mathbf{V}(n) - \mathbf{v}^*\|^2 = 0,$$

and convergence with probability 1.

$$P\left(\lim_{n\to\infty} \mathbf{V}(n) = \mathbf{v}^*\right) = 1.$$

Dvoretzky's theorem both states the very general allowed training procedure and gives the bounds on the stochastic term $\mathbf{R}_n$ and deterministic coefficients α_n, β_n, and γ_n. Our interest is limited to those cases where $\mathbf{t}_n$ is a function of $\mathbf{v}(n)$ but not of $\mathbf{v}$'s from earlier trials, i.e.,

$$\mathbf{t}_n[\mathbf{v}(n), \ldots, \mathbf{v}(n)] = \mathbf{t}_n[\mathbf{v}(n)] = E[\mathbf{V}(n+1) \mid \mathbf{V}(n) = \mathbf{v}(n)].$$

Discussion of Dvoretzky's Conditions. From constraints 1, 3, and 4, the deterministic coefficients γ_n and α_n must have the following two properties. First, γ_n represents the minimum decrease in distance to a $\mathbf{v}^*$ resulting from an expected correction. For large n, γ_n must therefore be chosen so that for expected corrections which decrease the distance to a $\mathbf{v}^*$, $\|\mathbf{v}^* - \mathbf{V}(n+1)\|$ is on the average decreasing sufficiently so that $\mathbf{V}(n)$ does not converge to some point short of a vector $\mathbf{v}^*$. Second, α_n is concerned with ensuring that, even for an expected correction which increases the distance to a vector $\mathbf{v}^*$, this increased distance will approach zero as n increases. Figure 4.16 illustrates γ_n and α_n in three dimensional $\mathbf{v}$ space. For α_n the condition is given by

$$\|\mathbf{t}_n - \mathbf{v}^*\| \leqslant \alpha_n, \tag{4.79}$$

so that

$$\lim_{n\to\infty} \alpha_n = 0 \tag{4.80}$$

must hold. This is Dvoretzky's condition 1. The condition on γ_n is first described by

$$\|\mathbf{t}_n - \mathbf{v}^*\| \leqslant \|\mathbf{v}(n) - \mathbf{v}^*\| - \gamma_n, \tag{4.81}$$

for which

$$\sum_{n=0}^{\infty} \gamma_n = \infty \tag{4.82}$$

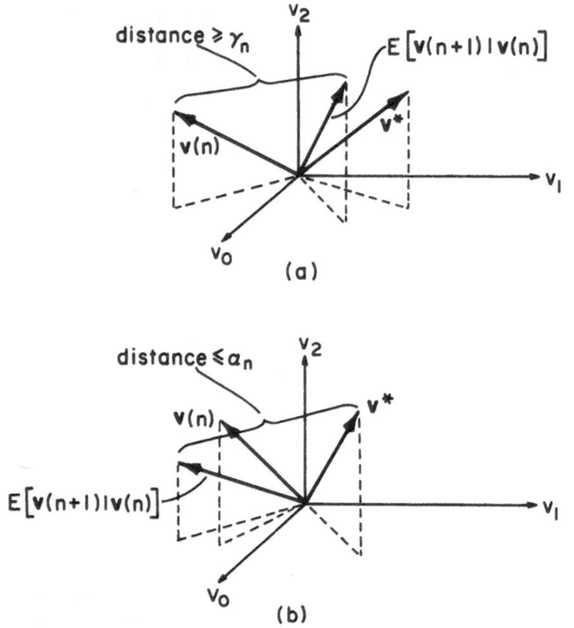

Figure 4.16. Illustration of (a) γ_n and (b) α_n.

must hold. This is Dvoretzky's condition 3. Actually, Dvoretzky allows a larger undershoot to occur in the early stages, as long as this tendency is overcome by the effect of γ_n. This generalization is obtained by the use of the parameter β_n, modifying Equation (4.81) to

$$\|\mathbf{t}_n - \mathbf{v}^*\| \leqslant (1 + \beta_n)\|\mathbf{v}(n) - \mathbf{v}^*\| - \gamma_n \tag{4.83}$$

where the constraint on β_n is

$$\sum_{n=0}^{\infty} \beta_n < \infty.$$

This is Dvoretzky's condition 2.

The conditions on the random term $\mathbf{R}_n$ are as follows. First condition: $\mathbf{R}_n$ must be unbiased, i.e.,

$$E(\mathbf{R}_n) = \mathbf{0}, \tag{4.84}$$

which means that the randomness must average to zero so that, on the average, the random errors tend to cancel themselves. This is the condition which has allowed us to write $\mathbf{t}_n = E[\mathbf{V}(n+1)|\mathbf{V}(n)]$. Second condition:

$$\sum_{n=0}^{\infty} E\|\mathbf{R}_n\|^2 < \infty, \tag{4.85}$$

which is Dvoretzky's condition 6. The following description of condition 6 is based upon a discussion by Chang [10]. For the condition of independent random errors $\mathbf{R}_n$, the variance of the sum of all the errors equals the sum of the average value of the square of the error magnitudes, i.e.,

$$E\left\|\sum_{n=0}^{\infty} \mathbf{R}_n\right\|^2 = \sum_{n=0}^{\infty} E\|\mathbf{R}_n\|^2. \tag{4.86}$$

But Equation (4.85) states that this sum must be finite. After a given number of trials, say m, the amount of variance left is

$$E\left\|\sum_{n=m+1}^{\infty} \mathbf{R}_n\right\|^2 = \sum_{n=0}^{\infty} E\|\mathbf{R}_n\|^2 - \sum_{n=0}^{m} E\|\mathbf{R}_n\|^2. \tag{4.87}$$

But as m increases, the second term on the right-hand side increases, since each of its sums is positive. Therefore the left-hand side must become indefinitely small as $m \to \infty$. This implies that, if condition 6 holds, the random fluctuations will tend to die out.

EXAMPLE 4.1: (cont.). Now let us apply Dvoretzky's theorem to Example 4.1 in order to find an appropriate form for $\rho(n)$ in the algorithm

$$V(n+1) = V(n) + \rho(n)Z(n)$$

and any required constraints on Z and V. For our example we will constrain $\beta_n = 0$ for all n, thereby satisfying Dvoretzky's condition 2.

Let

$$t_n = V(n) + \rho(n)G[V(n)] \tag{4.88}$$

and

$$R_n = \rho(n)\{Z(n) - G[V(n)]\}, \tag{4.89}$$

so that Dvoretzky's conditions 5 and 7 are satisfied by the algorithm. Next expand $|t_n - v^*|^2$ as follows:

$$|t_n - v^*|^2 = [V(n) - v^*]^2 - 2\rho(n)[v^* - V(n)]G[V(n)] + \rho^2(n)G^2[V(n)]. \tag{4.90}$$

We first consider the case where an expected correction *increases* the distance to v^*—i.e.,

$$|t_n - v^*| = |V(n) - v^*| + \varepsilon,$$

where ε is any positive real number. If $\rho(n) > 0$ and if $G[V(n)]$ has the same sign as $[v^* - V(n)]$, i.e., v^* is a local minimum of $J(v)$ in the region of all points $V(n)$, then Equation (4.90) can be weakened to

$$|t_n - v^*|^2 \leqslant [V(n) - v^*]^2 + \rho^2(n)G^2[V(n)].$$

This can be written

$$[|t_n - v^*| + |V(n) - v^*|][|t_n - v^*| - |V(n) - v^*|] \leqslant \rho^2(n)G^2[V(n)],$$

so that, for the case under discussion,

$$|t_n - v^*| \leqslant \frac{\rho^2(n)G^2[V(n)]}{\varepsilon} - |V(n) - v^*| \leqslant \rho^2(n)\frac{1}{\varepsilon}G^2[V(n)].$$

Now we select

$$\alpha_n = \rho^2(n) \frac{1}{\varepsilon} \sup_{V(n)} G^2[V(n)], \tag{4.91}$$

so that for this case Dvoretzky's condition 4 is satisfied. If $G^2[V(n)]$ is finite for all $V(n)$ then Equation (4.91) and condition 1 require that

$$\lim_{n\to\infty} \rho^2(n) = 0. \tag{4.92}$$

We next consider the case where an expected correction *decreases* the distance to v^*—i.e.,

$$|t_n - v^*| < |V(n) - v^*|.$$

By reference to Equation (4.90) we select a γ_n which satisfies the second part of Dvoretzky's condition 4. One such γ_n is

$$\gamma_n = \frac{1}{|V(n) - v^*|} \{\rho(n) \inf_\varepsilon([v^* - V(n)]G[V(n)]) - \tfrac{1}{2}\rho^2(n) \sup_{V(n)} G^2[V(n)]\} \tag{4.93}$$

where $0 < \varepsilon = |v^* - V(n)|$. This follows because

$$\begin{aligned}
[|V(n) - v^*| - \gamma_n]^2 &\geqslant |V(n) - v^*|^2 - 2|V(n) - v^*|\gamma_n \\
&\geqslant |V(n) - v^*|^2 - 2\rho(n)\inf_\varepsilon([v^* - V(n)]G[V(n)]) \\
&\quad + \rho^2(n) \sup_{V(n)} G^2[V(n)] \\
&\geqslant |t_n - v^*|^2,
\end{aligned}$$

so that $|t_n - v^*| \leqslant |V(n) - v^*| - \gamma_n$. Equations (4.92) and (4.93) and Dvoretzky's condition 3 require that

$$\sum_{n=0}^{\infty} \rho(n) = \infty. \tag{4.94}$$

We now investigate the effect of Dvoretzky's condition 6. By Equation (4.89),

$$\begin{aligned}
E|R_n|^2 &= \rho^2(n)E\{Z(n) - G[V(n)]\}^2 \\
&= \rho^2(n)E[E\{Z^2(n) - 2Z(n)G[V(n)] + G^2[V(n)] | V(n)\}] \\
&= \rho^2(n)E\{E[Z^2(n) | V(n)] - G^2[V(n)]\} \\
&\leqslant \rho^2(n)E[Z^2(n)].
\end{aligned}$$

Therefore, if $E[Z^2(n)]$ is finite, condition 6 requires that

$$\sum_{n=0}^{\infty} \rho^2(n) < \infty. \tag{4.95}$$

The stochastic approximation theorem has therefore placed the following constraints on Example 4.1. First, the solution point v^* must be a local minimum of $J(v)$ in the region of all points $V(n)$. Second, $G(v) = -dJ(v)/dv$ must be finite in the region of all points $v = V(n)$. Third, $E[Z^2(n)]$ must be finite. Finally, the constraints on $\rho(n)$ given by $\rho(n) > 0$, Equations (4.92), (4.94), and (4.95) must be satisfied. For example, the following form satisfies these constraints:

$$\rho(n) = \frac{1}{(1 + n)^\alpha} \rho_0,$$

where ρ_0 is a constant and $\frac{1}{2} < \alpha \leqslant 1$.

4.9 Gradients for Various Constituent Densities and Hyperplanes

In this section the effects of some plausible constituent density forms of the feature vector **x** are investigated. The investigation will be restricted to displaying the direction of the vector $-\nabla P(\text{error}|\mathbf{v})$ from Section 4.5.1 and the vector $-\nabla[M_1(\mathbf{v}) + M_2(\mathbf{v})]$ from Section 4.5.3 for specific hyperplanes. The principal aim of this investigation is to provide a general view of the forms and locations of constituent densities which will allow for the convergence of a sequence $\{\mathbf{V}(n)\}$ to an optimum vector $\mathbf{v}^*$ when utilizing certain stochastic approximation training procedures of Chapter 5. It is also desirable to describe possible regions in **v**-space from where this convergence either does not exist or is very slow.

For convenience of visualization, the illustrations will be restricted to two-dimensional feature space and cases which have a unique optimum decision hyperplane. This implies that **y** space will be three-dimensional and that the optimum set $\mathscr{V}^*$ contains only vectors $\mathbf{v}^*$ which have the same direction. Therefore $\mathbf{v}^*/\|\mathbf{v}\|$ is identical for all $\mathbf{v}^*$ in $\mathscr{V}^*$. In each of the figures of this section a heavy solid line represents the decision hyperplane in feature space described by the vector **v**. A heavy dashed line represents the desired hyperplane described by an optimum $\mathbf{v}^*$. The contours of the constituent densities illustrated in each figure are at a height equal to one-eighth of the peak value of $p(\mathbf{x})$. Their forms are smooth and unimodal. The (a) and (b) portions of these figures will represent two-dimensional projections of three-dimensional augmented feature space. Each figure is drawn so that portion (a) is the vertical projection of portion (b). The vectors **v** and $\mathbf{v}^*$ on the first set of illustrations, Figures 4.17–4.19, are restricted to an infinite cylinder whose axis is the y_0 axis and which has a radius of one, i.e., $\|\mathbf{w}\| = 1$. This normalization is used where the effects of $-\nabla P(\text{error}|\mathbf{v})$ are illustrated. The second set of illustrations, Figures 4.20–4.22, describe the direction of $-\nabla[M_1(\mathbf{v}) + M_2(\mathbf{v})]$ for the same constituent densities and hyperplanes illustrated in the first set. In this second set it is assumed that a guess of $v_2^* > 0$ has been made so that the second component of **v** and $\mathbf{v}^*$ is held constant at a positive value, i.e., $v_2 = v_2^* =$ positive constant.

Figure 4.17 depicts the contours of the constituent densities and a decision hyperplane for the first example. To visualize the direction of the gradient, express $\nabla P(\text{error}|\mathbf{v})$ by Equation (4.29). Note that $-\nabla P(\text{error}|\mathbf{v})$ is the difference of two vectors pointing toward the centroids of a planar slice of the constituent densities. These centroids are indicated by heavy dots in Figure 4.17. The lengths of the vectors are the products of the length of the position vector of the centroid and the area of the planar slice of the corresponding density. A dashed vector at the tip of **v** in Figure 4.17 indicates the direction of motion of **v** in the gradient descent training procedure. The same conventions are used in Figures 4.18 and 4.19.

Figure 4.17 illustrates the condition

$$(\mathbf{v}^* - \mathbf{v})^T[-\nabla P(\text{error}|\mathbf{v})] > 0,$$

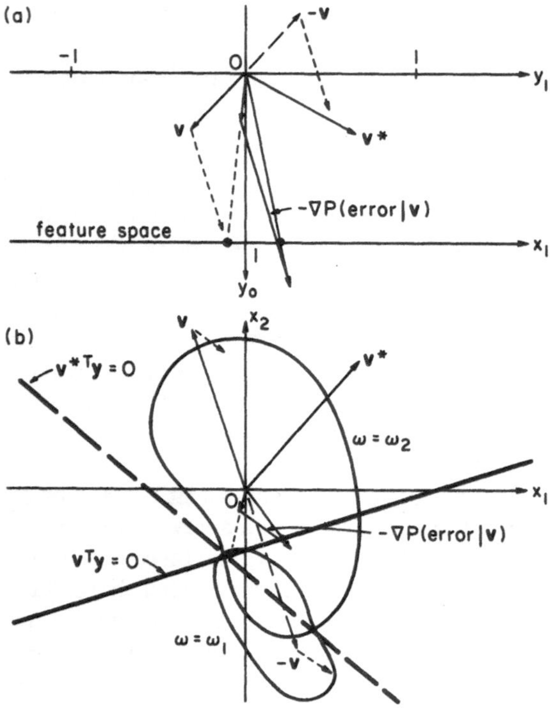

Figure 4.17. $-\nabla P(\text{error}|\mathbf{v})$ when densities and hyperplanes are close to origin.

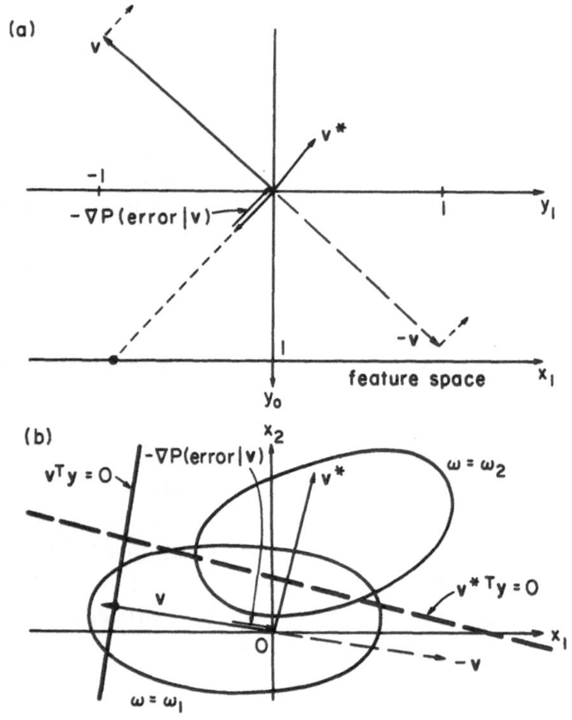

Figure 4.18. $-\nabla P(\text{error}|\mathbf{v})$ when hyperplane is far from one density and $\mathbf{w}$ passes through the hyperplane area centroid of the other.

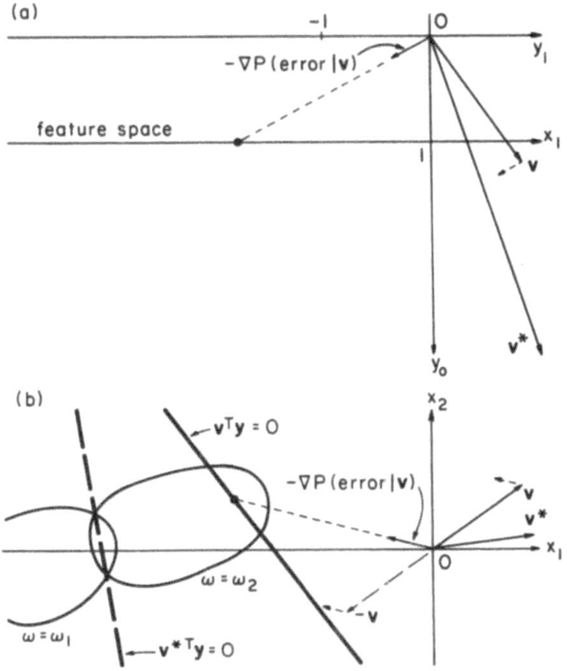

Figure 4.19. $-\nabla P(\text{error}\,|\,\mathbf{v})$ when densities are far from the origin.

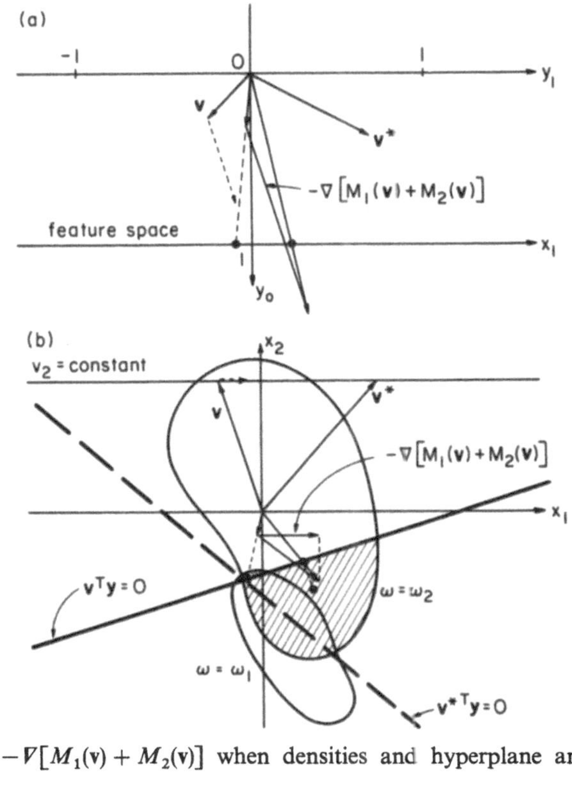

Figure 4.20. $-\nabla[M_1(\mathbf{v}) + M_2(\mathbf{v})]$ when densities and hyperplane are close to the origin.

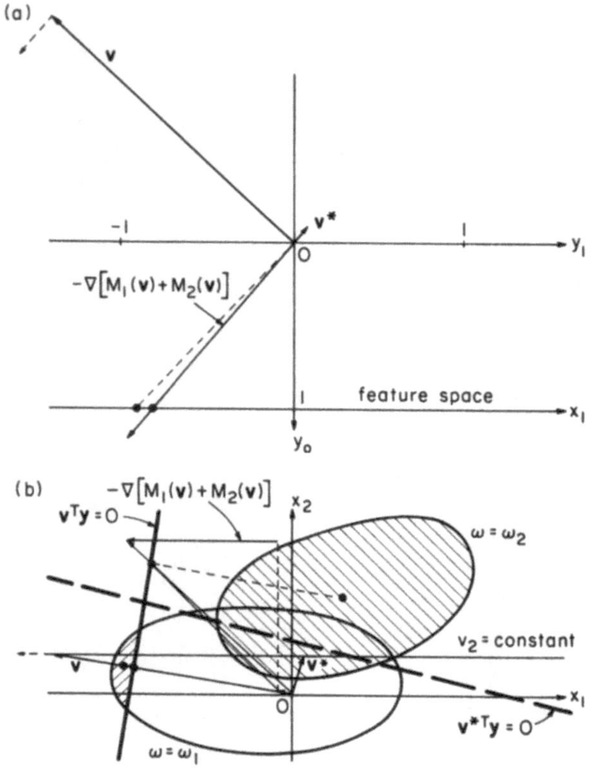

Figure 4.21. $-\nabla[M_1(\mathbf{v}) + M_2(\mathbf{v})]$ when hyperplane is far from one density and **w** passes through the centroid of the error tail of the other.

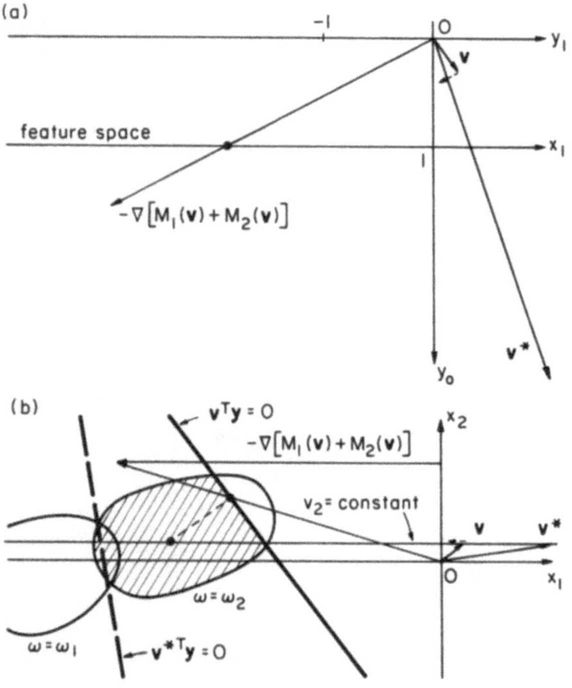

Figure 4.22. $-\nabla[M_1(\mathbf{v}) + M_2(\mathbf{v})]$ when densities are far from the origin.

which indicates that this gradient procedure moves $\mathbf{v}$ toward $\mathbf{v}^*$. This figure also illustrates that the decision hyperplane opposite to that of $\mathbf{v}$—namely the decision hyperplane of $-\mathbf{v}$ which chooses categories exactly opposite to that of $\mathbf{v}$—yields the same error gradient as that of $\mathbf{v}$:

$$\nabla P(\text{error}|\mathbf{v}) = \nabla P(\text{error}|-\mathbf{v}).$$

Here, too, we have

$$[\mathbf{v}^* - (-\mathbf{v})]^T[-\nabla P(\text{error}|\mathbf{v})] > 0.$$

Indeed, these relationships appear to be valid for almost all points $\mathbf{v}$ in this example if it is assumed that each constituent density is strictly positive for all $\mathbf{x}$. Two problems exist, however, which are better illustrated in Figures 4.18 and 4.19.

For the constituent densities in Figure 4.18 we again have

$$(\mathbf{v}^* - \mathbf{v})^T[-\nabla P(\text{error}|\mathbf{v})] > 0.$$

However, here a gradient descent procedure causes negligible rotation of the hyperplane, since the intersection of $f_2(\mathbf{x})$ with the hyperplane is negligibly small. Hence $\mathbf{v}$ passes through the centroid of the slice of the density of ω_1. This translates the hyperplane further from the origin in an attempt to further improve its classification of ω_1 samples. Therefore convergence to a desired $\mathbf{v}^*$ may not occur if the vector $\mathbf{v}$ is poorly positioned initially and the class densities are not strictly positive for all $\mathbf{x}$. If the class densities are strictly positive for all $\mathbf{x}$, the hyperplane will continue to translate away from the origin to the left, pass through infinity if increment corrections are made, and approach the region of the origin from the right. Note that both $\mathbf{v}$ and $-\mathbf{v}$ are approximately normal to $\mathbf{v}^*$. When $-\nabla P(\text{error}|\mathbf{v})$ is applied to $-\mathbf{v}$, the hyperplane moves in a preferred direction. This indicates the importance of a well chosen initial $\mathbf{v}$. In any case, when the hyperplane is in a region of low probability density the gradient may become so small as to give impractically slow convergence rates. An initial $\mathbf{v} = \mathbf{v}(0)$ should therefore be chosen close to $\mathbf{v}^*$ or, since the location of $\mathbf{v}^*$ is not known a priori, so that its hyperplane is in a region of relatively high probability density. As an example, a $\mathbf{v}(0)$ which appears to be appropriate for a large class of constituent density forms is one which positions the initial hyperplane between the centroids of the class densities. A sample approximation to such an initial vector can be determined from a training set $(\mathscr{X}_1, \mathscr{X}_2)$. Specific methods for finding such a $\mathbf{v}(0)$ are presented in Section 5.2 of Chapter 5.

Figure 4.19 describes a case where both of the constituent densities are far from the origin. Some of the earlier problems mentioned are accentuated here, and it is also not clear whether $(\mathbf{v}^* - \mathbf{v})^T[-\nabla P(\text{error}|\mathbf{v})] > 0$ is satisfied. This displays the property, discussed in Section 5.1.3, that gradient corrections to $\mathbf{v}$, and therefore convergence characteristics, are affected by the location of the constituent densities with respect to the origin. Indeed, this example illustrates that if convergence to $\mathbf{v}^*$ is desired, the region over which $\mathbf{v}(0)$ can vary from $\mathbf{v}^*$ becomes more restricted as the constituent densities are translated further from the origin. It therefore becomes desirable to translate the origin to the region between the means of the class densities. This can be done by defining a space translation vector **b**, a sample approximation of which can be found from the training set $(\mathscr{X}_1, \mathscr{X}_2)$; for example, **b** could be defined as the mean of all the feature vectors, i.e., the centroid of $p(\mathbf{x}) = f_1(\mathbf{x}) + f_2(\mathbf{x})$. As in the case of finding an appropriate initial hyperplane, the specific methods for finding an appropriate **b** are presented in Section 5.2 of Chapter 5.

Figures 4.20, 4.21, and 4.22 illustrate particular directions of $-\nabla[M_1(\mathbf{v}) + M_2(\mathbf{v})]$ for the same three constituent density forms given in Figures 4.17, 4.18, and 4.19, respectively. $\nabla[M_1(\mathbf{v}) + M_2(v)]$ is given by Equation (4.67). Note that the integral terms in this equation represent vectors directed from the origin toward the centroids of the projection onto the decision hyperplane of the $f_2(\mathbf{x})$ and $f_1(\mathbf{x})$ error tails. The magnitudes of these vectors are the magnitudes of the moments, taken about the origin, of the projections. Therefore, $-\nabla[M_1(\mathbf{v}) + M_2(\mathbf{v})]$ is found on the illustrations in a manner similar to that which was used for the earlier examples of $-\nabla P(\text{error}|\mathbf{v})$. One basic difference is the use of the projected error tail centroids and volumes rather than the centroids and slice areas of the constituent densities. The error tails are shown hatched on Figures 4.20, 4.21, and 4.22. Another basic difference is a result of the imposed constraint for these examples that $v_2 = \text{constant}$. In Figure 4.20, $-\nabla[M_1(\mathbf{v}) + M_2(\mathbf{v})]$ is obtained by first taking the difference of the two vectors which point toward centroids projected onto the hyperplane. However, a projection of this difference vector is then made so that its component in the v_2 direction is zero. $-\nabla[M_1(\mathbf{v}) + M_2(\mathbf{v})]$ is obtained for the examples of Figures 4.21 and 4.22 in the same manner. A dashed arrow with its tail at the tip of **v** indicates the direction of the correction. It is clear from Figure 4.20 that $(\mathbf{v}^* - \mathbf{v})^T\{-\nabla[M_1(\mathbf{v}) + M_2(\mathbf{v})]\} > 0$ for the given value of **v**. This is analogous to satisfying constraint 6 of Theorem 5.3 for the equalized error training procedure of Chapter 5. It appears that this condition is satisfied for almost all of the allowed points **v** *even though the constituent densities may not be strictly positive.* Some convergence problems do exhibit themselves in Figures 4.21 and 4.22. However, these problems do not appear to be as severe as those when $-\nabla P(\text{error}|\mathbf{v})$ is used.

It appears that for the example of Figure 4.21, $(\mathbf{v}^* - \mathbf{v})^T\{-\nabla[M_1(\mathbf{v}) + M_2(\mathbf{v})]\} > 0$ is not satisfied, so that the gradient descent is moving **v** away from $\mathbf{v}^*$. This again demonstrates the importance of a well chosen initial vector such as the one discussed earlier. A further investigation of this gradient correction reveals that although it is moving the decision hyperplane toward a vertical orientation, it is also translating the hyperplane toward the origin. Note that the orientation can never become vertical since v_k is constant. Once the hyperplane becomes close to the origin, the gradient descent training procedure will begin to rotate it toward its proper orientation.

For the example of Figure 4.22, the condition $(\mathbf{v}^* - \mathbf{v})^T\{-\nabla[M_1(\mathbf{v}) + M_2(\mathbf{v})]\} > 0$ is only marginally satisfied. Clearly in this example it would have been better to have chosen v_1 constant rather than v_2 constant. Some a priori knowledge of this better choice is likely, since the major portion of class ω_2 is clearly to the right of class ω_1. One of the major difficulties with this example is the large gradient, which continues to exist even when the hyperplane is in the region of the optimum hyperplane. This is due to large moment arms and can cause poor convergence properties as a result of overshooting. Again, a translation of the origin to a location between the means of the class densities would be helpful.

4.10 Conclusion

This chapter has investigated the general concept of loss functions for trainable classifiers, as well as the individual properties of several such loss functions. Gradient descent techniques and stochastic approximation received special attention as training procedures for these loss functions. Also presented were the conditions that ensure convergence to the minimum of a loss function when a gradient descent technique is used.

Since deterministic forms of loss functions are not generally available, a method for finding a sample gradient was presented. However, the calculation of this sample gradient is in general lengthy, so that methods of determining approximations to gradients of specific loss functions must often be used, as described in Chapter 5.

Section 4.4 discussed a modification of the gradient descent procedure. This modified procedure restricts the augmented weight vector $\mathbf{v}$ to a predetermined half-space. The half-space is chosen on the basis of some limited a priori knowledge of preferred categorization along a particular feature axis. When applicable, a restriction of this kind reduces the required constraints and often improves the convergence properties.

The specific loss functions presented were discussed as they applied to both linearly separable and linearly nonseparable cases. The more general linearly nonseparable case was illustrated in most of the examples.

The probability of error loss function discussed in Section 4.5.1 has a minimum equal to the least error rate. This is always the most desired value when correct classification costs are neglected and the costs of incorrect classification in each class are equal.

It was noted that the gradient of this loss function may be impractically small when the decision hyperplane lies only in regions of low class density. Thus training procedures based on this loss function may have convergence problems in certain regions of feature space. Consequently, other suboptimal loss functions were also discussed.

The first of these suboptimal loss functions was the mean square error loss function. Actually, with proper $\mathbf{y}$ transformations it approaches a minimum probability of error loss function. However the appropriate transformations are not known a priori, and furthermore they generally greatly increase the dimensionality of the problem. Therefore this loss function is often considered in its basic form without the transformation. We have described in some detail the minimum of this basic loss function.

Another suboptimal loss function discussed was based on magnitudes of the error tail moments. It was found that it could be viewed either as the sum of the magnitudes of the error tail moments or as the product of the probability of error and the average distance from the misclassified feature vectors to the decision hyperplane. The minimum of this loss function yields a region $\mathscr{V}^*$ where the error tails are equal and the moments of the projection of the error tails onto the decision hyperplane are equal.

The section on unequal costs presented a simple method of accounting for unequal costs of the types of misclassification. It was found that these costs could be used to define equivalent a priori class probabilities which could be substituted into any loss function or its gradient.

Stochastic approximation was presented as an extension of the gradient descent technique to the classification of noisy data. The definition of a regression function was given, Dvoretzky's theorem was presented, and the constraints of that theorem were discussed.

Finally, several examples of constituent density forms and decision hyperplanes were presented, together with the associated gradients. From these examples it appears that the size of the region within which an initial vector can be chosen which allows for convergence to the desired set $\mathscr{V}^*$ is larger when the origin is translated to a region between the constituent densities. This also supports the concern for a good choice of an initial weight vector.

In the following chapter the window, minimum mean square error, and equalized error training procedures will be presented. These are based upon the probability of error, mean square error, and error tail moment magnitude loss functions, respectively.

Exercises

4.1. For the constituent densities given in Exercise 1.9, plot $J(\mathbf{v})$, in terms of w_0, for the proportional increment training procedure in the region $-1 < \theta < 0$ when $\mathbf{v}$ is restricted so that $w_1 = +1$.

4.2. Refer to the constituent densities shown in Figure 4.23.

(a) Using these densities, find $J(\mathbf{v})$ for the proportional increment training procedure as given by Equation (4.8).

(b) Plot $J(\mathbf{v})$ as a function of w_0 while constraining $w_1 = \|\mathbf{w}\| = 1$. Recall that the threshold decision is made at $\theta = -w_0/w_1$.

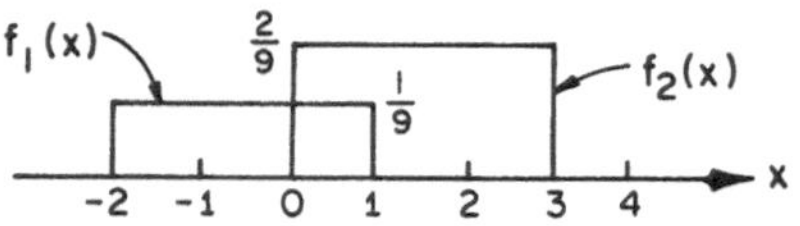

Figure 4.23. Illustration for Exercise 4.2.

(c) Indicate on the plot of $J(\mathbf{v})$ the value of $w_0 = w_0^*$ which gives the minimum $J(\mathbf{v})$ decision while keeping $\|\mathbf{w}\| = 1$.

4.3. Using Equation (4.8) or (4.11), plot $J(\mathbf{v})$ as a function of w_0 and w_1 while constraining $\|\mathbf{w}\| = 1$ for the constituent densities given in Exercise 4.2. (Be certain to consider both the $w_1 = +1$ and $w_1 = -1$ cases.)

4.4. Given $J(\mathbf{v}) = \int e^{-\mathbf{v}^T\mathbf{y}} f_2(\mathbf{x})\,d\mathbf{x} + \int e^{-\mathbf{v}^T\mathbf{y}} f_1(\mathbf{x})\,d\mathbf{x}$, find $\nabla J(\mathbf{v})$.

4.5. Find the sample gradient $\widehat{\nabla J}$, Equation (4.21), at $\mathbf{v} = [-1, 1]^T$ where $J(\mathbf{v})$ is the probability of misclassification, i.e., $J(\mathbf{v}) = P(\text{error}|\mathbf{v})$. The training set is given in Table 4.1. Use the $\{\mathbf{v}_k\}$ sample set of size $m = 4$ given by

$$\mathbf{v}_1 = \begin{bmatrix} -0.8 \\ 1.2 \end{bmatrix},$$

$$\mathbf{v}_2 = \begin{bmatrix} -1.2 \\ 1.2 \end{bmatrix},$$

$$\mathbf{v}_3 = \begin{bmatrix} -0.8 \\ 0.8 \end{bmatrix},$$

$$\mathbf{v}_4 = \begin{bmatrix} -1.2 \\ 0.8 \end{bmatrix}.$$

(These fall at the corners of an ε-hypercube neighborhood where $\varepsilon = 0.2$.)

Table 4.1. Training set for Exercise 4.5.

x values for $\omega = \omega_1$		x values for $\omega = \omega_2$
−2.3	.1	1.1
−1.8	.2	1.6
−1.7	.4	1.8
−1.4	.5	2.1
−1.1	.6	2.3
−.8	.7	2.4
−.6	.9	2.6
−.4	1.1	2.9
−.2	1.4	3.2
0	1.7	3.9

4.6. Find $\nabla J(\mathbf{v})|_{w_2=1}$ from Equation (4.25) for the $J(\mathbf{v})$ given in Exercise 4.4 when $\mathbf{x}$ is a two-dimensional feature vector.

4.7. Given the following constituent densities and vector $\mathbf{v}$:

$$f_2(\mathbf{x}) = \begin{cases} \frac{1}{8}, & \text{if } 0 < x_1 < 2, \quad 0 < x_2 < 2 \\ 0, & \text{otherwise} \end{cases}$$

$$f_1(\mathbf{x}) = \begin{cases} \frac{1}{8}, & \text{if } -1 < x_1 < 1, \quad -1 < x_2 < 1 \\ 0, & \text{otherwise} \end{cases}$$

$$\mathbf{v} = \begin{bmatrix} -1 \\ \frac{1}{2} \\ 1 \end{bmatrix},$$

find $\nabla J(\mathbf{v})$ for the probability of error loss function. Refer to Equation (4.29) and the comments which follow.

4.8. Find $\nabla J(\mathbf{v})$ for the error tail moment loss function given the constituent densities and vector $\mathbf{v}$ of Exercise 4.7. Refer to Equation (4.67) and the comments which follow.

4.9. For the constituent densities and vector $\mathbf{v}$ given in Exercise 1.10, find $\nabla J(\mathbf{v})$ for $J(\mathbf{v}) = P(\text{error}|\mathbf{v})$ using Equation (4.29).

4.10. Find the value of $\nabla J(\mathbf{v})$ for the constituent densities and $\mathbf{v}$ vector given in Exercise 1.10 for the loss function $J(\mathbf{v}) = M_1(\mathbf{v}) + M_2(\mathbf{v})$.

4.11. For the proportional increment loss function $J(\mathbf{v}) = E(-\mathbf{v}^T\mathbf{Z}|\mathbf{v})$, show that $\nabla J(\mathbf{v}) = -E(\mathbf{Z}|\mathbf{v})$. Note that $\mathbf{Z}$ is a function of $\mathbf{v}$. (*Hint:* Use the technique of differentiation of an integral expression as applied in Section 4.5.1.)

4.12. Find $\nabla J(\mathbf{v})$ for $J(\mathbf{v}) = E[(-\mathbf{v}^T\mathbf{Z}/\|\mathbf{w}\|)|\mathbf{v}]$, where $\mathbf{Z}$ is given by Equation (4.2).

4.13. Plot the discriminant function $g(\mathbf{x}) = P(\omega_2|\mathbf{x}) - P(\omega_1|\mathbf{x})$ for the constituent densities given in Exercise 4.2. (See Figure 4.7 for an example of $g(\mathbf{x})$.)

4.14. Prove the validity of Equation 4.46.

4.15. Assume the same definition of an error as in Example 4.1 of Section 4.8 but remove the restriction that the line must pass through the origin. Let $\mathbf{v} = [v_0, v_1, -1]^T$ and $\mathbf{Y} = [1, X_1, X_2]^T$.

(a) Show that $J(\mathbf{v}) = (1/2)E[(\mathbf{V}^T\mathbf{Y})^2|\mathbf{V} = \mathbf{v}]$ is proportional to the expected square of the error.

(b) Find an expression for the random vector $\mathbf{Z}$ obtained from a regression function of the form $\mathbf{G}(\mathbf{v}) = E[\mathbf{Z}|\mathbf{V} = v]$ where $\mathbf{G}(\mathbf{v}) = -\nabla J(\mathbf{v})$.

4.16. Let $\mathbf{V} = [V_0, V_1, \ldots, V_d]^T = [V_0, \mathbf{W}^T]^T$, $\mathbf{Y} = [1, X_1, X_2, \ldots, X_d]^T$ and an error ε_i is defined as the shortest distance of a vector $\mathbf{X}(i)$ from the hyperplane $\mathbf{V}^T\mathbf{Y} = 0$.

(a) Show that

$$J(\mathbf{v}) = \tfrac{1}{2}E\left[\left(\frac{1}{\|\mathbf{W}\|}\mathbf{V}^T\mathbf{Y}\right)^2\middle|\mathbf{V} = \mathbf{v}\right]$$

is one-half the expected square of the error.

(b) Find an expression for the random vector $\mathbf{Z}$ obtained from a regression function $G(\mathbf{v}) = E[\mathbf{Z}|\mathbf{V} = \mathbf{v}]$ where $\mathbf{G}(\mathbf{v}) = -\nabla J(\mathbf{v})$.

4.17. In Example 4.1 of Section 4.8, Dvoretzky's theorem establishes the constraints that both $G(\mathbf{v})$ and $E[Z^2(n)]$ must be finite. Express these constraints in terms of constraints on X_1 and X_2 for this example. Assume only finite v's and $V(n)$'s.

4.18. By applying Dvoretzky's stochastic approximation theorem to the algorithm $\mathbf{V}(n+1) = \mathbf{V}(n) + \rho(n)\mathbf{Z}(n)$ with the $\mathbf{Z}(n)$ from Exercise 4.15, find the constraints on $\mathbf{Z}(n)$ and $\rho(n)$. Assume only finite $\mathbf{v}$'s and $\mathbf{V}(n)$'s.

4.19. Repeat Exercise 4.18 with the $\mathbf{Z}(n)$ from Exercise 4.16.

4.20. Find the constraints on $\mathbf{Z}(n)$ and $\rho(n)$ when Dvoretzky's theorem is applied to the following algorithm:

$$\mathbf{V}(n+1) = \frac{\mathbf{V}(n) + \rho(n)\mathbf{Z}(n)}{\|\mathbf{V}(n) + \rho(n)\mathbf{Z}(n)\|}$$

with the $\mathbf{Z}(n)$ from Exercise 4.16. (The solution to this problem is lengthy and difficult to obtain.)

4.21. Assume that $\rho(n) = [n_0/(n_0 + n)]^\alpha \rho_0$ is to be used in the algorithm of Example 4.1 where ρ_0, n_0, and α are positive constants. Find the range of allowed values of α so that Equations (4.92), (4.94), and (4.95) will be satisfied.

4.22. Let $\mathscr{S}$ denote the decision hyperplane $\mathbf{v}^T\mathbf{y} = 0$. Let δ_{12} denote the distance of $\mathbf{x}$ to $\mathscr{S}$ when $\mathbf{x} \in \omega_1$ and $\mathbf{x}$ is misclassified by $\mathscr{S}$. Let δ_{21} denote the distance of $\mathbf{x}$ to $\mathscr{S}$ when $\mathbf{x} \in \omega_2$ and $\mathbf{x}$ is misclassified by $\mathscr{S}$.

(a) Show that

$$E(\delta|\mathbf{v}) = \frac{P(\omega_2, \mathbf{y} \in \mathscr{R}_1)E(\delta_{21}|\mathbf{v}) + P(\omega_1, \mathbf{y} \in \mathscr{R}_2)E(\delta_{12}|\mathbf{v})}{P(\omega_2, \mathbf{y} \in \mathscr{R}_1) + P(\omega_1, \mathbf{y} \in \mathscr{R}_2)}.$$

(b) Combining this equation with Equation (4.11), derive Equation (4.7).

References

1. S. S. Yau and J. M. Schumpert, Design of pattern classifiers with the updating property using stochastic approximation techniques. *IEEE Transactions on Computers*, **C-17** (September): 861–872 (1968).
2. S. S. Viglione, Applications of pattern recognition technology. In: *Adaptive Learning and Pattern Recognition Systems*, J. M. Mendel and K. S. Fu (eds.), Academic Press, New York, 1970, pp. 115–162.
3. B. B. Brick, Wiener's nonlinear expansion procedure applied to cybernetic problems. *IEEE Transactions on Systems Science and Cybernetics*, **SSC-1** (1): 67–74 (1965).
4. R. O. Duda and P. E. Hart, *Pattern Classification and Scene Analysis*. John Wiley & Sons, New York, 1973, Chapter 5.
5. A. Papoulis, *Probability, Random Variables and Stochastic Processes*. McGraw-Hill, New York, 1965.
6. H. Robbins and S. Monro, A stochastic approximation method. *The Annals of Mathematical Statistics*, **22**: 400–407 (1951).
7. J. Kiefer and J. Wolfowitz, Stochastic estimation of the maximum of a regression function. *The Annals of Mathematical Statistics*, **23**: 462–466 (1952).
8. J. R. Blum, Multidimensional stochastic approximation methods. *The Annals of Mathematical Statistics*, **25** (4): 737–744 (1954).
9. A. Dvoretzky, On stochastic approximation. *Proceedings of the Third Berkeley Symposium on Mathematical Statistics and Probability*, December, 1954 and June and July, 1955, pp. 39–55.
10. S. S. L. Chang, *Synthesis of Optical Control Systems*. McGraw-Hill, New York, 1961, pp. 289–93.

CHAPTER 5

Linear Classifiers for Nonseparable Classes

In this chapter we describe several training procedures based on the gradient and stochastic approximation techniques of the preceding chapter. Although our discussion in this chapter is restricted to two-class cases, the techniques may be extended to multiple-class cases by using the concepts of Section 2.6. The training procedures of this chapter apply to pairs of classes that are linearly nonseparable, i.e., that are not linearly separable, as well as those that are linearly separable. Linearly nonseparable classes include pairs of classes that are separable but not by a hyperplane, as well as pairs of classes that overlap.

The training procedures of this chapter have been selected for their robustness, their speed of convergence, and their asymptotic probabilities of error.

Before the training procedures are presented, a few advantageous modifications of a training procedure based strictly on the gradient of a loss function are discussed. Among these modifications, the first discussed is a modification of the gradient magnitude designed to overcome dependence on the weight vector magnitude. Next is a demonstration that the gradient, and therefore convergence properties, are affected by a translation of the origin. Also presented in this section is the relationship between two weight vectors, each of which define the same decision hyperplane, but one of which is defined with reference to a translated origin.

In the next few sections, methods for normalizing the feature vectors, for finding an advantageous origin translation, and for determining a good initial vector are discussed. Following these sections the training procedures are presented. The sections on training procedures include illustrative computer simulations of the training processes.

5.1 Modifications of Gradient Descent

5.1.1 Effects of Gradient Magnitude

As discussed in Chapters 2 and 4, under certain restrictions on the loss function one can find the region of minimum loss by always moving in a direction opposite to the gradient. The use of a digital computer for the training procedure demands that the training process move in discrete steps. The magnitude of these steps determines the rate of convergence, and indeed whether or not convergence can occur. As indicated in Equation (4.76) in a discussion on stochastic approximation, the step size $\rho(n)\mathbf{Z}(n)$ is partly specified by the deterministic sequence $\rho(n)$. The magnitude of the gradient of the loss function may also contribute to the step size. In fact, in a strict gradient descent procedure, the step size is proportional to the magnitude of the gradient. This can be helpful if the gradient magnitude is large when the weight vector is far from its optimum and small when it is near.

Even though the loss function may conform to the restrictions placed on it by the stochastic approximation method, if the gradient is small for $\mathbf{v}$'s which are significantly distant from the optimum weight vector, convergence may be impractically slow. There can also be a problem of convergence rate when the gradient is too small near the optimum weight vector. Another problem exists when the gradient is too large near the optimum weight vector causing continual overshooting. This can also make the convergence rate impractically small. If there is reason to believe that these problems might occur, the training procedure may be replaced by a form such as

$$\mathbf{V}(n+1) = \mathbf{V}(n) - \rho(n)\frac{\widehat{\nabla J}[\mathbf{V}(n)]}{\|\widehat{\nabla J}[\mathbf{V}(n)]\|}, \tag{5.1}$$

where the sample gradient $\widehat{\nabla J}$ has been used as an example. With this modification the magnitude of the gradient does not influence the step size.

Usually it is desirable to exploit some of the functional properties of the magnitude of a gradient while eliminating those which cause problems. Such might be the case when the length of the weight vector affects the size of the gradient correction at each step. Since the position of the decision hyperplane, and therefore the probability of error, is not a function of the magnitude of the weight vector, the dependency of the gradient magnitude correction on the length of the weight vector is usually undesirable. However, it may be desirable to take advantage of other gradient magnitude properties which do not depend on weight vector magnitude. The following Subsection describes a common form of dependence on the magnitude of a weight vector and some methods for eliminating this dependency.

5.1.2 Reducing Dependence on Magnitude of Weight Vector

Figure 5.1 illustrates two collinear augmented weight vectors $\mathbf{v}_1$ and $\mathbf{v}_2$ in two-dimensional augmented feature space, i.e., y-space. The line $y_0 = 1$ represents nonaugmented feature space. The intersection of this line with the lines which are perpendicular to the augmented weight vectors determines the decision threshold here denoted by $\theta_{1,2}$. Both $\mathbf{v}_1$ and $\mathbf{v}_2$ determine the same threshold $\theta_{1,2}$, since only their magnitudes differ. However, if the same vector $\Delta\mathbf{v}$ is added to each, and if $\Delta\mathbf{v}$ is not parallel to $\mathbf{v}_1$ and $\mathbf{v}_2$, the change in threshold differs. In Figure 5.1 the changes in threshold are denoted by $\Delta\theta_1$ and $\Delta\theta_2$. Analytically the change in threshold for some $\mathbf{v}$ is given by

$$\Delta\theta = \left(-\frac{v_0 + \Delta v_0}{v_1 + \Delta v_1}\right) - \left(-\frac{v_0}{v_1}\right) = \frac{(v_0/v_1)\,\Delta v_1 - \Delta v_0}{v_1 + \Delta v_1}, \tag{5.2}$$

or, in terms of the nonaugmented weight w_1,

$$\Delta\theta = \frac{(v_0/w_1)\,\Delta v_1 - \Delta v_0}{w_1 + \Delta v_1} \cong \frac{(v_0/w_1)\,\Delta v_1 - \Delta v_0}{w_1} \tag{5.3}$$

when $|\Delta v_1| \ll |w_1|$. For any two collinear $\mathbf{v}$ vectors, the weight w_1 varies in direct proportion to the length of the vector $\mathbf{v}$. However the ratio v_0/w_1

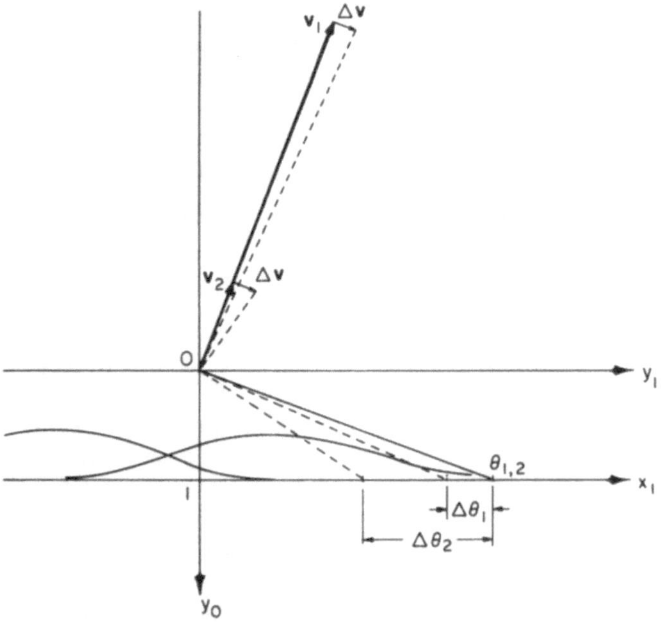

Figure 5.1. The dependence of threshold change on weight vector magnitude.

remains constant. Therefore Equation (5.3) demonstrates that when the step size is small the magnitude of threshold change varies approximately inversely with the magnitude $\|\mathbf{w}\|$ (which equals w_1 in this one-dimensional case) whenever $\operatorname{sgn}[w_1 + \Delta v_1] = \operatorname{sgn} w_1$. If $\operatorname{sgn}[w_1 + \Delta v_1] \neq \operatorname{sgn} w_1$, the decision regions become reversed as a result of the correction. This latter case is not of interest in the present discussion.

A similar situation exists in multidimensional space. Note that the distance of any vector $\mathbf{x}$ from $\mathscr{S}$ (the decision hyperplane determined by $\mathbf{v}$) is

$$d(\mathbf{x}, \mathscr{S}) = \mathbf{v}^T\mathbf{y}/\|\mathbf{w}\|. \tag{5.4}$$

Let $\mathscr{S}_n$ denote the decision hyperplane computed at the nth trial of the training process. By Equation (5.4),

$$\begin{aligned} d(\mathbf{x}, \mathscr{S}_{n+1}) - d(\mathbf{x}, \mathscr{S}_n) &= \frac{\mathbf{v}^T(n+1)\mathbf{y}}{\|\mathbf{w}(n+1)\|} - \frac{\mathbf{v}^T(n)\mathbf{y}}{\|\mathbf{w}(n)\|} \\ &\cong \frac{\Delta\mathbf{v}^T(n)\mathbf{y}}{\|\mathbf{w}(n)\|} \end{aligned} \tag{5.5}$$

when $\|\mathbf{w}(n+1)\| \cong \|\mathbf{w}(n)\|$.

For the case where the loss function is the probability of error, a form of gradient descent training procedure is

$$\mathbf{v}(n+1) = \mathbf{v}(n) - R_n \nabla J(\mathbf{v}(n)). \tag{5.6}$$

where R_n may vary with both n and $\|\mathbf{w}(n)\|$,

$$\nabla J(\mathbf{v}(n)) = -\frac{1}{\|\mathbf{w}(n)\|} \int_{\mathscr{S}_n} \mathbf{y}[f_2(\mathbf{x}) - f_1(\mathbf{x})]\, ds, \tag{5.7}$$

and $\mathscr{S}_n$ is the hyperplane determined by $\mathbf{v}(n)$.

Notice that $\nabla J(\mathbf{v}(n))$ is the product of $-\|\mathbf{w}(n)\|^{-1}$ and a function of $\mathbf{v}(n)$ that is independent of $\|\mathbf{v}(n)\|$. (This is because the hyperplane $\mathscr{S}_n$ is independent of $\|\mathbf{v}(n)\|$.) Thus

$$\mathbf{v}(n+1) = \mathbf{v}(n) + \frac{R_n}{\|\mathbf{w}(n)\|} \int_{\mathscr{S}_n} \mathbf{y}[f_2(\mathbf{x}) - f_1(\mathbf{x})]\, ds. \tag{5.8}$$

It is desirable that the hyperplanes represented by the $\mathbf{v}(n)$'s converge in feature space in a manner that is independent of $\|\mathbf{w}(n)\|$. It follows, therefore, from Equation (5.5) that it is desirable to design the training process so that the quantity $\Delta\mathbf{v}(n)/\|\mathbf{w}(n)\|$ is independent of $\|\mathbf{w}(n)\|$. By Equation (5.8),

$$\frac{\Delta\mathbf{v}(n)}{\|\mathbf{w}(n)\|} = \frac{R_n}{\|\mathbf{w}(n)\|^2} \int_{\mathscr{S}_n} \mathbf{y}[f_2(\mathbf{x}) - f_1(\mathbf{x})]\, ds. \tag{5.9}$$

To make $\Delta\mathbf{v}(n)/\|\mathbf{w}(n)\|$ independent of $\|\mathbf{w}(n)\|$, $R_n/\|\mathbf{w}(n)\|^2$ must be made independent of $\|\mathbf{w}(n)\|$. Hence we let

$$R_n/\|\mathbf{w}(n)\|^2 = \rho_n = a \text{ function of } n \text{ only},$$

which implies

$$R_n = \rho_n\|\mathbf{w}(n)\|^2. \tag{5.10}$$

Substituting Equation (5.10) into Equation (5.8) yields

$$\mathbf{v}(n+1) = \mathbf{v}(n) + \rho_n \|\mathbf{w}(n)\| \int_{\mathscr{S}_n} \mathbf{y}[f_2(\mathbf{x}) - f_1(\mathbf{x})\, d\mathbf{s}. \tag{5.11}$$

This equation may be generalized as follows:

$$\Delta\mathbf{v}(n) = \rho_n \mathbf{G}(\mathbf{v}(n)), \tag{5.12}$$

where

$$\mathbf{G}(\mathbf{v}) = -\|\mathbf{w}(n)\|^2 \nabla J(\mathbf{v}), \tag{5.13}$$

provided the loss function's gradient $\nabla J(\mathbf{v})$ varies inversely with $\|\mathbf{w}\|$.

In developing a stochastic approximation training procedure for such a loss function, we shall seek a random variable $\mathbf{Z}$ such that

$$\begin{aligned} \mathbf{G}(\mathbf{v}) &= E(\mathbf{Z} \mid \mathbf{v}) \\ &= \text{a regression function of } \mathbf{v}. \end{aligned}$$

We believe this choice of $\mathbf{Z}$ yields faster convergence than a $\mathbf{Z}$ satisfying $E(\mathbf{Z} \mid \mathbf{v}) = -\nabla J(\mathbf{v})$, and facilitates the mathematical demonstration of the convergence of $\mathbf{v}(n)$. A similar improvement in convergence is obtained by normalizing $\|\mathbf{w}(n)\|$ to a fixed quantity after each application of Equation (5.6). In both techniques, the change in position of $\mathscr{S}_n$ in $\mathbf{x}$-space is nearly independent of $\|\mathbf{w}(n)\|$.

5.1.3 Translating the Origin

For many training procedures the convergence properties are dependent on translations of the constituent densities in feature space; e.g., if the densities are $\{f_1(\mathbf{x} - \boldsymbol{\xi}), f_2(\mathbf{x} - \boldsymbol{\xi})\}$, the convergence properties generally vary with respect to $\boldsymbol{\xi}$. This is the case even if the initial decision hyperplane is fixed in $(\mathbf{x} - \boldsymbol{\xi})$-space as $\boldsymbol{\xi}$ varies.

Consider the case where the hyperplane is specified by

$$\mathbf{v}^T \left[\begin{array}{c} 1 \\ \hline \mathbf{x} - \boldsymbol{\xi} \end{array}\right] = 0$$

and the constituent densities, expressed in terms of $\boldsymbol{\xi}$, are $f_1(\mathbf{x} - \boldsymbol{\xi})$ and $f_2(\mathbf{x} - \boldsymbol{\xi})$. For this case, the hyperplane and constituent densities are translated together as $\boldsymbol{\xi}$ varies. The following discussion demonstrates that the corrections in the hyperplane position depend on $\boldsymbol{\xi}$ when these corrections are proportional to the gradient of the probability of error.

Let $\mathbf{v}^{(\boldsymbol{\xi})}$ denote an augmented weight vector defined by

$$\mathbf{v}^{(\boldsymbol{\xi})} = \left[\begin{array}{c} w_0 - \mathbf{w}^T\boldsymbol{\xi} \\ \hline \mathbf{w} \end{array}\right] \tag{5.14}$$

so that

$$\mathbf{v}^{(\boldsymbol{\xi})T}\mathbf{y} = \mathbf{v}^T \left[\begin{array}{c} 1 \\ \hline \mathbf{x} - \boldsymbol{\xi} \end{array}\right] = 0 \tag{5.15}$$

determines a hyperplane which is fixed in $(\mathbf{x} - \boldsymbol{\xi})$-space as $\boldsymbol{\xi}$ varies. The gradient with respect to $\mathbf{v}^{(\xi)}$ is given by

$$\nabla P(\text{error} \mid \mathbf{v}^{(\xi)}) = -\frac{1}{\|\mathbf{w}\|} \int_{\mathscr{S}} \left[\frac{1}{\mathbf{x} - \boldsymbol{\xi}}\right] [f_2(\mathbf{x} - \boldsymbol{\xi}) - f_1(\mathbf{x} - \boldsymbol{\xi})] \, ds, \tag{5.16}$$

where $\mathbf{x} \in \mathscr{S}$ iff $\mathbf{v}^{(\xi)T}\mathbf{y} = 0$. By Equation (5.15), $\mathscr{S}$ is a hyperplane which is fixed in $(\mathbf{x} - \boldsymbol{\xi})$-space as $\boldsymbol{\xi}$ varies. This demonstrates that $\nabla P(\text{error} \mid \mathbf{v}^{(\xi)})$ is independent of $\boldsymbol{\xi}$. In a gradient descent training procedure, a new augmented weight vector $\mathbf{v}^{(\xi)}(n + 1)$ is found by adding to $\mathbf{v}^{(\xi)}(n)$ the product of a constant ρ and the negative of the gradient. This is given by

$$\mathbf{v}^{(\xi)}(n + 1) = \mathbf{v}^{(\xi)}(n) - \rho \nabla P(\text{error} \mid \mathbf{v}^{(\xi)}(n)). \tag{5.17}$$

But since $\nabla P(\text{error} \mid \mathbf{v}^{(\xi)})$ is not a function of $\boldsymbol{\xi}$, and since $\mathbf{v}^{(\xi)}(n)$ has the functional form specified by Equation (5.14), it follows that $\mathbf{v}^{(\xi)}(n + 1)$ generally cannot have that form; i.e.,

$$\mathbf{v}^{(\xi)}(n + 1) \neq \frac{w_0(n + 1) - \mathbf{w}^T(n + 1)\boldsymbol{\xi}}{\mathbf{w}(n + 1)} \quad \text{for all } \boldsymbol{\xi}, \tag{5.18}$$

with $w_0(n + 1)$ and $\mathbf{w}(n + 1)$ independent of $\boldsymbol{\xi}$. Therefore gradient corrections based on loss functions which are proportional to the probability of error are in general dependent on $\boldsymbol{\xi}$. The following describes a method which will compensate for $\boldsymbol{\xi}$-dependent corrections.

Let $\mathbf{a}$ denote a vector which is fixed in $(\mathbf{x} - \boldsymbol{\xi})$-space. The vector $\mathbf{a}$ could, for example, be the centroid of $f_1(\mathbf{x} - \boldsymbol{\xi}) + f_2(\mathbf{x} - \boldsymbol{\xi})$. If some such $\mathbf{a}$ is either known or can be learned, a $\boldsymbol{\xi}$-independent gradient descent procedure can be established. The gradient for this procedure is found in terms of an augmented weight vector $\boldsymbol{v}$, analogous to $\mathbf{v}$, which describes a hyperplane in $\mathbf{y}$-space passing through a point $\boldsymbol{\beta}$, where

$$\boldsymbol{\beta} = \left[\frac{0}{\mathbf{a} + \boldsymbol{\xi}}\right].$$

Hereafter, we use $\mathbf{b}$ in place of $\mathbf{a} + \boldsymbol{\xi}$. Thus

$$\boldsymbol{\beta} = \left[\frac{0}{\mathbf{b}}\right]. \tag{5.19}$$

The general equation for this hyperplane is given by

$$\boldsymbol{v}^T[\mathbf{y} - \boldsymbol{\beta}] = 0. \tag{5.20}$$

As an example, the gradient of the probability of error loss function with respect to $\boldsymbol{v}$ is given by

$$\nabla_{\boldsymbol{v}} P(\text{error} \mid \boldsymbol{v}) = -\frac{1}{\|\mathbf{w}\|} \int_{\mathscr{S}} [\mathbf{y} - \boldsymbol{\beta}][f_2(\mathbf{x}) - f_1(\mathbf{x})] \, ds, \tag{5.21}$$

where as before $\mathbf{x} \in \mathscr{S}$ iff $\boldsymbol{v}^T\mathbf{y} = 0$. Note that $\nabla_{\boldsymbol{v}} P(\text{error} \mid \boldsymbol{v}) = \mathbf{0}$ for all values of $\mathbf{v}$ for which $\nabla P(\text{error} \mid \mathbf{v}) = \mathbf{0}$.

The vectors $\mathbf{v}$ and $\boldsymbol{\nu}$ are related so that they define the same separating hyperplane in feature space. This relationship is given by

$$\mathbf{v} = \boldsymbol{\nu} - \begin{bmatrix} \mathbf{w}^T\mathbf{b} \\ 0 \\ \vdots \\ 0 \end{bmatrix} = \boldsymbol{\nu} - \begin{bmatrix} \boldsymbol{\nu}^T\boldsymbol{\beta} \\ 0 \\ \vdots \\ 0 \end{bmatrix}. \tag{5.22}$$

The validity of Equation (5.22) is demonstrated by noting that $y_0 = 1$, and thus

$$\mathbf{v}^T\mathbf{y} = \left[\boldsymbol{\nu} - \begin{bmatrix} \boldsymbol{\nu}^T\boldsymbol{\beta} \\ 0 \\ \vdots \\ 0 \end{bmatrix}\right]^T \mathbf{y} = \boldsymbol{\nu}^T\mathbf{y} - \boldsymbol{\nu}^T\boldsymbol{\beta} = \boldsymbol{\nu}^T(\mathbf{y} - \boldsymbol{\beta}). \tag{5.23}$$

Therefore the hyperplane in $\mathbf{x}$-space described by $\mathbf{v}^T\mathbf{y} = 0$ is also described by $\boldsymbol{\nu}^T(\mathbf{y} - \boldsymbol{\beta}) = 0$ if Equation (5.22) holds.

Note that all but the first components of $\mathbf{v}$ and $\boldsymbol{\nu}$ are equal. When a gradient descent procedure which utilizes $\nabla_{\boldsymbol{\nu}} J(\boldsymbol{\nu})$ to make corrections on $\boldsymbol{\nu}$ is used, $\mathbf{v}$ can be found by the substitution of $\boldsymbol{\nu}$ into Equation (5.22).

Since $\boldsymbol{\nu}^T\boldsymbol{\beta} = \mathbf{w}^T\mathbf{b}$ as a result of the zero value of the first component of $\boldsymbol{\beta}$, it is also true that $\boldsymbol{\nu}^T\boldsymbol{\beta} = \mathbf{v}^T\boldsymbol{\beta}$. Therefore $\boldsymbol{\nu}$ can be found from $\mathbf{v}$ using the following equation obtained from Equation (5.22):

$$\boldsymbol{\nu} = \mathbf{v} + \begin{bmatrix} \mathbf{v}^T\boldsymbol{\beta} \\ 0 \\ \vdots \\ 0 \end{bmatrix}. \tag{5.24}$$

Thus a translation of the origin can be accounted for by replacing $\mathbf{v}$ by $\boldsymbol{\nu}$ and $\mathbf{y}$ by $\mathbf{y} - \boldsymbol{\beta}$. We refer to the space whose origin coincides with the point $\boldsymbol{\beta}$ as *$\boldsymbol{\beta}$-origin space.*

For convenience of notation in this book we avoid using $\boldsymbol{\nu}$ and $\boldsymbol{\beta}$ explicitly unless necessitated by the context.

5.2 Normalization, Origin Selection, and Initial Vector

In this section we describe a method for the normalization of feature vectors, methods for finding a suitable origin translation vector $\boldsymbol{\beta}$, and a method for finding a suitable initial augmented weight vector $\mathbf{v}(0)$. In each case the sample statistics of the feature vector training set $(\mathscr{X}_1, \mathscr{X}_2)$ will be used. The determination of a good $\boldsymbol{\beta}$ and $\mathbf{v}(0)$ usually greatly improves the convergence properties of training procedures.

5.2.1 Feature Vector Normalization

The training procedures presented in this chapter do not require any normalization of the feature vectors. However, in most cases normalization can provide several benefits:

1. More consistent convergence properties of a training procedure.
2. A guide in reducing the number of extracted features.
3. A simplified procedure for choosing the initial weight vector.

We describe two normalization procedures. The first procedure, *average variance normalization*, provides more consistent convergence of the training process. The second procedure, *mixture variance normalization*, is an effective guide in eliminating features that do not contribute effectively to the classification process.

Average Variance Normalization. In this procedure we scale the features so that if the features are uncorrelated, the class densities of the scaled features will more closely approximate unit hyperspheres than those of the unscaled features. Let

$$\tilde{\sigma}_i^2 = \tfrac{1}{2}(\hat{\sigma}_{1i}^2 + \hat{\sigma}_{2i}^2),$$

where $\hat{\sigma}_{ji}^2$ is the sample variance of the ith feature of class ω_j.

The average variance normalization process divides every feature x_i by $\tilde{\sigma}_i$. Thus

$$\begin{aligned} \tilde{x}_i &= \text{normalized } i\text{th feature} \\ &= x_i/\tilde{\sigma}_i. \end{aligned} \tag{5.25}$$

When $\hat{\sigma}_{1i} = \hat{\sigma}_{2i}$ for every i, the variance of $\tilde{x}_i$ in class ω_j is unity for every i and every j, thus yielding approximately hyperspherical class densities in the normalized feature space if the cross correlations among the $\tilde{x}_i$'s in each class are small. This normalization places the standard deviation $\tilde{\sigma}_{ji}$ of each normalized feature in class ω_j in the range

$$0 \leqslant \tilde{\sigma}_{ji} \leqslant \sqrt{2}.$$

Note that if normalization is desired, it must be applied to each feature vector in the training set before training and to every subsequent feature vector to be classified. After the training is completed, the scaling of the feature vectors may instead be replaced by an appropriately scaled augmented weight vector. (See Exercise 5.5.)

Two-dimensional examples illustrating the effect of average variance normalization are given in Figures 5.2–5.6. The constituent density contours shown in these figures represent a distance of one standard deviation, calculated radially, from the respective class means. The normalized form of either the (a) or (b) part of Figure 5.2 is given in Figure 5.3. The other figures illustrate the unnormalized form in the (a) portion of the figure with the

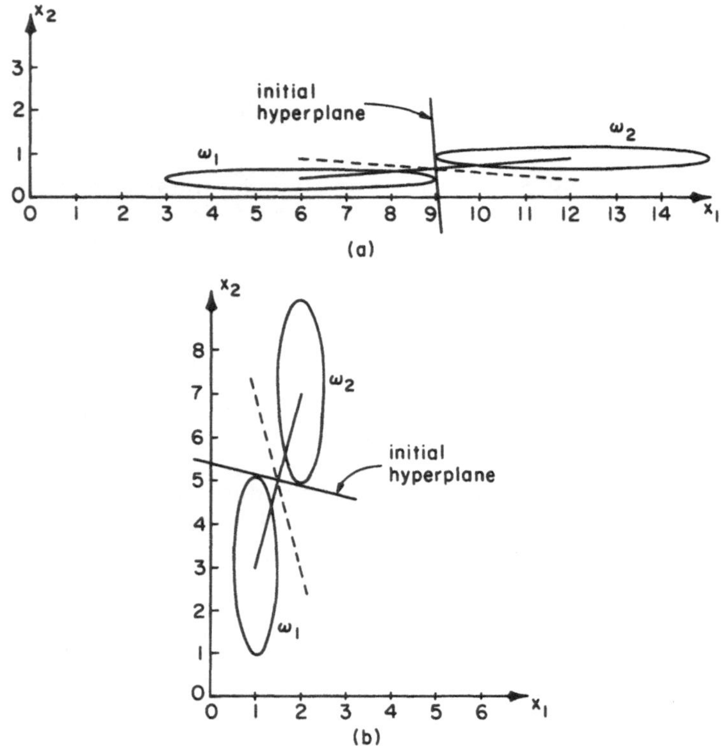

Figure 5.2. Examples of unnormalized features.

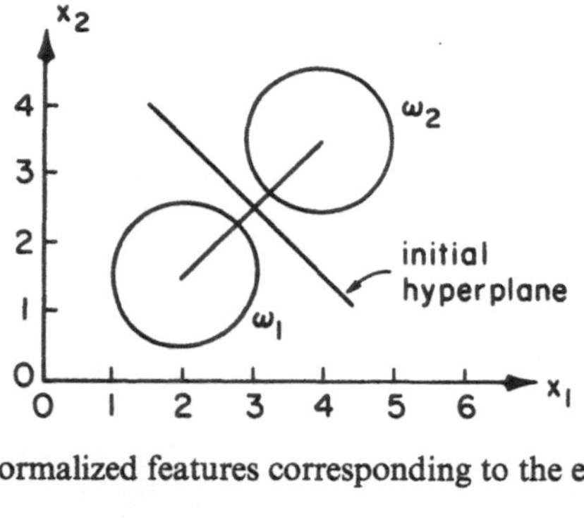

Figure 5.3. Normalized features corresponding to the examples in Figure 5.2.

corresponding normalized form in the (b) portion. Figures 5.5 and 5.6 each demonstrate contours where some cross correlation exists between the features. Often a suitable initial decision hyperplane after normalization is one which is normal to the line joining the means of the class densities and, in the case where $P(\omega_1) = P(\omega_2)$, located midway between the means. A line representing such a two-dimensional initial hyperplane is illustrated in each

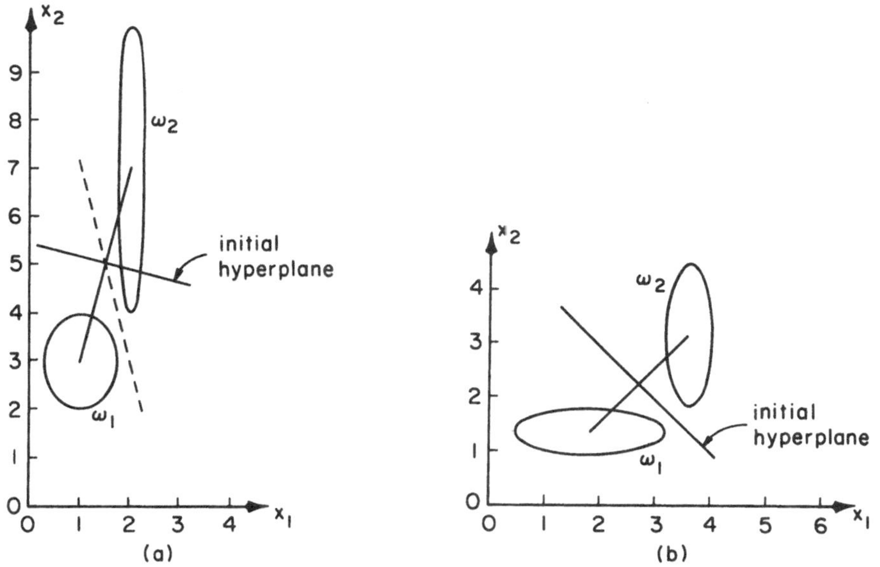

Figure 5.4. Example of (a) unnormalized features and (b) the same features normalized.

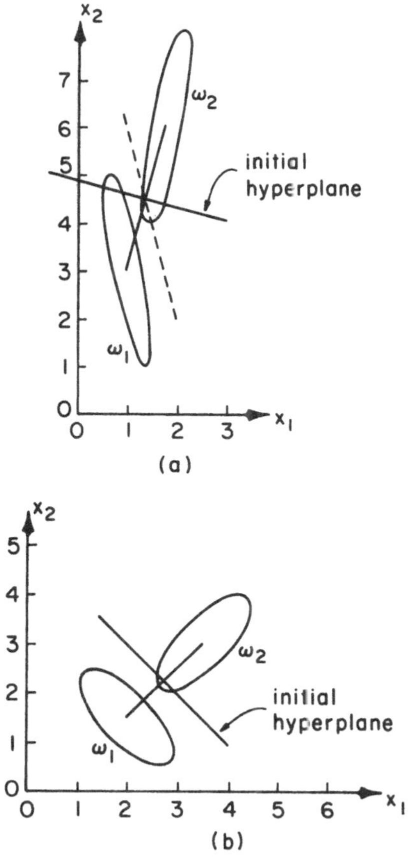

Figure 5.5. Features with some cross correlation: (a) unnormalized; (b) normalized.

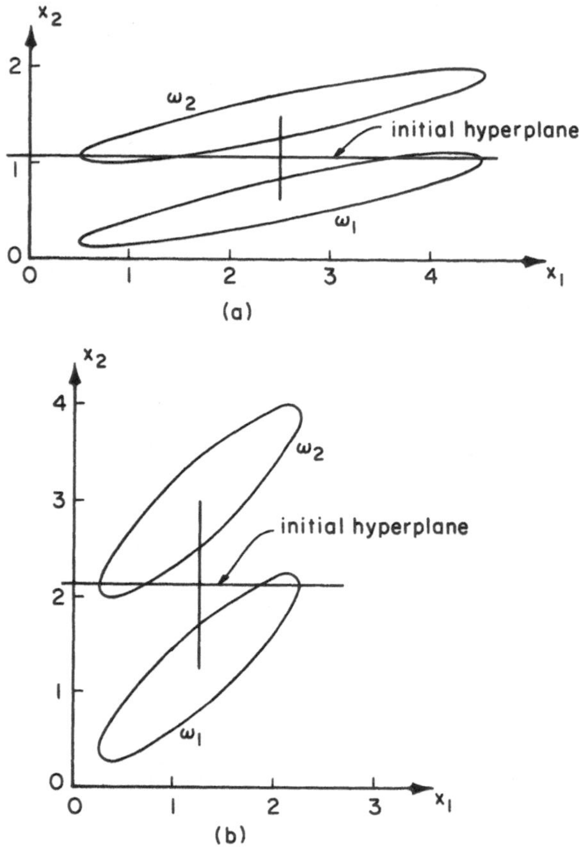

Figure 5.6. Features with some cross correlation and equal means of x_1 for each class: (a) unnormalized; (b) normalized.

of the figures. The examples are assumed to satisfy the condition of equal class probabilities. For reference, the unnormalized examples include a dotted initial hyperplane which corresponds to the simple hyperplane found after normalization. In each example, the normalization results in the determination of a reasonably good initial hyperplane.

Mixture Variance Normalization. The orientation of the optimum decision hyperplane in feature space can suggest which features may be deleted in order to economize on the cost of feature extraction. The mixture variance normalization scheme described here transforms the space occupied by the data so that the weakest feature tends to be the one associated with the axis of feature space most nearly parallel to the optimum hyperplane. This either (a) normalizes the variances of each of the features or (b) normalizes a measure of the spread of each feature. Let

$$\begin{aligned}\tilde{x}_i &= \text{normalized } i\text{th feature} \\ &= x_i/\hat{\sigma}_i.\end{aligned} \tag{5.26}$$

Here $\hat{\sigma}_i^2$ is the sample variance of the ith feature in the mixture of both classes.

The mixture variance normalization scheme described above produces a new mixture density which has unit-valued variances of each feature in the mixture. This type of normalization has the further advantage that features formed from products, e.g., $\hat{x}_i\hat{x}_j$, have unit variances as upper bounds, thereby making this transformation useful for quadratic as well as linear classifiers (using the technique of the $\Phi(\mathbf{x})$ transformation described in Section 3.2).

An alternative to the above scheme should also be considered. Since the mixture density is usually highly non-Gaussian, it may be effective to devise a measure of spread other than $\hat{\sigma}_i$ for each feature. For example, one may determine a spread s_i for each feature by specifying the areas of the tails of the mixture density of each feature outside of an interval of length $2s_i$. One then replaces $\hat{\sigma}_i$ in the above equations by s_i.

A demonstration of the relative effectiveness of these normalization procedures for feature selection is shown in Figure 5.7. In this figure part (a) illustrates the average variance normalization and part (b) illustrates the mixture variance normalization for the same constituent densities. The densities in this figure are uniform within the shaded regions, with equal a priori class probabilities: $P(\omega_i) = \frac{1}{2}$ for $i = 1, 2$. Here we see that the average variance normalization chooses x_2 as the more powerful feature (because of the

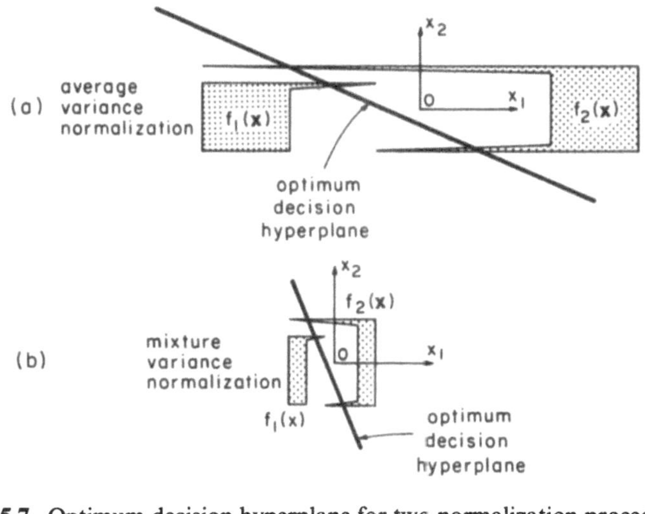

Figure 5.7. Optimum decision hyperplane for two normalization procedures.

orientation of the decision hyperplane), while the mixture variance normalization chooses x_1. An inspection of the figure shows that x_1 is the correct choice in this case.

5.2.2 Choosing an Origin

As indicated in Section 4.9, a translation of the origin of feature space, i.e., **x**-space, to a region close to the means of the class densities seems to improve the convergence properties of the gradient descent training procedures. It was suggested in Section 4.9 and subsection 5.1.3 that a translation of the origin could be made to the sample mean $\hat{\boldsymbol{\mu}}_{\mathbf{x}}$ of all the feature vectors in the training set. As before we denote the origin translation vector in **x**-space as **b**, so that in this case $\mathbf{b} = \hat{\boldsymbol{\mu}}_{\mathbf{x}}$. Although this choice of **b** lies on the line joining the class density means, it will lie closer to the mean $\boldsymbol{\mu}_{\mathbf{x}|\omega_j}$ of the class ω_j which has the larger class probability, $P(\omega_i) < P(\omega_j)$, $i, j = 1, 2$. However, it seems desirable to place **b** close to a point on the line through which the optimum decision hyperplane passes. In Subsection 5.2.4 we shall define the initial hyperplane as passing through the point **b**. It is advantageous to have the initial $\mathbf{v}(0)$ of this hyperplane near the optimum $\mathbf{v} = \mathbf{v}^*$. Since for a large class of problems the optimum decision hyperplane will pass closer to the mean of the density that has the *lower* class probability, a choice for **b** other than $\hat{\boldsymbol{\mu}}_{\mathbf{x}}$ is generally preferable. We shall describe three reasonable choices for **b**. The first of these choices for **b** is a vector directed from the origin and terminating at a point **b** on the line joining the centroids of the class densities such that the fractional distance from **b** to each centroid is proportional to the class probability $P(\omega_j)$, $j = 1, 2$. Such a **b** is illustrated in Figure 5.8 for an example in two-dimensional feature space. Clearly the point **b** must lie between the class means. If l_j, $j = 1, 2$, denotes the Euclidean distance or norm of the point **b** from the class mean $\boldsymbol{\mu}_{\mathbf{x}|\omega_j}$, then

$$\frac{l_j}{l_1 + l_2} = P(\omega_j)$$

or

$$\frac{l_1}{l_2} = \frac{P(\omega_1)}{P(\omega_2)}.$$

The expression for **b** in terms of the sample means of the training set is simply the vector sum of $\hat{\boldsymbol{\mu}}_{\mathbf{x}|\omega_1}$ and $(l_1/(l_1 + l_2))(\hat{\boldsymbol{\mu}}_{\mathbf{x}|\omega_2} - \hat{\boldsymbol{\mu}}_{\mathbf{x}|\omega_1})$:

$$\mathbf{b} = P(\omega_2)\hat{\boldsymbol{\mu}}_{\mathbf{x}|\omega_1} + P(\omega_1)\hat{\boldsymbol{\mu}}_{\mathbf{x}|\omega_2}. \tag{5.27}$$

The second of the choices for **b** includes considerations of the effects of different standard deviations of the class densities as well as different class probabilities. Occasionally the optimum decision hyperplane will not lie between the class means. This second method can produce initial hyperplanes with such a property. The constituent densities $f_1(\mathbf{x})$ and $f_2(\mathbf{x})$ will first be

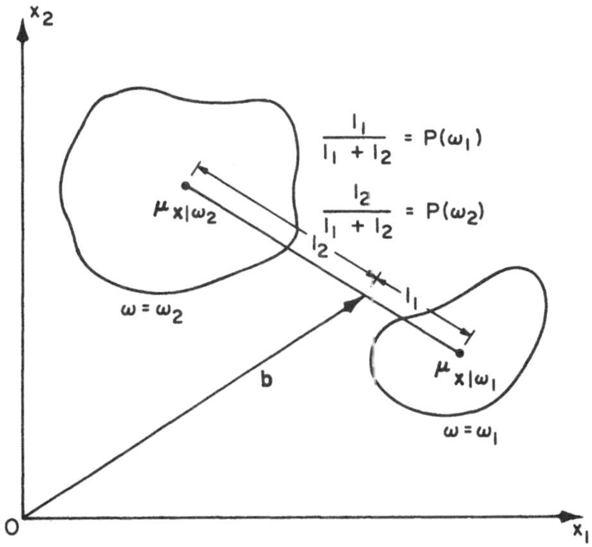

Figure 5.8. Illustration of the first suitable **b**.

modeled as Gaussian constituent densities whose class probabilities $P(\omega_j)$, $j = 1, 2$; means $\boldsymbol{\mu}_{\mathbf{x}|\omega_j}$, $j = 1, 2$; and covariance matrices $\boldsymbol{\Sigma}_j$ are equal to the sample statistics of their associated training sets. The origin translation vector **b**, which defines a point **b** in feature space, will then be defined as specifying the point of intersection of the modeled constituent densities along the line which joins their means. In general there will be two such intersections. The intersection chosen will be the one corresponding to the larger value of constituent density and will therefore be the intersection closest to a class mean. For a case where the class distributions are Gaussian with equal covariance matrices, the optimum surface will be a hyperplane which passes through **b**. This will be demonstrated in Subsection 5.2.4.

From Chapter 2, Gaussian constituent densities can be expressed as

$$f_j(\mathbf{x}) = \frac{P(\omega_j)}{(2\pi)^{d/2}|\boldsymbol{\Sigma}_j|^{1/2}} \exp\left[-\tfrac{1}{2}(\mathbf{x} - \hat{\boldsymbol{\mu}}_{\mathbf{x}|\omega_j})^T \boldsymbol{\Sigma}_j^{-1}(\mathbf{x} - \hat{\boldsymbol{\mu}}_{\mathbf{x}|\omega_j})\right], \tag{5.28}$$

where $|\boldsymbol{\Sigma}_j|$ denotes the determinant of the sample covariance matrix of class ω_j. The term $(\mathbf{x} - \boldsymbol{\mu}_{\mathbf{x}|\omega_j})^T \boldsymbol{\Sigma}_j^{-1}(\mathbf{x} - \boldsymbol{\mu}_{\mathbf{x}|\omega_j})$ is generally referred to as the squared Mahalanobis distance from **x** to $\boldsymbol{\mu}_{\mathbf{x}|\omega_j}$. To constrain **b** to values of **x** lying on the line joining the sample means, we express **b** as

$$\mathbf{b} = \hat{\boldsymbol{\mu}}_{\mathbf{x}|\omega_1} + k(\hat{\boldsymbol{\mu}}_{\mathbf{x}|\omega_2} - \hat{\boldsymbol{\mu}}_{\mathbf{x}|\omega_1}), \tag{5.29}$$

where k is to be determined from the constraint of the intersection of the model densities. This will be done by substituting **b** as given by Equation (5.29) for **x** in Equation (5.28) and equating the model constituent densities.

First note that

$$(\mathbf{b} - \hat{\boldsymbol{\mu}}_{\mathbf{x}|\omega_1})^T \boldsymbol{\Sigma}_1^{-1} (\mathbf{b} - \hat{\boldsymbol{\mu}}_{\mathbf{x}|\omega_1}) = k^2 (\hat{\boldsymbol{\mu}}_{\mathbf{x}|\omega_2} - \hat{\boldsymbol{\mu}}_{\mathbf{x}|\omega_1})^T \boldsymbol{\Sigma}_1^{-1} (\hat{\boldsymbol{\mu}}_{\mathbf{x}|\omega_2} - \hat{\boldsymbol{\mu}}_{\mathbf{x}|\omega_1})$$

and

$$(\mathbf{b} - \hat{\boldsymbol{\mu}}_{\mathbf{x}|\omega_2})^T \boldsymbol{\Sigma}_2^{-1} (\mathbf{b} - \hat{\boldsymbol{\mu}}_{\mathbf{x}|\omega_2}) = (1-k)^2 (\hat{\boldsymbol{\mu}}_{\mathbf{x}|\omega_1} - \hat{\boldsymbol{\mu}}_{\mathbf{x}|\omega_2})^T \boldsymbol{\Sigma}_2^{-1} (\hat{\boldsymbol{\mu}}_{\mathbf{x}|\omega_1} - \hat{\boldsymbol{\mu}}_{\mathbf{x}|\omega_2}).$$

Now let r_{ij}^2 denote the squared Mahalanobis distance from the sample mean $\hat{\boldsymbol{\mu}}_{\mathbf{x}|\omega_i}$ to the sample mean $\hat{\boldsymbol{\mu}}_{\mathbf{x}|\omega_j}$, i.e.,

$$r_{ij}^2 = (\hat{\boldsymbol{\mu}}_{\mathbf{x}|\omega_i} - \hat{\boldsymbol{\mu}}_{\mathbf{x}|\omega_j})^T \boldsymbol{\Sigma}_j^{-1} (\hat{\boldsymbol{\mu}}_{\mathbf{x}|\omega_i} - \hat{\boldsymbol{\mu}}_{\mathbf{x}|\omega_j}). \tag{5.30}$$

Equating the model densities gives

$$\frac{P(\omega_1)}{|\boldsymbol{\Sigma}_1|^{1/2}} e^{-(1/2)k^2 r_{21}^2} = \frac{P(\omega_2)}{|\boldsymbol{\Sigma}_2|^{1/2}} e^{-(1/2)(1-k)^2 r_{12}^2}, \tag{5.31}$$

which, after taking the natural logarithm of each side, can be expressed as

$$(r_{12}^2 - r_{21}^2)k^2 - 2r_{12}^2 k + r_{12}^2 + 2\ln\frac{P(\omega_1)}{P(\omega_2)} - \ln\frac{|\boldsymbol{\Sigma}_1|}{|\boldsymbol{\Sigma}_2|} = 0. \tag{5.32}$$

Therefore

$$k = \begin{cases} \frac{1}{2} + \dfrac{1}{r^2}\left[\ln\dfrac{P(\omega_1)}{P(\omega_2)} - \frac{1}{2}\ln\dfrac{|\boldsymbol{\Sigma}_1|}{|\boldsymbol{\Sigma}_2|}\right], & \text{for } r_{21}^2 = r_{12}^2 = r^2 \\ \dfrac{1}{r_{12}^2 - r_{21}^2}\left[r_{12}^2 - \left(r_{12}^2 r_{21}^2 - \left[2\ln\dfrac{P(\omega_1)}{P(\omega_2)} \right.\right.\right. & \\ \qquad \left.\left.\left. - \ln\dfrac{|\boldsymbol{\Sigma}_1|}{|\boldsymbol{\Sigma}_2|}\right][r_{12}^2 - r_{21}^2]\right)^{1/2}\right], & \text{otherwise,} \end{cases} \tag{5.33}$$

where the sign of the square root is chosen so that k specifies the model intersection closest to a sample mean.

If certain a priori knowledge exists of a desired direction of an initial weight vector $\mathbf{w} = \mathbf{w}(0)$, then a better $\mathbf{b}$ can be calculated. This knowledge might arise from the application of the techniques of Subsection 5.2.4 on initial vector calculation. First we express the marginal constituent densities in the direction of $\mathbf{w}$:

$$f_j\left(\frac{\mathbf{w}^T\mathbf{x}}{\|\mathbf{w}\|}\right) = \frac{P(\omega_j)}{\sqrt{2\pi}\sigma_{\lambda_j}} \exp\left\{-\frac{1}{2}\left[\frac{\mathbf{w}^T(\mathbf{x} - \hat{\boldsymbol{\mu}}_{\mathbf{x}|\omega_j})}{\|\mathbf{w}\|\sigma_{\lambda_j}}\right]^2\right\}, \quad j = 1, 2 \tag{5.34}$$

where $\lambda_j = [\mathbf{w}^T\mathbf{X}/\|\mathbf{w}\|\,|\omega_j]$, so that σ_{λ_j} can be found from the class covariance matrices by the relationship

$$\sigma_{\lambda_j}^2 = \left(\frac{\mathbf{w}}{\|\mathbf{w}\|}\right)^T \boldsymbol{\Sigma}_j \left(\frac{\mathbf{w}}{\|\mathbf{w}\|}\right), \quad j = 1, 2. \tag{5.35}$$

For simplicity of notation, let σ_j denote σ_{λ_j} in this section. Now substituting $\mathbf{b}$ given in Equation (5.29) for $\mathbf{x}$ in Equation (5.34) and equating constituent densities we obtain

$$\frac{P(\omega_1)}{\sigma_1}\exp\left\{-\tfrac{1}{2}\left[\frac{k\mathbf{w}^T(\hat{\boldsymbol{\mu}}_2-\hat{\boldsymbol{\mu}}_1)}{\|\mathbf{w}\|\sigma_1}\right]^2\right\} = \frac{P(\omega_2)}{\sigma_2}\exp\left\{-\tfrac{1}{2}\left[\frac{(k-1)\mathbf{w}^T(\hat{\boldsymbol{\mu}}_2-\hat{\boldsymbol{\mu}}_1)}{\|\mathbf{w}\|\sigma_2}\right]^2\right\}. \tag{5.36}$$

Again to simplify notation, we define a scalar s by

$$s = \frac{\mathbf{w}^T(\hat{\boldsymbol{\mu}}_2-\hat{\boldsymbol{\mu}}_1)}{\|\mathbf{w}\|}. \tag{5.37}$$

After taking the natural logarithm of each side, Equation (5.36) can now be expressed as

$$s^2\left(\frac{1}{\sigma_2^2}-\frac{1}{\sigma_1^2}\right)k^2 - 2\frac{s^2}{\sigma_2^2}k + \frac{s^2}{\sigma_2^2} + 2\ln\frac{P(\omega_1)}{P(\omega_2)} - \ln\frac{\sigma_1^2}{\sigma_2^2} = 0.$$

Therefore

$$k = \begin{cases} \frac{1}{2} + \dfrac{\sigma^2}{s^2}\ln\dfrac{P(\omega_1)}{P(\omega_2)}, & \text{for } \sigma_2^2 = \sigma_1^2 = \sigma^2 \\[2ex] \dfrac{1}{\left(\dfrac{1}{\sigma_2^2}-\dfrac{1}{\sigma_1^2}\right)}\left[\dfrac{1}{\sigma_2^2} - \left(\dfrac{1}{\sigma_1^2\sigma_2^2} - \dfrac{1}{s^2}\right.\right. \\ \quad \left.\left.\times\left[2\ln\dfrac{P(\omega_1)}{P(\omega_2)} - \ln\dfrac{\sigma_1^2}{\sigma_2^2}\right]\left[\dfrac{1}{\sigma_2^2}-\dfrac{1}{\sigma_1^2}\right]\right)^{1/2}\right], & \text{otherwise.} \end{cases} \tag{5.38}$$

If the direction of the optimum $\mathbf{w}$ for the Gaussian models is known, Equation (5.38) together with Equation (5.29) specify a point $\mathbf{b}$ through which the optimum decision hyperplane for the Gaussian models passes. Whenever it is assumed that the best initial hyperplane is normal to the line joining the means, $\mathbf{w}(0)$ has the same direction as $\boldsymbol{\mu}_2 - \boldsymbol{\mu}_1$. For this assumption Equation (5.37) simplifies to

$$s = \|\hat{\boldsymbol{\mu}}_2 - \hat{\boldsymbol{\mu}}_1\|. \tag{5.39}$$

One may also assume that the class densities may be replaced by two class densities having the same means $\boldsymbol{\mu}_i$ as the actual densities, and having covariance matrices which are the average of the actual covariance matrices:

$$\begin{aligned}\hat{\boldsymbol{\Sigma}}_1 = \hat{\boldsymbol{\Sigma}}_2 &= \text{estimated covariance matrix of each class} \\ &= P(\omega_1)\boldsymbol{\Sigma}_1 + P(\omega_2)\boldsymbol{\Sigma}_2,\end{aligned}$$

where $\boldsymbol{\Sigma}_i$ is the covariance matrix of the ith class. These estimated covariance matrices are then "whitened" by multiplying them by a whitening matrix $\mathbf{A}$ such that $\mathbf{A}\hat{\boldsymbol{\Sigma}}_i\mathbf{A}^T = \mathbf{I}$ for $i = 1, 2$. The perpendicular bisector of the straight line joining the transformed means $\mathbf{A}\boldsymbol{\mu}_i$ of the two classes is often a good first guess for a linear decision surface [11].

Since the calculation of all the sample covariances may be considered a rather lengthy task just for estimating an initial vector, statistical independence among the features may be assumed. The covariance matrices will then be diagonal matrices with components equal to the sample variances of the features. This also simplifies the calculation of the squared Mahalanobis distances or the $\sigma_j^2, j = 1,2$. After performing the normalization discussed in the last section, it may be found that the variance of each feature in each class is sufficiently close to unity to make the looser approximation

$$\boldsymbol{\Sigma}_j = \mathbf{I}, \quad j = 1, 2,$$

where $\mathbf{I}$ denotes the identity matrix. Then, either from Equations (5.33) and (5.30) or from Equations (5.38) and (5.39),

$$k = \tfrac{1}{2} + \frac{\ln(P(\omega_1)/P(\omega_2))}{\|\hat{\boldsymbol{\mu}}_{\mathbf{x}|\omega_2} - \hat{\boldsymbol{\mu}}_{\mathbf{x}|\omega_1}\|^2}. \tag{5.40}$$

For this special case, $\mathbf{b}$ can be written from Equation (5.29) as

$$\mathbf{b} = \frac{\hat{\boldsymbol{\mu}}_{\mathbf{x}|\omega_1} + \hat{\boldsymbol{\mu}}_{\mathbf{x}|\omega_2}}{2} + \frac{(\hat{\boldsymbol{\mu}}_{\mathbf{x}|\omega_2} - \hat{\boldsymbol{\mu}}_{\mathbf{x}|\omega_1})}{\|\boldsymbol{\mu}_{\mathbf{x}|\omega_2} - \boldsymbol{\mu}_{\mathbf{x}|\omega_1}\|^2} \ln \frac{P(\omega_1)}{P(\omega_2)}. \tag{5.41}$$

For any of the choices of $\mathbf{b}$ discussed in this section, the vector $\boldsymbol{\beta}$ which translates the origin in *augmented* space is defined as

$$\boldsymbol{\beta} = \left[\begin{array}{c} 0 \\ \hline \mathbf{b} \end{array} \right]. \tag{5.42}$$

5.2.3 Initial Spread from Hyperplane Point

In the application of training procedures, it is often useful to have a measure of the spread of the feature vector probability density around the point on the initial decision hyperplane closest to the mean of the entire training set of feature vectors. Let $\mathbf{a}$ denote this point on the hyperplane, so that

$$\mathbf{a} = \boldsymbol{\mu}_x - \frac{1}{\|\mathbf{w}(0)\|^2} \mathbf{v}(0)^T \left[\begin{array}{c} 1 \\ \hline \boldsymbol{\mu}_{\mathbf{x}} \end{array} \right] \mathbf{w}(0) \tag{5.43}$$

where

$$\boldsymbol{\mu}_{\mathbf{x}} = \boldsymbol{\mu}_1 P(\omega_1) + \boldsymbol{\mu}_2 P(\omega_2). \tag{5.44}$$

The initial window size for the window procedure (described later) and the initial step size $\rho(0)$ for any of the procedures can be better chosen if this spread is known at least approximately. We denote this spread in d-dimensional space by δ_d and define it to be the square root of the second moment of the probability density of $\mathbf{x}$ about the point $\mathbf{a}$. Alternatively δ_d can be viewed as the root mean square (RMS) distance of the $\mathbf{x}$'s from $\mathbf{a}$. Analytically,

$$\delta_d^2 = E(\|\mathbf{x} - \mathbf{a}\|^2). \tag{5.45}$$

Fortunately, it is possible to express δ_d as a relatively simple function of the class means and variances. First, δ_d^2 can be written as

$$\delta_d^2 = E\{E[\|\mathbf{x} - \mathbf{a}\|^2 \,|\, \omega_j]\}$$

or, after adding and subtracting $\boldsymbol{\mu}_j$,

$$\begin{aligned}\delta_d^2 &= E\{E[\|(\mathbf{x} - \boldsymbol{\mu}_j) - (\mathbf{a} - \boldsymbol{\mu}_j)\|^2 \,|\, \omega_j]\}\\ &= E\{E[\|\mathbf{x} - \boldsymbol{\mu}_j\|^2 - 2(\mathbf{x} - \boldsymbol{\mu}_j)^T(\mathbf{a} - \boldsymbol{\mu}_j) + \|\mathbf{a} - \boldsymbol{\mu}_j\|^2 \,|\, \omega_j]\}.\end{aligned}$$

After taking the expectation conditioned on $\omega = \omega_j$, we find

$$\delta_d^2 = E(\sigma^2_{\mathbf{x}|\omega_j} + \|\mathbf{a} - \boldsymbol{\mu}_j\|^2).$$

Now, taking the final expectation, we have

$$\delta_d^2 = (\sigma^2_{\mathbf{x}|\omega_1} + \|\mathbf{a} - \boldsymbol{\mu}_1\|^2)\,P(\omega_1) + (\sigma^2_{\mathbf{x}|\omega_2} + \|\mathbf{a} - \boldsymbol{\mu}_2\|^2)P(\omega_2). \tag{5.46}$$

Usually the class variances are known in terms of the coordinate class variances $\sigma^2_{x_i|\omega_j}$ rather than the total class variances $\sigma^2_{\mathbf{x}|\omega_j}$. However,

$$\sigma^2_{\mathbf{x}|\omega_j} = \sum_{i=1}^{d} \sigma^2_{x_i|\omega_j}. \tag{5.47}$$

Therefore δ_d can be expressed in terms of the $\sigma^2_{x_i|\omega_j}$ as

$$\begin{aligned}\delta_d = \Bigg[&\left(\sum_{i=1}^{d} \sigma^2_{x_i|\omega_1} + \|\mathbf{a} - \boldsymbol{\mu}_1\|^2\right)P(\omega_1)\\ &+ \left(\sum_{i=1}^{d} \sigma^2_{x_i|\omega_2} + \|\mathbf{a} - \boldsymbol{\mu}_2\|^2\right)P(\omega_2)\Bigg]^{1/2}.\end{aligned} \tag{5.48}$$

The window width for the window procedure is always one-dimensional, so that a modification of Equation (5.48) to reflect a one-dimensional spread is appropriate. We denote this one-dimensional spread by δ_1. Since $\sigma^2_{\mathbf{x}|\omega_j}$ or $\sum_{x=1}^{d} \sigma^2_{x_i|\omega_j}$ in expressions (5.46) or (5.48), respectively, represent d-dimensional class variances, replacing these with

$$\frac{1}{d}\sigma^2_{\mathbf{x}|\omega_j} \quad \text{or} \quad \frac{1}{d}\sum_{i=1}^{d} \sigma^2_{x_i|\omega_j}$$

will give an average one-dimensional variance. Therefore, δ_1 is defined as

$$\begin{aligned}\delta_1 = \Bigg[&\left(\frac{1}{d}\sum_{i=1}^{d} \sigma^2_{x_i|\omega_1} + \|\mathbf{a} - \boldsymbol{\mu}_1\|^2\right)P(\omega_1)\\ &+ \left(\frac{1}{d}\sum_{i=1}^{d} \sigma^2_{x_i|\omega_2} + \|\mathbf{a} - \boldsymbol{\mu}_2\|^2\right)P(\omega_2)\Bigg]^{1/2}.\end{aligned} \tag{5.49}$$

5.2.4 Calculation of Initial Vector

First we determine the decision surface for the case where the constituent densities are Gaussian and the optimum surface is a hyperplane. Since this is just the initial decision hyperplane of the training process, the Gaussian form need not be an accurate representation of the two classes. Equating Gaussian constituent densities whose equations are given by Equation (5.28) and taking the natural logarithm of each side, we find the intersection of their surfaces to be given by

$$(\mathbf{x} - \hat{\boldsymbol{\mu}}_{\mathbf{x}|\omega_1})^T \boldsymbol{\Sigma}_1^{-1} (\mathbf{x} - \hat{\boldsymbol{\mu}}_{\mathbf{x}|\omega_1}) - 2 \ln \frac{P(\omega_1)}{P(\omega_2)} + \ln \frac{|\boldsymbol{\Sigma}_1|}{|\boldsymbol{\Sigma}_2|} \\ = (\mathbf{x} - \hat{\boldsymbol{\mu}}_{\mathbf{x}|\omega_2})^T \boldsymbol{\Sigma}_2^{-1} (\mathbf{x} - \hat{\boldsymbol{\mu}}_{\mathbf{x}|\omega_2}). \tag{5.50}$$

Note that for this equation to represent a hyperplane, the quadratic terms must be equal, i.e. $\mathbf{x}^T \boldsymbol{\Sigma}_1^{-1} \mathbf{x} = \mathbf{x}^T \boldsymbol{\Sigma}_2^{-1} \mathbf{x}$. Therefore $\boldsymbol{\Sigma}_1 = \boldsymbol{\Sigma}_2 = \boldsymbol{\Sigma}$ represents the case where the optimum decision surface is a hyperplane. Denoting $\hat{\boldsymbol{\mu}}_{\mathbf{x}|\omega_j}$ by $\boldsymbol{\mu}_j$, Equation (5.50) can be written

$$\mathbf{x}^T \boldsymbol{\Sigma}^{-1} (\boldsymbol{\mu}_2 - \boldsymbol{\mu}_1) + \tfrac{1}{2}(\boldsymbol{\mu}_1^T \boldsymbol{\Sigma}^{-1} \boldsymbol{\mu}_1 - \boldsymbol{\mu}_2^T \boldsymbol{\Sigma}^{-1} \boldsymbol{\mu}_2) - \ln \frac{P(\omega_1)}{P(\omega_2)} = 0. \tag{5.51}$$

Recall that $\mathbf{y} = [1, \mathbf{x}]^T$, so that Equation (5.51) can be written

$$\left[\begin{array}{c} \tfrac{1}{2}(\boldsymbol{\mu}_1^T \boldsymbol{\Sigma}^{-1} \boldsymbol{\mu}_1 - \boldsymbol{\mu}_2^T \boldsymbol{\Sigma}^{-1} \boldsymbol{\mu}_2) - \ln(P(\omega_1)/P(\omega_2)) \\ \hdashline \boldsymbol{\Sigma}^{-1}(\boldsymbol{\mu}_2 - \boldsymbol{\mu}_1) \end{array}\right]^T \mathbf{y} = 0 \tag{5.52}$$

which is in the form of a hyperplane equation $\mathbf{v}^T \mathbf{y} = 0$. Now from Equations (5.29), (5.30), (5.33), and (5.42) of Subsection 5.2.2, the origin translation vector $\boldsymbol{\beta}$ for the case where $\boldsymbol{\Sigma}_1 = \boldsymbol{\Sigma}_2 = \boldsymbol{\Sigma}$ is

$$\boldsymbol{\beta} = \left[\begin{array}{c} 0 \\ \hdashline \dfrac{\boldsymbol{\mu}_1 + \boldsymbol{\mu}_2}{2} + \dfrac{\boldsymbol{\mu}_2 - \boldsymbol{\mu}_1}{\imath^2} \ln \dfrac{P(\omega_1)}{P(\omega_2)} \end{array}\right], \tag{5.53}$$

where $\imath^2 = \imath_{12}^2 = \imath_{21}^2 = (\boldsymbol{\mu}_2 - \boldsymbol{\mu}_1)^T \boldsymbol{\Sigma}^{-1} (\boldsymbol{\mu}_2 - \boldsymbol{\mu}_1)$. Recall from subsection 5.1.3 that an augmented weight vector $\boldsymbol{\nu}$ was defined which determines a hyperplane passing through the translated origin $\boldsymbol{\beta}$. The relationship between $\mathbf{v}$ and $\boldsymbol{\nu}$, such that they each determine the same decision hyperplane in $\mathbf{x}$-space, was given by Equation (5.24). Note that $\mathbf{v}$ and $\boldsymbol{\nu}$ differ only in their first component, where $\nu_0 = v_0 + \mathbf{v}^T \boldsymbol{\beta}$. From Equations (5.52) and (5.53) we find that $\mathbf{v}^T \boldsymbol{\beta}$ for the case where $\boldsymbol{\Sigma} = \boldsymbol{\Sigma}_1 = \boldsymbol{\Sigma}_2$ is given by

$$\mathbf{v}^T \boldsymbol{\beta} = -\left[\frac{\boldsymbol{\mu}_1^T \boldsymbol{\Sigma}^{-1} \boldsymbol{\mu}_1 - \boldsymbol{\mu}_2^T \boldsymbol{\Sigma}^{-1} \boldsymbol{\mu}_2}{2} - \ln \frac{P(\omega_1)}{P(\omega_2)}\right]. \tag{5.54}$$

Now from Equation (5.52), $\nu_0 = v_0 + \mathbf{v}^T \boldsymbol{\beta} = 0$. Therefore the decision hyperplane specified by Equation (5.52) is the same decision hyperplane specified by

$$\boldsymbol{\nu} = \left[\begin{array}{c} 0 \\ \hdashline \boldsymbol{\Sigma}^{-1}(\boldsymbol{\mu}_2 - \boldsymbol{\mu}_1) \end{array}\right] \tag{5.55}$$

in $\boldsymbol{\beta}$-origin space. Since $\nu_0 = 0$ and since hyperplanes specified by $\boldsymbol{\nu}$ must always pass through the origin of $\boldsymbol{\beta}$-origin space, this decision hyperplane must pass through the point $\mathbf{b}$ defined by Equations (5.29) and (5.33). In $\mathbf{x}$-space this hyperplane is normal to the vector $\mathbf{w}$ consisting of the components v_i, $i = 1, \ldots, d$ in $\mathbf{v}$, i.e., normal to $\mathbf{w} = \mathbf{\Sigma}^{-1}(\boldsymbol{\mu}_2 - \boldsymbol{\mu}_1)$.

As demonstrated by Equation (5.50), the optimum decision surface for Gaussian constituent density models where $\mathbf{\Sigma}_1 \neq \mathbf{\Sigma}_2$ is not a hyperplane. However, we wish to use these models to find a reasonably good initial hyperplane. Figure 5.6 illustrates a case where $\mathbf{\Sigma}_1 = \mathbf{\Sigma}_2$. For such cases, an initial hyperplane specified by Equation (5.55) is parallel to the major axis of the constituent densities. (The initial hyperplane illustrated in Figure 5.6 is not so oriented.) For cases where $\mathbf{\Sigma}_1 \neq \mathbf{\Sigma}_2$, it seems reasonable to choose an initial hyperplane which is parallel to the average of the orientation of the major axes of the two Gaussian constituent density models. We therefore define an average covariance matrix $\mathbf{\Sigma}_{\text{av}}$ by

$$\mathbf{\Sigma}_{\text{av}} = \tfrac{1}{2}(\mathbf{\Sigma}_1 + \mathbf{\Sigma}_2). \tag{5.56}$$

Note that $\mathbf{\Sigma}_{\text{av}}$ is the covariance matrix of a feature vector probability density obtained by both translating the class densities so that their means coincide and making the a priori class probabilities equal. In $\boldsymbol{\beta}$-origin space the initial augmented weight vector $\mathbf{v}(0)$ which specifies the desired initial hyperplane is given by

$$\mathbf{v}(0) = \left[\begin{array}{c} 0 \\ \hline \mathbf{\Sigma}_{av}^{-1}(\hat{\boldsymbol{\mu}}_{\mathbf{x}|\omega_2} - \hat{\boldsymbol{\mu}}_{\mathbf{x}|\omega_1}) \end{array}\right]. \tag{5.57}$$

As mentioned in Subsection 5.2.2, the task of calculating all the sample covariances can be a lengthy task. Therefore, as in Subsection 5.2.2, it may be desirable to assume statistical independence among the features. If this assumption is made and the average variance normalization presented in Subsection 5.2.1 has also been performed, $\mathbf{\Sigma}_{\text{av}}$ becomes the identity matrix. Under these conditions

$$\mathbf{v}(0) = \left[\begin{array}{c} 0 \\ \hline \hat{\boldsymbol{\mu}}_{\mathbf{x}|\omega_2} - \hat{\boldsymbol{\mu}}_{\mathbf{x}|\omega_1} \end{array}\right], \tag{5.58}$$

so that the decision hyperplane is normal to the line joining the class means and passes through the point $\mathbf{b}$ defined in Subsection 5.2.2.

If quadratic $\boldsymbol{\Phi}$-mapping is used, Equation (5.50) always represents a hyperplane in quadratic $\mathbf{x}_q$-space, where $\mathbf{x}_q$ is defined by

$$\mathbf{x}_q = [\mathbf{x}^T, x_1^2, x_2^2, \ldots, x_d^2, x_1x_2, \ldots, x_ix_j, \ldots, x_{d-1}x_d]^T, \quad i = 1, \ldots, d-1; \quad j = i+1, \ldots, d.$$

In this space, Equation (5.50) can be written

$$\mathbf{v}_q^T\mathbf{y}_q = 0.$$

For simplicity of notation in expressing $\mathbf{v}_q$ we first define a matrix $\mathbf{A} = \{a_{ij}\}$ as

$$\mathbf{A} = \mathbf{\Sigma}_1^{-1} - \mathbf{\Sigma}_2^{-1}$$

and then partition $\mathbf{v}_q$ as follows

$$\mathbf{v}_q = \begin{bmatrix} v_0 \\ \hline \mathbf{v}^{(x)} \\ \hline \mathbf{v}^{(xx)} \end{bmatrix}. \tag{5.59}$$

The $\mathbf{v}_q$ which describes the intersection of the surfaces given by Equation (5.50) is then given by

$$\begin{aligned} v_0 &= \boldsymbol{\mu}_1^T \boldsymbol{\Sigma}_1^{-1} \boldsymbol{\mu}_1 - \boldsymbol{\mu}_2 \boldsymbol{\Sigma}_2^{-1} \boldsymbol{\mu}_2 + \ln \frac{|\boldsymbol{\Sigma}_1|}{|\boldsymbol{\Sigma}_2|} - 2 \ln \frac{P(\omega_1)}{P(\omega_2)} \\ \mathbf{v}^{(x)} &= -2(\boldsymbol{\Sigma}_1^{-1} \boldsymbol{\mu}_1 - \boldsymbol{\Sigma}_2^{-1} \boldsymbol{\mu}_2) \\ \mathbf{v}^{(xx)} &= (a_{11}, a_{22}, \ldots, a_{dd}, 2a_{12}, \ldots, 2a_{ij}, \ldots, 2a_{d-1,d}) \quad i = 1, \ldots, d-1; \\ & \qquad j = i+1, \ldots, d. \end{aligned} \tag{5.60}$$

In the following sections several important training procedures will be discussed. The techniques just presented for finding a reasonable origin translation and initial vector will in general significantly improve the performance of these training procedures.

Hereafter we do not make explicit use of the notation $\boldsymbol{v},\boldsymbol{\beta}$ required for a translated origin. To obtain explicit expressions in terms of $\boldsymbol{v}$ and $\boldsymbol{\beta}$, replace $\mathbf{V}$ by $\boldsymbol{v}$ and $\mathbf{Y}$ by $\mathbf{Y} - \boldsymbol{\beta}$.

5.3 The Window Training Procedure

As discussed in Subsection 4.5.1, when the costs of the types of misclassifications are equal, usually the best decision hyperplane is the one which classifies feature vectors with a minimum probability of error. In this section we derive the *window training procedure*, which seeks a decision hyperplane that gives this minimum probability of classification error. A proof that the sequence $\{\mathbf{V}(n)\}$ determined by the window procedure converges to a weight vector in the set $\mathscr{V}^*$ in mean square and with probability one is presented in Appendix B. In this section $\mathscr{V}^*$ contains all the vectors $\mathbf{v}^*$ which specify a minimum probability of error decision hyperplane. In addition, training procedure design considerations and some simulation examples will be presented. Other special forms of the window training procedure are covered in Reference [1].

5.3.1 Derivation of the Regression Function

As discussed in Subsection 4.8.1, if $\mathbf{G}(\mathbf{v})$ denotes some deterministic function and if it is expressable in the form $\mathbf{G}(\mathbf{v}) = E[\mathbf{Z}(n) | \mathbf{V}(n) = \mathbf{v}]$, where $\mathbf{Z}(n)$ denotes a random variable, then $\mathbf{G}(\mathbf{v}) = E[\mathbf{Z}(n) | \mathbf{V}(n) = \mathbf{v}]$ is called a *vector regression function.* When applied to training procedures, $\mathbf{Z}(n)$ is a random variable as a result of its functional dependence on a random feature vector $\mathbf{X}(n)$. In this section we are concerned with finding a training procedure which will seek the minimum of the probability of error loss function as modified for weight vector magnitude compensation. Therefore, from Equations (4.29) and (5.13) we define

$$\mathbf{G}(\mathbf{v}) = \|\mathbf{w}\| \int_{\mathscr{S}} \mathbf{y}[f_2(\mathbf{x}) - f_1(\mathbf{x})]\, ds. \tag{5.61}$$

Unfortunately, it seems impossible to express $\mathbf{G}(\mathbf{v})$ directly as a vector regression function. However, a vector regression function $\hat{\mathbf{G}}(\mathbf{v}, c)$ can be defined which approximates $\mathbf{G}(\mathbf{v})$ and converges to $\mathbf{G}(\mathbf{v})$ in the limit as c approaches zero. This function is given in deterministic form as

$$\hat{\mathbf{G}}(\mathbf{v}, c) = \|\mathbf{w}\| \int \Psi\left(\frac{\mathbf{v}^T\mathbf{y}}{\|\mathbf{w}\|}, c\right)\left(\mathbf{y} - \frac{\mathbf{v}^T\mathbf{y}}{\|\mathbf{w}\|^2}\begin{bmatrix} 0 \\ \mathbf{w} \end{bmatrix}\right)[f_2(\mathbf{x}) - f_1(\mathbf{x})]\, d\mathbf{x} \tag{5.62}$$

where $\int d\mathbf{x}$ denotes integration over the entire feature space and where $\Psi(\mathbf{v}^T\mathbf{y}/\|\mathbf{w}\|, c)$ denotes a *Parzen window function* (see Section 3.6 and Reference [2]), with the property

$$\lim_{c \to 0} \Psi\left(\frac{\mathbf{v}^T\mathbf{y}}{\|\mathbf{w}\|}, c\right) = \delta\left(\frac{\mathbf{v}^T\mathbf{y}}{\|\mathbf{w}\|}\right). \tag{5.63}$$

The symbol $\delta(\xi)$ denotes the Dirac delta function [18] defined by

$$\int_{-\infty}^{\infty} g(\xi)\,\delta(\xi - \xi_0)\, d\xi_0 = g(\xi_0)$$

so that

$$\int_{-\infty}^{\infty} \delta(\xi)\, d\xi = \int_{-\varepsilon}^{\varepsilon} \delta(\xi)\, d\xi = 1 \quad \text{for all } \varepsilon > 0.$$

Examples of two Parzen window functions will be given later in this section. Recall from Section 1.5 and Figure 1.7 that the magnitude of $\mathbf{v}^T\mathbf{y}/\|\mathbf{w}\|$ is the shortest Euclidean distance from a point $\mathbf{x}$ to the hyperplane $\mathbf{v}^T\mathbf{y} = 0$ constrained to nonaugmented feature space, i.e., $\mathbf{x}$-space. Note from Figure 1.7 that $\mathbf{v}^T\mathbf{y}/\|\mathbf{w}\| > 0$ for $\mathbf{x}$ in the half-space on the side of the hyperplane to which $\mathbf{w}$ points. Conversely, $\mathbf{v}^T\mathbf{y}/\|\mathbf{w}\| < 0$ on the other side. The term

$$\mathbf{y} - \frac{\mathbf{v}^T\mathbf{y}}{\|\mathbf{w}\|^2}\begin{bmatrix} 0 \\ \mathbf{w} \end{bmatrix}$$

in Equation (5.62) is analagous to the $\mathbf{y}$ term in Equation (5.61). This analogy exists by noting that $\mathbf{y}$ must always lie on the hyperplane surface $\mathscr{S}$ in Equation (5.61) but a $\mathbf{y}$ term in Equation (5.62) would in general not lie on the hypersurface. This is a result of the nonzero window width, i.e., $c \neq 0$. Recall from Subsection 4.5.3 that

$$\mathbf{y} - \frac{\mathbf{v}^T\mathbf{y}}{\|\mathbf{w}\|^2}\begin{bmatrix} 0 \\ \mathbf{w} \end{bmatrix}$$

represents a projection of $\mathbf{y}$ in feature space, i.e., $\mathbf{x}$-space, onto the decision hyperplane, and that this projection is orthogonal to the hyperplane in $\mathbf{x}$-space. Therefore

$$\mathbf{y} - \frac{\mathbf{v}^T\mathbf{y}}{\|\mathbf{w}\|^2}\begin{bmatrix} 0 \\ \hdashline \mathbf{w} \end{bmatrix}$$

is always orthogonal to $\mathbf{v}$. (See the discussion following Equation (4.67).)

Now from (5.62), (5.63), and the definition of the Dirac delta function,

$$\lim_{c\to 0} \hat{\mathbf{G}}(\mathbf{v}, c) = \|\mathbf{w}\| \int_{\mathscr{S}} \mathbf{y}[f_2(\mathbf{x}) - f_1(\mathbf{x})]\, ds = \mathbf{G}(\mathbf{v}). \tag{5.64}$$

The validity of (5.64) can be demonstrated for any component of $\mathbf{y}$, say y_k, as follows. First note that for the kth component of $\hat{\mathbf{G}}$,

$$\lim_{c\to 0} \hat{G}_k(\mathbf{v}, c) = \lim_{c\to 0} \|\mathbf{w}\| \int_{-\infty}^{\infty} dx_1 \cdots \int_{-\infty}^{\infty} dx_d \times \Psi\left(\frac{\mathbf{v}^T\mathbf{y}}{\|\mathbf{w}\|}, c\right)\left(y_k - \frac{\mathbf{v}^T\mathbf{y}}{\|\mathbf{w}\|^2} w_k\right)[f_2(\mathbf{x}) - f_1(\mathbf{x})]. \tag{5.65}$$

But the Dirac delta function definition and

$$\lim_{c\to 0} \Psi(\xi - \xi_0, c) = \delta(\xi - \xi_0)$$

gives the general equality

$$\lim_{c\to 0} \int_{-\infty}^{\infty} d\xi\, \Psi(b\xi - \xi_0, c) g(\xi) = \int_{-\infty}^{\infty} d\xi\, \delta(b\xi - \xi_0) g(\xi) = \frac{1}{|b|}\, g(\xi_0/b), \tag{5.66}$$

where b denotes an arbitrary constant. Therefore

$$\begin{aligned}
\lim_{c\to 0} \hat{G}_k(\mathbf{v}, c) &= \|\mathbf{w}\| \int_{-\infty}^{\infty} dx_1 \cdots \int_{-\infty}^{\infty} dx_d\, \delta\left(\frac{\mathbf{v}^T\mathbf{y}}{\|\mathbf{w}\|}\right) \times \left[y_k - \frac{\mathbf{v}^T\mathbf{y}}{\|\mathbf{w}\|^2} w_k\right][f_2(\mathbf{x}) - f_1(\mathbf{x})] \\
&= \begin{cases}
\dfrac{\|\mathbf{w}\|^2}{|w_d|} \displaystyle\int_{-\infty}^{\infty} dx_1 \cdots \int_{-\infty}^{\infty} dx_{d-1}\, y_k \\
\quad \times [f_2(x_1, \ldots, x_{d-1}, x_d(\mathbf{v})) - f_1(x_1, \ldots, x_{d-1}, x_d(\mathbf{v}))], & k \neq d \\
\dfrac{\|\mathbf{w}\|^2}{|w_d|} \displaystyle\int_{-\infty}^{\infty} dx_1 \cdots \int_{-\infty}^{\infty} dx_{d-1}\, y_d(\mathbf{v}) \\
\quad \times [f_2(x_1, \ldots, x_{d-1}, x_d(\mathbf{v})) - f_1(x_1, \ldots, x_{d-1}, x_d(\mathbf{v}))], & k = d
\end{cases}
\end{aligned} \tag{5.67}$$

where $x_d(\mathbf{v})$ is defined by an expansion of $\mathbf{v}^T\mathbf{y}/\|\mathbf{w}\| = 0$ giving

$$x_d(\mathbf{v}) = -\frac{1}{w_d}\left(w_0 + \sum_{i=1}^{d-1} w_i x_i\right). \tag{5.68}$$

Finally note that (5.67) gives the same form as (4.28) multiplied by $-\|\mathbf{w}\|^2$.

Let $\hat{\mathbf{v}}^*(n)$, a function of $c(n)$, denote those zeros of $\hat{\mathbf{G}}(\mathbf{v}, c)$ such that

$$\lim_{c\to 0} \hat{\mathbf{G}}(\hat{\mathbf{v}}^*, c) = \mathbf{G}(\mathbf{v}^*),$$

and let $-\hat{\mathbf{v}}^*$ denote those zeros such that $\lim_{c\to 0} \hat{\mathbf{G}}(-\mathbf{v}^*, c) = \mathbf{G}(-\mathbf{v}^*)$, i.e.

$$\lim_{c(n)\to 0} \hat{\mathbf{v}}^*(c(n)) = \mathbf{v}^*. \tag{5.69}$$

If a vector random variable $\mathbf{Z}(\mathbf{X}, \mathbf{V}, \omega, c)$ is defined by

$$\mathbf{Z} = \begin{cases} 2c\|\mathbf{W}\|\left(\mathbf{Y} - \dfrac{\mathbf{V}^T\mathbf{Y}}{\|\mathbf{W}\|^2}\begin{bmatrix} 0 \\ \hdashline \mathbf{W} \end{bmatrix}\right)\Psi\left(\dfrac{\mathbf{V}^T\mathbf{Y}}{\|\mathbf{W}\|}, c\right), & \text{for } \omega = \omega_2, \\ -2c\|\mathbf{W}\|\left(\mathbf{Y} - \dfrac{\mathbf{V}^T\mathbf{Y}}{\|\mathbf{W}\|^2}\begin{bmatrix} 0 \\ \hdashline \mathbf{W} \end{bmatrix}\right)\Psi\left(\dfrac{\mathbf{V}^T\mathbf{Y}}{\|\mathbf{W}\|}, c\right), & \text{for } \omega = \omega_1, \end{cases} \tag{5.70}$$

then it can be seen from (5.62) that $\hat{\mathbf{G}}(\mathbf{v}, c)$ is a vector regression function and can be written

$$\hat{\mathbf{G}}(\mathbf{v}, c) = \frac{1}{2c} E(\mathbf{Z} \mid \mathbf{V} = \mathbf{v}). \tag{5.71}$$

Two Parzen window functions of particular interest are the rectangular and Gaussian windows.

Rectangular Window.

$$\Psi\left(\frac{\mathbf{v}^T\mathbf{y}}{\|\mathbf{w}\|}, c\right) = \begin{cases} \dfrac{1}{2c}, & \text{for } \dfrac{|\mathbf{v}^T\mathbf{y}|}{\|\mathbf{w}\|} \leqslant c, \\ 0, & \text{otherwise}, \end{cases} \tag{5.72}$$

$$\mathbf{Z} = \begin{cases} \|\mathbf{W}\|\left(\mathbf{Y} - \dfrac{\mathbf{V}^T\mathbf{Y}}{\|\mathbf{W}\|^2}\begin{bmatrix} 0 \\ \hdashline \mathbf{W} \end{bmatrix}\right), & \text{for } |\mathbf{V}^T\mathbf{Y}| \leqslant c\|\mathbf{W}\| \text{ and } \omega = \omega_2, \\ \mathbf{0}, & \text{for } |\mathbf{V}^T\mathbf{Y}| > c\|\mathbf{W}\|, \\ -\|\mathbf{W}\|\left(\mathbf{Y} - \dfrac{\mathbf{V}^T\mathbf{Y}}{\|\mathbf{W}\|^2}\begin{bmatrix} 0 \\ \hdashline \mathbf{W} \end{bmatrix}\right), & \text{for } |\mathbf{V}^T\mathbf{Y}| \leqslant c\|\mathbf{W}\| \text{ and } \omega = \omega_1. \end{cases} \tag{5.73}$$

An example of the rectangular window function for an arbitrary $\mathbf{v}$ is illustrated in two-dimensional feature space in Figure 5.9.

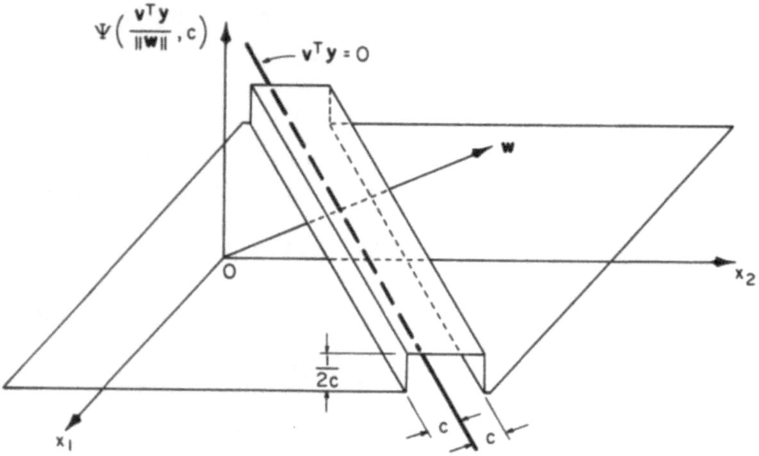

Figure 5.9. A rectangular window function.

Gaussian Window.

$$\Psi\left(\frac{\mathbf{v}^T\mathbf{y}}{\|\mathbf{W}\|}, c\right) = \frac{1}{\sqrt{2\pi}c}\exp\left[-\tfrac{1}{2}\left(\frac{\mathbf{v}^T\mathbf{y}}{\|\mathbf{W}\|c}\right)^2\right] \tag{5.74}$$

$$\mathbf{Z} = \begin{cases} \left(\dfrac{2}{\pi}\right)^{1/2}\|\mathbf{W}\|\left(\mathbf{Y} - \dfrac{\mathbf{V}^T\mathbf{Y}}{\|\mathbf{W}\|^2}\begin{bmatrix}0\\ \hdashline \mathbf{W}\end{bmatrix}\right)\exp\left[-\tfrac{1}{2}\left(\dfrac{\mathbf{V}^T\mathbf{Y}}{\|\mathbf{W}\|c}\right)^2\right], & \text{for } \omega = \omega_2, \\ -\left(\dfrac{2}{\pi}\right)^{1/2}\|\mathbf{W}\|\left(\mathbf{Y} - \dfrac{\mathbf{V}^T\mathbf{Y}}{\|\mathbf{W}\|^2}\begin{bmatrix}0\\ \hdashline \mathbf{W}\end{bmatrix}\right)\exp\left[-\tfrac{1}{2}\left(\dfrac{\mathbf{V}^T\mathbf{Y}}{\|\mathbf{W}\|c}\right)^2\right], & \text{for } \omega = \omega_1. \end{cases} \tag{5.75}$$

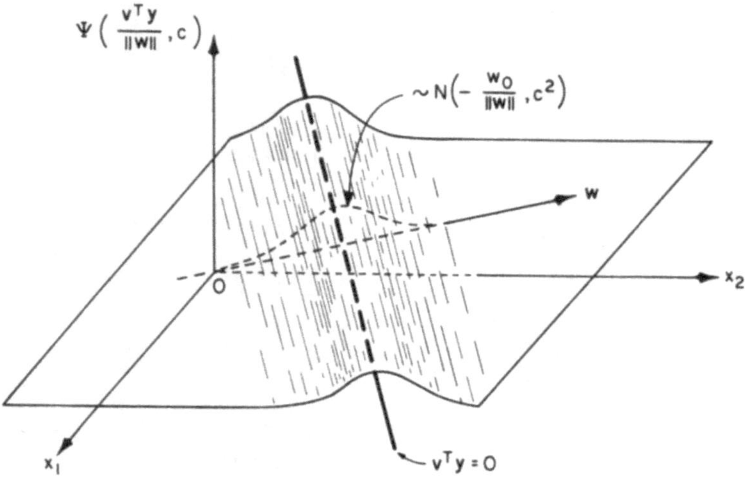

Figure 5.10. A Gaussian window function.

An example of the Gaussian window function for arbitrary $\mathbf{v}$ is illustrated in two-dimensional feature space in Figure 5.10.

When applied to (5.62), the effect of the rectangular or Gaussian window functions on the constituent densities may be viewed as an averaging or smoothing effect. This effect is less pronounced as c becomes smaller.

5.3.2 Window Training Procedures and Theorems

The first window procedure to be presented will take advantage of possible a priori knowledge of the spatial relations among the classes. A reasonable guess can then be made as to whether an assignment to class ω_2 is best for feature vectors whose component x_k is generally larger than the component x_k of those assigned to class ω_1 or whether the opposite assignment is best. As discussed in Section 4.4, this is indicated in the weight vector by the sign of v_k^*, $k \neq 0$. This will be denoted by $\operatorname{sgn}(v_k^*) = \pm 1$. Under these conditions the form of (4.32) is applicable.

Window Training Procedure: sgn (v_k^*) Known. This procedure is described by the sequence of recursively defined random vectors $\{\mathbf{V}(n)\}$:

$$\mathbf{V}(n+1) = \mathbf{V}(n) + \begin{cases} \rho(n)\mathbf{Z}(n), & \text{for } \|\mathbf{W}(n) + \rho(n)\mathbf{S}(n)\| < M, \\ \mathbf{0}, & \text{otherwise,} \end{cases} \tag{5.76}$$

where

$$\mathbf{Z}(n) = \begin{cases} 2c(n)\|\mathbf{W}(n)\|\left(\tilde{\mathbf{Y}}(n) - \dfrac{\mathbf{V}(n)^T\mathbf{Y}(n)}{\|\mathbf{W}(n)\|^2}\begin{bmatrix} 0 \\ \tilde{\mathbf{W}}(n) \end{bmatrix}\right) \\ \qquad \times \Psi\left(\dfrac{\mathbf{V}(n)^T\mathbf{Y}(n)}{\|\mathbf{W}(n)\|}, c(n)\right), \quad \text{for } \omega(n) = \omega_1, \\ -2c(n)\|\mathbf{W}(n)\|\left(\tilde{\mathbf{Y}}(n) - \dfrac{\mathbf{V}(n)^T\mathbf{Y}(n)}{\|\mathbf{W}(n)\|^2}\begin{bmatrix} 0 \\ \tilde{\mathbf{W}}(n) \end{bmatrix}\right) \\ \qquad \times \Psi\left(\dfrac{\mathbf{V}(n)^T\mathbf{Y}(n)}{\|\mathbf{W}(n)\|}, c(n)\right), \quad \text{for } \omega(n) = \omega_1, \end{cases} \tag{5.77}$$

$\mathbf{S}$ denotes the vector $[Z_1, Z_2, \ldots, Z_d]^T$,

$$V_k(0) = \begin{cases} +1, & \text{if } v_k^* > 0, \\ -1, & \text{if } v_k^* < 0, \end{cases} \tag{5.78}$$

where components of $\mathbf{V}(0)$ other than $V_k(0)$ are chosen arbitrarily, and where M is a large number chosen so that

$$M \gg \min \left\| \frac{\mathbf{v}^*}{v_k^*} \right\|. \tag{5.79}$$

The relationship between $\mathbf{V}(n)$ and $\mathbf{W}(n)$ is

$$\mathbf{V}(n) = \begin{bmatrix} W_0(n) \\ \hdashline \mathbf{W}(n) \end{bmatrix} \tag{5.80}$$

and, as described in Section 4.4, $\tilde{\mathbf{Y}}$ is defined by

$$\tilde{\mathbf{Y}} = \begin{bmatrix} 1 \\ \hdashline \tilde{\mathbf{x}} \end{bmatrix} = \begin{bmatrix} 1 \\ x_1 \\ \vdots \\ x_{k-1} \\ 0 \\ x_{k+1} \\ \vdots \\ x_d \end{bmatrix}, \tag{5.81}$$

i.e., the vector $\mathbf{Y}$ with its component x_k, $k \neq 0$, equal to zero. A choice $v_k^* > 0$ corresponds to a guess that the best assignment of a feature vector to class ω_2 is made for feature vectors whose x_k values are generally greater than the x_k values of feature vectors assigned to class ω_1. The opposite guess corresponds to $v_k^* < 0$.

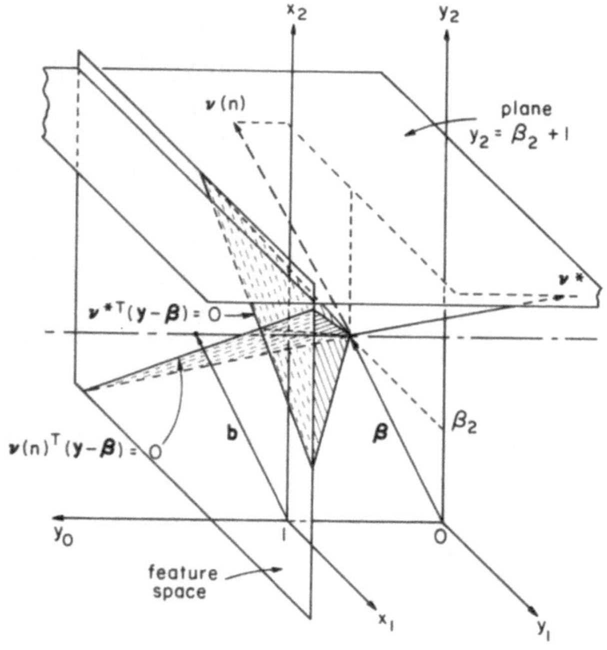

Figure 5.11. Typical locations of $v(n)$ and v^* for training procedures with $\mathrm{sgn}\,(v_k^*)$ known.

Note from (5.78) and since the z_k component of $\mathbf{Z}$ in Equation (5.77) is always zero, that $V_k(n) = V_k(0) = \pm 1$ for all n. If an origin translation has been made so that $\mathbf{y}$ is replaced by $(\mathbf{y} - \boldsymbol{\beta})$ and $\mathbf{V}$ is replaced by $\mathbf{v}$, then the requirement (5.78) becomes $v_k(n) = v_k(0) = \pm 1$. Therefore, the above training procedure restricts the vector $\mathbf{v}(n)$ to a vector originating at $\boldsymbol{\beta}$ and terminating on the hyperplane $y_k = \beta_k \pm 1$, where the sign is dependent on the choice of $v_k = \pm 1$. This is illustrated in augmented two-feature space in Figure 5.11, for the case where $v_k = +1$.

Theorem 5.1. *Suppose $\mathbf{X}(n)$ and $\omega(n)$ are stationary random sequences and $\rho(n)$ and $c(n)$ satisfy the constraints*

1. $$\rho(n) > 0, \qquad c(n) > 0$$
2. $$\rho(n) \to 0, \qquad c(n) \to 0$$
3. $$\sum_{n=0}^{\infty} \rho(n)c(n) = \infty$$
4. $$\sum_{n=0}^{\infty} \rho(n)^2 c(n) < \infty.$$

Suppose $\mathbf{X}(n)$, $\omega(n)$, and $f_j(\mathbf{x})$ satisfy the constraints

5. $$f_1(\mathbf{x}) + f_2(\mathbf{x}) < \infty, \quad \text{for all } \mathbf{x}$$
6. *and that for all $\varepsilon > 0$ there exists an $N(\varepsilon)$ such that*

$$\inf_{\varepsilon \leqslant \|\mathbf{v} - \hat{\mathbf{v}}^*\|} \frac{1}{c} (\hat{\mathbf{v}}^*(n) - \mathbf{v})^T E(\mathbf{Z}(n) \mid \mathbf{V}(n) = \mathbf{v}) > 0$$

for at least one $\hat{\mathbf{v}}^*(n)$ and for all $n > N(\varepsilon)$,

and suppose $\Psi(\xi, c)$ together with $f_j(\mathbf{x})$ satisfy

7. $$0 < \Psi(\xi, c) < \infty, \quad \text{for all } \xi \text{ and } c > 0$$
8. $$\lim_{c \to 0} \Psi(\xi, c) = \delta(\xi)$$
9. $$\int_{-\infty}^{\infty} \Psi(\xi, c)\, d\xi = 1, \qquad c \int_{-\infty}^{\infty} \Psi^2(\xi, c)\, d\xi < \infty$$
10. $$\sup_n \leqslant \left\| E\left\{ \left[\tilde{\mathbf{X}}(n) - \frac{\mathbf{V}^T(n)\mathbf{Y}(n)}{\|\mathbf{W}(n)\|^2} \tilde{\mathbf{W}}(n) \right] \times \boldsymbol{\Psi}\left(\frac{\mathbf{V}^T(n)\mathbf{Y}(n)}{\|\mathbf{W}(n)\|}, c(n) \right) \middle| \mathbf{V}(n), \omega(n) \right\} \right\| < \infty,$$

and

$$cE\left[\left\| \tilde{\mathbf{X}} - \frac{\mathbf{V}^T\mathbf{Y}}{\|\mathbf{W}\|^2} \tilde{\mathbf{W}} \right\|^2 \Psi^2\left(\frac{\mathbf{V}^T\mathbf{Y}}{\|\mathbf{W}\|}, c \right) \right] < \infty.$$

11. *If* $\mathbf{v}^+$ *denotes each value of* $\mathbf{v}$ *satisfying*

$$\int_{\mathscr{S}} f_1(\mathbf{x})d\mathbf{s} = \int_{\mathscr{S}} f_2(\mathbf{x})d\mathbf{s} = 0, \qquad \mathbf{v}^+ \notin \mathscr{V}^*,$$

then $\Psi(\mathbf{v}^T\mathbf{y}/\|\mathbf{w}\|, c)$ *must be chosen so that for* $0 < \varepsilon < \inf\|\mathbf{v}^+ - \mathbf{v}^*\|$, *there is an* $N(\varepsilon)$ *such that* $\|\hat{\mathbf{v}}^*(n) - \mathbf{v}^*\| \leqslant \inf\|\mathbf{v}^+ - \mathbf{v}^*\| - \varepsilon$ *for all* $c > 0$, $n > N(\varepsilon)$,

and when the $\hat{\mathbf{v}}^*$'s *are normalized so that* $|\hat{v}_k^*| = 1$,

$$\|\hat{\mathbf{v}}^*(n)\| < \infty, \quad \textit{for all } \hat{\mathbf{v}}^*, n$$

12. $$P(\|\mathbf{W}(n) + \rho(n)\mathbf{S}(n)\| < M \,|\, \mathbf{V}(n)) > 0, \quad \textit{for all allowed } \mathbf{V}(n),$$

then the random sequence $\{\mathbf{V}(n)\}$ *determined by* (5.76) *and* (5.77) *converges to a vector in the set* $\mathscr{V}^*$ *with probability one.*

The satisfaction of these constraints is often facilitated by a proper translation of the origin (Subsection 5.1.3).

INTERPRETATION OF CONSTRAINTS. If $\rho(n)$ and $c(n)$ are assumed to be of the form

$$\rho(n) = \left(\frac{n_\rho}{n_\rho + n}\right)^\alpha \rho(0), \qquad 0 < \rho(0) \tag{5.82}$$

$$c(n) = \left(\frac{n_c}{n_c + n}\right)^\gamma c(0), \qquad 0 < c(0) \tag{5.83}$$

where n_ρ, $\rho(0)$, n_c, and $c(0)$ are constants, then constraints 1–4 require that all of the following relationships hold:

$$0 < n_\rho \tag{5.84}$$

$$0 < n_c \tag{5.85}$$

$$0 < \alpha < 1 \tag{5.86}$$

$$1 - 2\alpha < \gamma \leqslant 1 - \alpha \tag{5.87}$$

$$0 < \gamma. \tag{5.88}$$

Constraint 2 establishes $0 < \alpha$ and $0 < \gamma$. Using constraint 3 and Inequalities (5.84) and (5.85) and letting $M \triangleq \max[n_\rho, n_c]$,

$$\frac{1}{n_\rho^\alpha n_c^\gamma \rho(0)c(0)} \sum_{n=0}^{\infty} \rho(n)c(n) \geqslant \sum_{n=0}^{\infty} \frac{1}{(M+n)^{\alpha\gamma}} = \sum_{n=M}^{\infty} \frac{1}{n^{\alpha\gamma}}$$

$$= \sum_{n=1}^{\infty} \frac{1}{n^{\alpha\gamma}} - \sum_{n=1}^{M} \frac{1}{n^{\alpha\gamma}} = \infty$$

for $\alpha\gamma \leqslant 1$. Similarly, using constraint 4 and Inequalities (5.84) and (5.85),

$$\frac{1}{n_\rho^{2\alpha} n_c^\gamma \rho^2(0)c(0)} \sum_{n=0}^{\infty} \rho^2(n)c(n) \leqslant \sum_{n=0}^{\infty} \frac{1}{(1+n)^{2\alpha\gamma}} = \sum_{n=1}^{\infty} \frac{1}{n^{2\alpha\gamma}} < \infty$$

for $2\alpha\gamma > 1$. Inequalities (5.86) and (5.87) are then established from $0 < \alpha$, $\alpha\gamma \leqslant 1$, and $2\alpha\gamma > 1$.

Constraint 6 most narrowly defines the forms of the constituent densities which will ensure convergence. $(\hat{\mathbf{v}}^* - \mathbf{v})$ represents the vector directed from the point $\mathbf{v}$ to one of the interim optimum weight vectors $\hat{\mathbf{v}}^*$.

$$\frac{1}{2c} E(\mathbf{Z}(n) \mid \mathbf{V}(n) = \mathbf{v}) = \hat{\mathbf{G}}(\mathbf{v}, c(n))$$

represents the approximation to $-\|\mathbf{W}\|^2 \nabla P(\text{error} \mid \mathbf{v})$ at trial n. Therefore, constraint 6 states that for every point $\mathbf{v}$ after trial N, both the vector function $\hat{\mathbf{G}}(\mathbf{v}, c(n))$ and at least one $(\hat{\mathbf{v}}^* - \mathbf{v})$ must be everywhere directed into some common half-space, i.e., $\hat{\mathbf{G}}(\mathbf{v}, c(n))$ corrections will move $\mathbf{v}$ "toward" some $\hat{\mathbf{v}}^*$. $\hat{\mathbf{G}}(\mathbf{v}, c(n))$ represents an approximation to the direction of maximum decrease of the probability of error. Constraint 6 also requires that the magnitude of $\hat{\mathbf{G}}(\mathbf{v}, c(n))$ be nonzero for values of $\mathbf{v}$ other than $\hat{\mathbf{v}}^*$. Therefore it states the requirement that a window function and/or initial values $\mathbf{V}(0)$, $\rho(0)$, $c(0)$ be chosen so that there is always a finite probability that after some $N(\varepsilon) < n$, corrections will on the average move $\mathbf{V}(n)$ toward some $\hat{\mathbf{v}}^*$. This is of concern whenever $p(\mathbf{x})$ is not strictly positive. When a priori knowledge of a range of $\hat{\mathbf{v}}^*$ is available, one can ameliorate the effect of constraint 6 by limiting the space in which $\mathbf{V}(n)$ is allowed to vary. Since for $n > N(\varepsilon)$ with $N(\varepsilon)$ large, $\hat{\mathbf{G}}(\mathbf{v}, c(n))$, is a good approximation to $-\|\mathbf{W}\|^2 \nabla P(\text{error} \mid \mathbf{v})$, the properties of $-\|\mathbf{W}\|^2 \nabla P(\text{error} \mid \mathbf{v})$ with respect to general forms of constituent densities are informative. Some of these properties have been discussed in Section 4.9.

Constraints 7, 8, and 9 are properties which define the principal Parzen window function requirements. These could be relaxed somewhat but seem to be sufficiently general in this form. These constraints are satisfied by the rectangular and Gaussian window functions described in Subsection 5.3.1.

Constraint 10 relates the window function to the constituent density forms. For well behaved window functions such as the rectangular or Gaussian, this constraint is similar to the requirement that the constituent densities have finite means and that the probability density $p(\mathbf{x})$ have a finite second moment.

Constraint 11 is concerned with those cases for which $p(\mathbf{x})$ is not strictly positive and states that a window function must be chosen so that the zeros of $\hat{\mathbf{G}}(\mathbf{v}, c(n))$ converge toward $\mathbf{v}^*$ and not toward the zeros of $\mathbf{G}(\mathbf{v})$, $\mathbf{v} \neq \mathbf{v}^*$, resulting from regions where $p(\mathbf{x}) = 0$. This restriction ensures that Equation (5.69) holds.

Constraint 12 is concerned with the degree of a priori knowledge of the direction of v_k^*. $\|\mathbf{w}^*/w_k^*\|$ is large only when the optimum decision hyperplane is nearly parallel to the x_k axis, which is indicative of weak a priori knowledge. However, if an M is chosen large enough to allow $\mathbf{W}(n)$ to reach its optimum value and constraint 6 is satisfied, constraint 12 will usually be satisfied.

The proof of Theorem 5.1, from which the constraints are determined, is given in Appendix B of this volume. Another proof of the convergence of window training procedures, based on Martingale sequences, is given in Reference [14].

The second window procedure assumes no knowledge of the sign of some v_k^*. In order to ensure the finiteness of the magnitude of the weight vector, it maintains $\|\mathbf{w}\| = 1$ by normalizing each newly corrected augmented weight vector.

Window Training Procedure: sgn (v_k^*) Unknown. This procedure is described by the following sequence of recursively defined random vectors:

$$\mathbf{V}(n+1) = \frac{\mathbf{V}(n) + \rho(n)\mathbf{Z}(n)}{\|\mathbf{W}(n) + \rho(n)\mathbf{S}(n)\|} \tag{5.89}$$

where

$$\mathbf{Z}(n) = \begin{cases} 2c(n)\left[\mathbf{Y}(n) - \mathbf{V}^T(n)\mathbf{Y}(n)\begin{pmatrix} 0 \\ \hdashline \mathbf{W}(n) \end{pmatrix}\right]\Psi(\mathbf{V}^T(n)\mathbf{Y}(n), c(n)), & \text{for } \omega(n) = \omega_2, \\ -2c(n)\left[\mathbf{Y}(n) - \mathbf{V}^T(n)\mathbf{Y}(n)\begin{pmatrix} 0 \\ \hdashline \mathbf{W}(n) \end{pmatrix}\right]\Psi(\mathbf{V}^T(n)\mathbf{Y}(n), c(n)), & \text{for } \omega(n) = \omega_1, \end{cases} \tag{5.90}$$

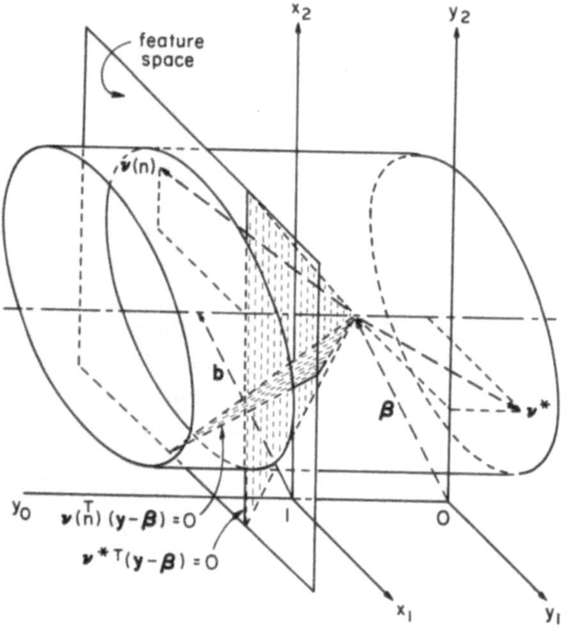

Figure 5.12. Typical locations of $v(n)$ and v^* for the training procedure with sgn (v_k^*) unknown.

$\mathbf{S}$ again denotes the vector $[Z_1, \ldots, Z_d]^T$, $\|\mathbf{W}(0)\| = 1$, and $W_0(0)$ is chosen arbitrarily.

Note from $\|\mathbf{W}(0)\| = 1$ and (5.89) that $\|\mathbf{W}(n)\| = 1$ for all n. When an origin translation has been made so that $\mathbf{Y}$ is replaced by $(\mathbf{Y} - \boldsymbol{\beta})$ and $\mathbf{V}$ is replaced by $\boldsymbol{v}$, then this training procedure restricts the vector $\boldsymbol{v}(n)$ to a vector originating at $\boldsymbol{\beta}$ and terminating on an infinitely long hypercylinder with radius one. The axis of the cylinder is parallel to the y_0-axis and passes through the point $\boldsymbol{\beta}$. This is illustrated in augmented two-feature space in Figure 5.12.

This training procedure utilizes a vector regression function which converges to $-\nabla P(\text{error}|\mathbf{v})$ as $c \to 0$. Although proof of the convergence of the sequence $\{\mathbf{V}(n)\}$ to the desired set $\mathscr{V}^*$ has not yet been developed, several simulations have supported the assumption that convergence will be obtained under constraints similar to those of Theorem 5.1.

5.3.3 Specialization to the Single-Feature Case

A specialization of the window training procedures for the case where the feature vector $\mathbf{x}$ is one-dimensional is made in this section. This will be done for both the case where $\operatorname{sgn}(v_k^*)$ is assumed known and when it is unknown.

For the training procedure where $\operatorname{sgn}(v_k^*)$ is known, we shall assume that we have a priori knowledge that the best choice of class ω_2 decisions is for features on the right, or increasing x, side of the threshold θ. We therefore choose $\operatorname{sgn}(v_k^*) = 1$ so that $v_k = +1$ for $k = 1$, i.e., $w_1 = +1$. The various variables of the procedure consequently take on the following values (note that several of the vector variables become one-dimensional, i.e., scalars):

$$\mathbf{X} = X$$

$$\mathbf{W} = w_1 = 1$$

$$\mathbf{V} = \begin{bmatrix} W_0 \\ 1 \end{bmatrix}$$

$$\mathbf{Y} = \begin{bmatrix} 1 \\ X \end{bmatrix}$$

$$\frac{\mathbf{V}^T\mathbf{Y}}{\|\mathbf{W}\|} = X + W_0$$

$$\tilde{\mathbf{X}} = 0$$

$$\tilde{\mathbf{Y}} = \begin{bmatrix} 1 \\ 0 \end{bmatrix}$$

$$\tilde{\mathbf{W}} = 0.$$

M is unspecified, since $\|\mathbf{W}(n)\| = 1$ for all n. When these values are substituted into Equations (5.76) and (5.77), the following recursively defined sequence is obtained:

$$\begin{bmatrix} W_0(n+1) \\ 1 \end{bmatrix} = \begin{bmatrix} W_0(n) \\ 1 \end{bmatrix} + \rho(n)\mathbf{Z}(n), \tag{5.91}$$

where

$$\mathbf{Z}(n) = \begin{cases} 2c(n)\begin{bmatrix} 1 \\ 0 \end{bmatrix} \Psi(X(n) + W_0(n), c(n)), & \text{for } \omega(n) = \omega_2, \\ -2c(n)\begin{bmatrix} 1 \\ 0 \end{bmatrix} \Psi(X(n) + W_0(n), c(n)), & \text{for } \omega(n) = \omega_1. \end{cases}$$

Now replace $-W_0$ by the threshold, θ, since $\theta = -W_0/\|\mathbf{W}\|$. Then (5.91) can be written

$$\theta(n+1) = \theta(n) - \rho(n)Z(n) \tag{5.92}$$

where

$$Z(n) = \begin{cases} 2c(n)\psi(X(n) - \theta(n), c(n)), & \text{for } \omega(n) = \omega_2, \\ -2c(n)\psi(X(n) - \theta(n), c(n)), & \text{for } \omega(n) = \omega_1. \end{cases}$$

This is the form of the one-dimensional minimum probability of error training procedure presented in Reference [3].

For the case where the direction which gives the best choice of ω_2 is *not* known, the training procedure with $\operatorname{sgn}(v_k^*)$ unknown is applicable. The various variables of this procedure take on the following values:

$$\mathbf{X} = X$$
$$\mathbf{W} = W_1 = \pm 1, \quad \text{since } \|\mathbf{W}\| = 1$$
$$\mathbf{V} = \begin{bmatrix} W_0 \\ W_1 \end{bmatrix}$$
$$\mathbf{Y} = \begin{bmatrix} 1 \\ X \end{bmatrix}$$
$$\mathbf{V}^T\mathbf{Y} = W_1 X + W_0$$
$$\mathbf{S} = Z_1.$$

Then (5.89) and (5.90) become

$$\mathbf{V}(n+1) = \frac{\mathbf{V}(n) + \rho(n)\mathbf{Z}(n)}{|W_1(n) + \rho(n)Z_1(n)|}, \tag{5.93}$$

where

$$\mathbf{Z}(n) = \begin{cases} 2c(n)\begin{bmatrix} 1 \\ -W_0(n)W_1(n) \end{bmatrix} \Psi(W_1(n)X(n) + W_0(n), c(n)), & \text{for } \omega(n) = \omega_2, \\ -2c(n)\begin{bmatrix} 1 \\ -W_0(n)W_1(n) \end{bmatrix} \Psi(W_1(n)X(n) + W_0(n), c(n)), & \text{for } \omega(n) = \omega_1. \end{cases} \tag{5.94}$$

From Equation (5.93)

$$\mathbf{V}(n+1) = \begin{bmatrix} W_0(n+1) \\ W_1(n+1) \end{bmatrix} = \frac{1}{|W_1(n) + \rho(n)Z_1(n)|} \begin{bmatrix} W_0(n) + \rho(n)Z_0(n) \\ W_1(n) + \rho(n)Z_1(n) \end{bmatrix}. \tag{5.95}$$

Note that W_1 is always normalized to ± 1. Now let θ denote the decision threshold on the X-axis, so that

$$\theta(n+1) = -\frac{W_0(n+1)}{W_1(n+1)}. \tag{5.96}$$

This single-feature window training procedure is illustrated in Figure 5.13. The heaviest lines represent decision hyperplanes in **y**-space. The example constituent densities are drawn in **x**-space, using the line $y_0 = 1$ as an abscissa. Notice that the constituent density $f_2(x)$ exceeds $f_1(x)$ when $x < \theta^*$. Notice also that $\mathbf{V}(n)$, the example augmented weight vector at trial n, is pointing into the right half-space, so that it is erroneously indicating that a decision of ω_2 should be made to the right of $\theta(n)$. However, if a sample from class ω_1 now occurs in the window (this is the most likely case) such as the $X(n) \in \omega_1$ illustrated, then an adjustment is made moving the weight vector to the $\mathbf{V}(n+1)$ indicated. This results in the new threshold $\theta(n+1)$. Note that a proper choice of class ω_2 to the left of $\theta(n+1)$ will now be made. The adjustment $\rho(n)\mathbf{Z}(n)$ is always normal to $\mathbf{V}(n)$, as can be seen from Equation (5.94). At the next trial the feature vector in the new window would most likely come from class ω_2, which would move the threshold in the direction of the optimum threshold θ^*.

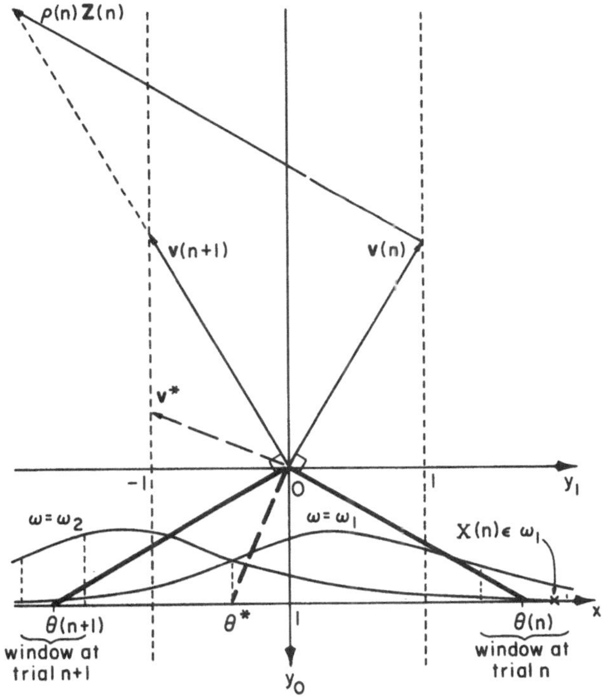

Figure 5.13. A possible configuration for the window training procedure, with sgn (v_k^*) unknown, applied to the single-feature case.

5.3.4 The Kiefer–Wolfowitz Approach*

In this section the Kiefer–Wolfowitz method [4] will be applied to the single-feature case to obtain a training procedure which will find a decision threshold giving the minimum probability of error. It will be assumed that a priori knowledge exists that a class ω_2 choice should be made to the right (or increasing x) side of the threshold.

As in the previous section, let θ denote a decision threshold. Then the probability of error conditioned on this threshold is given by

$$P(\text{error}\,|\,\theta) = \int_{-\infty}^{\theta} f_2(x)\,dx + \int_{\theta}^{\infty} f_1(x)\,dx. \tag{5.97}$$

Given a feature sample with N features, (5.97) can be approximated. Let $\hat{P}(\text{error}\,|\,\theta)$ denote this approximation and let $u(x)$ denote the unit step function. Then

$$\begin{aligned}\hat{P}(\text{error}\,|\,\theta) &= \frac{\text{Number of misclassifications in sample}}{\text{Total sample size}} \\ &= \frac{1}{N}\sum_{n=0}^{N-1} [\zeta_n(\omega_2)u(-X(n)+\theta) + \zeta_n(\omega_1)u(X(n)-\theta)],\end{aligned} \tag{5.98}$$

where the classification random variables $\zeta_n(\omega_j)$, $j = 1, 2$, are defined as

$$\zeta_n(\omega_j) = \begin{cases} 1, & \text{if } \omega(n) = \omega_j, \\ 0, & \text{otherwise.} \end{cases}$$

If the limit as $N \to \infty$ is taken in (5.98), then the finite averaging formula is replaced by an expectation, and the approximation is replaced by the function

$$P(\text{error}\,|\,\theta) = E[\zeta(\omega_2)u(-X+\theta) + \zeta(\omega_1)u(X-\theta)\,|\,\theta]. \tag{5.99}$$

Now since $u(-X+\theta) = 1 - u(X-\theta)$, (5.99) can be written

$$P(\text{error}\,|\,\theta) = E[\zeta(\omega_2) - [\zeta(\omega_2) - \zeta(\omega_1)]u(X-\theta)\,|\,\theta]. \tag{5.100}$$

Defining a new classification random variable

$$\zeta(\omega) = [\zeta(\omega_2) - \zeta(\omega_1)] = \begin{cases} 1, & \text{for } \omega = \omega_2, \\ -1, & \text{for } \omega = \omega_1, \end{cases}$$

and noting that $E[\zeta(\omega_2)\,|\,\theta] = P(\omega_2)$, (5.100) can be written

$$P(\text{error}\,|\,\theta) = P(\omega_2) - E[\zeta(\omega)u(X-\theta)\,|\,\theta]. \tag{5.101}$$

The expectation on the right-hand side of (5.101) describes a regression function whose maximum gives the minimum $P(\text{error}\,|\,\theta)$. The Kiefer–Wolfowitz procedure can now be applied to this regression function. The

* This section may be omitted without loss of continuity.

following recursive sequence describes the training procedure:

$$\theta(n+1) = \theta(n) + \frac{a(n)}{c(n)}[\zeta(\omega(n))u(X(n) - \theta(n) - c(n)) \\ - \zeta(\omega(n))(X(n) - \theta(n) + c(n))]. \tag{5.102}$$

Another classification random variable $Z(n)$ can now be defined as

$$Z(n) = \begin{cases} 1, & \text{for } |X(n) - \theta(n)| \leqslant c(n) \text{ and } \omega(n) = \omega_2, \\ 0, & \text{for } |X(n) - \theta(n)| > c(n), \\ -1, & \text{for } |X(n) - \theta(n)| \leqslant c(n) \text{ and } \omega(n) = \omega_1, \end{cases} \tag{5.103}$$

so that (5.102) can be written

$$\theta(n+1) = \theta(n) - \frac{a(n)}{c(n)} Z(n) \tag{5.104}$$

with the constraints

1. $a(n) > 0, \quad c(n) > 0$
2. $a(n) \to 0, \quad c(n) \to 0$
3. $\sum_{n=0}^{\infty} a(n) = \infty$
4. $\sum_{n=0}^{\infty} a(n)^2 c(n)^{-2} < \infty$
5. $|P(x_2 > \theta|\omega_2) - P(x_2 > \theta|\omega_1) - P(x_1 > \theta|\omega_1) + P(x_1 > \theta|\omega_2)| < A|x_2 - \theta^*| + B < \infty$

   for all $x_2 \neq x_1$, where θ^* denotes the optimum threshold and A and B are suitable constants,

Dvoretzky's conditions given in Subsection 4.8.2 are satisfied so that the sequence (5.104) converges to θ^* both in mean square and with probability one [5].

The training procedure described by Equations (5.103) and (5.104) is identical to the one-dimensional window training procedure given by (5.92) when a rectangular window is used and $\rho(n)$ is replaced by $a(n)/c(n)$. However, constraint 4 above is slightly more restrictive than its equivalent for the window training procedure.

The application of the Kiefer–Wolfowitz method to $P(\text{error}|\theta)$ is therefore a special case of the single-feature window training procedure when the optimum direction of category choice is known a priori.

5.3.5 Training Procedure Design Considerations

The primary design problem when using these training procedures is the choice of the sequences $\{\rho(n)\}$, $\{c(n)\}$ and the initial vector $\mathbf{V}(0)$. The forms which were used for our simulations are those given by (5.82) and (5.83)

with the accompanying constraints given by Equations (5.84)–(5.88). Once these forms are assumed, the design problem becomes one of selecting appropriate values of $\mathbf{V}(0)$, $\rho(0)$, n_ρ, α, $c(0)$, n_c, and γ for the training procedure. The term "appropriate" acknowledges that although convergence to an optimum vector has been proven for the training procedure, if $\mathbf{V}(0)$, $\rho(0)$, n_ρ, α, $c(0)$, n_c, and γ are poorly chosen the convergence rate will be impractically small. $\mathbf{V}(0)$'s which usually provide for good convergence properties have been discussed in Subsection 5.2.4.

The design problem for $c(n)$ is to ensure that the window size is sufficiently large to allow a significant number of samples to influence the motion of the hyperplane and yet small enough during the later trials of the training period to bring the vector $\mathbf{V}(n)$ reasonably close to $\mathbf{v}^*$. Simulations have shown that a value of n_c ranging from 20 for $N_t = 500$ trials up to 100 for an extremely large N_t and a $c(0)$ approximately equal to $\frac{1}{2}\delta_1$ produced good results. Recall that δ_1 denotes a measure of the one-dimensional spread of the $\mathbf{x}$'s around a point $\mathbf{a}$ and is given by Equation (5.49) in Subsection 5.2.3. The value of γ is chosen by selecting an appropriate final value $c(N_t)$ and then substituting $c(0)$, $c(N_t)$ and N_t into Equation (5.83). Simulation results have shown that a recommended minimum value of $c(N_t)$ is approximately $0.1\delta_1$ for $N_t = 1000$. Smaller values can be chosen for larger N_t.

The choice of appropriate $\rho(0)$, n_ρ, and α values involves minimizing the initial and final trial fluctuations while making certain that the fluctuations are sufficiently large to allow the sequence $\{\mathbf{V}(n)\}$ to reach a sufficiently small neighborhood of some $\mathbf{v}^*$. It was found from simulations on a digital computer (Subsection 5.3.6) that a value of n_ρ equal to n_c and $\alpha = 1 - \gamma$ were generally appropriate. So that α and N_t could influence the choice of $\rho(0)$, an approximation to $\sum_{n=0}^{N_t-1} \rho(n)$ was first defined. This approximation is denoted by D and is given by

$$D = \int_0^{N_t-1} \rho(n)\,dn = \rho(0)\int_0^{N_t-1} \left(\frac{n_\rho}{n_\rho + n}\right)^\alpha dn \tag{5.105}$$

or

$$D = \frac{n_\rho}{1-\alpha}\left[\left(\frac{N_t - 1}{n_\rho} + 1\right)^{1-\alpha} - 1\right]\rho(0). \tag{5.106}$$

Solving for $\rho(0)$ yields

$$\rho(0) = \frac{(1-\alpha)D}{n_\rho\left[\left(\dfrac{N_t - 1}{n_\rho} + 1\right)^{1-\alpha} - 1\right]}. \tag{5.107}$$

From simulations, a value of D approximately equal to $10\delta_d$, where δ_d is given by Equation (5.48), was generally found to be appropriate. Therefore, $\rho(0)$ was found by substituting $D = 10\delta_d$ into Equation (5.107).

In simulations where the number of trials exceeded 500, these design values produced adequate movement during the early trials and reasonably small fluctuations during later trials. The results were generally good even though the initial vector was sometimes poorly chosen. When a $\mathbf{V}(0)$ was calculated according to subsection 5.2.4, it was found that the above recommended values of $c(0)$ and $\rho(0)$ could be reduced.

5.3.6 Simulation Examples

Simulations were programmed in Fortran and executed on both an Interdata and a PDP-10 computer. Included in the program was a Monte Carlo method which transformed random numbers from an existing random number generator to various bivariate constituent density forms. Typical execution times were 12 seconds for 1000 trials. The results presented in this section are for the window training procedure with sgn (v_k^*) unknown using the Gaussian window. The procedure was modified so that no corrections were made when a sample was a distance greater than $\sqrt{10}c$ from the hyperplane. Simulations were also performed for the procedure with sgn (v_k^*) known with similar results. Prior to training the features were normalized by the average variance normalization procedure described in Subsection 5.2.1. The motion of the vector $\mathbf{V}(n)$ for normalized features was plotted by the computer. For the nonquadratic two-dimensional feature cases simulated, $\mathbf{V}(n)$ is augmented and therefore three dimensional. However, only the $V_0(n)$ and the $V_1(n)$ components were plotted. With the exception of the sign, this completely describes the motion for the simulations since the output was always normalized so that $\|\mathbf{W}(n)\| = (V_1(n)^2 + V_2(n)^2)^{1/2} = 1$ for all n. The printed outputs indicated that $V_2(n)$ was always positive for the simulations illustrated in this section. For the quadratic $\boldsymbol{\Phi}$-mapping simulation, $V_0(n)$ and $V_3(n)$ are plotted since $V_3(n)$ had the largest fluctuation.

The simulations are all for two-dimensional feature cases and represent typical results. For simulation examples of the one-dimensional feature case, see Reference [3]. Each simulation illustrated in this section was run for 2000 trials. For ease of visualization, only every 40th $\mathbf{V}(n)$ is plotted.

In both Figures 5.14 and 5.16 the minimum probability of error decision hyperplane is illustrated as a dashed line. The small arrows at the ends of each hyperplane indicate the side on which a category ω_2 decision is made.

Figure 5.14 illustrates contours of Gaussian constituent densities $f_1(\mathbf{x})$ and $f_2(\mathbf{x})$. The contours illustrated are for values of $\mathbf{x}$ where $f_1(\mathbf{x}) = f_2(\mathbf{x})$ and where the boundary of the union of these contours represents an approximate 80 precent sample occurrence probability boundary. The means are located at $\boldsymbol{\mu}_1 = [-0.5/\sqrt{2}, -0.5/\sqrt{2}]^T$ and $\boldsymbol{\mu}_2 = [1.5/\sqrt{2}, 1.5/\sqrt{2}]^T$. The covariance matrix of each is given by

$$\boldsymbol{\Sigma}_1 = \boldsymbol{\Sigma}_2 = \begin{bmatrix} 1 & 0 \\ 0 & 1 \end{bmatrix}.$$

The a priori probabilities are $P(\omega_1) = 0.25$ and $P(\omega_2) = 0.75$. The initial vector $\mathbf{V}(0)$ located the initial hyperplane along the x_1-axis as shown. The decision hyperplane found by the training procedure after 2000 trials is illustrated. The error probabilities of the two hyperplanes are: minimum probability of error = 0.127; error probability found after 2000 trials = 0.127. For this simulation,

$$c(n) = \left(\frac{40}{40+n}\right)^{0.76} (0.74), \quad \text{and} \quad \rho(n) = \left(\frac{40}{40+n}\right)^{0.24} (0.018).$$

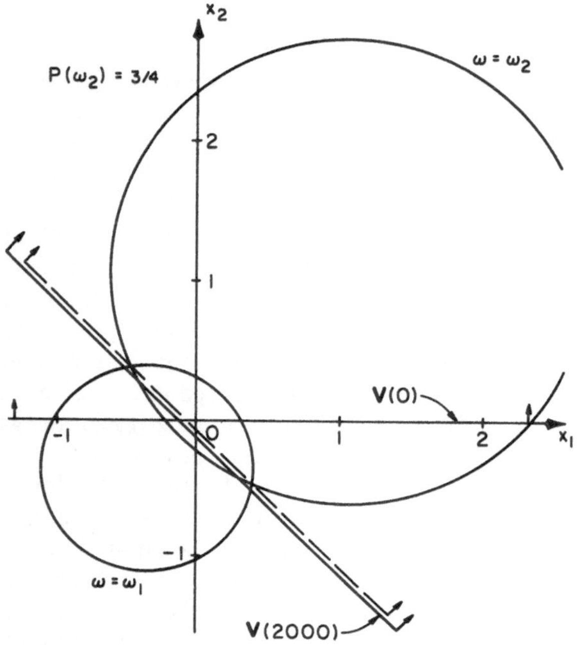

Figure 5.14. Simulation of Gaussian constituent densities.

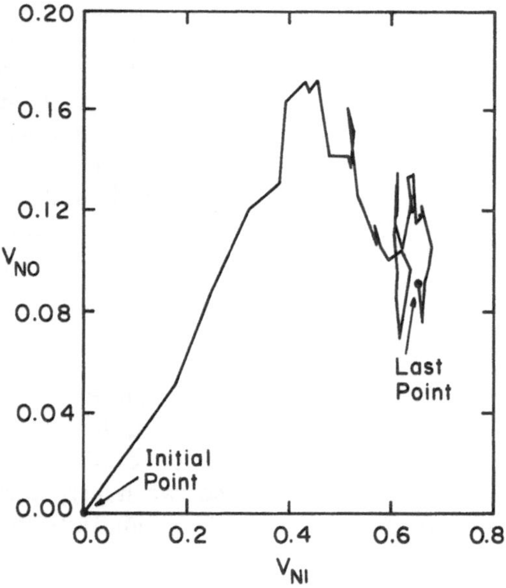

Figure 5.15. Motion of $\mathbf{V}_N(n)$ for the simulation of Figure 5.14.

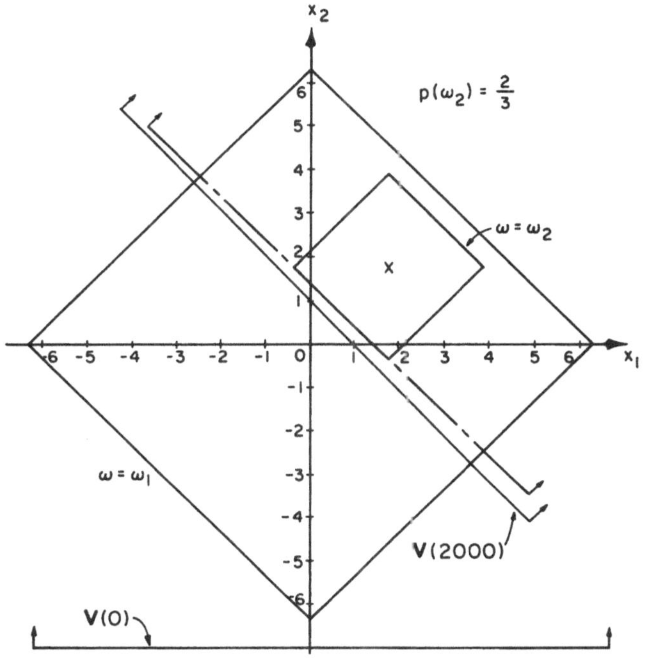

Figure 5.16. Simulation of bivariate uniform constituent densities.

Figure 5.15 displays the motion of the decision vector $\mathbf{V}_N(n)$ during training with normalized features. Every fortieth point is plotted.

Figure 5.16 illustrates bivariate uniform constituent densities. $f_1(\mathbf{x})$ is uniformly distributed over a 9-unit-sided square and $f_2(\mathbf{x})$ is uniformly distributed over a 3-unit-sided square. The means are located at $\boldsymbol{\mu}_1 = [0, 0]^T$ and $\boldsymbol{\mu}_2 = [2.5/\sqrt{2}, 2.5/\sqrt{2}]^T$. The a priori probabilities are $P(\omega_1) = \frac{1}{3}$ and $P(\omega_2) = \frac{2}{3}$. The initial hyperplane determined by $\mathbf{V}(0)$ was located as shown. The decision hyperplane found by the training procedure after 2000 trials is illustrated. The error probabilities of the two hyperplanes are: minimum probability of error = 0.130; error probability after 2000 trials = 0.142. For this simulation,

$$c(n) = \left(\frac{40}{40+n}\right)^{0.76} (2.16), \quad \text{and} \quad \rho(n) = \left(\frac{40}{40+n}\right)^{0.24} (0.044).$$

Figure 5.17 displays the motion of the decision vector during training with normalized features.

Figure 5.18 illustrates the constituent density regions for this third simulation. Class ω_2 is located within the ellipse, while class ω_1 exists within the rectangle but outside the ellipse. The equation of the ellipse is

$$\frac{(x_1 - 1)^2}{9} + \frac{(x_2 - 2)^2}{4} = 1.$$

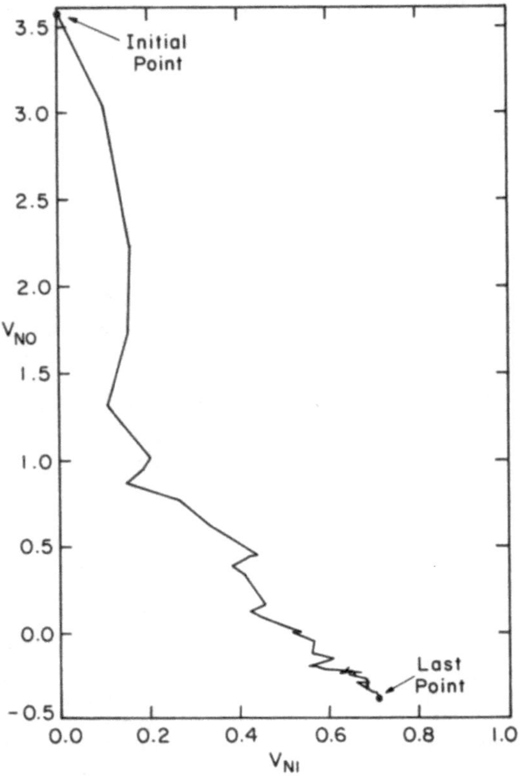

Figure 5.17. Motion of $\mathbf{V}_N(n)$ for the simulation of Figure 5.16.

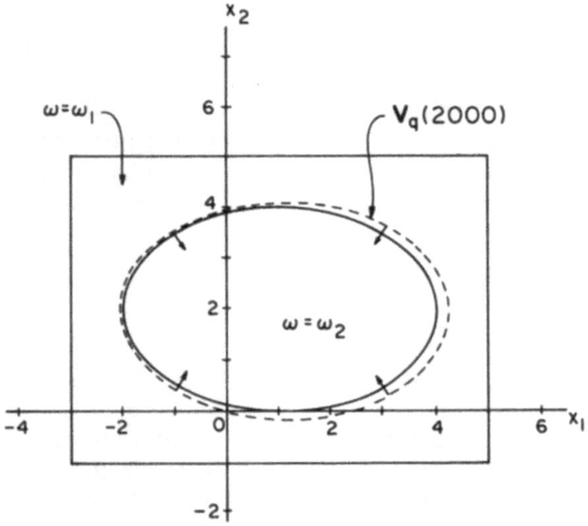

Figure 5.18. Simulation of uniformly distributed constituent densities, one inside and the other outside an ellipse.

The mean of each density is located at $[1,2]^T$. The constituent densities of both classes are uniformly distributed over their respective regions and have equal magnitudes of 1/48. The resulting a priori probabilities are $P(\omega_1) = 0.607$ and $P(\omega_2) = 0.393$. Quadratic $\boldsymbol{\Phi}$-mapping was used with a quadratic feature vector $\mathbf{x}_q$ defined by

$$\mathbf{x}_q = [x_1, x_2, x_1^2, x_1x_2, x_2^2]^T.$$

The initial vector $\mathbf{V}_q(0)$ was found using Equations (5.59) and (5.60). The quadratic surface found by the training procedure after 2000 trials is described by $\mathbf{V}_q(2000) = [-0.06, 0.25, 0.93, -0.11, -0.001, -0.24]^T$ and is illustrated as a dotted ellipse in Figure 5.18. The error probability found after 2000 trials = 0.029. For this simulation

$$c(n) = \left(\frac{40}{40+n}\right)^{0.76}(0.60), \quad \text{and} \quad \rho(n) = \left(\frac{40}{40+n}\right)^{0.24}(0.024).$$

Figure 5.19 displays the motion of the V_{qN0} and V_{qN3} components of the decision vector during training with normalized quadratic features.

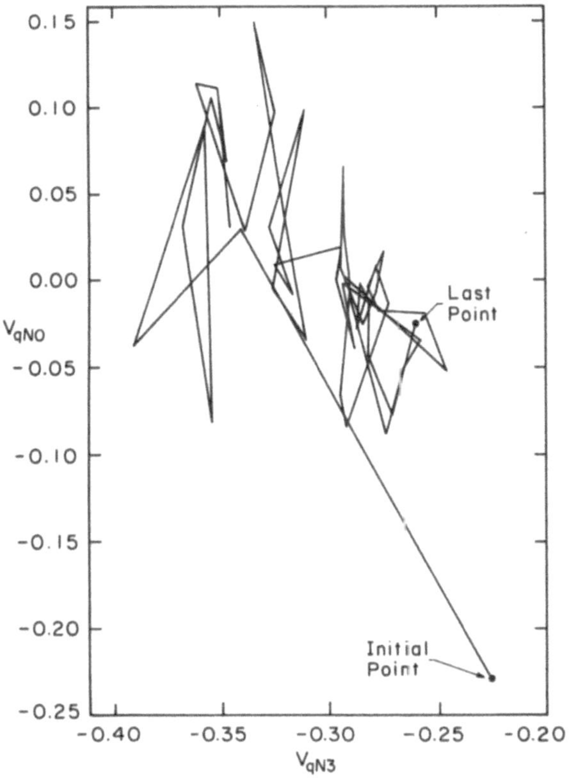

Figure 5.19. Motion of $\mathbf{V}_{qN}(n)$ for the simulation of Figure 5.18.

Each of the simulations supported the theory's prediction that the decision hyperplanes of the window training procedure approaches the minimum probability of error decision hyperplane.

5.4 The Minimum Mean Square Error Training Procedure

In this section, we present a training procedure based upon the mean square error loss function discussed in Subsection 4.5.2. As discussed in that subsection, this procedure will not in general seek the minimum probability of error decision hyperplane and is therefore a suboptimal procedure. However, it does have better convergence properties than the window procedure when the decision hyperplane is in a region of low probability density of feature vectors.

After presenting the basic minimum mean square error training procedure, we shall also present a modification based upon a second-order gradient descent technique. The use of this modified technique often results in better convergence properties. Also presented will be the design considerations and simulations for comparison with other training procedures.

5.4.1 The Regression Function

From Subsection 4.5.2, the deterministic form of the gradient of the mean square error loss function is given by Equation (4.48) and is repeated here for reference.

$$\nabla J(\mathbf{v}) = \int (\mathbf{v}^T\mathbf{y} + 1)\mathbf{y}f_1(\mathbf{x})\,d\mathbf{x} + \int (\mathbf{v}^T\mathbf{y} - 1)\mathbf{y}f_2(\mathbf{x})\,d\mathbf{x}. \tag{5.108}$$

The regression form of this gradient is found by defining the vector random variable $\mathbf{Z}(\mathbf{X}, \mathbf{V}, \omega)$ as

$$\mathbf{Z} = \begin{cases} -(\mathbf{V}^T\mathbf{Y} - 1)\mathbf{Y}, & \text{for } \omega = \omega_2, \\ -(\mathbf{V}^T\mathbf{Y} + 1)\mathbf{Y}, & \text{for } \omega = \omega_1. \end{cases} \tag{5.109}$$

Thus, the vector regression function $\mathbf{G}(\mathbf{v})$ becomes

$$\mathbf{G}(\mathbf{v}) = -\nabla J(\mathbf{v}) = E(\mathbf{Z} \mid \mathbf{V} = \mathbf{v}). \tag{5.110}$$

The demonstration that the $\nabla J(\mathbf{v})$ of (5.108) and (5.110) are equivalent is left to the reader as an exercise.

5.4.2 The Training Procedure

The basic minimum mean square error training procedure is then described by the following sequence of recursively defined random vectors.

$$\mathbf{V}(n + 1) = \mathbf{V}(n) + \rho(n)\mathbf{Z}(n) \tag{5.111}$$

where

$$\mathbf{Z}(n) = \begin{cases} -[\mathbf{V}^T(n)\mathbf{Y}(n) - 1]\mathbf{Y}(n), & \text{for } \omega(n) = \omega_2, \\ -[\mathbf{V}^T(n)\mathbf{Y}(n) + 1]\mathbf{Y}(n), & \text{for } \omega(n) = \omega_1, \end{cases} \tag{5.112}$$

and where $\mathbf{V}(0)$ is chosen arbitrarily. The relationship between $\mathbf{V}(n)$ and and $\mathbf{W}(n)$ for any n is

$$\mathbf{V}(n) = \begin{bmatrix} W_0(n) \\ \hline \mathbf{W}(n) \end{bmatrix}. \tag{5.113}$$

For this training procedure, the regression function $\mathbf{G}(\mathbf{v})$ equals the negative of the gradient. It is therefore given by (5.108), which can be written as

$$-\nabla J(\mathbf{v}) = -\|\mathbf{W}\| \left[\int \xi_1 \mathbf{y} f_1(\mathbf{x})\, d\mathbf{x} + \int \xi_2 \mathbf{y} f_2(\mathbf{x})\, d\mathbf{x} \right], \tag{5.114}$$

where ξ_1 and ξ_2 denote signed distances defined by (4.42). Since the factor $\|\mathbf{W}\|$ is already present as a multiplying factor in the regression function, $\|\mathbf{W}\|$ will not affect the magnitude of the correction.

Theorem 5.2. *Suppose $\mathbf{X}(n)$ and $\omega(n)$ are stationary random sequences and $\rho(n)$ satisfies the following constraints.*

1. $$\rho(n) > 0$$
2. $$\sum_{n=0}^{\infty} \rho(n) = \infty$$
3. $$\sum_{n=0}^{\infty} \rho^2(n) < \infty.$$

Suppose further that $\mathbf{X}$ and $f_j(\mathbf{x})$ satisfy the following constraints:

4. *$E(\mathbf{Y}\mathbf{Y}^T)$ and $E[(\mathbf{Y}\mathbf{Y}^T)^2]$ must exist and be positive definite.*
5. *$\int \mathbf{y} f_j(\mathbf{x})\, d\mathbf{x}$ and $\int \mathbf{y}\mathbf{y}^T \mathbf{y} f_j(\mathbf{x})\, d\mathbf{x}, j = 1, 2$, must exist.*

Then the random sequence $\{\mathbf{V}(n)\}$ determined by (5.111) and (5.112) converges to a vector in the solution set $\mathscr{V}^*$ both in mean square and with probability one.

The proof of this theorem is given by Blaydon [7, 8]. Other proofs of theorems closely related to this theorem are based on martingale sequences [13].

If $\rho(n)$ is assumed to be of the form

$$\rho(n) = \left(\frac{n_0}{n_0 + n} \right)^{\alpha} \rho(0), \qquad 0 < \rho(0), \tag{5.115}$$

then constraints 1–3 of the theorem require that the following relationships hold:

$$0 < n_0 \tag{5.116}$$

$$\tfrac{1}{2} < \alpha \leqslant 1. \tag{5.117}$$

For the minimum mean square error procedure, the following accelerated gradient descent technique is often proposed: the gradient correction is premultiplied by a finite-sample estimate of the matrix $[\nabla\nabla^T J(\mathbf{v})]^{-1}$ whenever the matrix $\nabla\nabla^T J(\mathbf{v})$ is nonsingular. From (5.108),

$$\nabla\nabla^T J(\mathbf{v}) = \int \mathbf{y}\mathbf{y}^T[f_1(\mathbf{x}) + f_2(\mathbf{x})]\,d\mathbf{x} = E(\mathbf{Y}\mathbf{Y}^T). \tag{5.118}$$

Let a finite-sample estimate of $\nabla\nabla^T J(\mathbf{v})$ based upon the N feature vectors in the training set $(\mathscr{X}_1, \mathscr{X}_2)$ be denoted by $\hat{\mathbf{H}}(N)$, and let $\hat{\mathbf{Q}}(N) = N\hat{\mathbf{H}}(N)$. Then

$$\hat{\mathbf{H}}(N) = \frac{1}{N}\sum_{i=0}^{N-1} \mathbf{Y}(i)\mathbf{Y}^T(i) = \frac{1}{N}\hat{\mathbf{Q}}(N). \tag{5.119}$$

A value of $\hat{\mathbf{Q}}(n)$ based upon the number of feature vectors which have been utilized at trial n can be computed recursively by

$$\hat{\mathbf{Q}}(n+1) = \hat{\mathbf{Q}}(n) + \mathbf{Y}(n)\mathbf{Y}^T(n). \tag{5.120}$$

The inverse of $\hat{\mathbf{Q}}(n)$ can also be calculated recursively (when it exists) by

$$[\hat{\mathbf{Q}}(n+1)]^{-1} = [\hat{\mathbf{Q}}(n)]^{-1} - \frac{[\hat{\mathbf{Q}}(n)]^{-1}\mathbf{Y}(n)\{[\hat{\mathbf{Q}}(n)]^{-1}\mathbf{Y}(n)\}^T}{1 + \mathbf{Y}^T(n)[\hat{\mathbf{Q}}(n)]^{-1}\mathbf{Y}(n)}. \tag{5.121}$$

The new algorithm with $[\mathbf{H}(n)]^{-1} = n[\hat{\mathbf{Q}}(n)]^{-1}$, is then given by

$$\mathbf{V}(n+1) = \mathbf{V}(n) + \rho(n)[\hat{\mathbf{H}}(n)]^{-1}\mathbf{Z}(n). \tag{5.122}$$

Often $[\hat{\mathbf{H}}(n)]^{-1}$ in Equation (5.122) is replaced by $[\hat{\mathbf{Q}}(n)]^{-1}$. This modified training procedure will also find a random sequence $\mathbf{V}(n)\}$ which converges to a vector in the solution set $\mathscr{V}^*$ both in mean square and with probability one under the constraints listed in Theorem 5.2.

For the case where the classes are finite and all members of the training set are equally likely, the loss function simplifies to

$$J(\mathbf{v}) = (\mathbf{A}\mathbf{v} - \mathbf{1})^T(\mathbf{A}\mathbf{v} - \mathbf{1}),$$

where $\mathbf{1}$ is the N-vector

$$\mathbf{1} = \begin{bmatrix} 1 \\ \vdots \\ 1 \end{bmatrix},$$

as discussed in Section 2.12. A gradient descent on this loss function produces the following training equation, which is equivalent to Equation (2.67) specialized to $\mathbf{b} = \mathbf{1}$.

$$\mathbf{V}(n+1) = \mathbf{V}(n) + \rho(n)\mathbf{Z}(n), \tag{5.123}$$

where

$$\mathbf{Z}(n) = -\mathbf{A}^T[\mathbf{A}\mathbf{V}(n) - \mathbf{1}].$$

The case where the size of the sample for estimating $\nabla\nabla^T J(\mathbf{v})$ is unity is of special interest. In this case only one feature vector is used at each trial of the training process. In the above finite-class training equation, $\mathbf{A}^T$ is replaced by $\boldsymbol{\eta}(n)$, the feature vector entering the nth trial. (Recall the definition of $\boldsymbol{\eta}$ in Equation (1.14).) This yields the following training equation.

$$\mathbf{V}(n+1) = \mathbf{V}(n) + \rho(n)[1 - \mathbf{V}(n)^T\boldsymbol{\eta}(n)]\boldsymbol{\eta}(n), \tag{5.124}$$

which is known as the *Widrow–Hoff rule* [12]. When $\rho(n) = c/n$, sequence $\mathbf{V}(n)\}$ generated by this rule has been observed to converge to the same $\mathbf{v}^*$ as for Equation (5.123), although a rigorous proof is not known. In this rule note that $\mathbf{V}(n)$ is incremented for every $\boldsymbol{\eta}(n)$ regardless of whether $\mathbf{v}^T\boldsymbol{\eta}(n)$ is positive or negative—i.e., regardless of whether the classifier misclassifies $\boldsymbol{\eta}(n)$.

A solution vector $\mathbf{v}^*$ for the finite-class case has an interesting interpretation. Here $\mathbf{v}^*$ not only minimizes $\|\mathbf{A}\mathbf{v} - \mathbf{1}\|^2$, but it also maximizes the function

$$t(\mathbf{v}) = \frac{[\mathbf{v}^T(\check{\boldsymbol{\mu}}_1 - \check{\boldsymbol{\mu}}_2)]^2}{\mathbf{v}^T(\check{\boldsymbol{\Sigma}}_1 + \check{\boldsymbol{\Sigma}}_2)\mathbf{v}}$$

(known as the *generalized Rayleigh quotient* in mathematical physics), where $\check{\boldsymbol{\mu}}_j$ and $\check{\boldsymbol{\Sigma}}_j$ are the sample mean and sample covariance matrix, respectively, of the augmented feature vectors $\{\mathbf{Y}\}$ in class ω_j.

The numerator of the above expression for $t(\mathbf{v})$ is a measure of the spread between classes in the direction of $\mathbf{v}$. More specifically, it is the mean square of the distances between projections of feature vectors in ω_1 and those in ω_2 on a straight line $L_\mathbf{v}$ oriented in the direction of the vector $\mathbf{v}$.

The denominator of the expression for $t(\mathbf{v})$ is proportional to the variance ot the projections of the $\mathbf{Y}$'s onto $L_\mathbf{v}$.

Thus $t(\mathbf{v})$ is the ratio of the interclass distance along $\mathbf{v}$ to the total dispersion of the projections of the feature vectors on $L_\mathbf{v}$. The solution vector $\mathbf{v}^*$ maximizes both this ratio and $\|\mathbf{A}\mathbf{v} - \mathbf{b}\|^2$ when $\mathbf{b} = [1, 1, \ldots, 1]^T$. (See Exercise 5.19.) $t(\mathbf{v})$ is a criterion function for Fisher's linear discriminant, which is widely used in discriminant analysis [2, 19].

5.4.3 Design Considerations

In this training procedure, the primary concern is the choice of the sequence $\{\rho(n)\}$ and the initial vector $\mathbf{V}(0)$. A recommended form of $\rho(n)$ is given by Equation (5.115) with the accompanying constraints (5.116) and (5.117). The choice of an appropriate initial vector $\mathbf{V}(0)$ has been discussed in Subsection 5.2.4.

When Equation (5.115) is used, $\rho(0)$, n_0 and α values must be determined. From (5.117), $\frac{1}{2} < \alpha \leqslant 1$. Simulations have indicated that a value of α close to $\frac{1}{2}$ appears to provide better convergence than values close to 1. Also from simulations, a value of 20 for n_0 seems to be appropriate for $N_t = 500$, with values approaching 100 for very large N_t. $\rho(0)$ can be determined from Equation (5.107) with D equal to approximately $2\delta_d$. Recall that δ_d is given by Equation (5.48) in Subsection 5.2.3. For simulations where $N_t > 500$ and/or the initial vector is known to be close to optimum, the value of n_0 should be increased and the value of $\rho(0)$ decreased. The simulation results were noticeably improved when N_t was 1000 or more. The results were generally unsatisfactory for $N_t < 500$.

5.4.4 Simulation Examples

The simulations were programed and plotted as discussed in Subsection 5.3.6. The outputs again were normalized so that $\|\mathbf{W}(n)\| = 1$ for all n. The three simulation examples to be presented had identical feature vector distributions and initial vectors as the examples in Subsection 5.3.6. Their specifications are given there. Each simulation illustrated in this section was run for 2000 trials. Typical execution times were in the 20-second range.

In each of Figures 5.20 and 5.22, the minimum mean square error decision hyperplane is illustrated as a dashed line. The small arrows at the ends of each hyperplane indicate the side on which a category ω_2 decision is made.

For the first simulation, which utilized Gaussian constituent densities, the initial vector $\mathbf{V}(0)$ again located the initial decision hyperplane along the x_1-axis as shown in Figure 5.20. The decision hyperplane found by the minimum mean square error training procedure after 2000 trials is illustrated. The error probabilities of the two hyperplanes illustrated are: minimum mean square error probability of error = 0.129; probability of error found after 2000 trials = 0.126. Figure 5.21 displays the motion of two components of the three-dimensional vector $\mathbf{V}_N(n)$ during training with normalized features. Every 40th point is plotted. For this simulation,

$$\rho(n) = \left(\frac{40}{40 + n}\right)^{1/2} (0.007).$$

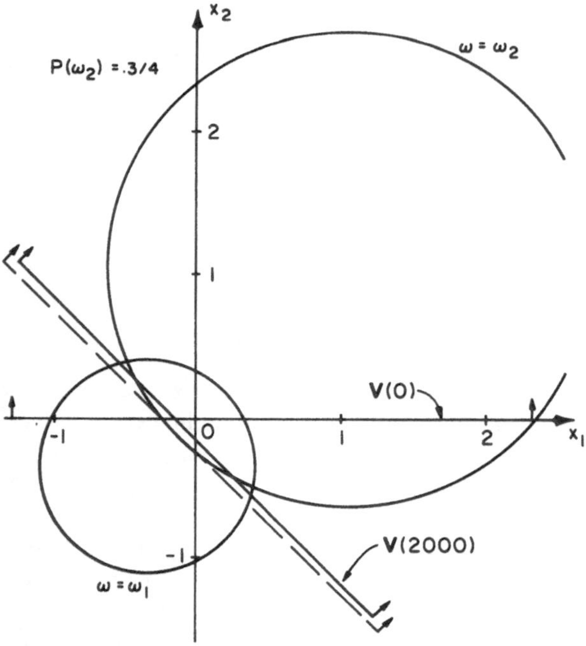

Figure 5.20. Simulation of Gaussian constituent densities for the minimum mean square error training procedure.

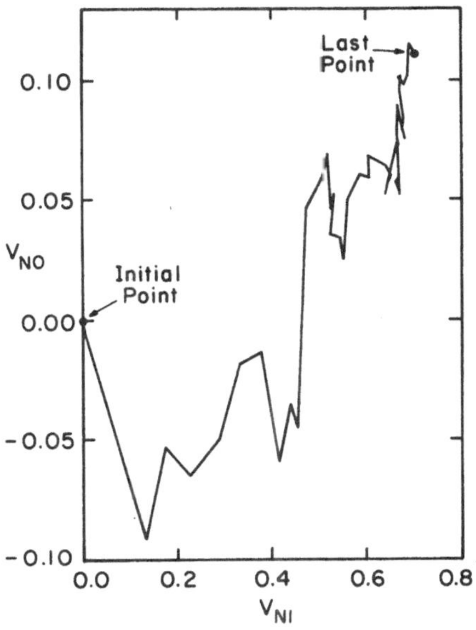

Figure 5.21. Motion of $\mathbf{V}_N(n)$ for the simulation of Figure 5.20.

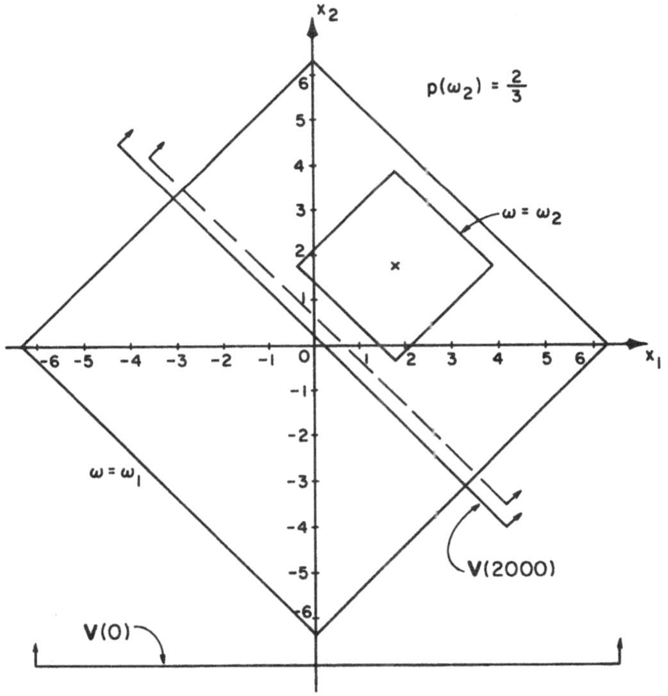

Figure 5.22. Simulation of bivariate uniform constituent densities for the minimum mean square error training procedure.

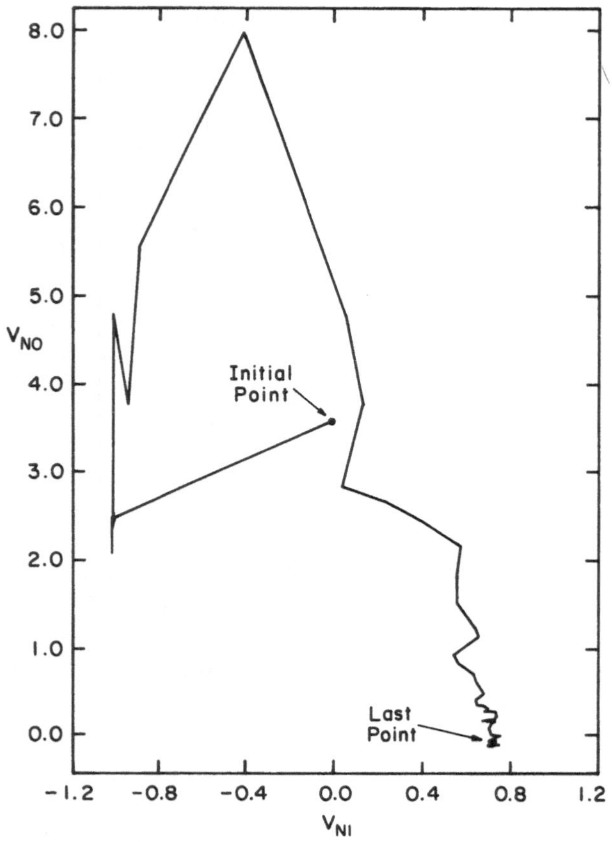

Figure 5.23. Motion of $\mathbf{V}_N(n)$ for the simulation of Figure 5.22.

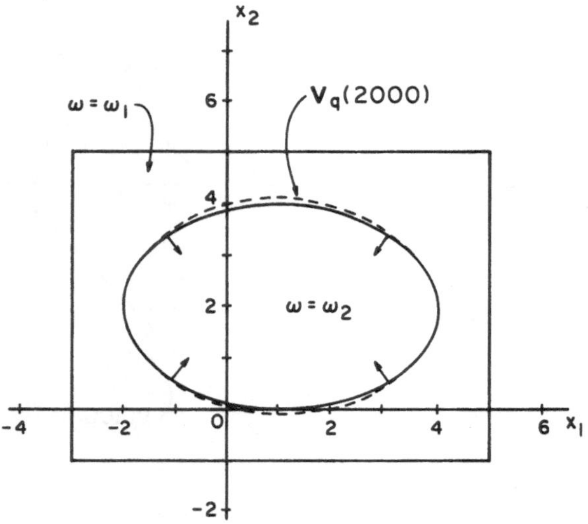

Figure 5.24. The ellipse found by the minimum mean square error procedure after 2000 trials is shown as a broken curve. The solid curve is the ellipse in the simulation.

The second simulation, for which the constituent densities were uniformly distributed, located the initial decision hyperplane along the line $x_2 = -7$. This is illustrated in Figure 5.22. The decision hyperplane found by the minimum mean square error training procedure after 2000 trials is illustrated. The error probabilities of the two hyperplanes illustrated are: minimum mean square error probability of error = 0.151; probability of error found after 2000 trials = 0.170. Figure 5.23 displays the motion of two components of $\mathbf{V}_N(n)$ during training with normalized features. Again, only every 40th point is plotted. For this simulation,

$$\rho(n) = \left(\frac{40}{40 + n}\right)^{1/2} (0.018).$$

As before, the third simulation utilized constituent densities which were uniformly distributed either within or outside an ellipse. Quadratic $\boldsymbol{\Phi}$-mapping was again used, and the initial $\mathbf{V}_q$ vector was found using Equations (5.59) and (5.60). The ellipse found by the training procedure after 2000 trials is described by $\mathbf{V}_q(2000) = [0.03, 0.23, 0.94, -0.12, -0.002, -0.24]^T$ and is illustrated in Figure 5.24. The probability of error associated with this $\mathbf{V}_q$ is 0.034. Figure 5.25 displays the motion of $V_{qN0}(n)$ and $V_{qN3}(n)$ during training with normalized features. Excepting $V_{qN0}(n)$, the component with the largest

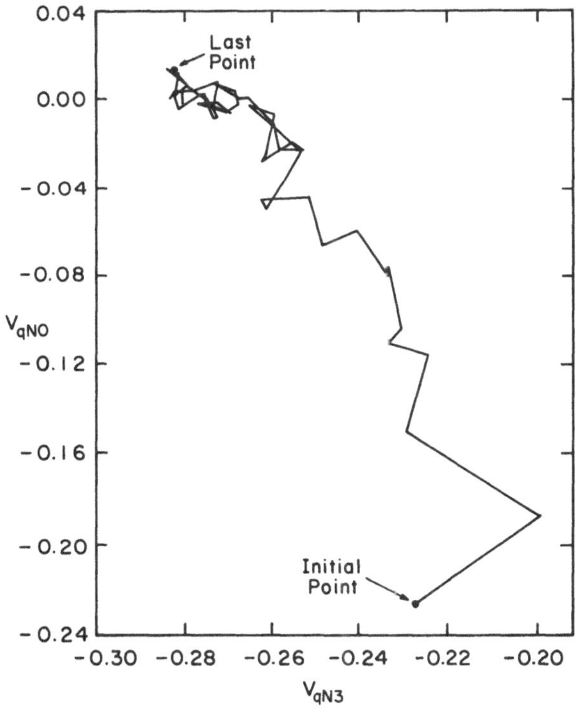

Figure 5.25. Motion of $\mathbf{V}_{qN}(n)$ for the simulation of Figure 5.24.

fluctation was $V_{qN3}(n)$. For this simulation,

$$\rho(n) = \left(\frac{40}{40 + n}\right)^{1/2} (0.01).$$

5.5 The Equalized Error Training Procedure

The equalized error training procedure presented in this section seeks a decision hyperplane which minimizes the error tail moment magnitude loss function. Since this hyperplane does not, in general, equal the minimum probability of error hyperplane, this procedure is asymptotically suboptimal. However, the convergence properties of this procedure are superior to the window procedure whenever the location of the decision hyperplane is in a region of low probability density of the feature vectors. Furthermore, the optimum hyperplane sought by the procedure will classify feature vectors with a probability of error close to the minimum for a large class of constituent densities.

In this section, we derive the equalized error training procedure, present a proof that the sequence $\{\mathbf{V}(n)\}$ determined by this procedure converges to a weight vector in the set $\mathscr{V}^*$ in mean square and with probability one, discuss the design considerations, and present some simulation examples. In this section, $\mathscr{V}^*$ contains all vectors $\mathbf{v}^*$ which specify a hyperplane which mimizes the error tail moment magnitude loss function.

For other special forms of the equalized error training procedure, see References [1] and [10].

5.5.1 Derivation of the Regression Function

The gradient to be utilized for this regression function is derived from the loss function based on the error tail moment magnitudes. This was discussed in Subsection 4.5.3. As discussed there the most appropriate form of the gradient is the form in which some w_k is maintained constant, since the gradient approaches infinity as $|\mathbf{w}_0|/\|\mathbf{w}\| \to \infty$. This gradient is given by Equation (4.70) together with Equation (4.2), and is rewritten here with $\mathbf{Z}$ redefined:

$$\nabla J(\mathbf{v})\big|_{w_k = \text{const.}} = -\frac{1}{\|\mathbf{w}\|^2} E(\mathbf{Z}|\mathbf{V} = \mathbf{v}), \tag{5.125}$$

where $\mathbf{Z}$ has now become

$$\mathbf{Z} = \begin{cases} \|\mathbf{W}\|\left(\tilde{\mathbf{Y}} - \dfrac{\mathbf{V}^T\mathbf{Y}}{\|\mathbf{W}\|^2}\left[\dfrac{0}{\tilde{\mathbf{W}}}\right]\right), & \text{for } \mathbf{V}^T\mathbf{Y} < 0 \text{ and } \omega = \omega_2, \\ -\|\mathbf{W}\|\left(\tilde{\mathbf{Y}} - \dfrac{\mathbf{V}^T\mathbf{Y}}{\|\mathbf{W}\|^2}\left[\dfrac{0}{\tilde{\mathbf{W}}}\right]\right), & \text{for } \mathbf{V}^T\mathbf{Y} > 0 \text{ and } \omega = \omega_1, \\ \mathbf{0}, & \text{otherwise,} \end{cases} \tag{5.126}$$

and where $\tilde{\mathbf{Y}}$ and $\tilde{\mathbf{W}}$ are respectively the $\mathbf{Y}$ vector and $\mathbf{W}$ vector with X_k and W_k, $k \neq 0$, equal to zero. Recall that the choice of k and the sign of v_k are based on some priori knowledge, such as the known relative values of the class means of ω_1 and ω_2 for a feature x_k. The vector regression function for the equalized error training procedure is simply (5.125) with weight vector magnitude compensation. This was discussed in Subsection 5.1.2. and is given by (5.13). The regression function $\mathbf{G}(\mathbf{v})$ is therefore

$$\mathbf{G}(\mathbf{v}) = E(\mathbf{Z} \mid \mathbf{V} = \mathbf{v}), \tag{5.127}$$

where $\mathbf{Z}$ is given by (5.126).

5.5.2 The Procedure and a Theorem

The equalized error training procedure is described by the following sequence of recursively defined random vectors:

$$\mathbf{V}(n+1) = \mathbf{V}(n) + \begin{cases} \rho(n)\mathbf{Z}(n), & \text{for } \|\mathbf{V}(n) + \rho(n)\mathbf{Z}(n)\| < M, \\ \mathbf{0}, & \text{otherwise}, \end{cases} \tag{5.128}$$

where

$$Z(n) = \begin{cases} \|\mathbf{W}(n)\| \left\{ \tilde{\mathbf{Y}}(n) - \dfrac{\mathbf{V}^T(n)\mathbf{Y}(n)}{\|\mathbf{W}(n)\|^2} \begin{bmatrix} 0 \\ \hline \tilde{\mathbf{W}}(n) \end{bmatrix} \right\}, & \text{for } \mathbf{V}^T(n)\mathbf{Y}(n) < 0 \text{ and } \omega = \omega_2, \\ -\|\mathbf{W}(n)\| \left\{ \tilde{\mathbf{Y}}(n) - \dfrac{\mathbf{V}^T(n)\mathbf{Y}(n)}{\|\mathbf{W}(n)\|^2} \begin{bmatrix} 0 \\ \hline \tilde{\mathbf{W}}(n) \end{bmatrix} \right\}, & \text{for } \mathbf{V}^T(n)\mathbf{Y}(n) > 0 \text{ and } \omega = \omega_1, \\ \mathbf{0}, & \text{otherwise}, \end{cases} \tag{5.129}$$

and

$$V_k(0) = \begin{cases} +1, & \text{if guess is } v_k^* > 0, \\ -1, & \text{if guess is } v_k^* < 0, \end{cases} \tag{5.130}$$

with $k \neq 0$, where the remaining components of $\mathbf{V}(0)$ are chosen arbitrarily and where M is a large number chosen so that $M \gg \min \|\mathbf{v}^*\|/|v_k^*|$. The relationship between $\mathbf{V}(n)$ and $\mathbf{W}(n)$ is

$$\mathbf{V}(n) = \begin{bmatrix} W_0(n) \\ \hline \mathbf{W}(n) \end{bmatrix}. \tag{5.131}$$

If there is insufficient confidence in a guess, the training can be performed twice using a different choice of the sign of v_k^* each time. The results can then be compared and the linear classifier which gives the least error rate based on the training feature vector sample can then be chosen.

$V_k(n)$ never changes from its initial value $V_k(0) = \pm 1$. If an origin translation has been made so that $\mathbf{Y}$ is replaced by $(\mathbf{Y} - \boldsymbol{\beta})$ and $\mathbf{V}$ is replaced by $\mathbf{v}$, then $v_k(n) = \pm 1$ for all n. The training procedure then restricts the vector $\mathbf{v}(n)$ to a vector originating at $\boldsymbol{\beta}$ and terminating on the hyperplane $y_k = \beta_k \pm 1$. This has been illustrated in Figure 5.11 for the case $y_k = \beta_k + 1$ resulting from the choice $v_k = +1$.

Theorem 5.3. *Suppose* $\mathbf{X}(n)$ *and* $\omega(n)$ *are stationary random sequences satisfying, together with* $\rho(n)$, *the constraints*

1. $$\rho(n) = 0$$
2. $$\sum_{n=0}^{\infty} \rho(n) = \infty$$
3. $$\sum_{n=0}^{\infty} \rho(n)^2 < \infty$$
4. $$E\|\mathbf{X}(n)\|^2 < \infty$$
5. $$P(\|\mathbf{V}(n) + \rho(n)\mathbf{Z}(n)\| < M) > 0 \text{ for all } n,$$
6. *for all* $\varepsilon > 0$ *there must exist at least one* $\mathbf{v}^*$ *such that*
$$\inf_{\varepsilon \leqslant \|\mathbf{v}-\mathbf{v}^*\|} (\mathbf{v}^* - \mathbf{v})^T E(\mathbf{Z}|\mathbf{V} = \mathbf{v}) > 0$$
and when the $\mathbf{v}^*$*'s are normalized so that* $|v_k^*| = 1$, $\|\mathbf{v}^*\| < \infty$ *for all* $\mathbf{v}^*$,

then the random sequence $\{\mathbf{V}(n)\}$ *determined by* (5.128) *and* (5.129) *converges to a vector in the set* $\mathscr{V}^*$ *in mean square and with probability one.*

The proof of Theorem 5.3 is given in Appendix C of this volume.

INTERPRETATION OF CONSTRAINTS. If $\rho(n)$ is assumed to be of the form

$$\rho(n) = \left(\frac{n_0}{n_0 + n}\right)^{\alpha} \rho(0), \qquad \rho(0) > 0, \tag{5.132}$$

then constraints 1–3 require that

$$0 < n_0 \tag{5.133}$$

$$\tfrac{1}{2} < \alpha \leqslant 1. \tag{5.134}$$

Assuming a finite mean of $\mathbf{x}$, constraint 4 requires that the variance of the magnitude of the feature vectors be finite.

Constraint 5 is interpreted identically to constraint 12 of Theorem 5.1. This constraint has been discussed in Subsection 5.3.2.

In constraint 6 $(\mathbf{v}^* - \mathbf{v})$ represents a vector directed from the point $\mathbf{v}$ to any one of the desired weight vectors $\mathbf{v}^*$. $E(\mathbf{Z}|\mathbf{V} = \mathbf{v})$ represents the vector function $\mathbf{G}(\mathbf{v})$ defined by (5.127) with $G_k(\mathbf{v})$ equal to zero. Constraint 6 therefore requires that the constituent densities describing $p(\mathbf{x})$ be restricted to forms such that for at least one $\mathbf{v}^*$, $(\mathbf{v}^* - \mathbf{v})$ and $E(\mathbf{Z}|\mathbf{V} = \mathbf{v})$ are everywhere directed into some common half-space.

5.5.3 Specialization to the Single-Feature Case

When the feature vector $\mathbf{X}$ is one dimensional, then the training procedure given by (5.128) simplifies to the form

$$W_0(n + 1) = W_0(n) + \rho(n)Z(n) \tag{5.135}$$

where

$$Z(n) = \begin{cases} 1, & \text{for } W_0(n) + W_1 X(n) < 0 \text{ and } \omega(n) = \omega_2, \\ -1, & \text{for } W_0(n) + W_1 X(n) > 0 \text{ and } \omega(n) = \omega_1, \\ 0, & \text{otherwise,} \end{cases} \tag{5.136}$$

and

$$W_1 = \begin{cases} 1, & \text{for } v_1^* > 0, \\ -1, & \text{for } v_1^* < 0, \end{cases} \tag{5.137}$$

for all n.

If it is assumed that $v_1^* > 0$, then $W_1 = +1$. Then if $\theta(n)$ denotes the decision threshold defined by

$$\theta(n) = -W_0(n)/W_1 \tag{5.138}$$

the training procedure can be written

$$\theta(n+1) = \theta(n) - \rho(n)Z(n) \tag{5.139}$$

where

$$Z(n) = \begin{cases} 1, & \text{for } X(n) < \theta(n) \text{ and } \omega(n) = \omega_2, \\ -1, & \text{for } X(n) > \theta(n) \text{ and } \omega(n) = \omega_1, \\ 0, & \text{otherwise.} \end{cases} \tag{5.140}$$

But this is simply a form of the well-known one-dimensional error correction training procedure. Here the stochastic approximation restrictions on $\rho(n)$ are given by Theorem 5.3 to be

$$\textstyle\sum_n \rho(n) = \infty, \qquad \sum_n \rho(n)^2 < \infty. \tag{5.141}$$

5.5.4 Design Considerations

The design problem for this training procedure is the choice of the sequence $\rho(n)$ and the initial weight vector $\mathbf{V}(0)$. For the form of $\rho(n)$ given by Equation (5.132), a choice of n_0 ranging from 20 for $N_t = 500$ trials up to 100 for very large N_t, $\alpha \approx 0.5$, and $\rho(0)$ calculated from Equation (5.107) with $D = 6\delta_d$ generally produced good results even when the initial vector was poorly chosen. It was again found that $\rho(0)$ could be reduced if a $\mathbf{V}(0)$ is chosen which has been calculated according to Subsection 5.2.4. Simulations where $N_t < 500$ appear to be generally unsatisfactory and simulations where N_t is 1000 or more have given significantly improved results.

5.5.5 Simulation Examples

The three examples of this section have the same feature vector distributions and initial vectors as the examples for the window training procedure of Subsection 5.3.6 and the examples for the minimum mean square error training procedure of Subsection 5.4.4. Their specifications are given in Subsection 5.3.6.

In order that the results could be better compared with other simulations, the vector $\mathbf{V}(n)$ was normalized in magnitude before plotting so that $\|\mathbf{W}(n)\| = 1$ for all n. The simulation examples were run for 2000 trials. Execution times were usually of the order of 20 seconds.

In both Figures 5.26 and 5.28, the minimum error tail moments decision hyperplane is illustrated as a dashed line. The small arrows at the end of the hyperplane indicate the side on which ω_2 decisions are made.

Figure 5.26 illustrates the first example. Here the constituent densities are bivariate Gaussian. The initial hyperplane specified by $\mathbf{V}(0)$ was again located along the x_1-axis as shown. The decision hyperplane found by the equalized error training procedure after 2000 trials is illustrated. The error probabilities of the two hyperplanes illustrated are: 0.131 for the minimum error tail moments and 0.129 for the equalized error procedure after 2000 trials. To check whether the simulations converge toward an equal error point, the error probabilities contributed by each class were computed at $N_t = 2000$. These error probabilities were found to be 0.067 and 0.062 for classes ω_1 and ω_2, respectively. For this simulation, the sequence $\{p(n)\}$ was specified by

$$\rho(n) = \left(\frac{40}{40 + n}\right)^{1/2} (0.022).$$

Figure 5.27 displays the motion of the $V_{N0}(n)$ and $V_{N1}(n)$ components of $\mathbf{V}_N(n)$ during training with normalized features. Every 40th point is plotted.

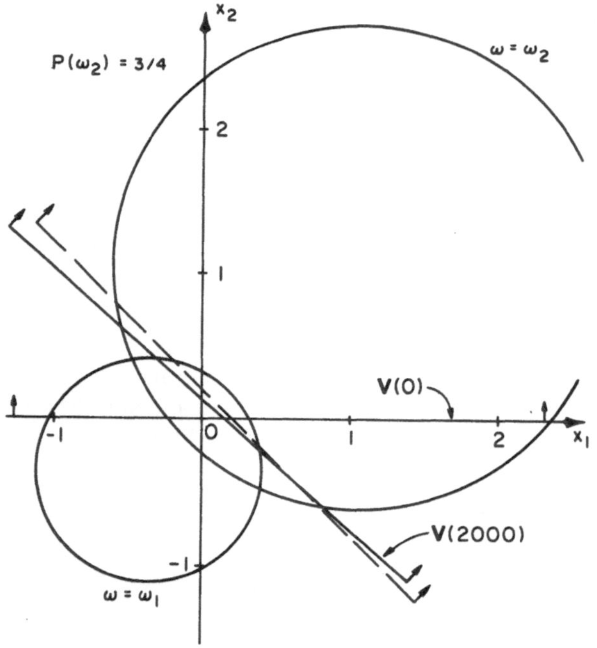

Figure 5.26. Simulation of Gaussian constituent densities for the equalized error training procedure.

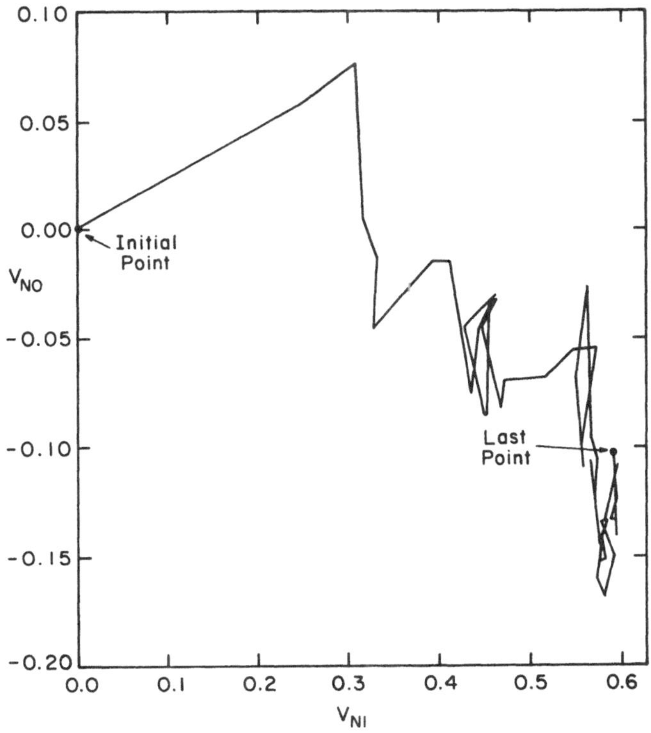

Figure 5.27. Motion of $\mathbf{V}_N(n)$ for the simulation of Figure 5.26.

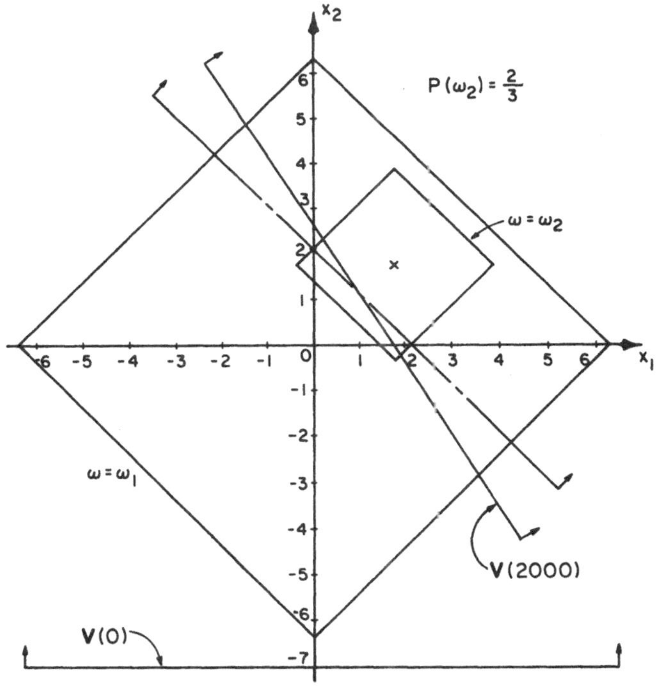

Figure 5.28. Simulation of bivariate uniform constituent densities for the equalized error training procedure.

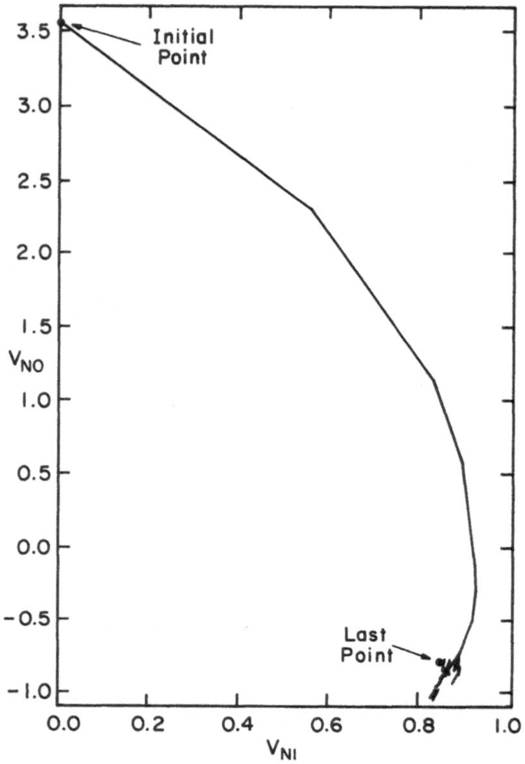

Figure 5.29. Motion of $\mathbf{V}_N(n)$ for the simulation of Figure 5.28.

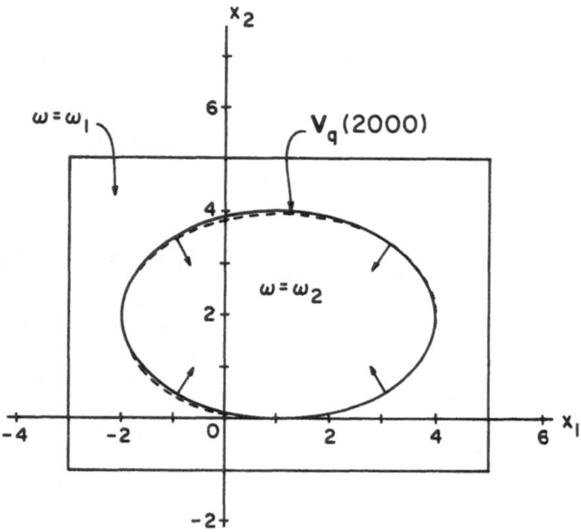

Figure 5.30. The ellipse found by the equlaized error procedure after 2000 trials is shown as a broken curve. The solid curve is the ellipse in the simulation.

Figure 5.28 illustrates bivariate uniform constituent densities which are identical to those of Figures 5.16 and 5.22. The initial hyperplane was located at $x_1 = -7$ as shown. The decision hyperplane found by this training procedure after 2000 trials is illustrated. The error probabilities for the two hyperplanes illustrated are: 0.222 for the minimum error tail moments and 0.222 for the equalized error procedure after 2000 trials. The error probabilities contributed by each class were 0.126 and 0.096 for class ω_1 and ω_2, respectively. For this simulation,

$$\rho(n) = \left(\frac{40}{40 + n}\right)^{1/2} (0.054).$$

Figure 5.29 displays the motion of the vector $\mathbf{V}_N(n)$ for normalized features. Every 40th point is plotted.

Again, the third simulation used constituent densities which were uniformly distributed either within or outside of an ellipse. The specifications are listed in Subsection 5.3.6. Quadratic $\boldsymbol{\Phi}$-mapping was used and the initial $\mathbf{V}_q$ vector was found using Equations (5.59) and (5.60). The ellipse found by the equalized error training procedure after 2000 trials is described by the vector $\mathbf{v}_q(2000) = [-0.093, 0.22, 0.94, -0.11, 0.003, -0.24]^T$ and is illustrated in Figure 5.30. The probability of error associated with this decision surface is 0.002. Figure 5.31 displays the motion of $V_{qN0}(n)$ and $V_{qN3}(n)$ during

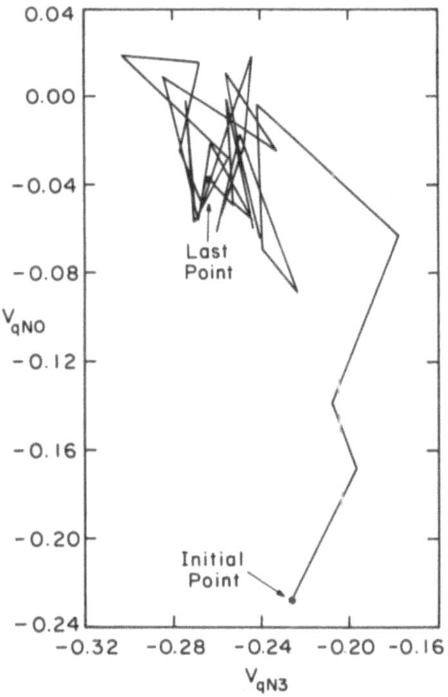

Figure 5.31. Motion of $\mathbf{V}_{qN}(n)$ for the simulation of Figure 5.30.

training with normalized features. With the exception of $V_{qN0}(n)$, the component with the largest fluctuation was again $V_{qN3}(n)$. For this simulation,

$$\rho(n) = \left(\frac{40}{40+n}\right)^{1/2} (0.030).$$

Each of the simulations supported the theory's prediction that the decision hyperplanes of the equalized error training procedures approached its suboptimal (minimal-loss) decision hyperplane.

5.6 Accounting for Unequal Costs

A common goal associated with the application of the training procedures presented in this chapter is the finding of a $\mathbf{v}^*$ which specifies a hyperplane giving a minimum probability of error decision. A simple modification of these procedures can be made which accounts for cases where the costs of the types of misclassification are *not* equal. For these cases, the goal is transformed to finding a $\mathbf{v}^*$ which specifies a decision hyperplane giving a *minimum cost*.

As discussed in Section 4.7, we first find equivalent a priori class probabilities from Equation (4.74). This equation is repeated here for reference:

$$P_{\text{eq}}(\omega_j) = \frac{c_{ij}}{c_{12}P(\omega_2) + c_{21}P(\omega_1)} P(\omega_j), \qquad j = 1, 2;\ i \neq j. \qquad (5.142)$$

Recall that c_{ij} denotes the cost of choosing class ω_i when the correct class is ω_j. The training procedures then need to be modified so that a feature vector selected from the training set has a probability $P_{\text{eq}}(\omega_j)$ of being from class ω_j. This can be accomplished in one of two ways.

In the first method, we assume that the feature vectors have been labeled with their class and are being selected at random from a training set. The ratio of the number of class ω_j feature vectors in the training set to the total number of feature vectors can then be increased or decreased so that these ratios are equal to $P_{\text{eq}}(\omega_j)$, $j = 1, 2$.

The second method is superior to the first if several combinations of costs are to be tried. For this method, the set of class ω_1 feature vectors are first separated from the set of class ω_2 feature vectors. Then, perhaps through the use of a random number generator, the selection of each feature vector is made after first choosing a particular class set. The class set is chosen according to the rule: Choose class ω_j, $j = 1, 2$ with probability $P_{\text{eq}}(\omega_j)$.

5.7 An Application

Window training, equalized error training, and minimum mean square error training algorithms have been applied to the classification of radar signatures [16] and the classification of radiographic images of human breasts [15].

Table 5.1. Partitioning of feature vectors of breast tissue xeromammograms.

Type of test set	Number of test sets of this type	Number of feature vectors per test set	Number of feature vectors in each class for each test set	
			Normal	Ductal
1	110	10	4	6
2	16	11	5	6
3	1	8	4	4

Application of these training techniques to the classification of nodules in radiographic images of lung tissue is in progress at this writing. Below we summarize the results of the work on the classification of breast tissue.

Each of 32 xeromammograms (electrostatically printed x-ray images of the human female breast) were divided into 64 square sections, each containing an image of diagnostically interesting tissue. Each section was approximately one centimeter in width and height, and consisted of 128×128 picture elements ("pixels"). Under the guidance of an expert mammographer*, our computer tagged each of 2048 of these sections either as containing one or more specified abnormalities or as normal. This tagging process yielded 524 normal sections and 760 ductal sections—a total of 1284 sections for the design and test of our classifier. (The ductal regions are often described as containing "increased ducting" or "prominent ducting"—which has the appearance of a plowed field in the xeromammogram). Dr. John Wolfe's research suggests that breast tissue containing a sufficiently large quantity of ducting indicates high risk of subsequent contraction of breast cancer [17].

A list of 30 features were formed from statistics of the gray levels in each square section. These features were ranked in the order of the minimum error rates obtained when each feature is used individually. Histograms of these features and correlations of pairs of these features were examined. Using this ranking and our examination of the histogram and correlations in combination with the normalization and feature reduction procedures described in Subsection 5.2.1, we reduced these features to a set of twelve. Preliminary window training applied to subsets of these twelve features and their squares and cross products yielded a final set of 35 features consisting of eleven of the original set and 23 square and cross product features.

A window training procedure applied to the resulting design set of 1284 35-dimensional feature vectors yielded a classifier having a resubstitution error rate of 0.158. To estimate the accuracy of this error rate we used the same data in a rotation schedule for design and test. In this schedule the 1284 feature vectors of our 1284 sections were partitioned into 127 test sets, as shown in Table 5.1. This yielded 127 classifier designs, one for each test

* Dr. G. Frankl of the Kaiser Permanente Hospital, 1505 N. Edgemont, Los Angeles, CA 90027.

set, obtained by removing one of the test sets from the data, and applying the window training procedure to the remaining data. The accumulated errors for these tests yielded a rotation error rate of 0.188. This differs from the resubstitution error rate by 0.030. The actual error in the resubstitution error rate of this window-trained classifier is probably less than 0.030, because the design sets in the rotation are smaller than that in the resubstitution.

The relative sizes of the components of the weight vectors obtained in the resubstitution window-trained classifier indicated that five features might be removed with little effect on the error rate. Of these five, the removal of any one among a subset of three had negligible effect on the error rate [15].

Using the above design set, the minimum mean square error training algorithm yielded an error rate of 0.24, and the equalized error procedure yielded an error rate of 0.20. This demonstrated the superiority of the window training algorithm for this data.

These training procedures were implemented on an Interdata 7/32 minicomputer. This minicomputer had a core memory of 64,000 32-bit words. We found that convergence of our training processes was quite satisfactory even when the dimensionality of feature space was relatively large. In particular, we found that in a 35-dimensional feature space our relatively slow computer (15-microsecond floating point multiplication time) completed 5000 trials in about 8 minutes.

5.8 Summary

The principal emphasis of this chapter has been the presentation of training procedures (i.e., algorithms) which determine linear classifiers that operate efficiently even for pattern classification problems where the class distributions overlap. These training procedures were based upon a modified gradient descent technique.

In a discussion of the modifications to a strict gradient descent technique, some of the effects of weight vector magnitude and the position of the distributions in feature space were described. Suggestions for reducing the unfavorable effects of these factors were discussed.

As a preparation for the application of the training procedures, techniques of distribution normalization, origin translation, and initial vector calculation were presented. Although distribution normalization is not necessary for the application of the training procedures, it aids in the choice of training procedure parameters and in the recognition of the least important features. The origin translation techniques presented can significantly improve convergence properties, as can the initial vector calculation techniques presented.

The window training procedure was next developed. This procedure finds an augmented weight vector which asymptotically approaches the vector describing the decision hyperplane giving the minimum probability of error.

The constraints necessary when using this procedure were presented and described. Suggested parameter values necessary for practical applications were given and several simulation examples were presented.

Two suboptimal training procedures were also described. These were the minimum mean square error and the equalized error procedures. Although these procedures do not in general find the minimum probability of error decision hyperplane, their convergence properties and robustness make them attractive. In each case, the necessary constraints were presented and described, some suggested practical parameter values were given, and several simulation examples were presented.

Two methods were described which modify the application of any one of the training procedures so that unequal costs of misclassifications can be taken into account.

Finally, a discussion was presented of an application of the window training procedure and the minimum mean square error training procedure to the classification of x-ray images of breast tissue into two classes, normal and ductal.

Exercises

5.1. For the case where $J(\mathbf{v}) = P(\text{error}|\mathbf{v})$ and $\mathbf{G}(\mathbf{v})$ is given by Equation (5.13), demonstrate that the amount of translation of a decision hyperplane in $\mathbf{x}$-space is in general larger for larger distances of the decision hyperplane from the origin when the training procedure is $\mathbf{V}(n+1) = \mathbf{V}(n) + \rho(n)\mathbf{G}[\mathbf{V}(n)]$ and $\int_{\mathscr{S}} [f_2(\mathbf{x}) - f_1(\mathbf{x})] ds = \text{constant}$.

5.2. Show that $\nabla_{\mathbf{v}} P(\text{error}|\mathbf{v}) = \mathbf{0}$ for all values of $\mathbf{v}$ for which $\nabla P(\text{error}|\mathbf{v}) = \mathbf{0}$. Use Equations (4.29) and (5.21).

5.3. Assume a two-dimensional feature example where the origin in $\mathbf{x}$-space has been translated to the point $\mathbf{b} = [2, 3]^T$ during training. Further assume that the augmented weight vector $\boldsymbol{v}$, referenced to $\boldsymbol{\beta}$-origin space, found from the training is given by $\boldsymbol{v} = [-2, 4, 5]^T$. Find the augmented weight vector $\mathbf{v}$, referenced to $\mathbf{x}$-space without origin translation, which describes the same hyperplane as does $\boldsymbol{v}$.

5.4. Assume that the following *sample means* and *sample variances* have been determined from the training set of a two-feature problem having sample size of 100 for class ω_1 and size 300 for class ω_2.

$$\hat{\mu} = -0.01, \quad \hat{\sigma}^2 = 0.1, \quad \text{for } x_1 \text{ samples from class } \omega_1$$
$$\hat{\mu} = 0.5, \quad \hat{\sigma}^2 = 144, \quad \text{for } x_2 \text{ samples from class } \omega_1$$
$$\hat{\mu} = 1.01, \quad \hat{\sigma}^2 = 0.09, \quad \text{for } x_1 \text{ samples from class } \omega_2$$
$$\hat{\mu} = 16.2, \quad \hat{\sigma}^2 = 130, \quad \text{for } x_2 \text{ samples from class } \omega_2.$$

Find the sample normalized standard deviations for each feature of each class using:

(a) Average variance normalization.
(b) Mixture variance normalization.

5.5. If the normalization indicated by Equation (5.25) has been applied during training, find the adjustment which must be applied to the augmented weight vector found from the training so that it can be used to classify unnormalized feature vectors.

Answer: If $\mathbf{V}_N = [V_{NO}, \mathbf{W}_N^T]^T$ denotes the augmented weight vector for normalized features, then for use with unnormalized feature vectors,

$$\mathbf{V} = \begin{bmatrix} V_{NO} \\ \mathbf{W}_N^T \boldsymbol{\Sigma}_N^{-1} \end{bmatrix},$$

where

$$\boldsymbol{\Sigma}_N = \begin{bmatrix} \tilde{\sigma}_1 & & & 0 \\ & \tilde{\sigma}_2 & & \\ & & \ddots & \\ 0 & & & \tilde{\sigma}_d \end{bmatrix}$$

5.6. Given the following constituent densities:

$$f_1(\mathbf{x}) = \begin{cases} 1/12, & \text{if } 0 < x_1 < 2, 0 < x_2 < 4, \\ 0, & \text{otherwise} \end{cases}$$

$$f_2(\mathbf{x}) = \begin{cases} 1/24, & \text{if } 0 < x_1 < 4, 0 < x_2 < 2, \\ 0, & \text{otherwise,} \end{cases}$$

find the origin translation vector $\mathbf{b}$ as given by Equation (5.29). Use actual means and variances in place of their sample approximations.

5.7. Repeat Exercise 5.6, assuming a priori that the best initial $\mathbf{w}$ is given by

$$\mathbf{w} = [1, -1]^T.$$

5.8. Using Equations (5.29), (5.30), (5.33) and (5.42), derive Equation (5.53) for the case where $\boldsymbol{\Sigma}_1 = \boldsymbol{\Sigma}_2 = \boldsymbol{\Sigma}$.

5.9. Demonstrate the validity of Equation (5.55) by using Equations (5.52) and (5.54).

5.10. Show that $\boldsymbol{\Sigma}_{\text{av}}$ as given by Equation (5.56) becomes the identity matrix if the features are statistically independent and the average variance normalization presented in Subsection 5.2.1 has been performed.

5.11. Demonstrate the validity of Equations (5.59) and (5.60) with reference to the equation $\mathbf{v}_q^T \mathbf{y}_q = 0$ and the definition of $\mathbf{x}_q$.

5.12. Given that $\mathbf{V}(2) = [0, 1, 1]^T$, $\rho(n) = 1/\sqrt{n}$, and $c(n) = 3/\sqrt{n}$, assume $M = 10^{10}$ and $\boldsymbol{\beta} = \mathbf{0}$. Also assume that you know that $v_1^* > 0$. Find $\mathbf{V}(3)$ using the window training procedure with the rectangular window when the second training feature vector is:

(a) $\mathbf{X}(2) = [-1, -1]^T$ from class ω_2.
(b) $\mathbf{X}(2) = [1, 0]^T$ from class ω_2.
(c) $\mathbf{X}(2) = [0, 1]^T$ from class ω_1.

5.13. Repeat exercise 5.12 with $c(n) = 1.5/\sqrt{n}$.

5.14. Referring to Equation (5.96), find $\theta(n + 1)$ in terms of W_0, W_1, ρ, Z_0, and Z_1.

5.15. By reference to Equations (5.93) and (5.95), demonstrate that the adjustment $\rho(n)\mathbf{Z}(n)$ is always normal to $\mathbf{V}(n)$.

5.16. Demonstrate that $\nabla J(\mathbf{v})$ as given by Equations (5.108) and (5.110) are equivalent.

5.17. Repeat Exercise 5.12 using the minimum mean square error training procedure.

5.18. Demonstrate the validity of Equation (5.121).

5.19. Show that the weight vector $\mathbf{v}^*$ that minimizes

$$(\mathbf{A}\mathbf{v} - \mathbf{b})^T(\mathbf{A}\mathbf{v} - \mathbf{b}),$$

where $\mathbf{b} = [1, 1, \ldots, 1]^T$, maximizes the function $t(\mathbf{v})$ defined in Subsection 5.4.2.

5.20. Repeat Exercise 5.12 using the equalized error training procedure.

5.21. Show that

$$\frac{(\mathbf{V}^T\mathbf{Y})^2}{\|\mathbf{W}\|^2} - 2\frac{(\mathbf{V}^T\mathbf{Y})(\mathbf{W}^T\mathbf{X})}{\|\mathbf{W}\|^2} = \frac{W_0^2 - (\mathbf{W}^T\mathbf{X})^2}{\|\mathbf{W}\|^2}.$$

which is used in Appendix C.

5.22. Suppose you have a training set of feature vectors where 200 are labeled as class ω_1 and 300 are labeled as class ω_2. Assume that this class ratio reflects the ratio of natural sampling. If the cost of misclassifying an ω_1 feature vector is \$50 and misclassifying an ω_2 feature vector is \$100, describe the adjusted training set and the method of selection during training which takes the costs into account

(a) when the first method of Section 5.7 for accounting for unequal costs is applied.

(b) when the second method is applied.

References

1. G. N. Wassel, Training a linear classifier to optimize the error probability. Ph.D. Dissertation, UCI Pattern Recognition Project, School of Engineering, University of California, Irvine, California, Technical Report TP-72-5, December 1972.
2. R. O. Duda and P. E. Hart, *Pattern Recognition and Scene Analysis*. John Wiley & Sons, New York, 1973.
3. G. N. Wassel and J. Sklansky, Training a one-dimensional classifier to minimize the probability of error. *IEEE Trans. on Systems, Man, and Cybernetics*, **SMC-2**: 533–541 (1972).
4. J. Kiefer and J. Wolfowitz, Stochastic estimation of the maximum of a regression function. *The Annals of Mathematical Statistics*, **23**: 462–466 (1952).
5. A. Dvoretzky, On stochastic approximation. In: *Proc. Third Berkeley Symp. Mathematical Statistics and Probability*, December 1954, June-July 1955, pp. 39–55.
6. J. R. Blum, Multidimensional stochastic approximation methods. *The Annals of Mathematical Statistics*, **25**, (4): 1954.
7. C. Blaydon, Recursive algorithms for pattern classification. Division of Engineering and Applied Physics, Harvard University, Cambridge, Massachusetts, Technical Report 520, March 1967.
8. C. C. Blaydon and Y. C. Ho, Recursive algorithms for pattern classification. *Proceedings of 1966 National Electronics Conference*.
9. Y. C. Ho and R. L. Kashyap, An algorithm for linear inequalities and its applications. *IEEE Trans. on Electronic Computers*, **EC-14**: 683–688 (1965).
10. G. N. Wassel and J. Sklansky, An adaptive nonparametric linear classifier. *Proc. of the IEEE*, **64**, (8): 1162–1171 (1976).

11. K. Fukunaga, *Introduction to Statistical Pattern Recognition*. Academic Press, New York, 1972.

12. B. Widrow and M. E. Hoff, Adaptive switching circuits. *1960 IRE WESCON Convention Record*, Part 4, August 1960, pp. 96–104.

13. W. C. Miller, A modified mean-square-error criterion for use in unsupervised learning. Technical Report No. 6778-2, Stanford University, August 1967.

14. H. Do-Tu and M. Installe, Learning algorithms for nonparametric solution to the minimum error classification problem. *IEEE Trans. on Computers, C-27*: 648–659 (1978).

15. C. Kimme-Smith, G. Wassel, G. Frankl, and J. Sklansky, Toward a computerized estimate of the index of risk for developing breast cancer. UCI Technical Report TP-79-3, University of California, School of Engineering, Irvine, California, 1979.

16. C. Hightower, K. Rowe, and J. Sklansky, Target pattern correlation for BMD threat discrimination. *Proceedings of the Twenty-Third Tri-Service Radar Symposium*, West Point, New York, July 1977.

17. J. N. Wolfe, Breast pattern as an index of risk for development breast cancer. *American Journal of Roentgenology*, **126**: 1130–1139 (1976).

18. M. J. Lighthill, *Introduction to Fourier Analysis and Generalized Functions*. Cambridge University Press, New York, 1958.

19. R. A. Fisher, The use of multiple measurements in taxonomic problems. *Annals of Eugenics*, **7** (Part II): 179–188 (1936). Reprinted in: *Contributions to Mathematical Statistics*, John Wiley & Sons, New York, 1950.

CHAPTER 6

Markov Chain Training Models for Nonseparable Classes

6.1 Introduction

In Chapter 2, we derived the convergence under separability property of the proportional increment training procedure. As interesting as this property is, convergence is usually of less practical importance than a good estimate of the performance of the classifier as a function of training length. Indeed, it is sometimes possible to obtain a noncovergent training procedure that performs better over a finite training period than a convergent training procedure on the same input data. This situation has motivated much of the research on the learning dynamics of trainable classifiers.

Another shortcoming of the convergence under separability theorem is that it applies only to linearly separable pairs of class regions. But nonseparable classes and nonlinearly separable* classes occur quite often in practice, notable examples occurring in weather prediction and medical diagnosis.

This chapter and Chapter 7 are devoted to modeling and analyzing the learning dynamics of trainable linear classifiers operating on nonseparable classes. Both chapters are based on Markov models. The applications of such models include both the design of automatic classifiers as well as the modeling of human learning.

* "Nonlinearly separable" does not mean "not linearly separable." It means "separable by nonlinear decision surfaces".

In this chapter we first review the modeling of a single-feature trainable classifier operating on a pair of nonseparable classes (i.e., a pair of classes whose class regions overlap). We devote subsequent portions of this chapter to Markov chain models of the training of decision surfaces of linear classifiers whose class regions are (a) not linearly separable and (b) described statistically. Because of the large number of states in typical Markov chain models of multiple-feature classifiers, we give special attention to single-feature classifiers.

6.2 The Problem of Analyzing a Stochastic Difference Equation

The most common method of describing the class regions of a classifier when the class regions overlap is by probability density functions—i.e., by $p_{\mathbf{x}}(\mathbf{x}|\omega)$, the conditional probability density of feature vector $\mathbf{X}$, given the class ω; and by $P(\omega)$, the a priori probability of occurrence of class ω. Unfortunately, it is difficult to analyze the learning dynamics of most training procedures whose input data are described in this way. The reason for this difficulty is that the motion of the weight vector $\mathbf{v}(n)$ is usually governed by a difference equation in which one of the variables is a random vector. Equation (2.13), reproduced below, illustrates this situation:

$$\mathbf{v}(n+1) = \begin{cases} \mathbf{v}(n) + \rho\boldsymbol{\eta}(n), & \rho > 0, \quad \text{if } \mathbf{v}^T(n)\boldsymbol{\eta}(n) \leqslant 0, \\ \mathbf{v}(n), & \text{otherwise.} \end{cases} \tag{6.1}$$

When the input data are random variables, $\mathbf{v}(n)$ and $\boldsymbol{\eta}(n)$ must be replaced by random vectors—which we denote by $\mathbf{V}(n)$ and $\mathbf{H}(n)$, respectively. Then Equation (6.1) becomes

$$\mathbf{V}(n+1) = \begin{cases} \mathbf{V}(n) + \mathrm{o}\mathbf{H}(n), & \rho > 0, \quad \text{if } \mathbf{V}^T(n)\mathbf{H}(n) \leqslant 0, \\ \mathbf{V}(n), & \text{otherwise.} \end{cases} \tag{6.2}$$

Here the augmented weight vector is determined by a vector difference equation in which the independent variable is a random vector $\mathbf{H}(n)$. We may express $\mathbf{V}(n)$ explicitly in terms of $\mathbf{H}(n)$ by expressing $\mathbf{V}(n)$ as the sum of $\mathbf{V}(j+1) - \mathbf{V}(j)$ over j from $j = 0$ to $j = n - 1$, and using Equation (6.2). This yields

$$\mathbf{V}(n) = \mathbf{V}(0) + \rho \sum_{j=0}^{n-1} \zeta(j)\mathbf{H}(j), \tag{6.3}$$

where $\zeta(j) = 1$ when $\mathbf{V}^T(j)\ {}^{\mathbf{I}}(j) \leqslant 0$, $\xi(j) = 0$ when $\mathbf{V}^T(j)\ {}^{\mathbf{I}}(j) > 0$.

Since $\mathbf{H}(n)$ is a random vector, Equation (6.2) is a *stochastic difference equation*, and Equation (6.3) is a summed form of that difference equation. Note that even though the $\mathbf{H}(j)$'s are often independently and identically distributed, the random vectors $\{\zeta(j)\mathbf{H}(j)\}$ are neither independently nor identically distributed. For this reason the application of the central limit theorem to Equation (6.3) for finding the mean and variance of $\mathbf{V}(n)$ is not possible. This has made the derivation of exact, closed form expressions for the mean and variance of $\mathbf{V}(n)$ very difficult. In fact such expressions have not been obtained for most cases of interest.

It has proved fruitful to deal with these difficulties by two methods. One method is to find trainable classifiers that can be analyzed by Markov chain models. An example of such a classifier is a single-feature classifier with constant increment training, to be described below. Although Markov chain models have been shown to be useful for single-feature trainable classifiers, their usefulness is much more limited for multiple-feature classifiers. The reason for this is that the number of states of the models of multiple-feature classifiers is usually large. Nevertheless Markov chain models can be of at least qualitative rather than quantitative utility for multiple-feature classifiers. Markov chain models for both single-feature and multiple-feature classifiers are discussed in this chapter.

A second method of dealing with the difficulties of analyzing the learning dynamics of trainable classifiers is to approximate the difference equation (6.1) by an appropriate differential equation. This second method is often more useful than Markov chain models, since the approximation is often quite good, and since the method retains the ability to analyze multiple-feature classifiers without an excessive increase in computational complexity over that required for single-feature classifiers. This second method is discussed in Chapter 7.

When it is applicable, the Markov chain method is often more accurate than the differential equation method, and the mathematical manipulations are easier.

6.3 Examples of Single-Feature Classifiers

Although most pattern classifiers analyze multidimensional feature vectors, there exist classifiers that analyze only one-dimensional feature vectors—i.e., a single scalar feature. These classifiers have special interest for us because their learning dynamics are easier to analyze than those of multiple-feature classifiers.

We describe below two examples of such classifiers.

EXAMPLE 6.1: (Signal detection in radar, electronic communication and human hearing.) Consider a pulsed data communication link in which the transmitted data is a sequence $\Omega(n)$ of 1's and 2's. We may view 1 as a label for ω_1, and 2 as a label for ω_2. These 1's and 2's are transmitted at a uniform rate in waveform $U(t)$ consisting of a sequence of rectangular pulses of unit height, all pulses having the same width. A 2 is transmitted as a pulse of unit height and a 1 is transmitted as a pulse of zero height. This is illustrated in Figure 6.1 for the transmitted data sequence

$$\{\Omega(n) \mid n = 0, 1, 2, 3, \ldots\} = 2, 1, 1, 2, 2, 2, \ldots.$$

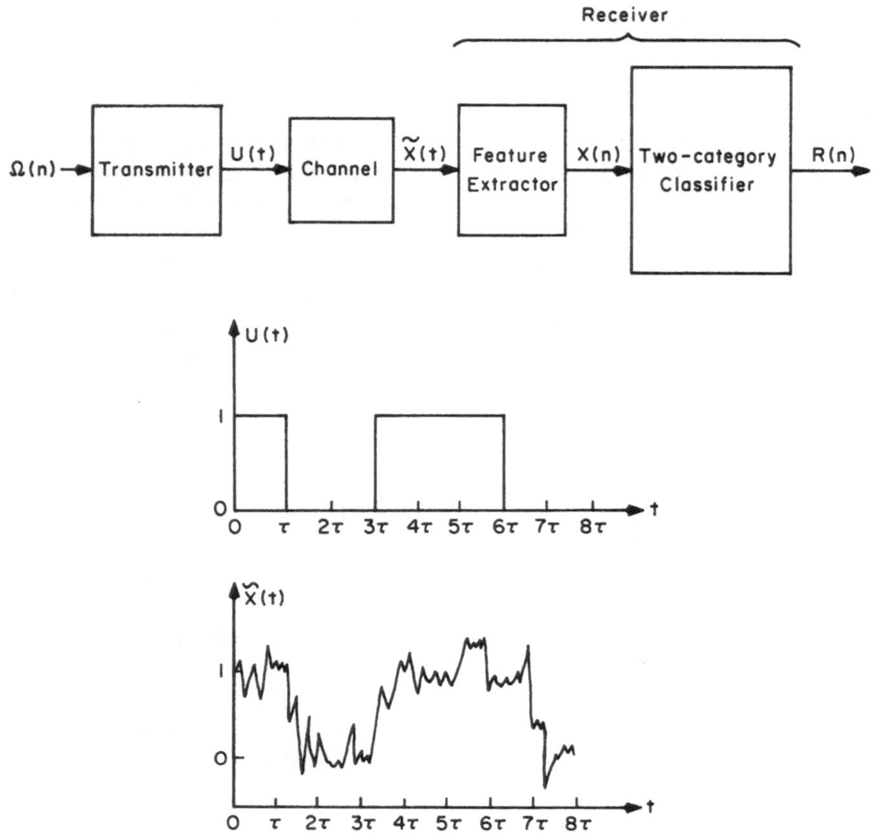

Figure 6.1. Signal detection, an example of single-feature pattern classification.

The transmitted waveform for this data is

$$U(t) = \Omega(n) - 1, \quad \text{for } n\tau < t \leqslant (n + 1)\tau.$$

The channel mixes $U(t)$ with noise, producing the noisy waveform $\tilde{X}(t)$. The job of the receiver is to produce a guess $R(n)$ equal to 0 or 1 such that the probability $P[R(n) = \Omega(n)]$ is at or close to its maximum possible value. This model of signal detection occurs in radar, in digital-data telephone lines, and in human hearing.

Let us suppose that we are faced with the problem of designing an optimum or near-optimum receiver for the signal detection model described above. This problem can be approached from the viewpoint of pattern classification. If we view $\Omega(n)$ as a class to which $\tilde{X}(t)$ belongs for $n\tau < t \leqslant (n + 1)\tau$, then the job of the receiver is to behave as a two-category classifier. The receiver includes a device for extracting one or more features from $\tilde{X}(t)$, as well as a device for classifying the features. Often just one feature is sufficient, namely, the integrated square of $\tilde{X}(t)$ over a unit interval. Thus, if

$$X(n) = \int_{n\tau}^{(n+1)\tau} \tilde{X}^2(t)\, dt,$$

then the single feature $X(n)$ is often adequate for guessing $\Omega(n)$.

EXAMPLE 6.2: (Edge detection in continuous-tone photographs.) In recent years several special-purpose computer facilities for analyzing continous-tone photographs (as differentiated from photographs having just a few discrete gray levels) have been constructed. In one case the photographs are pictures of the surfaces of planets and the moon; in other cases they are aerial photographs of agricultural crops, radiographs, or photomicrographs of biological cells.

In all of these computer facilities one of the uses of the computer is to enhance the visual quality of the photograph—for example, between normal and abnormal human tissue. For this enhancement process the computer sometimes is programed to detect the boundary or edge between significantly different regions of the photograph [1].

One way of enhancing the edges in continuous-tone photographs is to compare the average gray levels in small areas of neighboring regions. When the magnitude of the difference between these average gray levels is sufficiently large, an edge is detected; otherwise no edge is detected. Thus in this case the magnitude of the difference $X(n)$ between average gray levels may be used as a single feature for classifying the nth observed point of a picture into one of two categories: *edge* and *nonedge*. (Definitions of greater sophistication for $X(n)$ have been proposed. See, in particular, Chow and Kaneko [2].) Then the classifier assigns every observed $X(n)$ to one of two regions, labeled according to the two categories. If the classifier is trainable, then the boundary between these two regions in feature space may be adjusted as a function of n in response to a training sequence.

We now describe how the learning dynamics of a single-feature classifier with a constant increment training procedure may be analyzed.

6.4 A Single-Feature Classifier with Constant Increment Training

A block diagram of a single-feature classifier is shown in Figure 6.2. Here the classifier generates $R(n) = 2$ (representing the guess that $\omega(n) = \omega_2$) if $X(n) > \Theta(n)$, where $X(n)$ is the feature and $\Theta(n)$ is a scalar weight at trial n; otherwise $R(n) = 1$. The sign detector produces 1 if its input is positive; otherwise it produces -1. As is shown in Figure 6.2, the sign detector's input is $X(n) - \Theta(n)$.

The weight $\Theta(n)$ is subject to adjustment during the training procedure. It is desirable to choose a training procedure that will bring $\Theta(n)$ close to a

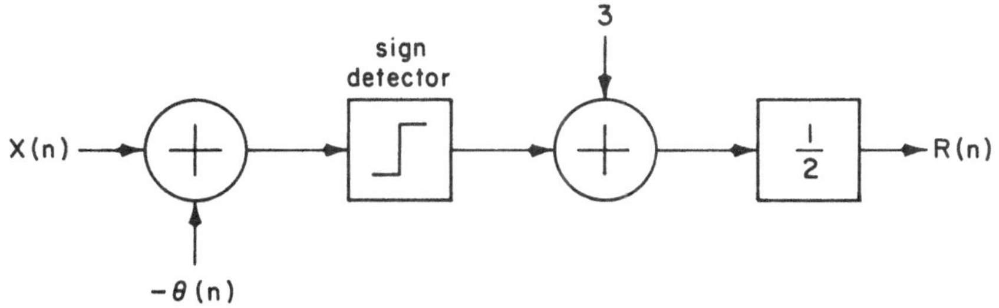

Figure 6.2. A simple single-feature classifier.

value θ^* such that the probability

$$P[\Omega(n) = R(n) \mid \Theta(n) = \theta^*],$$

where $\Omega(n) = 1$ or 2 whenever $\omega(n) = \omega_1$ or ω_2, respectively, is a maximum.

Since in this classifier the output $R(n)$ is 2 only when $X(n) > \Theta(n)$, and $R(n)$ is otherwise 1, the weight $\Theta(n)$ has sometimes been referred to as a *threshold.* This nomenclature, in turn, has led to the name *threshold learning process* or TLP for this classifier [3].

To illustrate the basic concepts of training this classifier, we shall assume that the following procedure is used:

$$\Theta(n+1) = \Theta(n) - \rho Z(n), \qquad \rho > 0,$$

where $Z(n) = \Omega(n) - R(n)$ and therefore is equal to 1 if an ω_2 feature is incorrectly classified, is equal to -1 if an ω_1 feature is incorrectly classified, and is equal to 0 for a correct classification. Hence all of the nonzero adjustments of $\Theta(n)$ during the training phase are equal in size. For this reason, we refer to this training procedure as *constant increment.**

In this training procedure, $\Theta(n)$ is increased or decreased by a constant amount only when an error takes place. (Note that an error occurs when $|Z(n)| = |\Omega(n) - R(n)| = 1$.) Otherwise $\Theta(n+1) = \Theta(n)$. During the working phase, the weight $\Theta(n)$ remains fixed at the value it reached at the end of the preceding training phase.

6.5 Basic Properties of Learning Dynamics

In this section we describe a few basic properties associated with the learning dynamics of classifiers, and show how these properties are analyzed in a single-feature classifier undergoing constant increment training.

6.5.1. Performance Measures

The basic measure of performance of a trainable classifier is the *success probability* at trial n, i.e.,

$$z(n) = P[\Omega(n) = R(n)] = P[Z(n) = 0] \equiv P[\text{success} \mid n]. \tag{6.4}$$

If the weight at trial n is known or given, then the probability of success is the *conditional success probability* defined by

$$S(\theta) = P[\Omega(n) = R(n) \mid \Theta(n) = \theta] = P[Z(n) = 0 \mid \Theta(n) = \theta]. \tag{6.5}$$

The variable n does not appear in $S(\theta)$ because when θ is given, the knowledge of n does not change the conditional success probability.

* The constant increment training procedure should not be confused with the fixed increment procedure referred to by Nilsson and others [4]. *Fixed increment* is a misnomer for that procedure, and is replaced in this book by *proportional increment.*

In the process of choosing a training procedure it is desirable to know the success probability as a function of n, taking account of the fact that $\Theta(n)$ is unknown. This success probability $z(n)$ is the expectation of $S[\Theta(n)]$. Thus

$$z(n) = E[S(\Theta(n))]. \tag{6.6}$$

If Θ is absolutely continuous (see R. B. Ash [5]) then from Equation (6.6)

$$z(n) = \int_{-\infty}^{\infty} p_{\Theta(n)}(\theta)S(\theta)\,d\theta \tag{6.7}$$

where $p_{\Theta(n)}$ is the probability density of $\Theta(n)$. If Θ is discrete, then

$$z(n) = \sum_{\theta=-\infty}^{\infty} P[\Theta(n) = \theta]S(\theta). \tag{6.8}$$

We refer to the curve of $z(n)$ versus n as a *learning curve*.

Some reflection on the use of $z(n)$ alone as a performance measure will show that it is inadequate. Consider that at the end of a training period of length N, the value of the conditional success probability $S[\Theta(N)]$ is a random variable rather than a deterministic quantity, because $\Theta(N)$ is a random variable. It is $S[\Theta(N)]$ which describes the expected performance during the working phase after an initial training phase of length N. The quantity $z(n)$ gives us only the mean value of $S[\Theta(N)]$.

Hence, in addition to $z(n)$, it is useful to know the variance of $S[\Theta(n)]$. We denote this by $\tilde{q}(n)$, and define it as

$$\tilde{q}(n) = E\{[S(\Theta(n)) - z(n)]^2\}. \tag{6.9}$$

We refer to the curve of $\tilde{q}(n)$ versus n as a *variance curve*. Ideally, the training of a classifier will stop when the learning curve exceeds its specified minimum acceptable value by a tolerance determined by the variance curve.

Other useful performance measures are the error probability and the expected cost. Let $e(n)$ denote the event of an error (i.e., an incorrect classification) at trial n. The *error probability* at trial n is

$$P[e(n)] = P[\Omega(n) \neq R(n)] = 1 - z(n). \tag{6.10}$$

The *expected cost* is an elaboration on the error probability. In a two-category classifier, there are two types of errors: $(\omega = \omega_2, R = 1)$ and $(\omega = \omega_1, R = 2)$. These are often referred to as *false negative* (or *Type 1*) and *false positive* (or *Type 2*), respectively. The total error probability may be expressed as a sum of the probability of a false positive and the probability of a false negative:

$$P[e(n)] = P(\Omega \neq R) = P(\Omega = 1, R = 2) + P(\Omega = 2, R = 1). \tag{6.11}$$

Sometimes the Type 1 error and the Type 2 error have unequal values or costs. For example, in medical diagnosis where class ω_1 might denote normal and class ω_2 abnormal, a Type 2 error (false positive) is often (but not always) less dangerous to the patient than the Type 1 error (false negative). Hence in place of Equation (6.11) we write the *expected cost* as

$$E[C(n)] = c_{21}P(\Omega = 1, R = 2) + c_{12}P(\Omega = 2, R = 1), \tag{6.12}$$

where c_{21}, c_{12} are the costs associated with false positives and false negatives, respectively. As in earlier chapters, let c denote the expected cost if every feature were incorrectly classified, i.e.,

$$c = c_{12}P(\omega_2) + c_{21}P(\omega_1).$$

Then $(1/c)E[C(n)]$ has the properties of a probability, so that

$$\frac{1}{c}E[C(n)] = \lambda_{21}P(\Omega = 1, R = 2) + \lambda_{12}P(\Omega = 2, R = 1),$$

where $\lambda_{ij} = c_{ij}/c$, can replace $P[e(n)]$ in the equations of this chapter when it is desirable to account for costs. Another useful quantity contributing to an evaluation of the training process is the expected value of $\Theta(n)$, namely,

$$\mu(n) = E[\Theta(n)].$$

The distance of $\mu(n)$ from its asymptotic value is an indirect indicator of when to stop training ($z(n)$ is a direct indicator). The curve of $\mu(n)$ versus n is referred to as a *centroid curve.*

6.5.2 Constituent Densities and Optimum Weights

It is often assumed that the input data of a classifier arises from a set of conditional densities $p(\mathbf{x}|\omega_j)$ and $P(\omega_j)$, where $p_{\mathbf{X}}(\mathbf{x}|\omega_j)$ is the conditional probability of $\mathbf{X}$ when the class is ω_j, and where $P(\omega_j)$ is the a priori probability of the event $\omega = \omega_j, j = 1, 2$.

Recall the definition of the *constituent density* $f_j(\mathbf{x})$:

$$\begin{aligned} f_j(\mathbf{x}) &= P(\omega_j)p_{\mathbf{X}}(\mathbf{x}|\omega_j) \\ &\equiv p_{\mathbf{X},\omega}(\mathbf{x}, \omega_j). \end{aligned}$$

Thus in the single feature case $f_j(x)\,dx$ is the joint probability of the events $X \in (x, x + dx)$ and $\omega = \omega_j$.

If an estimate of the constituent densities is available, and if these densities are reasonably well behaved, one can estimate the learning curves and variance curves for various training procedures, as well as estimate a set of optimum decision surfaces. We illustrate this by analyzing the single-feature classifier shown in Figure 6.2. Here we have two constituent densities, $f_1(x)$ and $f_2(x)$, where

$$f_1(x) = P(\omega = \omega_1)P_X(x|\omega = \omega_1) \equiv P(\Omega = 1)p(x|\Omega = 1) \tag{6.13}$$

$$f_2(x) = P(\omega = \omega_2)P_X(x|\omega = \omega_2) \equiv P(\Omega = 2)p(x|\Omega = 2). \tag{6.14}$$

These densities are illustrated in Figure 1.12. In this figure the value of the threshold θ is marked by the dashed line.

We now can compute the conditional error probability, given θ:

$$P(e|\theta) = P[\Omega \neq R|\theta] = P(\Omega = 2, R = 1) + P(\Omega = 1, R = 2) \tag{6.15}$$

$$P(\Omega = 2, R = 1) = \int_{-\infty}^{\theta} f_2(x)\,dx \tag{6.16}$$

$$P(\Omega = 1, R = 2) = \int_{\theta}^{\infty} f_1(x)\,dx. \tag{6.17}$$

The false negative and false positive probabilities in Equations (6.16) and (6.17) are indicated respectively by the diagonally and horizontally shaded regions in Figure 1.12. Thus

$$P(e|\theta) = P[\Omega \neq R|\theta] = \int_{-\infty}^{\theta} f_2(x)\,dx + \int_{\theta}^{\infty} f_1(x)\,dx. \tag{6.18}$$

The error probability is given by

$$P[e(n)] = E\{P[e|\Theta(n)]\} = \sum_{\theta=-\infty}^{\infty} P(\theta)P(e|\theta) \tag{6.19}$$

if $\Theta(n)$ and θ are discrete variables. Thus to compute $P[e(n)]$ as well as $z(n)$ we must first find the probability function $P(\theta)$, i.e., the probability that $\Theta(n) = \theta$, for all θ. In the next section we show how this may be done for the constant increment training procedure.

More generally, we may define $E(C|\theta)$ as the *conditional expected cost*, given θ. By Equations (6.12), (6.16), and (6.17) we get

$$E(C|\theta) = c_{12} \int_{-\infty}^{\theta} f_2(x)\,dx + c_{21} \int_{\theta}^{\infty} f_1(x)\,dx. \tag{6.20}$$

From Equation (6.20) one can find a value of θ which minimizes $E(C|\theta)$. To do this one first differentiates both members of Equation (6.20) and sets the derivative of $E(C|\theta)$ equal to zero. This yields

$$c_{21} f_1(\theta^*) = c_{12} f_2(\theta^*). \tag{6.21}$$

where θ^* is a value of θ for which $E(C|\theta)$ is a local minimum or a local maximum. In particular, if $c_{12} = c_{21}$ and if $f_1(x)$ and $f_2(x)$ have a single point of intersection at $x = \theta^*$, then the minimum error probability is achieved when $\theta = \theta^*$.

It can be shown (see Exercise 6.2) that $E(C|\theta)$ is a local minimum at $\theta = \theta^*$ even if $E(C|\theta)$ has a finite discontinuity at $\theta = \theta^*$, provided

$$[c_{21} f_1(x) - c_{12} f_2(x)](x - \theta^*) < 0 \quad \text{for all } x \neq \theta^*. \tag{6.22}$$

If there is just one value of θ^* that satisfies Equations (6.21) and (6.22) it is clearly the optimum value of θ. If there is more than one such value, say $\{\theta_i^*\}$, an optimum value of θ, say $\theta = \theta_m^*$, is one for which

$$E(C|\theta_m^*) \leqslant E(C|\theta_i^*) \quad \text{for all } \theta_i^* \neq \theta_m^*.$$

6.5.3 Markov Motion of the Threshold $\Theta(n)$

In the preceding section we showed how the error probability at trial n may be computed in terms of the probability function of the random variable $\Theta(n)$, namely $P(\theta)$. In this section we show how this function may be computed when the training procedure is the following variation of a constant increment procedure.

Let the allowed values of $\Theta(n)$ be bounded from below and above by a and b, respectively, and let $\Theta(n+1)$ be determined by

$$\Theta(n+1) = \begin{cases} \Theta(n) - \rho Z(n), & \text{if } a \leqslant [\Theta(n) - \rho Z(n)] \leqslant b, \\ \Theta(n), & \text{otherwise.} \end{cases} \tag{6.23}$$

For this training procedure, we see that $\Theta(n)$ can only take on a finite set of values. Denote this set by

$$\{\theta^{(i)} \mid i = 1, \ldots, m\}, \tag{6.24}$$

where

$$\theta^{(i)} = \theta^{(i-1)} + \rho. \tag{6.25}$$

To simplify the notation in the subsequent discussion, let a vector $\mathbf{r}(n)$ be defined by

$$\mathbf{r}(n) = [r_1(n), \ldots, r_m(n)]^T, \tag{6.26}$$

where

$$r_i(n) = P[\Theta(n) = \theta^{(i)}], \tag{6.27}$$

and let

$$p_{ij} = P[\Theta(n+1) = \theta^{(j)} \mid \Theta(n) = \theta^{(i)}]. \tag{6.28}$$

Here we assume that the chain of $\Theta(n)$ has m states.

We now show that p_{ij} is independent of n, which simplifies our analysis still further. Note that $Z(n)$ can only take on the values -1, 0, and 1. From Equation (6.23) we see that

$$p_{ij} = \begin{cases} 0, & \text{if } |i-j| \geqslant 2, \\ P[Z(n) = 1 \mid \Theta(n) = \theta^{(i)}], & \text{if } j = i-1, \\ P[Z(n) = 0 \mid \Theta(n) = \theta^{(i)}], & \text{if } j = i, \\ P[Z(n) = -1 \mid \Theta(n) = \theta^{(i)}], & \text{if } j = i+1. \end{cases}$$

Thus

$$\begin{aligned} p_{i,i-1} &= P[\omega(n) = \omega_2, r(n) = 1 \mid \Theta(n) = \theta^{(i)}] \\ &= P[\omega(n) = \omega_2, X(n) < \theta^{(i)} \mid \Theta(n) = \theta^{(i)}]. \end{aligned}$$

Note that $\omega(n)$ and $X(n)$ are independent of $\Theta(n)$. Hence

$$\begin{aligned} p_{i,i-1} &= P[\omega(n) = \omega_2, X(n) < \theta^{(i)}] \\ &= \int_{-\infty}^{\theta(i)} P[\omega(n) = \omega_2, x \leqslant X \leqslant x + dx]\, dx \\ &= \int_{-\infty}^{\theta(i)} f_2(x)\, dx. \end{aligned}$$

Hence $p_{i,i-1}$ is independent of n. Similarly, $p_{i,i+1}$ is independent of n. Furthermore since (a) $p_{ij} = 0$ if $|i-j| \geqslant 2$ and (b) $\sum_{j=1}^{m} p_{ij} = 1$, it follows that p_{ii} is independent of n. Hence p_{ij} is independent of n for all i, j.

Note from Equations (6.27) and (6.28) that

$$r_j(n+1) = \sum_{i=1}^{m} p_{ij} r_i(n). \tag{6.29}$$

Hence the vectors $\mathbf{r}(n)$ and $\mathbf{r}(n+1)$ are related by

$$\mathbf{r}^T(n+1) = \mathbf{r}^T(n)\mathbf{P}, \tag{6.30}$$

where $\mathbf{P}$ is the matrix defined by

$$\mathbf{P} = [p_{ij}]. \tag{6.31}$$

From Equation (6.30) we deduce

$$\mathbf{r}^T(n) = \mathbf{r}^T(0)\mathbf{P}^n. \tag{6.32}$$

Thus $\mathbf{r}(n)$ depends only on the initial vector $\mathbf{r}(0)$ and the *transition matrix* $\mathbf{P}$.

We see in Equation (6.32) that the computation of $\mathbf{P}^n$ is basic to the computation of $\mathbf{r}(n)$. Fortunately the computation of $\mathbf{P}^n$ is a well developed technique in the theory of Markov chains [7, 9].

Note that all nonzero elements of $\mathbf{P}$ lie in the principal diagonal and the two minor diagonals. All the remaining elements of $\mathbf{P}$ are zero. Thus $\mathbf{P}$ is a tridiagonal matrix, as illustrated in part (a) of Figure 6.3, and $\theta(n)$ is the special form of Markov chain known as a one-dimensional *random walk*.

Often $\mathbf{P}$ is represented by a transition graph, as illustrated in part (b) of Figure 6.3.

We are now in a position to compute the error probability and the success probability as functions of n. Let

$$h_i = P[e \mid \theta^{(i)}], \tag{6.33}$$

(a)

$$\mathbf{P} = \begin{bmatrix} p_{11} & p_{12} & 0 & 0 & 0 & \cdots & 0 \\ p_{21} & p_{22} & p_{23} & 0 & 0 & \cdots & 0 \\ 0 & p_{32} & p_{33} & p_{34} & 0 & \cdots & 0 \\ 0 & 0 & p_{43} & p_{44} & p_{45} & \cdots & 0 \\ \vdots & \vdots & \vdots & \vdots & \vdots & & \vdots \\ 0 & 0 & 0 & 0 & 0 & \cdots & p_{mm} \end{bmatrix}$$

(b)

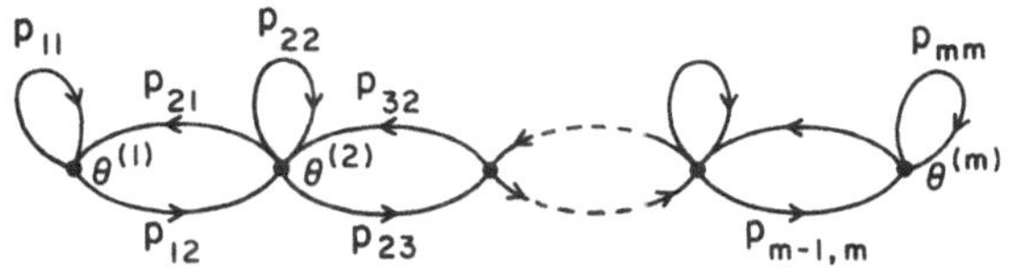

Figure 6.3. Transition probability matrix and transition graph for a finite state, constant increment training procedure.

where $P(e|\theta)$ is defined by Equation (6.18). Let $\mathbf{h} = \{h_i\}$. Then, by Equations (6.19) and (6.27) we have

$$P[e(n)] = \mathbf{r}^T(n)\mathbf{h}. \tag{6.34}$$

By Equation (6.32),

$$P[e(n)] = \mathbf{r}^T(0)\mathbf{P}^n\mathbf{h}. \tag{6.35}$$

The Learning Curve. The success probability is, by Equation (6.10),

$$z(n) = 1 - \mathbf{r}^T(0)\mathbf{P}^n\mathbf{h}. \tag{6.36}$$

One can also write $z(n)$ in the form

$$z(n) = \mathbf{r}^T(0)\mathbf{P}^n(\mathbf{1} - \mathbf{h}), \tag{6.37}$$

where $\mathbf{1} = [1, \ldots, 1]^T$. This may be derived from the fact that the present Markov chain is finite and irreducible, and that for such a Markov chain $\mathbf{r}(n)$ satisfies

$$\mathbf{r}^T(n)\mathbf{1} = 1 \quad \text{for all } n. \tag{6.38}$$

(An *irreducible* Markov chain is one which has no proper subset of states such that escape from the subset is impossible [7]. Thus the present random walk is irreducible.)

It is sometimes convenient to express Equation (6.37) in the form

$$z(n) = \mathbf{r}^T(0)\mathbf{P}^n\mathbf{s}, \tag{6.39}$$

where

$$\mathbf{s} = (\mathbf{1} - \mathbf{h}) = \{S(\theta^{(i)})\},\ i = 1, \ldots, m, \tag{6.40}$$

and where

$$\begin{aligned} S(\theta) &= P(Z(n) = 0|\theta) = 1 - P(e|\theta) \\ &= \int_{-\infty}^{\theta} f_1(x)\,dx + \int_{\theta}^{\infty} f_2(x)\,dx. \end{aligned}$$

The Variance Curve. The variance $\tilde{q}(n)$ is

$$\begin{aligned} \tilde{q}(n) &= E\{[S(\Theta(n)) - z(n)]^2\} \\ &= E\{S^2(\Theta(n))\} - z^2(n). \end{aligned} \tag{6.41}$$

Let

$$\mathbf{s}^{(2)} = \{S^2(\theta^{(i)})\}, \qquad i = 1, \ldots, m. \tag{6.42}$$

Then

$$\begin{aligned} E\{S^2(\Theta^{(n)})\} &= \sum_{i=1}^{n} S^2(\theta^{(i)})P(\Theta(n) = \theta^{(i)}) \\ &= \mathbf{r}^T(n)\mathbf{s}^{(2)} \\ &= \mathbf{r}^T(0)\mathbf{P}^n\mathbf{s}^{(2)}. \end{aligned} \tag{6.43}$$

Hence

$$\tilde{q}(n) = \mathbf{r}^T(0)\mathbf{P}^n\mathbf{s}^{(2)} - [\mathbf{r}^T(0)\mathbf{P}^n\mathbf{s}]^2. \tag{6.44}$$

6.5.4 The Memory of the Training Procedure

Although we have assumed a training procedure and constituent densities that yield a finite Markov chain—i.e., a Markov chain with a finite number of states—infinite Markov chains are often obtained under somewhat weaker assumptions. For example, if in Equation (6.23) we use

$$a = -\infty, \qquad b = \infty,$$

then we obtain an infinite-state random walk in which successive states of the random walk can be mapped one-to-one onto successive elements of the sequence of all integers from $-\infty$ to ∞. Much of the theory of Markov chains discussed below is applicable to these infinite-state Markov chains as well as the finite-state chains. Where the theory is restricted to finite Markov chains, this fact will be indicated.

Sometimes we limit the number of states in the Markov chain model of a training procedure for computational reasons, since the required information storage or memory capacity of the classifier is larger when the number of states in the Markov chain is larger. If the classifier is implemented by a digital computer or a mechanism similar to a digital computer, the number of bits of memory devoted to the training procedure in that computer must be proportional to the logarithm of the number of states. Specifically, if M is the number of states of the Markov chain model of the training procedure, we say that $\log_2 M$ is the *size* of the training procedure's memory.* Thus if the Markov chain model is finite, we say that the training procedure or the learning process has *finite memory*. If the Markov chain has only one state, the process modeled by the chain is said to have *no memory* or *zero memory*.

6.5.5 Transient and Steady-State Behavior

The theory of Markov chains facilitates the analysis of the dynamics of training in single-feature classifiers, and provides insights into these dynamics. In this section we describe how this theory facilitates the computation of the transient and asymptotic portions of $\mathbf{P}^n$ versus n, as well as the related learning curves and variance curves, of a single-feature classifier undergoing constant increment training.

First we define

$$\boldsymbol{\Phi}(n) = \{\Phi_{ij}(n)\} = \mathbf{P}^n. \tag{6.45}$$

Call

$$\underset{\sim}{\boldsymbol{\Phi}}(x) = \sum_{n=0}^{\infty} \boldsymbol{\Phi}(n)x^n = \sum_{n=\infty}^{\infty} \mathbf{P}^n x^n = (\mathbf{I} - \mathbf{P}x)^{-1} \tag{6.46}$$

* This definition of the size of the memory differs somewhat from that of Cover and Hellman, who define the size of the memory as M [6].

the *generating function* or *x-transform* of $\boldsymbol{\Phi}(n)$. (Note that if $\boldsymbol{\Phi}(n) = \mathbf{K}\xi^n$, where $\mathbf{K}$ is a constant matrix, then $\underset{\sim}{\boldsymbol{\Phi}}(x) = \mathbf{K}(1 - \xi x)^{-1}$.) We now describe how to find $\boldsymbol{\Phi}(n)$ as an explicit function of n.

Write $\underset{\sim}{\boldsymbol{\Phi}}(x)$ in the form

$$\begin{aligned} \underset{\sim}{\boldsymbol{\Phi}}(x) = (\mathbf{I} - \mathbf{P}x)^{-1} &= \frac{1}{|\mathbf{I} - \mathbf{P}x|}\underset{\sim}{\mathbf{H}}(x) \\ &= \sum_{i=0}^{m} \frac{\mathbf{A}_i}{1 - (x/\xi_i)}, \end{aligned} \tag{6.47}$$

where $\mathbf{I}$ is a unit matrix, $|\mathbf{I} - \mathbf{P}x|$ is the determinant of $\mathbf{I} - \mathbf{P}x$, $\{\xi_i\}$ are the roots of $|\mathbf{I} - \mathbf{P}x|$, $\underset{\sim}{\mathbf{H}}(x)$ is a matrix whose elements are polynomials in x, and $\{\mathbf{A}_i\}$ are the matrix coefficients of the partial fraction expansion $\underset{\sim}{\boldsymbol{\Phi}}(x)$. To find $\underset{\sim}{\mathbf{H}}(x)$, use

$$\underset{\sim}{\mathbf{H}}(x) = [|\mathbf{I} - \mathbf{P}x|(\mathbf{I} + \mathbf{P}x + \cdots + \mathbf{P}^m x^m)]_m, \tag{6.48}$$

where $[\cdot]_m$ denotes the truncated power series of the enclosed expression, starting with the 0th power of x and ending with x^m. Once $\underset{\sim}{\mathbf{H}}(x)$ is obtained, find the $\mathbf{A}_i$'s as follows.

First find the zeros of $|\mathbf{I} - \mathbf{P}x|$. Denote them by $\xi_0, \ldots, \xi_m$ arranged so that $\xi_0 < \xi_1 < \cdots < \xi_m$. (In all Markov chains, $\xi_0 = 1$.) Then

$$\mathbf{A}_i = \lim_{x \to \xi_i}\left[\underset{\sim}{\mathbf{H}}(x)\frac{1 - (x/\xi_i)}{|\mathbf{I} - \mathbf{P}x|}\right] \tag{6.49}$$

for $i = 0, 1, \ldots, m$. Substitute the ξ_i's and the $\mathbf{A}_i$'s in the following expression for $\boldsymbol{\Phi}(n)$ as an explicit function of n:

$$\boldsymbol{\Phi}(n) = \sum_{i=0}^{m} \mathbf{A}_i \xi_i^{-n}. \tag{6.50}$$

This expression is derived from Equation (6.47) by taking the inverse x-transform of each partial fraction. It is possible to reduce the computational effort in Equation (6.49) by exploiting certain properties of the matrices $\{\mathbf{A}_i\}$. These properties are described in the chapter on algebraic treatment of Markov chains in Reference [7].

Thus the learning curve $z(n)$ of a constant increment training procedure on a single-feature classifier is

$$z(n) = \mathbf{r}^T(0)\boldsymbol{\Phi}(n)\mathbf{s}, \tag{6.51}$$

where $\boldsymbol{\Phi}(n)$ is given by Equation (6.50). The variance curve is obtained by substituting Equation (6.50) for $\mathbf{P}^n$ in Equation (6.44).

6.6 Ergodicity and Stability in the Large

In this section we demonstrate the following interesting property of both finite-state and infinite-state constant increment training procedures on single-feature classifiers: $\mathbf{r}(\infty)$ exists, is unique and positive, and is independent of $\mathbf{r}(0)$. (Here we permit ourselves to use vector notation even for infinite-dimensional vectors.) It follows immediately that for these training processes both $z(\infty)$ and $\tilde{q}(\infty)$ are unique and independent of $\mathbf{r}(0)$, $z(0)$ and $\tilde{q}(0)$. This is a form of stability in the large in the following sense: regardless of the initial value $\Theta(0)$, there is precisely one probability function of $\Theta(\infty)$. We derive this property below. (In Chapter 7 we show that stability in the large often does not exist in multiple-feature training procedures. In particular, the well established proportional increment training procedure is unstable in the large for certain multiple-feature constituent densities.)

First we need a few definitions. Using Feller's nomenclature [7], we say that a state i is *persistent* if the probability Ψ_{ii} of ever returning to state i, starting from state i, is unity. State i is *transient* if Ψ_{ii} is less than 1. The *mean delay* δ_{ij} is the average number of transitions consumed in moving from state i to an occurrence of state j for the first time after the start at i. State i is *null* if $\Psi_{ii} = 1$ and $\delta_{ii} = \infty$. Let d denote the greatest common divisor of the set of values of n such that $\Phi_{ii}(n) > 0$; d is called the *period*. If $d > 1$, state i is *periodic*. If $d = 1$, state i is *aperiodic*. State i is *ergodic* if it is persistent, aperiodic, and non-null. We say that a Markov chain is ergodic if all of its states are ergodic.

We prove the existence, uniqueness, and positiveness of the asymptotic state probabilities of single-feature constant increment training procedures in three stages. First we show that an irreducible Markov chain is aperiodic if all the lengths of simple loops (defined below) in the state transition graph are relatively prime. Second we show that if a Markov chain is irreducible and aperiodic then either (a) $\mathbf{r}(n) \to \mathbf{0} \equiv [0, \ldots, 0]^T$ as $n \to \infty$ or (b) the chain is ergodic, in which case $\mathbf{r}(n)$ has a unique positive asymptote as $n \to \infty$. Third, using the results of the preceding two stages, we show that, under certain loose constraints on the constituent densities in a single-feature constant increment training procedure, the Markov chain of the motion of the weight is ergodic and irreducible.

We digress for a definition and a statement of a well known basic theorem of Markov chains.

A *simple loop* in a transition graph is a sequence of branches from the transition graph forming a closed, non-self-intersecting path with no regressions in direction along the loop. For example, in Figure 6.4 *abc* and *adec* are simple loops, but *abfg* and *abcgfc* are not simple loops. The *length* of a

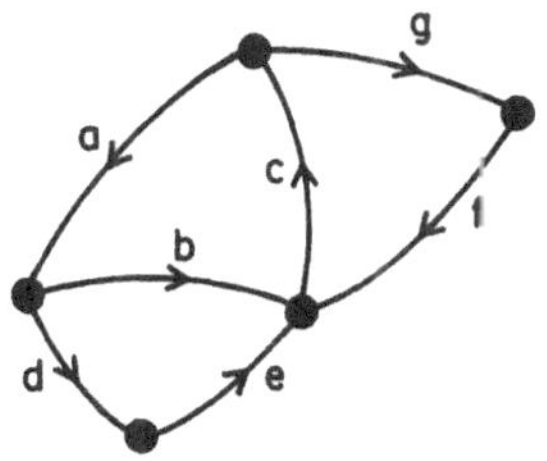

Figure 6.4. Transition graph.

simple loop is the number of branches in the loop. Let $\{l_i\}$ denote the lengths of the complete set of simple loops of a Markov chain. Clearly the length L of any closed path in the transition graph is an integer weighted sum of the l_i's, i.e., $L = \sum m_i l_i$, where the m_i's are nonnegative integers.

For convenience we state here without proof a portion of the well known equivalence class theorem of Markov chains. (For a proof, see W. Feller [7].)

Theorem 6.1. *In an irreducible Markov chain all states belong to the same* equivalence class: *they are all transient, all persistent and null, or all persistent and non-null. In every case they have the same period. Moreover, every state can be reached from every other state. If a Markov chain is not irreducible, the persistent states can be divided uniquely into closed sets* $\mathscr{C}_1, \mathscr{C}_2, \ldots$ *such that states from* $\mathscr{C}_i$ *cannot be reached from any state in* $\mathscr{C}_j$ *for* $i \neq j$.

$\mathscr{C}_1, \mathscr{C}_2, \ldots$ are referred to as *irreducible sets*.

It follows from this theorem that a finite Markov chain has at least one persistent state [7].

We are now in a position to prove the following theorem. (For further discussion of this and related theorems, see G. Lendaris [8].)

Theorem 6.2. *An irreducible Markov chain is aperiodic if and only if the lengths* $\{l_i\}$ *of all simple loops in its transiting graph are relatively prime* (i.e., *the largest integer-valued common divisor of the* l_i's *is* 1).

PROOF. Let d denote the largest common divisor of $\{l_i\}$.

(a) Suppose the members of $\{l_i\}$ are not relatively prime, i.e., $d > 1$. Then the length of every closed path is a multiple of d. In that case every state has a period $\geqslant d$, so that the chain is periodic.

(b) Suppose the members of $\{l_i\}$ are relatively prime. If any state of the chain is periodic with period $d > 1$, then by Theorem 6.1 every state of the chain is periodic with period d, since the chain is irreducible. Hence, the length of a closed loop in the chain is an integer multiple of d. Hence all the simple loops are integer multiples of d. But then the members of $\{l_i\}$ are not relatively prime, contradicting the hypothesis. Hence the chain is aperiodic, and the theorem is proved. □

We will use this theorem in the proof of Theorem 6.3 to show that the Markov chain of a single-feature constant increment training procedure satisfying the constraints stated in Theorem 6.3 is irreducible and aperiodic. It then follows from Theorem 6.1 that either (a) this Markov chain is ergodic or (b) all the asymptotic state probabilities of this Markov chain are zero, i.e., $\boldsymbol{\Phi} = \mathbf{0}$, where $\boldsymbol{\Phi}$ is defined as a probability vector whose jth component is the asymptotic probability of the occurrence of state j. A principal objective of Theorem 6.3 is to show that $\boldsymbol{\Phi} > \mathbf{0}$ (i.e., $\Phi_i > 0$ for every i), and hence that the Markov chain of any member of the stated class of training procedures is irreducible and ergodic.

Now, the *ergodic theorem* for Markov chains tells us that in every irreducible aperiodic Markov chain $\Phi_{ik}(n) \to 1/\delta_{kk}$ as $n \to \infty$ for all i, k, where δ_{kk} is the mean delay or *mean recurrence time* of state k [9]. If the chain is ergodic, $\delta_{kk} < \infty$. In that case $\delta_{ik}(\infty) > 0$. Thus in an irreducible ergodic

Markov chain there exists a unique positive stationary probability vector

$$\boldsymbol{\Phi} = \{\delta_{kk}^{-1}\} = \{\Phi_{ik}(\infty)\} \quad \text{for any } i$$

such that $\boldsymbol{\Phi} > \mathbf{0}$, $\boldsymbol{\Phi}^T = \boldsymbol{\Phi}^T\mathbf{P}$, and

$$\mathbf{P} = \begin{bmatrix} \boldsymbol{\Phi}^T \\ \boldsymbol{\Phi}^T \\ \vdots \\ \boldsymbol{\Phi}^T \end{bmatrix}.$$

(A state probability vector $\boldsymbol{\alpha}$ is said to be *stationary* if and only if $\boldsymbol{\alpha}^T = \boldsymbol{\alpha}^T\mathbf{P}$.) From this property we deduce that every state probability vector $\mathbf{r}(n)$ of the chain of Theorem 6.3 approaches a unique positive stationary probability vector $\boldsymbol{\Phi}$ as $n \to \infty$. We return to this matter after the proof of Theorem 6.3.

Thus in an ergodic Markov chain, $\lim_{n\to\infty} \mathbf{r}(n)$ is independent of $\mathbf{r}(0)$. Let us view $\mathbf{r}(n)$ as the expected value of the stochastic process $\mathbf{R}(n)$, where

$$R_i(n) = \begin{cases} 1, & \text{if state } i \text{ is occupied at trial } n, \\ 0, & \text{if state } i \text{ is not occupied at trial } n. \end{cases}$$

The time average of $\mathbf{R}(n)$ is

$$\lim_{n\to\infty} \left[\frac{1}{n} \sum_{k=1}^{n} \mathbf{R}(k)\right] = \mathbf{r}(\infty).$$

The ensemble average of $\mathbf{R}(n)$ is $E[\mathbf{R}(n)] = \mathbf{r}(n)$. In the limit as $n \to \infty$, the ensemble average of $\mathbf{R}(n)$ equals the time average of $\mathbf{R}(n)$. $\mathbf{R}(n)$ is an *ergodic process*—defined as a stochastic process in which the time average equals the ensemble average.

Because ergodicity of a Markov chain implies the existence of a unique asymptotic state probability vector, ergodicity of Markov chains is equivalent to the concept of stability in the large as used in the theory of the dynamics of systems governed by differential equations [10]. In the next chapter we expand upon this concept of stability of training processes.

The following definitions are needed in Theorem 6.3 for describing the constraints on the constituent densities. The *support* Ωf of $f(x)$ is the closure of the set of points $\Omega' f$ such that $x \in \Omega' f$ if and only $f(x) > 0$. A set of points $\mathscr{S}$ is *convex* if every straight line segment joining a pair of points in $\mathscr{S}$ lies wholly in $\mathscr{S}$. The *convex hull* of any set of points $\mathscr{R}$ is the smallest convex set of points containing $\mathscr{R}$. Let $\mathscr{C}\mathscr{R}$ denote the convex hull of $\mathscr{R}$.

Theorem 6.3. *If, for a single-feature constant increment training procedure in a two-category classifier, the constituent densities $f_1(x)$, $f_2(x)$ satisfy*

1. $$0 < \int_{-\infty}^{\infty} f_j(x)\,dx < 1, \qquad j = 1, 2$$

and

2. $$\mathscr{C}\Omega f_1 \cap \mathscr{C}\Omega f_2 \neq \varnothing,$$

where $\varnothing$ is the empty set, then the Markov chain of the motion of the weight vector $\Theta(n)$ is ergodic and irreducible.

PROOF. The equation describing the single-feature constant increment training procedure is

$$\Theta(n+1) = \Theta(n) - \rho Z(n)$$

where $\Theta(n)$ is the weight at trial n, and $Z(n) = 1, -1$, or 0 for an incorrectly classified ω_2 feature, an incorrectly classified ω_1 feature, or a correctly classified feature, respectively. Let $\theta^{(i)}$ denote the set of possible values of Θ, where $\theta^{(i)} < \theta^{(i+1)}$.

Consider the case where

$$\begin{aligned} \mathscr{C}\Omega f_1 \cap \mathscr{C}\Omega f_2 &= (-\infty, \infty) \\ &= \text{entire real line.} \end{aligned} \tag{6.52}$$

Let

$$a_i = \int_{\theta^{(i)}}^{\infty} f_1(x)\,dx \tag{6.53}$$

be the probability of moving rightward, given that $\Theta(n) = \theta^{(i)}$;

$$b_i = \int_{-\infty}^{\theta^{(i)}} f_2(x)\,dx \tag{6.54}$$

be the probability of moving leftward, given that $\Theta(n) = \theta^{(i)}$;

$$c_i = 1 - a_i - b_i \tag{6.55}$$

be the probability that $\Theta(n+1) = \Theta(n)$, given that $\Theta(n) = \theta^{(i)}$; and

$$r_i(n) = P[\Theta(n) = \theta^{(i)}].$$

The equation of motion of $\Theta(n)$ is governed by

$$r_i(n+1) = r_{i-1}(n)a_{i-1} + r_i(n)(1 - a_i - b_i) + r_{i+1}(n)b_{i+1}, \qquad i = 0, \pm 1, \pm 2, \ldots.$$

The transition graph associated with this equation is shown in Figure 6.5. Here $\Theta(n)$ follows a doubly infinite random walk because of the assumption in Equation (6.52). Cases of semi-infinite and finite random walks, corresponding respectively to replacing the right-hand side of Equation (6.52) by (M, ∞) and (M_1, M_2), where M, M_1, and M_2 are finite numbers, can be analyzed in a manner similar to that presented here.

The complete set of lengths of simple loops in Figure 6.5 is $\{1,2\}$. Hence, by Theorem 6.2, the Markov chain of this training procedure is irreducible and aperiodic. By Theorem 6.1 there exists at least one vector $\boldsymbol{\Phi}$ such that $\boldsymbol{\Phi}^T = \boldsymbol{\Phi}^T\mathbf{P}$, and either $\boldsymbol{\Phi} > \mathbf{0}$ or $\boldsymbol{\Phi} = \mathbf{0}$. We will show that $\boldsymbol{\Phi} > \mathbf{0}$ and that $\boldsymbol{\Phi}$ is unique.

Since $\boldsymbol{\Phi}^T = \boldsymbol{\Phi}^T\mathbf{P}$,

$$\Phi_i = \Phi_{i-1}a_{i-1} + \Phi_i(1 - a_i - b_i) + \Phi_{i+1}b_{i+1}. \tag{6.56}$$

Rearranging terms, we get

$$\Phi_i a_i - \Phi_{i+1}b_{i+1} = \Phi_{i-1}a_{i-1} - \Phi_i b_i = K \tag{6.57}$$

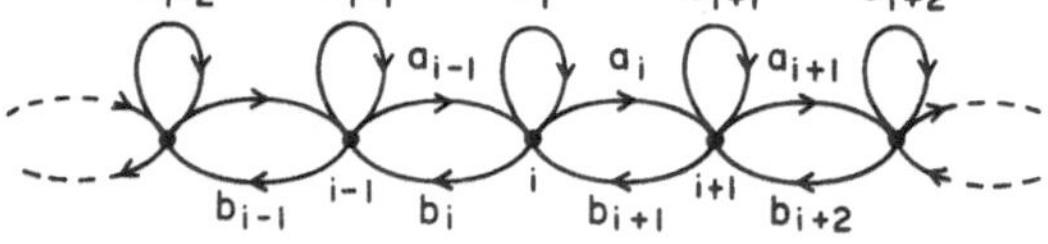

Figure 6.5. State transition graph of doubly infinite Markov chain of single-feature constant increment training.

where K is a constant. Note that

$$\sum_{i=-\infty}^{\infty} |\Phi_{i-1}a_{i-1} - \Phi_i b_i| < \infty \tag{6.58}$$

since $\sum_{i=-\infty}^{\infty} \Phi_i = 1$, $0 \leqslant a_i \leqslant 1$, $0 \leqslant b_i \leqslant 1$. Hence $K = 0$, and

$$\Phi_{i+1} = h_i \Phi_i, \tag{6.59}$$

where $h_i = a_i / b_{i+1}$.

Note from Equations (6.53) and (6.54) that $a_i > a_{i+1}$ and $b_i < b_{i+1}$ for all i. Further, note that hypotheses 1 and 2 in the statement of the theorem imply that $0 = b_{-\infty} < a_{-\infty} < 1$ and $0 = a_\infty < b_\infty < 1$. Hence there exists an integer m such that $a_i \leqslant b_{i+1}$ for $i > m$ and $a_i \geqslant b_{i+1}$ for $i < m$. Also, $h_i \leqslant 1$ for $i > m$, $h_i \geqslant 1$ for $i < m$. Hence, by Equation (6.59)

$$\Phi_{m+k} = \Phi_m \prod_{j=0}^{k-1} h_{m+j}. \tag{6.60}$$

Similarly,

$$\Phi_i = h_i^{-1} \Phi_{i+1} \tag{6.61}$$

and

$$\Phi_{m-k} = \Phi_m \prod_{j=1}^{k} h_{m-k}^{-1}. \tag{6.62}$$

Summing Equations (6.60) and (6.62) over k from $k = 1$ to $k = \infty$, and noting that

$$\sum_{k=-\infty}^{\infty} \Phi_{m+k} = 1, \tag{6.63}$$

we obtain

$$\Phi_m \left\{ 1 + \sum_{k=1}^{\infty} \left[\prod_{j=1}^{k} h_{m-j}^{-1} + \prod_{j=0}^{k-1} h_{m+j} \right] \right\} = 1. \tag{6.64}$$

By hypotheses 1 and 2, h_{m-j}^{-1} for $j \geqslant 1$ and h_{m+j} for $j \geqslant 0$ are positive and less than or equal to 1. For convenience we assume here that h_{m-j}^{-1} for $j \geqslant 1$ and h_{m+j} for $j \geqslant 0$ are strictly less than 1, since the proof can be modified without difficulty (but with greater complication of expression) to include the possibility of $h_{m-j}^{-1} = 1$ and for $h_{m+1} = 1$ (see Exercise 6.7). Under these assumptions there exists a number H such that $0 < H < 1$, $h_{m-j}^{-1} < H$ for $j \geqslant 1$, and $h_{m+j} < H$ for $j \geqslant 0$. Hence, by Equation (6.64)

$$\Phi_m \geqslant \frac{1}{1 + 2\sum_{k=1}^{\infty} H^k} = \frac{1}{1 + 2H(1-H)^{-1}} > 0. \tag{6.65}$$

Since the chain is irreducible, all the remaining elements of $\boldsymbol{\Phi}$ are also positive, i.e., $\boldsymbol{\Phi} > \mathbf{0}$. It follows from Theorem 6.1 and the irreducibility of this chain that the chain is ergodic. □

By the ergodic theorem, $\boldsymbol{\Phi}$ is unique and

$$\boldsymbol{\Phi}(n) \to \begin{bmatrix} \boldsymbol{\Phi}^T \\ \boldsymbol{\Phi}^T \\ \vdots \\ \boldsymbol{\Phi}^T \end{bmatrix} \quad \text{as } n \to \infty.$$

Furthermore, any state probability vector $\mathbf{r}(n)$ satisfies $\mathbf{r}^T(n) = \mathbf{r}^T(0)\boldsymbol{\Phi}(n)$. Then, since

$$\sum_{i=-\infty}^{\infty} r_i(0) = 1,$$

it follows that

$$\mathbf{r}^T(n) \to \mathbf{r}^T(0)\begin{bmatrix} \boldsymbol{\Phi}^T \\ \boldsymbol{\Phi}^T \\ \vdots \\ \boldsymbol{\Phi}^T \end{bmatrix} = \boldsymbol{\Phi} \quad \text{as } n \to \infty.$$

Thus the state probability vector of the Markov chain of Theorem 6.3 approaches $\boldsymbol{\Phi} > \mathbf{0}$ as $n \to \infty$, regardless of $\mathbf{r}(0)$.

6.7 Train–Work Schedules: Two-Mode Classes

The use of training procedures for classifiers is usually justified on the basis that the best set of decision parameters (i.e., the best decision surface) is unknown. A simple example occurs in the transmission of binary-valued signals through a noisy communication link. In this case the training procedure converges to a detection threshold where the error rate (or probability of error) is close to or at a minimum.

In some cases the statistics underlying the class regions change randomly as a function of time. Here the best decision surface changes whenever the class regions change. The classifier, however, has no instantaneous access to perfect information on the changing statistical parameters of the class regions. A simple example of this situation is the Gilbert or two-mode channel [11]. This channel transmits information in one of two modes: $\mathscr{A}$ and $\mathscr{B}$. In mode $\mathscr{B}$ it acts as a zero-memory channel with relatively poor transmission characteristics. In the classical Gilbert channel, which is encountered in telephone communication, the transitions between $\mathscr{A}$ and $\mathscr{B}$ are described by a two-state Markov chain—each state associated with one of the modes. This is an example of a randomly time-varying channel. (Tsetlin refers to such a channel as a *composite medium* [12].)

A single training phase followed by a single working phase is not, as a rule, a satisfactory way of training a classifier whose input data arrives from a time-varying channel. The reason for this is that a working phase restricted to a single time-invariant decision surface is certain to be nonoptimal over extended periods of time if the channel is time-varying. Since the classifier accepting data from such a channel has only imperfect access to information on the state of the channel—namely, through the training procedure—some aspects of training must be introduced into the working phase on a continuous or intermittent basis. One approach is through training without a teacher—i.e., without the use of training sequences in which the classes of the feature vectors are identified. A second approach is through a periodic or quasi-periodic train–work schedule. A Markov chain model for this second approach is discussed below.

6.7.1 Markov Chain Model of a Trainable Classifier Operating on a Two-Mode Channel

Suppose a single-feature classifier receives its input data from the output of a two-mode Gilbert channel, and suppose correctly labeled data is available for use during one or more training phases. During a training phase, a sequence of noise-free binary-valued inputs are entered into the channel, the channel converts each of these inputs into a random variable X (X is the scalar feature), and the classifier guesses the value of the input that produced X by comparing X to an adjustable threshold Θ.

Suppose the training procedure is of constant increment type. Then the state of the classifier follows a random walk as long as the mode is fixed. When the mode moves at random between two values, however, the state of the classifier can no longer be described by a random walk. This classifier's state follows a Markov chain which, in a sense, is a combination of two random walks—one random walk for each mode.

To derive this Markov chain, we must first define the state of a classifier trained through a two-mode Gilbert channel. Consider the sequence of events shown in Figure 6.6, and refer to Figures 6.1 and 6.2. Assume that only 0's and 1's are entered into the channel in the form of rectangular pulses of heights 0 and 1, respectively. To generate its guess at trial n, the classifier receives a clock pulse $c(n, t)$ which instructs the classifier to guess whether

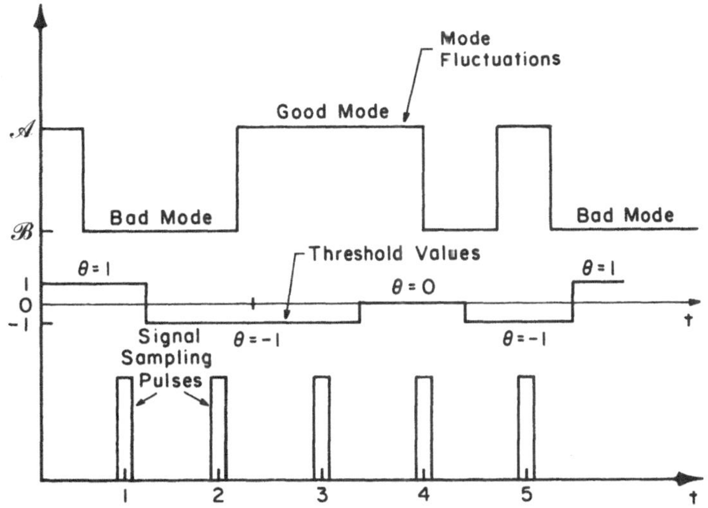

Figure 6.6. Timing relations among mode-to-mode transitions, threshold shifts, and sampling pulses.

the most recently observed feature $X(n)$ was caused by $\Omega(n) = 1$ or $\Omega(n) = 2$ associated with class ω_1 and ω_2, respectively. In response to $c(n, t)$ and $X(n)$, the threshold of the classifier is set at $\Theta(n) = \theta^{(k)}$. During the clock pulse $c(n, t)$, the channel is either in mode $\mathscr{A}$ or mode $\mathscr{B}$. If it is in $\mathscr{A}$, the classifier is said to be in state A_k from the time that the threshold takes on the value $\Theta(n) = \theta^{(k)}$ until the next time a threshold transition can occur. (Note that there must be a slight delay between each clock pulse and the resultant threshold transition.) Similarly, if the channel is in mode $\mathscr{B}$ when $c(n, t)$ occurs, and if $\Theta(n) = \theta^{(k)}$, the classifier is in state B_k during the interval between the pulses $c(n, t)$ and $c(n + 1, t)$. The timing relations among the mode-to-mode transitions, the threshold adjustments, and the clock pulses are illustrated in Figure 6.6 for a three-threshold classifier. (Note that by this definition of A_k and B_k the classifier may be in state A_k even though the channel may have shifted to mode $\mathscr{B}$. Similarly the state may be B_k even though the mode may have shifted to $\mathscr{A}$. This situation has no adverse effect on the Markov chain model.)

Let us assume that the training procedure yields a transition matrix

$$\mathbf{A} = [a_{ij}]$$

when the channel is fixed at mode A and a transition matrix

$$\mathbf{B} = [b_{ij}]$$

when the channel is fixed at mode B. The quantities a_{ij} and b_{ij} are defined as

$$\begin{aligned} a_{ij} &= P[\Theta(n+1) = \theta^{(j)} \,|\, \Theta(n) = \theta^{(i)}, \mathscr{A}] \\ b_{ij} &= P[\Theta(n+1) = \theta^{(j)} \,|\, \Theta(n) = \theta^{(i)}, \mathscr{B}]. \end{aligned}$$

Let $\boldsymbol{\Gamma}$ denote the mode-to-mode transition matrix:

$$\boldsymbol{\Gamma} = \begin{bmatrix} 1 - \delta & \delta \\ \varepsilon & 1 - \varepsilon \end{bmatrix},$$

Where

$$\begin{aligned} \delta &= P[\text{next mode} = \mathscr{B} \,|\, \text{present mode} = \mathscr{A}] \\ 1 - \delta &= P[\text{next mode} = \mathscr{A} \,|\, \text{present mode} = \mathscr{A}] \\ \varepsilon &= P[\text{next mode} = \mathscr{A} \,|\, \text{present mode} = \mathscr{B}] \\ 1 - \varepsilon &= P[\text{next mode} = \mathscr{B} \,|\, \text{present mode} = \mathscr{B}] \end{aligned}$$

The state transition graphs associated with these matrices are shown in Figure 6.7.

From our definitions of the states $\{A_i\}$ and $\{B_j\}$, it follows that a transition from $\theta^{(i)}$ to $\theta^{(j)}$ depends only on $\theta^{(i)}$ and the mode at the clock pulse immediately preceding the transition. Hence every transition from A_i to B_j is the result of a transition from $\mathscr{A}$ to $\mathscr{B}$ followed by a transition from B_i to B_j. Since the Markov chains for the mode-to-mode transitions and the threshold transitions within each mode are statistically independent, it follows that

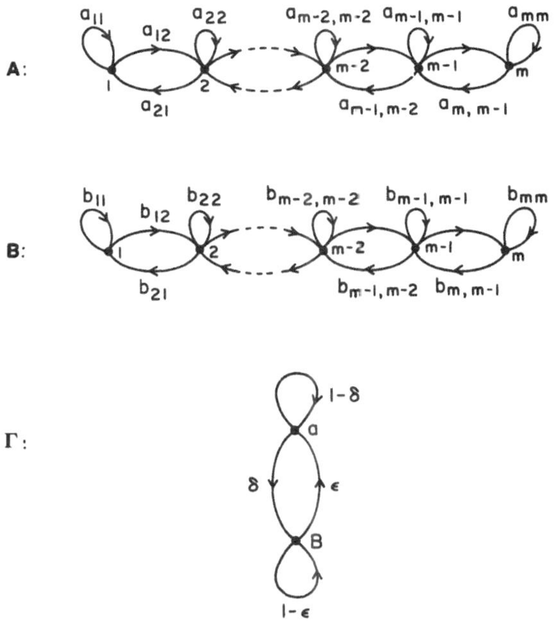

Figure 6.7. State transition graphs associated with **A**, **B**, and $\boldsymbol{\Gamma}$.

the probability of a transition from state A_j to B_j is as follows. Let $\mathscr{S}(n)$ denote a random variable whose value at trial n is the currently occupied state among $\{A_i\} \cup \{B_j\}$. Let $\mathscr{M}(n)$ denote the random variable whose value is the mode of the channel at trial n. Then

$$\begin{aligned}
P[\mathscr{S}(n+1) = B_j | \mathscr{S}(n) = A_i] \\
&= P[\mathscr{M}(n+1) = \mathscr{B} | \mathscr{S}(n) = A_i] P[\mathscr{S}(n+1) = B_j | \mathscr{S}(n) = B_i] \\
&= P[\mathscr{M}(n+1) = \mathscr{B} | \mathscr{M}(n) = \mathscr{A}] P[\mathscr{S}(n+1) = B_j | \mathscr{S}(n) = B_i] \\
&= \delta b_{ij}.
\end{aligned}$$

By similar reasoning,

$$\begin{aligned}
P[\mathscr{S}(n+1) = B_i | \mathscr{S}(n) = B_j] \\
&= P[\mathscr{M}(n+1) = \mathscr{B} | \mathscr{M}(n) = \mathscr{B}] P[\mathscr{S}(n+1) = B_i | \mathscr{S}(n) = B_j] \\
&= (1 - \varepsilon) b_{ij}
\end{aligned}$$

$$P[\mathscr{S}(n+1) = A_j | \mathscr{S}(n) = A_i] = (1 - \delta) a_{ij}$$
$$P[\mathscr{S}(n+1) = A_j | \mathscr{S}(n) = B_i] = \varepsilon a_{ij}.$$

The resultant Markov chain has the state transition graph shown in Figure 6.8.

It is possible to represent this graph more completely by representing the set of states $\{A_i | i = 1, \ldots, m\}$ by an m-element vector **a** and $\{B_i | i = 1, \ldots, m$ by an m-element vector **b**. The transition probabilities $\{\delta b_{ij}\}$, $\{\varepsilon a_{ij}\}$, $\{(1-\delta) b_{ij}\}$

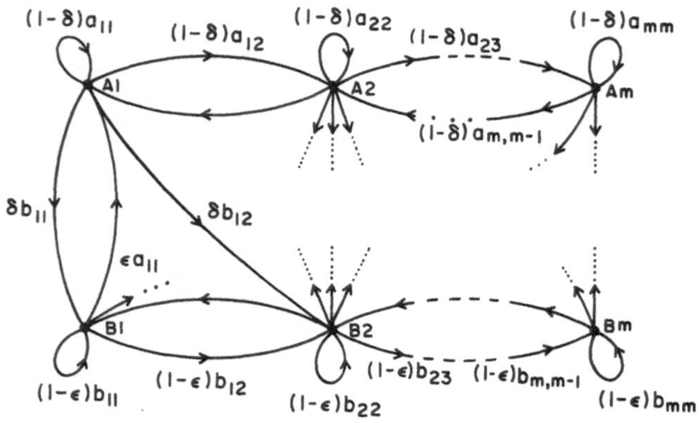

Figure 6.8. State transition graph of Markov chain model of two-mode training.

and $\{(1-\varepsilon)a_{ij}\}$ are then replaced by the matrices $\delta\mathbf{B}$, $\varepsilon\mathbf{A}$, $(1-\delta)\mathbf{B}$, and $\varepsilon\mathbf{A}$. The probabilities $\{r_{Ai}(n)\}$ of occupying the states $\{A_i\}$ are then represented by the vector $\mathbf{r}_A(n)$. Similarly $\mathbf{r}_B(n)$ represents $\{r_{Bi}(n)\}$. Thus, the scalar state transition graph of Figure 6.8 is equivalent to the vector state transition graph of Figure 6.9.

From the vector state transition graph, one may obtain the following equations determining the vector state probabilities $\mathbf{r}_A(n)$ and $\mathbf{r}_B(n)$:

$$\left.\begin{aligned}\mathbf{r}_A^T(n+1) &= \mathbf{r}_A^T(n)(1-\delta)\mathbf{A} + \mathbf{r}_B^T(n)\varepsilon\mathbf{A}\\ \mathbf{r}_B^T(n+1) &= \mathbf{r}_A^T(n)\delta\mathbf{B} + \mathbf{r}_B^T(n)(1-\varepsilon)\mathbf{B}\end{aligned}\right\} \tag{6.66}$$

This pair of equations may be made more compact by the notation

$$\mathbf{P} = \begin{bmatrix}\mathbf{A} & \mathbf{0}\\ \mathbf{0} & \mathbf{B}\end{bmatrix}$$

$$\mathbf{r}^T(n) = [\mathbf{r}_A^T(n), \mathbf{r}_B^T(n)]^T.$$

Then Equation (6.66) becomes

$$\mathbf{r}^T(n+1) = \mathbf{r}^T(n)\Gamma\mathbf{P}, \tag{6.67}$$

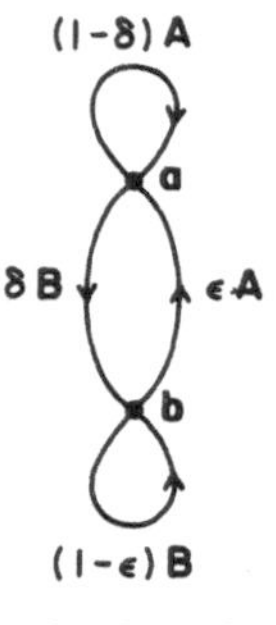

Figure 6.9. Vector state transition graph of Markov chain model of two-mode training.

where $\boldsymbol{\Gamma}$ is defined as before. The solution of this equation is

$$\mathbf{r}^T(n) = \mathbf{r}^T(0)(\boldsymbol{\Gamma}\mathbf{P})^n. \tag{6.68}$$

Note that $(\boldsymbol{\Gamma}\mathbf{P})^n \neq \boldsymbol{\Gamma}^n\mathbf{P}^n$, except in trivially degenerate cases.

Using the above model we are in a position to find mathematical expressions for the learning curves in the training phase and the working phase. We will derive such expressions below. We will also show how one may design train–work schedules that are matched to **A**, **B**, and $\boldsymbol{\Gamma}$.

6.7.2 The Training Phase

The success probability for the training phase may be found by the following formula:

$$z(n) = \mathbf{r}_A^T(n)\mathbf{s}_A + \mathbf{r}_B^T(n)\mathbf{s}_B, \tag{6.69}$$

where

$$z(n) = P[\text{success} \mid n]$$

$$\mathbf{s}_A = [P(\text{success} \mid A_j)], \qquad j = 1, \ldots, m, \text{ an } m\text{-dimensional vector}$$

$$\mathbf{s}_B = [P(\text{success} \mid B_j)], \qquad j = 1, \ldots, m, \text{ an } m\text{-dimensional vector.}$$

If ε and δ are sufficiently small, the channel tends to behave like a single-mode Markov chain for long periods of time. We refer to such a two-mode channel as *weakly coupled.* The learning curve on a weakly coupled two-mode channel can be approximated as a linear sum of the component single-mode learning curves. To see this, note that the initial probabilities of mode A and mode B are the steady-state probabilities of the modes:

$$P[\mathscr{A} \mid n = 0] = \varepsilon/(\delta + \varepsilon) \tag{6.70}$$

$$P[\mathscr{B} \mid n = 0] = \delta/(\delta + \varepsilon). \tag{6.71}$$

Thus if $\mathbf{i}$ is the initial probability vector of the thresholds $\{\theta^{(j)}\}$, i.e., if

$$\mathbf{i} = \{P[\theta^{(j)} \mid n = 0]\}, \qquad j = 1, \ldots, m \tag{6.72}$$

then

$$\mathbf{r}_A(0) = \gamma\mathbf{i} \tag{6.73}$$

$$\mathbf{r}_B(0) = (1 - \gamma)\mathbf{i}, \tag{6.74}$$

where

$$\gamma = \varepsilon/(\delta + \varepsilon). \tag{6.75}$$

By the above theory, one can show that, for the weakly coupled condition, i.e., for $0 < \varepsilon \ll 1$, $0 < \delta \ll 1$, the learning curve during the training phase is given by the approximation

$$z(n) \cong \gamma z_{Ao}(n) + (1 - \gamma)z_{Bo}(n),$$

where $z_{Ao}(n)$ (read Ao as "A only") is the learning wave of a classifier when the channel is fixed in mode A, and $z_{Bo}(n)$ is the learning curve when the channel is fixed in mode B.

6.7.3 The Working Phase

During the working phase of a classifier trained through a one-mode channel, the expected rate of success (i.e., the success probability) is constant. This constant is equal to the success probability achieved at the end of the training phase.

When the classifier is trained through a two-mode Gilbert channel, there is a nonzero probability that the value of the weight θ during the working phase is not optimum for the particular mode the channel happens to occupy. This mismatch probability is relatively small at the beginning of the working phase (i.e., the end of the training phase) and becomes greater as time progresses through the working phase. Hence, with a two-mode channel, the classifier's success probability during the working phase diminishes monotonically from a peak at the beginning of the working phase, and levels off at an asymptotic value ζ_f where the subscript f denotes the fact that the working-phase threshold is fixed.

The state transition graph of the Markov chain of the working phase consists of disjoint two-state components, each component corresponding to a fixed threshold $\theta^{(j)}$. The distribution of state probabilities at the beginning of the working phase, $\mathbf{r}(m)$, is the same as at the end of the preceding training phase.

The formula for the learning curve during the working phase is

$$z_f(m) = \zeta_f + [z_f(0) - \zeta_f](1 - \delta - \varepsilon)^m, \tag{6.76}$$

where m is the number of working-phase trials, ζ_f is the asymptotic value of $z_f(m)$, and $z_f(0)$ is the initial value of $z_f(m)$. This formula is derived from the fact that each component of the graph yields a transient response of the form $c_1 + c_2(1 - \delta - \varepsilon)^n$, where c_1 and c_2 are constants. The quantities ζ_f and $z_f(0)$ are determined by the state probabilities at $n = 0$, namely, $\mathbf{r}_A(0)$ and $\mathbf{r}_B(0)$. Specifically,

$$\zeta_f = \gamma^2 \Psi_A + (1 - \gamma)^2 \Psi_B + \gamma(1 - \gamma)(\Psi_{AB} + \Psi_{BA}) \tag{6.77}$$

$$z_f(0) = \gamma \Psi_A + (1 - \gamma)\Psi_B \tag{6.78}$$

where

$$\Psi_A = \frac{1}{\gamma}\mathbf{r}_A^T(0)\mathbf{s}_A$$

$$\Psi_B = \frac{1}{1-\gamma}\mathbf{r}_B^T(0)\mathbf{s}_B$$

$$\Psi_{AB} = \frac{1}{\gamma}\mathbf{r}_A^T(0)\mathbf{s}_B$$

$$\Psi_{BA} = \frac{1}{1-\gamma}\mathbf{r}_B^T(0)\mathbf{s}_A.$$

Equation (6.76) displays monotonic reduction in success probability during the working phase. That this reduction is brought about by the mode-to-mode fluctuations is displayed by the fact that the transient portion of Equation (6.76) is determined solely by δ and ε. Another factor determining the extent of reduction of success probability during the working phase is the degree of mismatch that occurs when a weight optimized for one mode coexists with the other mode. This may be seen by an inspection of Equations (6.77) and (6.78). If modes $\mathscr{A}$ and $\mathscr{B}$ were identical, then ζ_f would equal $z_f(0)$. The degree to which ζ_f and $z_f(0)$ differ depends on Ψ_{AB} and Ψ_{BA}.

6.7.4 Train–Work Cycles

We have seen in the preceding discussion that the mode-to-mode fluctuations of a two-mode channel induce an exponential decay in the working-phase section of the classifier's learning curve. Hence an alternating sequence of training and working phases will be needed if the lowest acceptable success probability, denoted by ζ_L, is greater than the working phase's asymptotic success probability ζ_f. This is necessary in spite of the fact that the success probability at the beginning of the working phase (namely, $z_f(0)$) exceeds ζ_L. Under these conditions, it is possible that the success probability $z_f(0)$ at the beginning of the working phase exceeds the lowest acceptable success probability ζ_L, but that ζ_L exceeds the asymtotic success probability of the working phase $\zeta_f = z_f(\infty)$. To overcome this reduction of $z_f(n)$ in the working phase, one may use a sequence of training and working phases. (In contrast, if the channel were a single-mode channel, then $z_f(0) = z_f(n)$ for all n. Thus, for a single-mode channel the training phase can often bring the success probability permanently to a desired value.) We describe below how knowledge of the component single-mode dynamics permits us to construct a reasonably good train–work schedule.

An exaggerated example of a train–work schedule is illustrated in Figure 6.10. During the training phase the state probability vector is determined by

$$\mathbf{r}^T(n) = \mathbf{r}^T(0)(\Gamma \mathbf{P})^n. \tag{6.79}$$

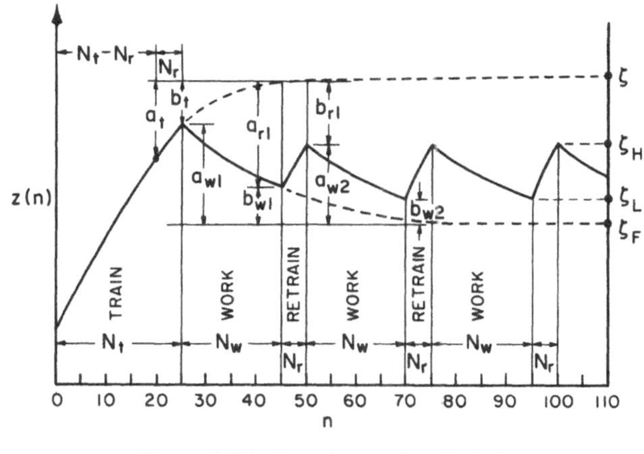

Figure 6.10. A train–work schedule.

Suppose the available thresholds are $\{\theta^{(j)} | j = 1, \ldots, K\}$. Let N_t denote the number of trials in the training phase. At the end of the training phase the $2K$ elements of $\mathbf{r}(N_t)$ are grouped into pairs, each pair consisting of the two joint probabilities associated with a threshold $\theta^{(j)}$. These pairs are then used in the following equation to find the variation of these pairs throughout the working phase:

$$\mathbf{r}_j^T(N_t + m) = [r_{Aj}(N_t), r_{Bj}(N_t)] \begin{bmatrix} 1 - \delta & \delta \\ \varepsilon & 1 - \varepsilon \end{bmatrix}^m \tag{6.80}$$

for $m = 1, 2, \ldots, N_w$, where N_w is the number of trials in the working phase, $r_{Aj}(N_t)$ is the joint probability of mode $\mathscr{A}$ and threshold $\theta^{(j)}$ at trial N_t. At the end of the working phase the pairs $((r_{Aj}(N_t + m), r_{Bj}(N_t + m)))$ are regrouped into a $2K$-dimensional vector $\mathbf{r}(N_t + N_w)$. This vector is used as the initial condition for the retraining phase, and used in Equation (6.79) with a proper shift of time origin to obtain a formula for the state probability throughout the retraining phase.

6.7.5 Numerical Example

We illustrate the above techniques by a numerical example in which the two-mode channel produces the constituent densities shown in Figure 6.11. We assume the following:

1. The frequencies of occurrence of 0's and 1's in mode $\mathscr{A}$ are equal to those in mode $\mathscr{B}$.
2. The constituent densities in mode $\mathscr{A}$ do not overlap.

In Figure 6.11, note that mode $\mathscr{A}$ yields perfect performance and mode $\mathscr{B}$ yields imperfect performance under the optimum threshold θ_B^* whenever $\alpha_B < 1$. Thus $\mathscr{A}$ is a relatively good mode and $\mathscr{B}$ is a relatively bad mode whenever $\alpha_B < 1$.

Let ρ_A be the probability of sending a 0 in mode $\mathscr{A}$, ρ_B be the probability of sending a 0 in mode $\mathscr{B}$,

$$\alpha_A = \rho_A^{-1} f_{1A}(1.5),$$

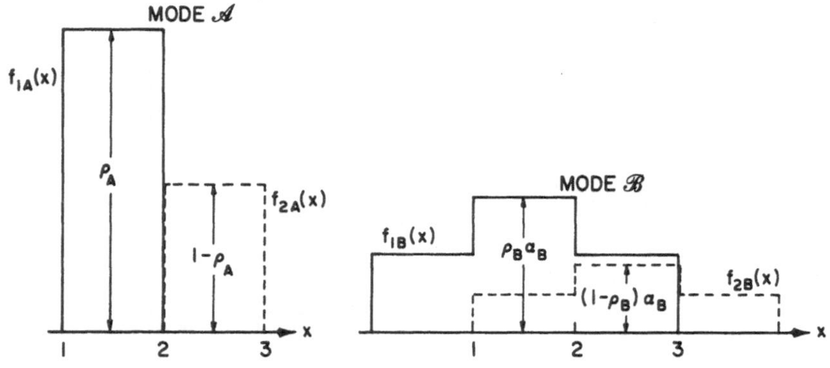

Figure 6.11. Constituent densities for the numerical example.

and

$$\alpha_B = \rho_B^{-1} f_{1B}(1.5),$$

where $f_{iA}(x)$, $f_{iB}(x)$ are the constituent densities for class ω_i in modes $\mathscr{A}$ and $\mathscr{B}$, respectively. In this notation, assumptions 1 and 2 are equivalent to

(a) $$\rho_A = \rho_B$$

(b) $$\alpha_A = 1.$$

Thus the four constituent densities in modes $\mathscr{A}$ and $\mathscr{B}$ can be specified by just two parameters, ρ_A and α_B. Let

$$\rho = \rho_A = \rho_B \tag{6.81}$$

$$\alpha = \alpha_B. \tag{6.82}$$

If we assume $\delta \ll 1$ and $\varepsilon \ll 1$, the learning curves are approximately specified by ρ, α, and γ, as defined in Equations (6.81), (6.82), and (6.75). Suppose

$$\alpha = 0.6, \rho = 0.9, \delta = \varepsilon = 0.01.$$

For this case,

$$\mathbf{A} = \begin{bmatrix} 0.1 & 0.9 & 0 \\ 0 & 1 & 0 \\ 0 & 0.1 & 0.9 \end{bmatrix} \tag{6.83}$$

$$\mathbf{B} = \begin{bmatrix} 0.280 & 0.720 & 0 \\ 0.020 & 0.800 & 0.180 \\ 0 & 0.080 & 0.920 \end{bmatrix}. \tag{6.84}$$

To see how **A** and **B** are computed, consider for example the element a_{11} at the first row and first column of **A**. This element is

$$\begin{aligned} a_{11} &= P[\Theta(n+1) = 1 \mid \Theta(n) = 1] \\ &= 1 - (\rho_A/2)(1 + \alpha_A) = 0.1. \end{aligned}$$

Assume that $\mathbf{r}_A(n)$ and $\mathbf{r}_B(n)$ each have just three elements, corresponding to $\{\theta^{(j)}\} = \{1, 2, 3\}$, and suppose that the threshold at $n = 0$ is $\Theta(0) = 1$. Then

$$\begin{aligned} \mathbf{r}_A(0) &= \mathbf{r}_B(0) = \mathbf{i} = [\tfrac{1}{2}, 0, 0]^T \\ \mathbf{r}^T(n) &= [\mathbf{r}_A^T(n), \mathbf{r}_B^T(n)] \\ &= (\mathbf{i}^T, \mathbf{i}^T) \left(\begin{bmatrix} 0.9 & 0.1 \\ 0.1 & 0.9 \end{bmatrix} \begin{bmatrix} \mathbf{A} & \mathbf{0} \\ \mathbf{0} & \mathbf{B} \end{bmatrix} \right)^N \\ z(n) &= \mathbf{r}_A^T(n)\mathbf{s}_A + \mathbf{r}_B^T(n)\mathbf{s}_B. \end{aligned}$$

Suppose the length of the training phase is N_t. The learning curve for the working phase is computed by finding $\mathbf{r}_A(N_t)$ and $\mathbf{r}_B(N_t)$ and applying Equations (6.80) and (6.69). Suppose the length of the working phase is N_w. The learning curve for the retraining phase is then given by

$$\mathbf{r}^T(N_t + N_w + k) = \mathbf{r}^T(N_t + N_w)(\boldsymbol{\Gamma}\mathbf{P})^k,$$

Table 6.1. Selected train–work schedules and the associated work ratios and smallest success probabilities.

Train–work triplet (N_t, N_w, N_r)	Work ratio (N_w/N_r)	ζ_L
(6,6,1)	6	0.9201
(7,10,2)	5	0.9202
(12,18,7)	2.57	0.9215
(15,19,10)	1.90	0.9221

etc. One may then keep the lengths of the working phases and retraining phases at N_w and N_r, respectively. Under these conditions one achieves a minimum success probability

$$\zeta_L = \min_n \; z(n)$$

for each triplet (N_t, N_w, N_r). The results for a number of such triplets are given in Table 6.1. In this table we list ζ_L and the "work ratio" N_w/N_r. The work ratio is a measure of the duty cycle. The higher the work ratio, the greater a percentage of the time is devoted to work. The quantity ζ_L is a measure of the worst performance, i.e., the smallest success probability during all of the working phases.

6.7.6 Synthesis in the Face of Ignorance

The need for training a classifier arises from the designer's or experimenter's ignorance of the statistics of the class regions and/or the statistics of the mode-to-mode fluctuations. In cases where the number of modes in the channel may be greater than two, the designer may be ignorant of the number of modes. In these cases one may examine the train–work behavior over the entire range of unknown parameters (e.g.: ρ, α, etc.). For each point in parameter space one may choose the optimum train–work schedule and then, among these optimum schedules, choose a composite train–work schedule (N_t^*, N_w^*, N_r^*) as follows. Let

$$(N_t(\mathbf{p}), N_w(\mathbf{p}), N_r(\mathbf{p}))$$

denote the train–work triplets as a function of the parameter vector $\mathbf{p}$, and let $\zeta_L(\mathbf{p})$ denote the associated values of minimum success probability

$$\begin{aligned} N_t^* &= \max_p \; N_t(\mathbf{p}) \\ N_w^* &= \min_p \; N_w(\mathbf{p}) \\ N_r^* &= \max_p \; N_r(\mathbf{p}). \end{aligned} \tag{6.85}$$

Let

$$\zeta_L^* = \min_{p} \zeta_L(\mathbf{p}).$$

Then, because of the typically montonic increasing behavior of $z(n)$ in the training and retraining phases, and the monotonically decreasing behavior of $z(n)$ in the working phases, the schedule (N_t^*, N_w^*, N_r^*) will in most cases yield a train–work learning curve for which ζ exceeds ζ_L^*.

6.8 Optimal Finite Memory Learning

6.8.1 Introduction

In the constant increment single-feature training procedure described in preceding sections, it is assumed that the location of the best decision surface is unknown, but that the direction of that decision surface (i.e., the direction of the normal to the surface pointing to the region where the decision $X \in \omega_2$ is made) is known. For example, in that procedure one may assume that $X \in \omega_2$ is a better decision than $X \in \omega_1$, on average, when $X > \theta$. If one omits the assumption of knowing the better of the two possible directions of the decision surface, so that both the location and the direction of the decision surface are unknown, one may partition the training procedure into two levels: at one level the classifier learns the *location* of the decision surface, and at the second level the classifier learns the *direction* of the decision surface.

We have seen how a finite Markov chain can describe the learning of the location of the decision surface in response to a constant increment training procedure. In the following section, we describe a two-level training procedure for learning both the location and the direction of the decision surface for a single-feature classifier. Since in this type of classifier there are only two possible directions of the decision surface, we may formulate the problem of choosing the better of these directions as a two-hypothesis testing problem.

We shall describe an optimal finite memory training procedure for learning the better of these two directions, based on the work of Cover and Hellman for optimal finite memory testing of two hypotheses. An approach to extending this finite memory training procedure to learning the better of two other training procedures will also be described.

6.8.2 Two-Level Training

In Section 6.5 we developed a Markov chain model of constant increment training of classifiers whose class regions ω_1 and ω_2 are described statistically by the constituent densities $f_1(x)$ and $f_2(x)$. It was assumed that the constituent densities overlap, as in Figure 1.12, and that there exists at least one optimum threshold θ^* such that $f_2(x) > f_1(x)$ when $x > \theta^*$ and $f_2(x) < f_1(x)$ when $x < \theta^*$. This is equivalent to assuming that the optimum decisions assign x to ω_2 when $x > \theta^*$, which is equivalent to assuming knowledge of the better of two opposite choices of the direction of the weight vector which describes the optimum decision surface.

In this section, we weaken these assumptions by eliminating knowledge of the half-space direction of the weight vector which describes the optimum decision surface. Specifically, we are given a class $C(\theta^*)$ of pairs of constituent densities $\{(p_i(x), q_i(x))\}$ such that $p_i(x) > q_i(x)$ for $x < \theta^*$ and $p_i(x) < q_i(x)$ for $x > \theta^*$. We are given that either $(f_1(x), f_2(x)) \in C(\theta^*)$ or $(f_2(x), f_1(x)) \in C(\theta^*)$. (Clearly, both pairs cannot belong to $C(\theta^*)$.) Let $H = \{H_0, H_1\}$ where H_0, H_1, denote the two hypotheses:

$$\begin{aligned} H_0&: (f_1(x), f_2(x)) \in C(\theta^*) \\ H_1&: (f_2(x), f_1(x)) \in C(\theta^*). \end{aligned} \tag{6.86}$$

(We assume that the a priori probabilities of H_0 *and* H_1, are π_0 and π_1, respectively.) For each of these hypotheses we build a trainable machine: L_0 for H_0 and L_1 for H_1. The training procedures in L_0, L_1 are designed for hypotheses H_0 and H_1, respectively. For example, if constant increment training procedures are used for the first level, the following training procedure appears in $L_k (k = 0, 1)$:

$$\Theta(n+1) = \Theta(n) + \begin{cases} -\rho, & \text{for } X(n) > \Theta(n) \text{ and } \omega(n) = \omega_{1+k}, \\ \rho, & \text{for } X(n) < \Theta(n) \text{ and } \omega(n) = \omega_{2-k}, \\ 0, & \text{otherwise,} \end{cases} \tag{6.87}$$

where $X(n)$ is an observed value of x at trial n. Both L_0 and L_1 receive the same observations $\Xi(n)$, where

$$\Xi(n) = (\omega(n), X(n)), \qquad \omega(n) \in \{\omega_1, \omega_2\}. \tag{6.88}$$

$\omega(n)$ and $X(n)$ are statistially independent random variables. L_k generates a decision $R_k(n)$, where $R_k(n) \in (1, 2)$. $R_k(n) = 1$ denotes the assignment of x to ω_1; $R_k(n) = 2$ denotes the assignment of x to ω_2. L_k has an adjustable threshold $\Theta_k(n)$ which determines $R_k(n)$ by

$$R_0(n) = \begin{cases} 1, & \text{for } x < \Theta_0(n), \\ 2, & \text{for } x > \Theta_0(n), \end{cases}$$

$$R_1(n) = \begin{cases} 2, & \text{for } x < \Theta_1(n), \\ 1, & \text{for } x > \Theta_1(n). \end{cases}$$

A decision process chooses $R(n)$ to be either $R_0(n)$ or $R_1(n)$. This decision process depends on $\omega(n)$, $R_0(n)$, and $R_1(n)$ for all n. Let $z(n) = P[\bigcup_i \{R(n) = i, \omega(n) = \omega_i\}]$ be the success probability of this system at trial n, and let $z_k(n) = P[\bigcup_i \{R_k(n) = i, \omega(n) = \omega_i\}]$ be the success probability of L_k at trial n.

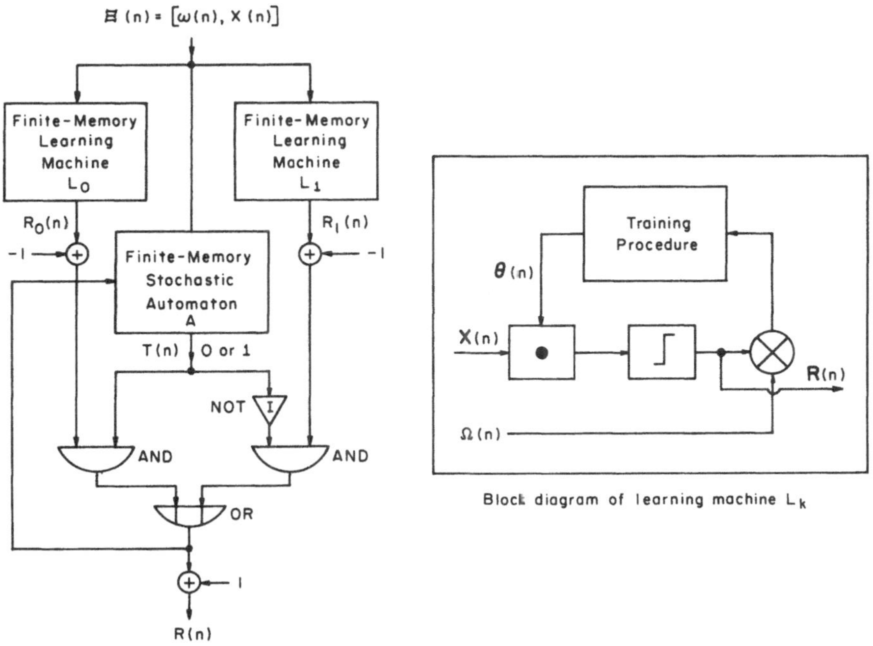

Figure 6.12. Block diagram of a scheme for learning the better of two trainable classifiers.

Our first problem is to design the training procedures in L_k and to choose k so that the success probabilities $\{z_k(n) | k = 0, 1\}$ are maximized. We assume that L_k is so designed that $z_k(n)$ approaches a value not far from optimum as $n \to \infty$ whenever H_k is true. When we solve this problem, we are left with the problem of finding a procedure for choosing $R(n)$ between $R_0(n)$ and $R_1(n)$.

An approach to solving this second problem makes use of a finite-state learning automaton A that observes the successes and errors in both L_0 and L_1 simultaneously. The output of A is a binary-valued signal that chooses $R(n)$ among $R_0(n)$ and $R_1(n)$. This approach is illustrated in block diagram form in Figure 6.12.

The learning machines L_0, L_1, and A constitute a two-level training procedure: L_0, L_1 at the first level and A at the second level.

6.8.3 The Single-Threshold Problem (Single-State L_i's)

To simplify our discussion as much as possible in a nontrivial manner, we restrict our analysis at first to the case where the threshold is known but the direction of the better decision surface is unknown; i.e., we assume that the alphabets of the thresholds of L_0 and L_1 are both limited to θ^*. The threshold θ^* could be any real number. Where possible, however, we will want to set θ^* equal to the optimum threshold under H_0 and H_1. The learning automaton A must choose the machine $L_k (k = 0, 1)$ that yields the higher success probability when the threshold is fixed at θ^* in both L_0 and L_1.

Let $p(x)$, $q(x)$ be defined as a pair of probability densities having the following two properties:

$$(p(x), q(x)) \in C(\theta^*) \tag{6.89a}$$

$$\begin{aligned}(f_1(x|H_0), f_2(x|H_0)) &= (f_2(x|H_1), f_1(x|H_1)) \\ &= (p(x), q(x))\end{aligned} \tag{6.89b}$$

where $f_i(x|H_j)$ denotes the constituent density of class ω_i under hypothesis H_j. Let

$$S_k(n) = \begin{cases} 1, & \text{for } R_k(n) = i \text{ and } \omega(n) = \omega_i;\ i = 0, 1 \\ 0, & \text{otherwise} \end{cases} \tag{6.90}$$

We refer to $S_k(\mathrm{n})$ as the *success* or the *outcome* of L_k at trial n. Let

$$z_{jk}(s) = P[S_j(n) = s|H_k], \qquad s \in \{0, 1\}. \tag{6.91}$$

Thus $z_{jk}(1)$ is the success probability at trial n of L_j under hypothesis H_k. Since in the present case $\Theta_0(n)$ and $\Theta_1(n)$ are both fixed at θ^*, $z_{jk}(s)$ is not a function of n. This is not so in more general situations, in which case $z_{jk}(s)$ would have to be replaced by $z_{jk}(n, s)$.

By Equations (6.91) and 6.89),

$$\begin{aligned} z_{jk}(1) &= \begin{cases} \int_{-\infty}^{\theta^*} f_1(x|H_k)\,dx + \int_{\theta^*}^{\infty} f_2(x|H_k)\,dx, & \text{if } j = 0, \\ \int_{-\infty}^{\theta^*} f_2(x|H_k)\,dx + \int_{\theta^*}^{\infty} f_1(x|H_k)\,dx, & \text{if } j = 1, \end{cases} \\ &= \begin{cases} \int_{-\infty}^{\theta^*} p(x)\,dx + \int_{\theta^*}^{\infty} q(x)\,dx, & \text{if } j = k, \\ \int_{-\infty}^{\theta^*} q(x)\,dx + \int_{\theta^*}^{\infty} p(x)\,dx, & \text{if } j \neq k. \end{cases} \end{aligned} \tag{6.92}$$

Letting

$$\zeta = \int_{-\infty}^{\theta^*} p(x)\,dx + \int_{\theta^*}^{\infty} q(x)\,dx, \tag{6.93}$$

Equation (6.92) becomes

$$z_{jk}(1) = \begin{cases} \zeta, & \text{if } j = k, \\ 1 - \zeta, & \text{if } j \neq k. \end{cases} \tag{6.94}$$

By Equation (6.91),

$$z_{jk}(0) = 1 - z_{jk}(1). \tag{6.95}$$

Hence

$$z_{jk}(s) = 1 - z_{jk}(1 - s) = \zeta \tag{6.96}$$

for $j + k + s = 1$ modulo 2.

The automaton A observes L_0 and L_1 under a sequence of experiments or tests $\{T(n)\}$, $T(n) \in \{0, 1\}$ for all n. $T(n)$ selects either $R(n) = R_0(n)$ or

$R(n) = R_1(n)$ by the formula

$$R(n) = R_0(n)T(n) + R_1(n)(1 - T(n)).$$

This is indicated in Figure 6.12 by equivalent Boolean operations. At each test, A observes $R(n)$ and $\omega(n)$ and computes the success indicator $S(n)$ by

$$S(n) = \begin{cases} 1, & \text{if } R(n) = i \text{ and } \omega(n) = \omega_i, \\ 0, & \text{otherwise.} \end{cases}$$

The choice of $T(n)$ is governed by a stochastic function that depends on the state of A. The state of A, in turn, is determined by a recursive process in which the state of A at trial $n + 1$ depends on the state at trial n and the pair $(T(n), S(n))$.

We assume that L_j is matched to H_j so that the success probability $z_{jk}(1)$ exceeds $\frac{1}{2}$ when $j = k$. Hence, by Equation (6.94),

$$\zeta > \tfrac{1}{2}.$$

We are faced with the problem of designing A so that it will choose $R(n) = R_k(n)$ most of the time if $H = H_k$. This problem is mathematically equivalent to the following "two-armed bandit problem" [6, 13–15]: Let L_0 and L_1 denote two coins with biases ζ and $1 - \zeta$ toward heads, with $\zeta > \frac{1}{2}$. (The bias of a coin is defined here as the probability of tossing a head.) It is not known whether L_0 or L_1 has the bias ζ. The problem is to find a strategy that most of the time tosses the coin whose bias is ζ.

This problem can be solved with an asymptotic performance close to optimum by a class of finite-state automata described by Cover and Hellman [6]. Each member of this class of automata, which we refer to here as a *C–H ε-optimal automaton*, may be modeled by an m-state random walk of the form shown in Figure 6.13. The number of states m, referred to as the *size of the memory*, depends on the desired approximation to optimum performance. Let

$$\xi(n) = (T(n), S(n)).$$

(Cover and Hellman refer to $\xi(n)$ as an *observation* associated with automaton A [6].) Sometimes we use the short notation

$$\xi = (T, S),$$

where the variable n is omitted. Since the values of both $T(n)$ and $S(n)$ are members of $\{0, 1\}$, the alphabet of $\xi(n)$ consists of the four pairs (0, 0), (0, 1), (1, 0), and (1, 1). Among this alphabet are observations yielding maximum information in favor of the hypothesis $H = H_0$. Other observations yield maximum information in favor of the hypothesis $H = H_1$. The automaton

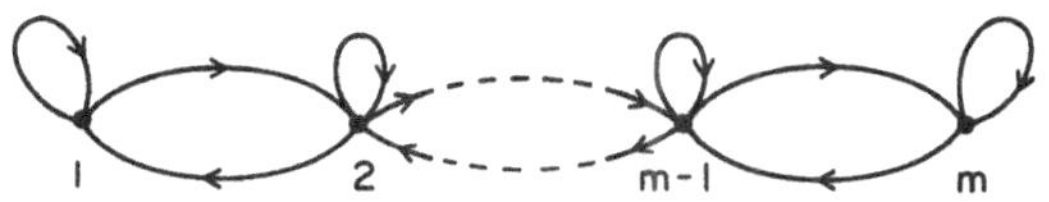

Figure 6.13. An m-state random walk.

is constructed so that transitions from state i to state j, for $i \neq j$, take place only on maximum-information observations, and in a direction toward one of two *ε-trapping states*. These ε-trapping states are the two extreme states, state 1 and state m, of the random walk. When state 1 is occupied, the probability of $P[T(n) = 0]$ is high, and when state m is occupied the probability $P[T(n) = 1]$ is high. Furthermore, the probability of leaving either of the two ε-trapping states is low, and becomes arbitrarily small as $m \to \infty$. Thus the automaton A is constructed so that maximum-information observations favoring $H = H_k$ tend to produce experiments that match H_k, i.e., $T(n) = k$. Before describing the rules for the transitions among the states, we will derive the maximum-information observations and the experiments that they favor.

Let $l_0(s)$ denote the likelihood ratio for the event $S(n) = s$. Under the experiment $T(n) = 0$, this likelihood ratio is

$$l_0(s) = \frac{z_{00}(s)}{z_{01}(s)} = \begin{cases} (1-\zeta)/\zeta, & \text{if } s = 0, \\ \zeta/(1-\zeta), & \text{if } s = 1. \end{cases} \tag{6.97}$$

The larger of these two likelihood ratios, for $T(n) = 0$, is

$$\begin{aligned} \hat{l}_0 &= \max\,[l_0(0), l_0(1)] = \max\left(\frac{1-\zeta}{\zeta}, \frac{\zeta}{1-\zeta}\right) \\ &= l_0(1) = \zeta/(1-\zeta). \end{aligned} \tag{6.98}$$

This is the maximum likelihood ratio with respect to s when $T(n) = 0$.

By similar reasoning, the maximum likelihood ratio with respect to s when $T(n) = 1$ is

$$\hat{l}_1 = l_1(0) = \zeta/(1-\zeta). \tag{6.99}$$

Let

$$\hat{l} = \max(\hat{l}_0, \hat{l}_1) = \zeta/(1-\zeta). \tag{6.100}$$

(In the present case, $\hat{l} = \hat{l}_0 = \hat{l}_1$. In more general cases, however, $\hat{l}_0 \neq \hat{l}_1$.) Let $\hat{\xi} = (\hat{T}, \hat{S})$ denote an observation for which the likelihood ratio of S is $\hat{l}$. Then $\hat{\xi}$ is a maximum-information observation favoring the hypothesis $H = H_0$, and hence favoring the decision $r(n) = r_0(n)$. In the present case, an examination of Equations (6.98) and (6.99) reveals that $\hat{\xi} = ((0,1), (1,0))$. Hence $(0,1)$ and $(1,0)$ are maximum-information observations favoring the experiment $T = 0$.

In a similar manner, we obtain the minimum likelihood ratio of S for any $T(n)$:

$$\begin{aligned} \check{l} = \check{l}_0 = \check{l}_1 &= l_0(0) = l_1(1) \\ &= (1-\zeta)/\zeta. \end{aligned} \tag{6.101}$$

(Note that $\hat{l} = 1/\check{l}$.)

Let $\check{\xi} = (\check{T}, \check{S})$ denote an observation for which the likelihood ratio of S is $\check{l}$. By Equations (6.98) and (6.99), $\check{\xi} = ((0,0), (1,1))$. Hence $(0,0)$ and $(1,1)$ are maximum-information observations favoring the experiment $T = 1$. The maximum-information observations $\check{\xi}$ and $\hat{\xi}$ generate stochastically near-optimal state transitions in the ε-optimal automaton. To provide an understanding of how this automaton achieves near-optimal behavior, we describe the rules for state transitions in such an automaton.

Suppose the number of states in the automaton is m. When the automaton occupies state 1, it chooses experiment 0 (i.e., $T = 0$) with probability $1 - \beta\delta$, and experiment 1 (i.e., $T = 1$) with probability $\beta\delta$, where δ is arbitrarily small, and where β is given by

$$\beta = \frac{(\hat{l})^{m-1}(\pi_0\pi_1)^{1/2} - \pi_0}{2[(\hat{l})^{m-1}\pi_0 - (\pi_0\pi_1)^{1/2}]}. \tag{6.102}$$

(Note that $\beta = \frac{1}{2}$ when $\pi_0 = \pi_1 = \frac{1}{2}$.) When the automaton occupies one of the intermediate states ($i = 2, \ldots, m-1$) it chooses $T = 0$ and $T = 1$ with equal probability. When the automaton occupies state m, it chooses $T = 0$ with probability δ and $T = 1$ with probability $1 - \delta$. Thus, in a sense, states $(2, \ldots, m-1)$ are training states and states $(1, m)$ are working states. The automaton moves from state i to state $i + 1$ whenever $\xi = \check{\xi} = (1,1)$, for $i = 1, \ldots, m-1$ (or, equivalently, whenever $\xi = \check{\xi} = (0,0)$, for $i = 1, \ldots, m-1$). It moves from state j to state $j - 1$ whenever $\xi = \hat{\xi} = (0,1)$, for $j = 2, \ldots, m$ (or, equivalently, whenever $\xi = \hat{\xi} = (1,0)$, for $j = 2, \ldots, m$).

For this automaton, the transition probabilities are as follows. For $i < m$:

$$\begin{aligned}
p_{i,i+1} &= P[\xi = (1,1)] = P(T = 1)P(S = 1 \mid T = 1) \\
&= \begin{cases} \frac{1}{2} z_{1_k}(1), & \text{if } 1 < i < m, \\ \beta\delta z_{1_k}(1), & \text{if } i = 1, \end{cases} \\
&= \begin{cases} \frac{1}{2}\zeta, & \text{if } k = 1,\ 1 < i < m, \\ \beta\delta\zeta, & \text{if } k = 1,\ i = 1, \\ \frac{1}{2}(1 - \zeta), & \text{if } k = 0,\ 1 < i < m, \\ \rho\delta(1 - \zeta), & \text{if } k = 0,\ i = 1, \end{cases} \\
p_{i,i-1} &= P[\xi = (0,1)] = P(T = 0)P(S = 1 \mid T = 0) \\
&= \begin{cases} \frac{1}{2} z_{0_k}(1), & \text{if } 1 < i < m, \\ \delta z_{0_k}(1), & \text{if } i = m, \end{cases} \\
&= \begin{cases} \frac{1}{2}\zeta, & \text{if } k = 0,\ 1 < i < m, \\ \delta\zeta, & \text{if } k = 0,\ i = m, \\ \frac{1}{2}(1 - \zeta), & \text{if } k = 1,\ 1 < i < m, \\ \delta(1 - \zeta), & \text{if } k = 1,\ i = m. \end{cases}
\end{aligned}$$

The state transition graph of this automaton is shown in Figure 6.14.

Let

$$b = \lim_{n \to \infty} P[H = H_{T(\mathbf{n})}]. \tag{6.103}$$

Thus b is the asymptotic probability that A will choose the output of the better of (L_0, L_1). Let

$$\hat{b}_m = \text{highest value of } b \text{ attainable by any } m\text{-state automaton}. \tag{6.104}$$

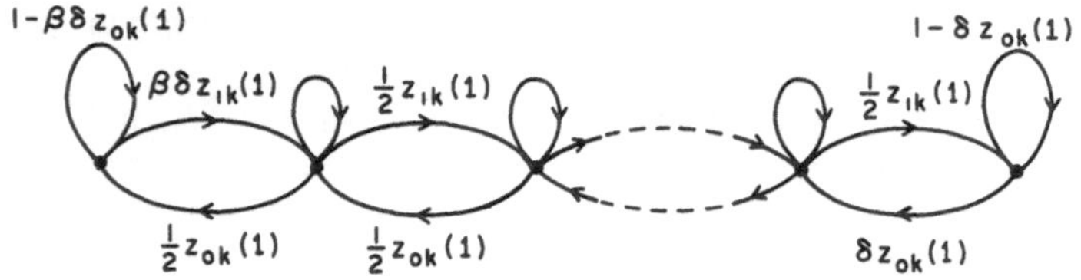

Figure 6.14. Transition graph of the *C–H* ε-optimal automaton for hypothesis H_k.

In the next section we show that

$$b \to \hat{b}_m \quad \text{as } \delta \to 0. \tag{6.105}$$

We also show in that section that, for the automaton A derived in this section, assuming $\pi_0 = \pi_1 = \frac{1}{2}$,

$$\hat{b}_m = \frac{1}{1 + (\hat{l})^{-(m-1)}}. \tag{6.106}$$

Since $\zeta > \frac{1}{2}$, it follows that $\hat{l} > 1$. Hence, by Equation (6.106),

$$\hat{b}_m \to 1 \quad \text{as } m \to \infty.$$

Thus we choose m so as to obtain b_m within specifications, and we choose δ so as to obtain b as close to $\hat{b}_m$ as possible within a reasonable length of training. The smaller δ is, the longer is the time taken by the automaton to approach a steady state condition, but the closer b comes to b_m. In fact when $\delta = 0$, the learning time is infinite, so that we never choose $\delta = 0$ unless we already know the solution to the problem.

6.8.4 Derivation of Basic Theorems for Automaton A*

In this section we derive three theorems describing the basic properties of automaton A. To derive these theorems we need the following additional nomenclature, definitions, and assumptions:

$$Q_n = \text{the state of } A \text{ at trial } n$$

$$r_i^k(n) = P[Q_n = i \mid H_k]$$

$$\mathbf{r}(n) = \text{vector } \{r_i^k(n)\}$$

$$\boldsymbol{\Phi}^k = \lim_{n \to \infty} \mathbf{r}^k(n)$$

$$p_{ij}^k = P[Q_n = j \mid Q_{n-1} = i, H_k], \qquad k = 0, 1$$

$$\xi = (T, S) = \text{an observation for } A \text{, consisting of a test } T \text{ and an outcome } S$$

$$p_{ij}(\xi) = P[Q_n = j \mid Q_{n-1} = i, \xi].$$

* This section may be omitted at the first reading of this chapter.

Assumptions:

(a) Test at trial $n = T_n = T(Q_{n-1})$, which is a function (possibly stochastic) of Q_{n-1}.

(b) Q_n is a function (possibly stochastic) of Q_{n-1} and ξ.

(c) $$Q_n \in \{1, \ldots, m\}$$

(d) $$T(n) \in \{0, 1\}$$

(e) $$S(n) \in \{0, 1\}$$

(f) A is an irreducible m-state automaton. (Actually we need not restrict A to irreducible automata. We do so because irreducible automata are inherently better for our purposes than reducible automata [6].)

Theorem 6.4. *If A is irreducible, then*

$$1/\hat{l} \leqslant p_{ij}^0/p_{ij}^1 \leqslant \hat{l}.$$

(*Note:* If p_{ij}^0 and p_{ij}^1 are zero, then their ratio is undefined.)

PROOF.

$$\begin{aligned} p_{ij}^k &= P[Q_n = j \mid Q_{n-1} = i, H_k] \\ &= \sum_T P[Q_n = j \mid Q_{n-1} = i, H_k, T] P[T \mid Q_{n-1} = i, H_k] \\ &= \sum_T P[Q_n = j \mid Q_{n-1} = i, H_k, T] P[T \mid Q_{n-1} = i]. \end{aligned}$$

Let

$$\alpha_i = P(T = 0 \mid Q_{n-1} = i).$$

Then

$$1 - \alpha_i = P(T + 1 \mid Q_{n-1} = i)$$

$$p_{ij}^k = \alpha_i P[Q_n = j \mid Q_{n-1} = i, H_k, T = 0] + (1 - \alpha_i) P[Q_n = j \mid Q_{n-1} = i, H_k, T = 1]$$

$$\begin{aligned} P[Q_n = j \mid Q_{n-1} = i, H_k, T] &= \sum_S P[Q_n = j \mid Q_{n-1} = i, T, S] P[S \mid H_k, T] \\ &= \sum_s p_{ij}(T, s) z_{Tk}(s), \end{aligned} \tag{6.107}$$

where $z_{Tk}(s)$ is defined in Equation (6.91). (Note that $P[S_j(n) = s \mid H_k] = P[S(n) = s \mid H_k, T(n) = j]$.) Thus

$$\begin{aligned} p_{ij}^k = \alpha_i \{ & P[Q_n = j \mid Q_{n-1} = i, T = 0, s = 0] z_{0k}(0) \\ & + P[Q_n = j \mid Q_{n-1} = i, T = 1, s = 1] z_{0k}(1) \} \\ & + (1 - \alpha_i) \{ P[Q_n = j \mid Q_{n-1} = i, T = 1, s = 0] z_{1k}(0) \\ & + P[Q_n = j \mid Q_{n-1} = i, T = 1, s = 1] z_{1k}(1) \}. \end{aligned} \tag{6.108}$$

Recall that

$$l_0(s) = z_{00}(s)/z_{01}(s), \qquad l_1(s) = z_{10}(s)/z_{11}(s). \tag{6.109}$$

Using Equations (6.109), replace $z_{00}(s)$ by $l_0(s) z_{01}(s)$ and $z_{10}(s)$ by $l_1(s) z_{11}(s)$ in Equation (6.108), with $k = 0$. Note that $z_{00}(s) \leqslant \hat{l} z_{01}(s)$. Then we obtain

$$p_{ij}^0 \leqslant \hat{l} p_{ij}^1.$$

In a similar manner, we may show that

$$p_{ij}^1 \leqslant \hat{l} p_{ij}^0.$$

Hence

$$1/\hat{l} \leqslant p_{ij}^0/p_{ij}^1 \leqslant \hat{l},$$

which completes the proof. □

Let

$$\lambda_i = \Phi_i^0/\Phi_i^1, \qquad i = 1, \ldots, m$$

be the state likelihood ratio.

Theorem 6.5. *If A is irreducible and the λ_i's are arranged so that $\lambda_1 \leqslant \cdots \leqslant \lambda_m$, then*

$$1 \leqslant \lambda_{i+1}/\lambda_i \leqslant (\hat{l})^2 \quad \text{for all } i.$$

PROOF. Since $\lambda_i \leqslant \lambda_{i+1}$ for all i, it follows that $1 \leqslant \lambda_{i+1}/\lambda_i$. This demonstrates the validity of the lower bound.

To establish the upper bound, partition the state space into two sets C and C'. In the steady state the probability of a transition from C to C' must equal the probability of a transition from C' to C. This must hold separately for each hypothesis, H_0 and H_1. Hence

$$\sum_{i \in C} \sum_{j \in C'} \Phi_i^k p_{ij}^k = \sum_{i \in C'} \sum_{j \in C} \Phi_i^k p_{ij}^k, \qquad k = 0, 1. \tag{6.110}$$

Suppose there exists r such that

$$\lambda_{r+1}/\lambda_r < (\hat{l})^2. \tag{6.111}$$

We shall prove that this supposition is false.

Let

$$c = \lambda_r. \tag{6.112}$$

Then we have

$$\lambda_1 \leqslant \cdots \leqslant \lambda_r = c \leqslant c(\hat{l})^2 < \lambda_{r+1} \leqslant \cdots \leqslant \lambda_m. \tag{6.113}$$

Let

$$C = \{1, \ldots, i\}, \qquad C' = \{i+1, \ldots, m\}. \tag{6.114}$$

Then

$$\Phi_i^0/\Phi_i^1 \leqslant c \quad \text{for } i \in C \tag{6.115}$$

$$\Phi_i^0/\Phi_i^1 > c(\hat{l})^2 \quad \text{for } i \in C'. \tag{6.116}$$

Setting $k = 0$ in Equation (6.110) and using Theorem 6.4 and Equation (6.115), we have

$$\sum_{i \in C} \sum_{j \in C'} \Phi_i^0 p_{ij}^0 \leqslant c\hat{l} \sum_{i \in C} \sum_{j \in C'} \Phi_i^1 p_{jk}^1. \tag{6.117}$$

Similarly, setting $k = 1$ in Equation (6.110) and using Theorem 6.4 and Equation (6.116), we have

$$\sum_{i \in C'} \sum_{j \in C} \Phi_i^0 p_{ij}^0 > c\hat{l} \sum_{i \in C} \sum_{j \in C'} \Phi_i^1 p_{jk}^1. \tag{6.118}$$

By setting $k = 0$ in Equation (6.110), the left-hand sides of Equations (6.117) and (6.118) are equal. By setting $k = 1$ in Equation (6.111), the right-hand sides of Equations (6.117) and (6.118) are equal, thereby obtaining a contradiction of supposition (6.111) and proving the theorem. □

Recall the definition of b:

$$b = \lim_{n \to \infty} P[H = H_{T(n)}],$$

i.e., b is the asymptotic probability that A chooses the output of the better of (L_0, L_1).

Theorem 6.6. *If A is an m-state automaton, b is bounded from above by $\hat{b}_m$,* where

$$\hat{b}_m = \max\left[\frac{(\hat{l})^{2(m-1)} - 2(\pi_0\pi_1)^{1/2}(\hat{l})^{(m-1)}}{(\hat{l})^{2(m-1)} - 1}, \pi_0, \pi_1\right].$$

Note that when $\pi_0 = \pi_1 = \frac{1}{2}$, $\hat{b}_m$ becomes

$$\hat{b}_m = \frac{(\hat{l})^{(m-1)}}{(\hat{l})^{(m-1)} + 1} = \frac{1}{1 + (\hat{l})^{-(m-1)}}.$$

PROOF. Let

$$b_k = \lim_{n \to \infty} P[T(n) = k \mid H = H_k], \qquad k = 0, 1 \tag{6.119}$$

be the asympototic probability of choosing the output of the better of (L_0, L_1) when hypothesis H_k holds ($k = 0, 1$). Then

$$b = \pi_0 b_0 + \pi_1 b_1, \tag{6.120}$$

where

$$\pi_k = P[H = H_k], \qquad k = 0, 1.$$

$$\begin{aligned} P[T(n) = 0 \mid H_0] &= \sum_i P(T = 0 \mid H_0, Q_i, n) P(Q_i \mid H_0, n) \\ &= \sum_i P(T = 0 \mid Q_i) P(Q_i \mid H_0, n). \end{aligned} \tag{6.121}$$

Recall that

$$\Phi_i^0 = \lim_{n \to \infty} P(Q_i \mid H_0, n). \tag{6.122}$$

Let

$$\alpha_i = P(T = 0 \mid Q_i). \tag{6.123}$$

Then by Equation (6.121)

$$b_0 = \lim_{n \to \infty} P(T(n) = 0 \mid H_0) = \sum_{i=1}^{m} \alpha_i \Phi_i^0. \tag{6.124}$$

Similarly,

$$b_1 = \sum_{i=1}^{m} (1 - \alpha_i)\Phi_i^1 = 1 - \sum_{i=1}^{m} \alpha_i \Phi_i^1. \tag{6.125}$$

By Theorem 6.2,

$$\lambda_m \leqslant \lambda_{m-1}(\hat{l})^2 \leqslant \lambda_{m-2}(\hat{l})^4 \leqslant \cdots \leqslant \lambda_1(\hat{l})^{2(m-1)}. \tag{6.126}$$

Since

$$\lambda_i \leqslant \lambda_m \quad \text{for } 1 \leqslant i \leqslant m, \tag{6.127}$$

it follows that

$$\lambda_i = \Phi_i^0/\Phi_i^1 \leqslant \lambda_1(\hat{l})^{2(m-1)} \quad \text{for } 1 \leqslant i \leqslant m. \tag{6.128}$$

Using Equation (6.128) and (6.125) in Equation (6.124) we obtain

$$\begin{aligned} b_0 &\leqslant \lambda_1(\hat{l})^{2(m-1)} \sum_i \alpha_i \Phi_i \\ &= \lambda_1(\hat{l})^{2(m-1)}(1 - b_1). \end{aligned} \tag{6.129}$$

Similarly,

$$b_1 \leqslant (1/\lambda_1)(1 - b_0). \tag{6.130}$$

Multiplying (6.129) and (6.130) yields

$$b_0 b_1 \leqslant (\hat{l})^{2(m-1)}(1 - b_0)(1 - b_1), \tag{6.131}$$

since b_0 and b_1 are nonnegative. Maximizing $\pi_0 b_0 + \pi_1 b_1$ subject to the constraint (6.131) can be performed by Lagrange extremization techniques. This yields

$$\hat{b}_m = \max\left[\frac{(\hat{l})^{2(m-1)} - 2(\pi_0\pi_1)^{1/2}(\hat{l})^{m-1}}{(\hat{l})^{2(m-1)} - 1}, \pi_0, \pi_1\right] \tag{6.132}$$

as the maximum possible value of b for an m-state automaton A. This maximum value is achieved when the inequality in Equation (6.131) is replaced by the equality

$$b_0 b_1 = (\hat{l})^{2(m-1)}(1 - b_0)(1 - b_1) \tag{6.133}$$

and all parameters in b_0 and b_1 are adjusted so as to maximize $\pi_0 b_0 + \pi_1 b_1$. □

6.8.5 Asymptotic Optimality of Automaton A

In this section we show that the asymptotic performance of automaton A approaches the best possible performance as $\delta \to 0$.

Let

$$\alpha_i = P[T(n) = 0 \mid Q_n = i].$$

Thus, by the description of A in Subsection 6.8.4,

$$\alpha_i = \begin{cases} 1 - \beta\delta, & \text{if } i = 1, \\ \frac{1}{2}, & \text{if } 2 \leqslant i \leqslant m, \\ \delta, & \text{if } i = m. \end{cases}$$

To determine the steady-state or asympototic performance of A, we may use the fact that in the steady state the probability of a transition from any subset of states of A to the set of remaining states is equal to the probability of a transition from the second set to the first. In particular, if the first set of states is $\{1, \ldots, i\}$ and the second set is $\{i + 1, \ldots, m\}$, then we have

$$\mu_i^k(1 - \alpha_i)z_{1k}(1) = \mu_{i+1}^k \alpha_{i+1} z_{0k}(1). \tag{6.134}$$

Thus,

$$\mu_{i+1}^k = \left(\frac{1 - \alpha_i}{\alpha_{i+1}}\right)\frac{z_{1k}(1)}{z_{0k}(1)}\mu_i^k, \qquad k = 0, 1. \tag{6.135}$$

Using Equations (6.94) and (6.100), Equation (6.135) becomes

$$\mu_{i+1}^0 = \frac{1 - \alpha_i}{\alpha_{i+1}} \hat{l}^{-1} \mu_i^0, \tag{6.136}$$

and

$$\mu_{i+1}^1 = \frac{1-\alpha_i}{\alpha_{i+1}} \hat{l} \mu_i^1, \tag{6.137}$$

where $\hat{l} = \zeta/(1-\zeta)$. (Recall that we assume that $\zeta > \frac{1}{2}$.)

Now we shall derive expressions for b_0 and b_1, where b_k is the steady-state probability that $T(n) = k$ under hypothesis H_k. Clearly,

$$b_0 = \sum_{i=1}^{m} \alpha_i \mu_i^0 \tag{6.138}$$

and

$$b_1 = \sum_{i=1}^{m} (1-\alpha_i)\mu_i^1 \tag{6.139}$$

$$= 1 - \sum_{i=1}^{m} \alpha_i \mu_i^1 \tag{6.140}$$

since

$$\sum_{i=1}^{m} \mu_i^k = 1, \qquad k = 0, 1. \tag{6.141}$$

By Equations (6.135) and (6.136) we obtain

$$\begin{aligned} \mu_i^0 &= 2\beta\delta\hat{l}^{-(i-1)}\mu_1^0, \qquad 2 \leqslant i \leqslant m-1 \\ \mu_m^0 &= \beta\hat{l}^{-(m-1)}\mu_1^0 \end{aligned} \tag{6.142}$$

and

$$\begin{aligned} \mu_i^1 &= 2\beta\delta\hat{l}^{(i-1)}\mu_1^1, \qquad 2 \leqslant i \leqslant m-1 \\ \mu_m^1 &= \beta\hat{l}^{m-1}\mu_1^1. \end{aligned} \tag{6.143}$$

Thus, by Equations (6.138) and (6.142),

$$b_0 = \mu_1^0 \left\{1 + \beta\delta\left[-1 + \sum_{i=2}^{m} \hat{l}^{-(i-1)}\right]\right\}. \tag{6.144}$$

Similarly,

$$b_1 = 1 - \mu_1^1 \left\{1 + \beta\delta\left[-1 + \sum_{i=2}^{m} \hat{l}^{i-1}\right]\right\}. \tag{6.145}$$

Let

$$\bar{b}_k = \lim_{\delta\to 0} b_k, \qquad \bar{b} = \lim_{\delta\to 0} b, \quad \text{and } \bar{\mu}_i^{-k} = \lim_{\delta\to 0} \mu_i^k. \tag{6.146}$$

From Equations (6.142) and (6.143), it is easy to show that

$$\bar{\mu}_1^0 + \bar{\mu}_m^0 = 1 \quad \text{and} \quad \bar{\mu}_1^1 + \bar{\mu}_m^1 = 1 \tag{6.147}$$

as a result of the constraints

$$\sum_{i=1}^{m} \mu_i^k = 1 \quad \text{for } k = 0, 1. \tag{6.148}$$

By Equations (6.143) and (6.144),

$$\bar{b}_0 = \bar{\mu}_1^0, \qquad \bar{b}_1 = 1 - \bar{\mu}_1^1. \tag{6.149}$$

By Equations (6.148) and (6.149),

$$\bar{b}_0 = 1 - \bar{\mu}_m^0, \qquad \bar{b}_1 = \bar{\mu}_m^1. \tag{6.150}$$

Hence

$$\lim_{\delta \to 0} \frac{b_0 b_1}{(1 - b_0)(1 - b_1)} = \frac{\bar{b}_0 \bar{b}_1}{(1 - \bar{b}_0)(1 - \bar{b}_1)} = \frac{\bar{\mu}_1^0 \bar{\mu}_m^1}{\bar{\mu}_m^0 \bar{\mu}_1^1}. \tag{6.151}$$

By Equations (6.142), (6.143), and (6.151),

$$\frac{\bar{b}_0 \bar{b}_1}{(1 - \bar{b}_0)(1 - \bar{b}_1)} = \hat{l}^{2(m-1)}. \tag{6.152}$$

But this equation is the condition for achieving the maximum possible value of $\bar{b}$, as specified by Equation (6.133). Thus choosing β so as to maximize $\bar{b}$ will yield a maximum possible $\bar{b}$ for an m-state automaton D.

By (6.142) and (6.148),

$$\bar{\mu}_1^0 = 1/(1 + \beta \hat{l}^{-(m-1)}), \qquad \bar{\mu}_m^0 = \beta \hat{l}^{-(m-1)} \bar{\mu}_1^0. \tag{6.153}$$

By (6.142) and (6.147),

$$\bar{\mu}_1^1 = 1/(1 + \beta \hat{l}^{m-1}), \qquad \mu_m^1 = \beta \hat{l}^{m-1} \bar{\mu}_1^1. \tag{6.154}$$

By Equations (6.149), (6.150), (6.153), and (6.154),

$$\bar{b} = \pi_0 b_0 + \pi_1 b_1 = \frac{\pi_0}{1 + \beta \hat{l}^{-(m-1)}} + \frac{\pi_1 \beta \hat{l}^{m-1}}{1 + \beta \hat{l}^{m-1}}.$$

Choosing β so as to maximize $\bar{b}$ in the above equation yields

$$\beta = \frac{\hat{l}^{m-1}(\pi_0 \pi_1)^{1/2} - \pi_0}{2[\hat{l}^{m-1}\pi_0 - (\pi_0 \pi_1)^{1/2}]},$$

as specified in Equation (6.102).

6.8.6 Extension to Multistate L_i's

One may extend the above technique to choosing the better among two multistate L_i's. In particular, suppose that each of the L_i's uses a constant increment training procedure, and that each of these training procedures is matched to a distinct hypothesis H_k as specified in Equation (6.86). Then, as $n \to \infty$, each of the training procedures will produce a success probability

$$z_{jk}(1) = \lim_{n \to \infty} P[S = 1 \mid n, L_j, H_k].$$

The automaton A will then move through state transitions only at maximum-information observations $\xi = (T, S)$, as described in subsection 6.8.3. In this case, however, there may be no symmetry between L_0 and L_1, in the sense that it is likely that $z_{10}(1) \neq z_{01}(1)$ and $z_{00}(1) \neq z_{11}(1)$.

The mathematical derivations for this case are extensions of our discussion in the preceding sections.

6.9 Multidimensional Feature Space

Most of the preceding material in this chapter is concerned with Markov chain training models of single-feature classifiers. The use of such models for multiple-feature classifiers is less attractive than for single-feature classifiers because in most such cases the Markov process followed by the weight vector has either a countably infinite or a very large finite number of states. Thus Markov chain models for multiple-feature classifiers are usually used for qualitive rather than quantitive purposes. Examples of the qualitative uses of Markov chains for such classifiers are given below.

Consider the case of proportional increment training on finite linearly separable classes. For this case, the weight vector follows a finite Markov chain. To see this, recall that the weight vector in this case terminates at a solution vector within a finite number of trials. Furthermore, since the classes are finite and since, from Equation (2.13),

$$\mathbf{V}(n+1) = \mathbf{V}(n) + \rho\varepsilon(n)\boldsymbol{\eta}(n)$$

where $\varepsilon(n) = 1$ in the event of an error and $\varepsilon(n) = 0$ in the event of no error, it follows that

$$\mathbf{V}(n) = \mathbf{V}(0) + \rho \sum_{j=0}^{n-1} \varepsilon(j)\boldsymbol{\eta}(j). \tag{6.155}$$

Since the number of combinations of $\{\mathbf{X}(j)\}$ and therefore $\{\boldsymbol{\eta}(j)\}$ is finite, it follows from Equation (6.155) that the number of possible values of $\mathbf{V}(n)$ is finite. Hence $\mathbf{V}(n)$ must follow a finite Markov chain. Furthermore, because $\mathbf{V}(n)$ in this case must terminate at a solution vector in a finite number of trials, the Markov chain is an *absorbing* Markov chain. (An absorbing Markov chain is a Markov chain in which every path terminates in a finite time at a state from which escape is impossible. Such a state is an *absorbing state.*) In such a chain all states except the absorbing states are transient states. Since the chain is finite, the probability of occupying a transient state at trial n decays exponentially with n. This can be seen from Equation (6.50).

In contrast, if the class regions are not linearly separable, no error-free hyperplane decision surface exists for such a classifier. Hence the weight vector generated by the proportional increment training procedure for such a classifier cannot follow an absorbing Markov chain.

In many artificial classifiers the weight vectors are of finite dimensionality and the component weights are integers. An example of such a classifier is one in which the weight vectors are generated by a digital computer. In this case each component of each weight vector is represented by a finite set of digits or bits, and the weight vector has a finite number of components. Thus for these classifiers the number of states in the Markov chain of the weight vector is necessarily finite, but for multidimensional weight vectors the number of states is usually very large. For these classifiers the Markov chain may or may not be ergodic. Ergodicity here depends on whether or not the chain is irreducible. (If the chain is not irreducible, then it has more than one irreducible set of persistent states, where "irreducible set" is defined in the sense of Theorem 6.1.)

If a Markov chain is finite, it must have at least one persistent state. In an error-correcting trainable classifier modeled by a finite Markov chain, at least one of these persistent states must be aperiodic. To see this, suppose a given feature vector $\mathbf{X}(n)$ results in an error at trial $n = m$, and $\mathbf{V}(m)$ belongs to a persistent set. Let $\mathbf{X} = \mathbf{X}(m)$. Suppose $\mathbf{X}$ is repeated for $n = m, m + 1, \ldots$. Then eventually a $\mathbf{V}(n^*)$ will be reached, at which either no error occurs, resulting in a self-transition to $\mathbf{V}(n^*)$, or members of both classes appear on at least one side of the hyperplane. In the latter situation a self-transition to $\mathbf{V}(n^*)$ must exist, because if two feature vectors are taken from different classes on the same side of the hyperplane, one of them must result in a correct classification and hence a self-transition. But by the definition of an aperiodic state in Section 6.5, a persistent state with a self-transition must be aperiodic. Hence if the Markov chain of an error-corrected trainable classifier is finite, the chain must have at least one persistent aperiodic state.

A proportional increment training procedure may yield more than one irreducible set of persistent states. To see this, suppose that all of the components of every feature vector in the training set are integers, and suppose the adjustment coefficient ρ is 2, i.e.,

$$\mathbf{V}(n+1) = \begin{cases} \mathbf{V}(n) + 2\boldsymbol{\eta}(n), & \text{if an error occurs at trial } n, \\ \mathbf{V}(n) & \text{if no error occurs at trial } n. \end{cases}$$

Then if the components of $\mathbf{V}(0)$ are all odd, the components of $\mathbf{V}(n)$ are all odd for every n. Similarly, if the components of $\mathbf{V}(0)$ are all even, the components of $\mathbf{V}(n)$ are all even for every n. Thus there must be at least two irreducible sets of persistent states in this case—one set consisting of even-valued $\mathbf{V}(n)$'s, and the other of odd-valued $\mathbf{V}(n)$'s.

In some trainable classifiers one can be certain that there is only one irreducible set of persistent states. In particular, Duda and Munson have shown, "if the training sequence is formed by randomly selecting from a finite set of [feature vectors] and their negatives, so that some of the elements of the training sequence are mislabeled, and if the [proportional increment training procedure] produces a finite-state Markov chain, then the resulting Markov chain has exactly one [irreducible] set of persistent states" [16]. In other words, a finite Markov chain produced from a partially mislabeled training sequence has precisely one irreducible set of persistent states.

Exercises

6.1. Show that Equation (6.6) follows from Equations (6.4) and (6.5).

6.2. (a) Find an expression for $dE(C|\theta)/d\theta$, where $E(C|\theta)$ is given by Equation (6.20).

(b) Derive Equation (6.21).

(c) Let $f'_u(\theta) = df_u/d\theta$. Show that $c_{21}f'_1(\theta^*) < c_{12}f'_2(\theta^*)$ implies that $E(C|x)$ has a local minimum at $x = \theta^*$.

(d) Derive Equation (6.18).

6.3. Consider the following procedure. Let $\Theta = \{\theta^{(1)}, \theta^{(2)}, \ldots, \theta^{(m)}\}$, such that $\theta^{(i)} < \theta^{(j)}$ wherever $i < j$. If $\Theta(n) = \theta^{(i)}$, then

$$\Theta(n+1) = \begin{cases} \theta^{(i+1)}, & \text{if } \Omega(n) < R(n) \text{ and } \theta^{(i)} < X(n), \\ \theta^{(i-1)}, & \text{if } \Omega(n) > R(n) \text{ and } \theta^{(i)} > X(n), \\ \theta(n), & \text{otherwise.} \end{cases}$$

Show that
(a) p_{ij} is independent of n;
(b) $\mathbf{r}^T(n) = \mathbf{r}^T(0)\mathbf{P}^n$;
where $\mathbf{r}(n)$ and $\mathbf{P}$ are defined by Equations (6.26) and (6.31).

6.4. Let

$$B(n) = 1 - |\Omega(n) - R(n)|,$$

where $\Omega(n)$ and $R(n)$ specify the input and output classes, respectively, of a single-feature trainable classifier. Show that the expected value of $B(n)$ is the success probability, i.e., that

$$E[B(n)] = E[S(\theta(n))].$$

6.5. Let

$$R(\theta) = P[Z = -1 | \theta]$$
$$S(\theta) = P[Z = 0 | \theta]$$
$$L(\theta) = P[Z = 1 | \theta],$$

where $Z = \Omega - R$. Consider a one-dimensional classifier undergoing constant increment training.
(a) Show that $R(\theta) + S(\theta) + L(\theta) = 1$ for all θ.
(b) Let

$$F_i(\theta) = \int_{-\infty}^{\theta} f_i(x)\, dx,$$

where $\{f_i(x) | i = 1, 2\}$ are the constituent densities of the channel and the input data. Let $\rho = F_1(\infty)$. Show that

$$R(\theta) = \rho - F_1(\theta)$$
$$S(\theta) = 1 - \rho + F_1(\theta) - F_2(\theta)$$
$$L(\theta) = F_2(\theta).$$

(c) Show that $R(\theta)$ is a nonincreasing function of θ, and that $L(\theta)$ is a nondecreasing function of θ.
(d) Show that if $f_1(x)$, $f_2(x)$ are differentiable, if $f_1(x) - f_2(x)$ has precisely one zero at $X = \theta^*$, and if $f'_1(\theta^*) - f'_2(\theta^*) < 0$, then $S(\theta)$ is single-peaked (i.e., $S(\theta)$ has precisely one local maximum).

6.6. Prove the following: An irreducible Markov chain is ergodic if and only if there exists a vector $\boldsymbol{\alpha}$ such that $\boldsymbol{\alpha}^T = \boldsymbol{\alpha}^T\mathbf{P}$ and $\boldsymbol{\alpha} > \mathbf{0}$, where $\mathbf{P}$ is the transition probability matrix of the chain.

6.7. In the proof of Theorem 6.3, allow for the possibility that h_{m-j}^{-1} and h_{m+j}^{-1} are equal to 1 for some values of j. Show that in this case, as in the given proof,

$$\Phi_m > 0.$$

Hint: Note that $h_{m-j}^{-1} = 1$ and/or $h_{m+i} = 1$ only over a finite range of i and j.

6.8. Find an expression for Φ_{m+k} for $k > 0$, where $\boldsymbol{\Phi}$ is the stationary probability vector of Theorem 6.3.

6.9. (a) Show that the states of the random walk having the state transition graph shown in Figure 6.15 are null.

(b) If the weight $\Theta(n)$ of a single-feature training procedure follows the Markov chain shown in Figure 6.15, does the state probability vector $\mathbf{r}(n)$ converge as $n \to \infty$ to a unique asymptotic vector independently of $\mathbf{r}(0)$? Explain.

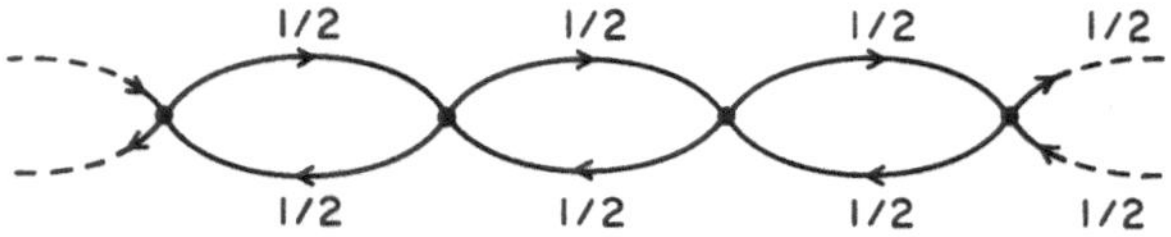

Figure 6.15. Illustration of random walk for Exercise 6.9.

6.10. Show that the learning curve of the training phase for a weakly coupled two-mode channel is given by

$$z(n) \cong \gamma z_{Ao}(n) + (1 - \gamma) z_{Bo}(n),$$

where $z_{Ao}(n)$ and $z_{Bo}(n)$ are the component single-mode learning curves, and γ is given by Equation (6.75).

6.11. Show that the success probability in a working phase of a classifier on a two-mode Gilbert channel cannot be less than one-half the success probability at the beginning of that working phase.

6.12. Derive the entries in Table 6.1 with the help of the weak-coupling approximation.

6.13. Consider the following automaton. It is an m-state random walk, choosing experiments 0, 1 according to the rule in Subsection 6.8.3. It moves from state i to state $i + 1$ whenever $\xi = \bar{\xi} =$ either $(0, 0)$ or $(1, 1)$ for states $i = 2, \ldots, m$. It moves from state 1 to state 2 whenever $\xi = (0, 1)$.

(a) Find the elements of the transition probability matrix of the Markov chain model of this automaton.

(b) Show that this automaton is an ε-optimal automaton.

References

1. J. Sklansky, Boundary detection in medical radiographs. In: *Digital Processing of Biomedical Images*, K. Preston and M. Onoe (eds.), Plenum Press, New York, 1976, pp. 309–322.

2. C. K. Chow and T. Kaneko, Boundary detection of radiographic images by a threshold method. In: *Proc. IFIP Congress 1971*, North-Holland, Amsterdam, 1972, Section TA-7, pp. 130–134.

3. J. Sklansky, Threshold training of two-mode signal detection. *IEEE Trans. on Information Theory*, **IT-11** (3): 353–362 (1965).

4. N. J. Nilsson, *Learning Machines*. McGraw-Hill, New York, 1965.

5. R. B. Ash, *Basic Probability Theory*. John Wiley & Sons, New York, 1970.

6. T. M. Cover and M. E. Hellman, The two-armed bandit problem with time-invariant finite memory. *IEEE Trans. on Information Theory*, **IT-16** (2): 185–195 (1970).
7. W. Feller, *An Introduction to Probability Theory and its Applications*, 3d edition. John Wiley & Sons, New York, 1968.
8. G. Lendaris, Two theorems concerning Markov chains. Institute of Engineering Research, Series No. 60, Issue No. 347, Electronics Research Laboratory, University of California, Berkeley, California, February 1961.
9. D. R. Cox and H. D. Miller, *The Theory of Stochastic Processes*. Methuen & Co., London, 1965.
10. P. M. De Russo, R. J. Roy, and C. M. Close, *State Variables for Engineers*. John Wiley & Sons, New York, 1966.
11. E. N. Gilbert, Capacity of a burst-noise channel. *Bell System Technical Journal*, **39**: 1253–1266 (1960).
12. M. L. Tsetlin, On the behavior of finite automata in random media. *Avtomatika i Telemekhanika*, **22**, (10): 1345–1354 (1961). (English translation in *Automation and Remote Control*, pp. 1210–1219 (1962).)
13. H. Robbins, Some aspects of the sequential design of experiments. *Bulletin of the American Mathematical Society*, **58**: 529–532 (1952).
14. R. N. Bradt, S. M. Johnson, and S. Karlin, On sequential designs for maximizing the sum of *n* observations. *Annals of Mathematical Statistics*, **27**: 1060–1074 (1956).
15. R. N. Bradt and S. Karlin, On the design and comparison of certain dichotomous experiments. *Annals of Mathematical Statistics*, **27**: 390–409 (1956).
16. R. O. Duda, Linear machines and Markov processes. In: *Pattern Recognition*, L. Kanal (ed.), Thompson Book Co., Washington, D.C., 1968, pp. 251–282.

CHAPTER 7

Continuous-State Models

7.1 Introduction

In Chapter 6 we showed how the transient and steady-state behavior of single-feature training procedures can be analyzed by Markov chains. However in most training procedures the state space is multidimensional and the density of states is indefinitely large, thereby making finite Markov chain models inappropriate. Consider, for example, the proportional increment training procedure described by Equations (2.13) and (2.14) in Chapter 2. (In this chapter, we replace $r(n)$ by the random variable $R(n)$.) The augmented weight vector $\mathbf{V}(n)$ takes the role of a state, in the sense that the future statistical behavior of $\mathbf{V}(n + M)$, $m \geqslant 0$, is completely determined from a knowledge of the present weight vector (i.e., $\mathbf{V}(n)$), the constituent densities, and the training procedure. The motion of $\mathbf{V}(n)$ is similar to Brownian motion, as illustrated in Figure 7.1, so that the set of possible values of $\mathbf{V}(n)$ is infinitely dense. Thus $\mathbf{V}(n)$ cannot be a Markov chain in this case. Yet $\mathbf{V}(n)$ is a Markov process, i.e.,

$$P[\mathbf{V}(n + 1) | \mathbf{V}(0), \ldots, \mathbf{V}(n)] = P[\mathbf{V}(n + 1) | \mathbf{V}(n)],$$

since (a) the proportional increment training procedure generates $\mathbf{V}(n + 1)$ from $\omega(n)$, $R(n)$, and $\mathbf{V}(n)$, (b) the statistics of $R(n)$ depend on $\mathbf{V}(n)$ and $\omega(n)$, and (c) the random variables $\{\omega(n)\}$ are statistically independent. In fact we come to the same conclusion—i.e., that $\mathbf{V}(n)$ is a Markov process—if the proportional increment training procedure is generalized to the following

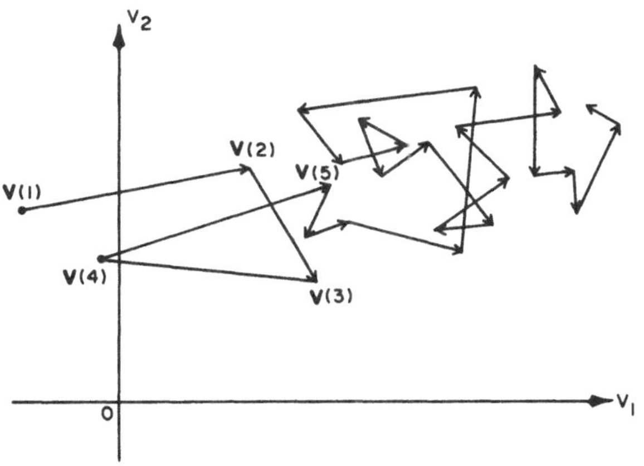

Figure 7.1. An example of Brownian motion.

form:

$$\mathbf{V}(n+1) = \mathbf{V}(n) + \rho_n \mathbf{Z}(\Gamma(n), \mathbf{Y}(n)), \tag{7.1}$$

where $\Gamma(n) = \Omega(n) - R(n)$ and where $\Omega(n) = i$ when $\omega(n) = \omega_i$. (Unless otherwise specified, the feature vectors and weight vector are assumed to be in augmented form.) We refer to $\rho_n \mathbf{Z}(\Gamma(n), \mathbf{Y}(n))$ in Equation (7.1) as the *step* at trial *n*. A simple illustrative example of Equation (7.1) is the one-dimensional time-varying increment procedure:

$$\Theta(n+1) = \Theta(n) + \rho_n(-\Gamma(n)). \tag{7.2}$$

Here $\Theta(n) = -V_0(n)$, $V_1(n) = 1$, for all n, $Z_0(\Gamma(n), Y(n)) = \Gamma(n)$. When $\rho_n =$ constant, this procedure reduces to the constant increment procedure discussed in the preceding chapter.

To analyze the dynamic behavior of the type of multiple feature training process specified in Equation (7.1), one may assume that the ρ_n's are so small that the motion of $\mathbf{V}(n)$ through weight space can be approximately represented by differential rather than difference equations. To obtain these equations, we define a function $g(t)$ such that

$$\rho_t = g(t).$$

Then we replace ρ_n by $\rho g(n\rho)$ and let $\rho \to 0$ while $n\rho = t$. We refer to this approximation as a *continuous-state Markov* model. One of the earliest examples of the continuous-state approximation is the classical analysis of Brownian motion by Albert Einstein [2].

In this chapter, continuous-state equations are derived for the mean of $\mathbf{V}(n)$ and for the covariance matrix of $\mathbf{V}(n)$. Specifically, let

$$\boldsymbol{\mu}(n) = E[\mathbf{V}(n)] = \text{mean of } \mathbf{V}(n) \tag{7.3}$$

$$\begin{aligned}\boldsymbol{\Sigma}(n) &= E\{[\mathbf{V}(n) - \boldsymbol{\mu}(n)][\mathbf{V}(n) - \boldsymbol{\mu}(n)]^T\} \\ &= \text{covariance matrix of } \mathbf{V}(n).\end{aligned} \tag{7.4}$$

Continuous-state equations are derived for $\boldsymbol{\mu}(n)$ and $\boldsymbol{\Sigma}(n)$, $n \leqslant 0$, in terms of the constituent densities and the training procedure. This is done by expressing $\boldsymbol{\mu}(n)$ and $\boldsymbol{\Sigma}(n)$ in terms of $\boldsymbol{\eta}(\mathbf{v})$ and $\boldsymbol{\Omega}(\mathbf{v})$, where

$$\boldsymbol{\eta}(\mathbf{v}) = E\left[\frac{\Delta\mathbf{V}(n)}{\rho_n}\middle|\mathbf{V}(n) = \mathbf{v}\right], \tag{7.5}$$

$$\boldsymbol{\Omega}(\mathbf{v}) = E\left[\left(\frac{\Delta\mathbf{V}(n)}{\rho_n} - \boldsymbol{\eta}(\mathbf{v})\right)\left(\frac{\Delta\mathbf{V}(n)}{\rho_n} - \boldsymbol{\eta}(v)\right)^T\middle|\mathbf{V}(n) = \mathbf{v}\right], \tag{7.6}$$

which in turn may be expressed in terms of the constituent densities and the training procedure. (ρ_n in the above equations is a coefficient proportional to the step size.) From $\boldsymbol{\mu}(n)$ and $\boldsymbol{\Sigma}(n)$ it is then possible to estimate the learning curves and variance curves.

By replacing ρ_n by $\rho g(n\rho)$ and letting $\rho \to 0$ while $n\rho = t$, we obtain continuous-state approximations for $\boldsymbol{\mu}(n)$ and $\boldsymbol{\Sigma}(n)$. These approximations are sometimes referred to as *small-step* approximations, since in theory they are valid only when ρ is sufficiently small. Nevertheless, in practice they are often good even when ρ is only moderately small.

We also show in this chapter how the continuous-state model makes possible an extension of the concept of *stability in the large*, originally developed for automatic control systems, to training theory. And we show how the continuous-state model leads to restrictions on the possible shapes of the learning curves and variance curves in one-dimensional training.

7.2 The Centroid Equation

In this section we show how a continuous-state model leads to an approximate differential equation for $\boldsymbol{\mu}(n)$. We refer to $\boldsymbol{\mu}(n)$ as the *centroid* of $\mathbf{V}(n)$. We refer to the trajectory of $\boldsymbol{\mu}(n)$ in weight (state) space as a *centroid curve*, and the approximating differential equation for $\boldsymbol{\mu}(n)$ as a *centroid equation*.

In the approximation leading to the centroid equation the discrete motion of $\mathbf{V}(n)$ is replaced by a continuous motion. It is for this reason that we refer to this approximation as a *continuous-state* model.

Recall from Equation (7.1) that ρ_n is the coefficient of proportionality of the step size at trial n. Consider a set of proportional increment training procedures which all are alike except for differing values of ρ; i.e., replace ρ_n by $\rho g(n\rho)$ and consider a set of proportional increment training procedures the members of which have the sequence $\{\rho g(n\rho)\}$ with different values of ρ. For example, suppose

$$\rho_n = \frac{1}{1 + \sqrt{n}} + \frac{2}{n^2}. \tag{7.7}$$

If

$$g(t) = \frac{1}{1 + \sqrt{t}} + \frac{2}{t^2}, \tag{7.8}$$

it follows that $\rho g(n\rho) = \rho_n$ when $\rho = 1$. Hence we may choose $g(t)$ as in Equation (7.8). Thus, in place of the single training procedure with the step size given by Equation (7.7), we now consider a set of training procedures in which the step sequence is given by

$$\rho g(n\rho) = \frac{\rho}{1 + \sqrt{n\rho}} + \frac{2}{n^2 \rho}, \tag{7.9}$$

each with a different value of ρ. For each of these training procedures,

$$\begin{aligned} \Delta\boldsymbol{\mu}(n) &= \boldsymbol{\mu}(n+1) - \boldsymbol{\mu}(n) = E[\mathbf{V}(n+1) - \mathbf{V}(n)] \\ &= \rho g(n\rho) E\{E[\Delta\mathbf{V}(n)/\rho g(n\rho) \mid \mathbf{V}(n)]\}. \end{aligned} \tag{7.10}$$

Hence

$$\Delta\boldsymbol{\mu}(n) = \rho g(n\rho) E\{\boldsymbol{\eta}[\mathbf{V}(n)]\}, \tag{7.11}$$

where $\boldsymbol{\eta}[\mathbf{V}(n)]$ is defined by Equation (7.5). Equation (7.11) determines $\boldsymbol{\mu}(n)$ exactly. Unfortunately this equation is generally quite difficult to solve. The reason for this difficulty arises from the fact that $E\{\boldsymbol{\eta}[\mathbf{V}(n)]\}$ depends on $p_{\mathbf{V}(n)}(\mathbf{v})$, which in multidimensional state space is difficult to compute for all but very small values of n. For example, in the one-dimensional training process described by Equation (7.2), $p_{\Theta(n)}(\theta)$ is determined by the following difference equation:

$$\begin{aligned} p_{\Theta(n+1)}(\theta) &= p_{\Theta(n)}(\theta - \rho_n)R(\theta - \rho_n) + p_{\Theta(n)}(\theta)[1 - L(\theta) - R(\theta)] \\ &\quad + p_{\Theta(n)}(\theta + \rho_n)L(\theta + \rho_n) \end{aligned} \tag{7.12}$$

with the initial condition

$$p_{\Theta(0)}(\theta) = \delta[\theta - \theta(0)],$$

where $\delta[\theta - \theta(0)]$ denotes a Dirac delta function and where

$$\begin{aligned} L(\theta) &= P[\omega(n) = \omega_2, r(n) = 1 \mid \Theta(n) = \theta] \\ R(\theta) &= P[\omega(n) = \omega_1, r(n) = 2 \mid \Theta(n) = \theta]. \end{aligned}$$

A general solution of equations of the from of Equation (7.12) is not known.

To overcome the difficulty of computing $p_{\mathbf{V}(n)}(\mathbf{v})$, expand $\boldsymbol{\eta}(\mathbf{V}(n))$ about $\boldsymbol{\mu}(n)$ in a Taylor series:

$$\begin{aligned} \boldsymbol{\eta}(\mathbf{V}(n)) &= \boldsymbol{\eta}(\boldsymbol{\mu}(n)) + \frac{d\boldsymbol{\eta}}{d\boldsymbol{\mu}}(\mathbf{V}(n) - \boldsymbol{\mu}(n)) \\ &\quad + (\mathbf{V}(n) - \boldsymbol{\mu}(n))^T \frac{d^2\boldsymbol{\eta}}{d\boldsymbol{\mu}^2}(\mathbf{V}(n) - \boldsymbol{\mu}(n)) + \cdots, \end{aligned} \tag{7.13}$$

where

$$\begin{aligned} \frac{d^2\boldsymbol{\eta}}{d\boldsymbol{\mu}^2} &= \left\{\frac{d^2\eta_i}{d\boldsymbol{\mu}^2}\right\} = \left\{\left(\frac{d}{d\boldsymbol{\mu}}\right)\left(\frac{d}{d\boldsymbol{\mu}}\right)^T \eta_i\right\} \\ &= \left(\frac{d}{d\boldsymbol{\mu}}\right)\left(\frac{d}{d\boldsymbol{\mu}}\right)^T \boldsymbol{\eta} \end{aligned}$$

is a vector of matrices. Applying the expectation operator to Equation (7.13) we get

$$E\boldsymbol{\eta}[\mathbf{V}(n)] = \boldsymbol{\eta}(\boldsymbol{\mu}(n)) + \mathcal{O}(\boldsymbol{\Sigma}(n))\mathbf{1}, \tag{7.14}$$

where

$$\begin{aligned} \boldsymbol{\Sigma}(n) &= E[(\mathbf{V}(n) - \boldsymbol{\mu}(n))(\mathbf{V}(n) - \boldsymbol{\mu}(n))^T], \\ \mathcal{O}(\boldsymbol{\Sigma}(n)) &= \mathcal{O}\left[\max_{i,j} \Sigma_{ij}(n)\right]. \end{aligned} \tag{7.15}$$

and $\mathbf{1} = [1, 1, \ldots, 1]^T$.* In Section 7.3, we show under the assumptions of boundedness of $d\boldsymbol{\eta}/d\mathbf{v}$ and $\boldsymbol{\Omega}$, and differentiability of $\boldsymbol{\eta}(\mathbf{v})$, that

$$\boldsymbol{\Sigma}(n) = \mathcal{O}(\rho)\mathbf{U} \quad \text{whenever } n\rho \leqslant t < \infty, \tag{7.16}$$

where $\mathbf{U}$ denotes a matrix whose elements all are equal to one. In other words, there exist $C > 0$, $M(t) > 0$ such that $\max_{i,j} |\Sigma_{ij}| < \rho M(t)$ whenever $\rho \leqslant C$ and $n\rho \leqslant t \leqslant \infty$.
Hence

$$E\boldsymbol{\eta}(\mathbf{V}(n)) = \boldsymbol{\eta}(\boldsymbol{\mu}(n)) + \mathcal{O}(\rho)\mathbf{1} \tag{7.17}$$

for $n\rho \leqslant t < \infty$. By Equations (7.11) and (7.17),

$$\Delta\boldsymbol{\mu}(n) = \rho g(n\rho)\boldsymbol{\eta}[\boldsymbol{\mu}(n)] + \mathcal{O}(\rho^2)\mathbf{1}, \qquad n\rho \leqslant t < \infty. \tag{7.18}$$

Hereafter, we let $t = n\rho$. Let $\tilde{\boldsymbol{\mu}}(t)$ and $g(t)$ denote differentiable functions which satisfy

$$\begin{aligned} \tilde{\boldsymbol{\mu}}(n\rho) &= \boldsymbol{\mu}(n), \quad \text{for } n = 0, 1, \ldots \\ g(n) &= \rho_n, \quad \text{for } n = 0, 1, \ldots. \end{aligned} \tag{7.19}$$

By Equation (7.19), Equation (7.18) becomes

$$(\tilde{\boldsymbol{\mu}}(n\rho + \rho) - \tilde{\boldsymbol{\mu}}(n\rho))/\rho = g(n\rho)\,\boldsymbol{\eta}[\tilde{\boldsymbol{\mu}}(n\rho)] + \mathcal{O}(\rho)\mathbf{1}. \tag{7.20}$$

Note that, by a Taylor expansion of $\tilde{\boldsymbol{\mu}}(t)$ about $t = n\rho$,

$$\begin{aligned} (\tilde{\boldsymbol{\mu}}(n\rho + \rho) - \tilde{\boldsymbol{\mu}}(n\rho))/\rho &= \tilde{\boldsymbol{\mu}}'(n\rho) + (\rho/2)\tilde{\boldsymbol{\mu}}''(n\rho) + \cdots \\ &= \tilde{\boldsymbol{\mu}}'(n\rho) + \mathcal{O}(\rho)\mathbf{1}, \end{aligned} \tag{7.21}$$

where $\tilde{\boldsymbol{\mu}}'(t)$ and $\tilde{\boldsymbol{\mu}}''(t)$ denote the first and second derivatives of $\tilde{\boldsymbol{\mu}}(t)$ with respect to t. By Equations (7.20) and (7.21)

$$\begin{aligned} \tilde{\boldsymbol{\mu}}'(t) &= g(t)\boldsymbol{\eta}[\tilde{\boldsymbol{\mu}}(t)] + \mathcal{O}(\rho)\mathbf{1}, \quad \text{for } t = n\rho \\ \tilde{\boldsymbol{\mu}}(0) &= \mathbf{V}(0). \end{aligned} \tag{7.22}$$

Let $\hat{\boldsymbol{\mu}}(t)$ denote the solution to

$$\hat{\boldsymbol{\mu}}'(t) = g(t)\boldsymbol{\eta}[\hat{\boldsymbol{\mu}}(t)], \qquad \hat{\boldsymbol{\mu}}(0) = \mathbf{V}(0). \tag{7.23}$$

* A positive sequence $\{b_k\}$ is said to be *of the order of a positive sequence* $\{c_k\}$, i.e., $b_k = \mathcal{O}(c_k)$, if for some K there exists a constant M, M independent of K, such that $b_k/c_k < M$ for all $k < K$.

Comparing Equations (7.22) and (7.23), it is plausible that

$$\tilde{\boldsymbol{\mu}}(t) \cong \hat{\boldsymbol{\mu}}(t) \quad \text{for small } \rho, \tag{7.24}$$

from which we obtain

$$\boldsymbol{\mu}(n) \cong \hat{\boldsymbol{\mu}}(n\rho) \quad \text{for small } \rho. \tag{7.25}$$

In fact, a detailed analysis [1, 3] has shown that

$$\tilde{\boldsymbol{\mu}}(t) = \hat{\boldsymbol{\mu}}(t) + \mathcal{O}(\rho)\mathbf{1} \tag{7.26}$$

under certain weak constraints on $g(t)$, $\boldsymbol{\eta}(\mathbf{v})$, and $\boldsymbol{\Omega}(\mathbf{v})$. The following theorem states this result precisely.

Theorem 7.1 (Centroid Theorem). *Suppose the training procedure of a stationary-input pattern classifier is of the form*

$$\mathbf{V}(n+1) = \mathbf{V}(n) + \rho g(n)\mathbf{Z}(\Gamma(n), \mathbf{Y}(n)), \tag{7.27}$$

where $\mathbf{Y}(n)$ *and* $\mathbf{V}(n)$ *are the augmented feature vector and the augmented weight vector in* $(d+1)$*-space at trial* n. *Suppose* $\boldsymbol{\eta}(\mathbf{v})$, $d\boldsymbol{\eta}/d\mathbf{v}$, $d^2\boldsymbol{\eta}/d\mathbf{v}^2$, *and* $\boldsymbol{\Omega}(\mathbf{v})$ *are bounded, where we define* $d\boldsymbol{\eta}/d\mathbf{v}$ *as the matrix*

$$d\boldsymbol{\eta}/d\mathbf{v} = \{\partial\eta_i/\partial v_j\}. \tag{7.28}$$

Suppose $g(t)$ *is a nonincreasing function of* t. *Then*:

(a)
$$\boldsymbol{\mu}(n) = \hat{\boldsymbol{\mu}}(n\rho) + \mathcal{O}(\rho)\mathbf{1}, \tag{7.29}$$

i.e., there exist a bounded function $M(t)$ *and a positive constant* ρ_0 *such that*

$$\|\hat{\boldsymbol{\mu}}(n\rho) - \boldsymbol{\mu}(n)\| < \rho M(n\rho) \tag{7.30}$$

whenever $|\rho| < \rho_0$, *and* $n\rho \leqslant t < \infty$.

(b) $\hat{\boldsymbol{\mu}}(t)$ *satisfies*

$$d\hat{\boldsymbol{\mu}}(t)/dt = g(t)\boldsymbol{\eta}[\hat{\boldsymbol{\mu}}(t)] \tag{7.31}$$

with the initial condition

$$\hat{\boldsymbol{\mu}}(0) = \mathbf{V}(0) = \textit{initial augmented weight vector}. \tag{7.32}$$

We refer to Equation (7.31) as the continuous-state *centroid equation.* Note that is is equivalent to Equation (4.14) for a continuous gradient descent. Equation (7.31), together with Equation (7.29), states that the vector velocity of the centroid of $\mathbf{V}(n)$ is approximately proportional to the expectation of the change in $\mathbf{V}(n)$.

Equation (7.31) is solved much more easily than Equation (7.11) since $\boldsymbol{\eta}(\mathbf{v})$ can be determined from the constituent densities and the training procedure.

7.3 Proof that $\boldsymbol{\Sigma}(n) = \mathcal{O}(\rho)\mathbf{U}$ for $n\rho \leqslant t < \infty$*.

When we say that $\boldsymbol{\Sigma}(n) = \mathcal{O}(\rho)\mathbf{U}$ for $n\rho \leqslant t < \infty$, we mean that there exist a positive constant C and a positive function $M(t)$ such that $\max_{i,j} |\Sigma_{ij}(n)| < \rho M(t)$ whenever $\rho \geqslant C$ and $n\rho \leqslant t < \infty$. This is a mathematical statement of the property that the variance of the weight vector approaches zero at least as rapidly as ρ when ρ approaches zero while maintaining $n\rho = t$.

In this section we prove that $\boldsymbol{\Sigma}(n) = \mathcal{O}(\rho)\mathbf{U}$ for $n\rho \leqslant t < \infty$ under the assumptions that $\boldsymbol{\eta}(\mathbf{v})$ is differentiable, that $\Omega(\mathbf{v})$ and $d\boldsymbol{\eta}(\mathbf{v})/d\mathbf{v}$ are bounded, and that $\rho_0 \geqslant \rho_n$ for all n.

First we define some new notation:

$$\operatorname{Var} Y = E[(Y - EY)^2], \tag{7.33}$$

$$\operatorname{Cov}(X, Y) = E[(X - EX)(Y - EY)], \tag{7.34}$$

$$\boldsymbol{\Sigma}(n) = \{\Sigma_{ij}(n)\}. \tag{7.35}$$

For convenience we will often omit the explicit display of the argument n when such omission causes no confusion; e.g., $\Sigma_{ii} \equiv \Sigma_{ii}(n)$, $\boldsymbol{\mu} \equiv \boldsymbol{\mu}(n)$, etc.

By Schwartz's inequality, $\Sigma_{ii}\Sigma_{jj} \geqslant \Sigma_{ij}^2$, so that we need only show that $\Sigma_{ii} = \mathcal{O}(\rho)$ for all i. Recalling the definition of $\boldsymbol{\Sigma}(n)$ in Equation (7.4) note that

$$\begin{aligned}\Sigma_{ii}(n+1) &= \operatorname{Var}[V_i(n) + \Delta V_i(n)] \\ &= \Sigma_{ii}(n) + 2\operatorname{Cov}[V_i(n), \Delta V_i(n)] + \operatorname{Var}[\Delta V_i(n)].\end{aligned} \tag{7.36}$$

Since

$$E\left(\frac{\Delta V_i}{\rho_n}\right) = E\left\{E\left(\frac{\Delta V_i}{\rho_n}\middle|\mathbf{V}\right)\right\} = E\eta_i(\mathbf{V}), \tag{7.37}$$

$$\begin{aligned}\operatorname{Var}\Delta V_i &= \rho_n^2 E\left[E\left\{\left[\frac{\Delta V_i}{\rho_n} - E\eta_i(\mathbf{V})\right]^2\middle|\mathbf{V}\right\}\right] \\ &= \rho_n^2 E\left[E\left\{\left[\frac{\Delta V_i}{\rho_n} - \eta_i(\mathbf{V}) + \eta_i(\mathbf{V}) - E\eta_i(\mathbf{V})\right]^2\middle|\mathbf{V}\right\}\right] \\ &= \rho_n^2 E\left[E\left[\left[\frac{\Delta V_i}{\rho_n} - \eta_i(\mathbf{V})\right]^2\middle|\mathbf{V}\right] + E\left\{\left[\eta_i(\mathbf{V}) - E\eta_i(\mathbf{V})\right]^2\middle|\mathbf{V}\right\} \\ &\leqslant \rho_n^2 E[\Omega_{ii}(\mathbf{V}) + E\{[\eta_i(\mathbf{V}) - \eta_i(\boldsymbol{\mu})]^2|\mathbf{V}\}]\end{aligned} \tag{7.38}$$

from the property

$$E[(Y - E(Y))^2] \leqslant E[(Y - a)^2] \tag{7.39}$$

* This section may be omitted on a first reading of this chapter.

for any random variable Y and any constant a. Hence

$$\begin{aligned}\operatorname{Var} \Delta V_i &\leqslant \rho_n^2[\Omega_{ii}(\mathbf{V})] + \rho_n^2 E\left\{(V_i - \mu_i)^2 \left(\frac{\eta_i(\mathbf{V}) - \eta_i(\boldsymbol{\mu})}{V_i - \mu_i}\right)^2 \middle| \mathbf{V}\right\} \\ &\leqslant \rho_n^2 \alpha_{ii} + \rho_n^2 \Sigma_{ii} \gamma_{ii}^2, \end{aligned} \tag{7.40}$$

where we define

$$\alpha_{ij} = \sup |\Omega_{ij}(\mathbf{v})| \tag{7.41}$$

$$\gamma_{ij} = \sup \left| \frac{\partial \eta_i(\mathbf{v})}{\partial v_j} \right|. \tag{7.42}$$

Thus

$$\begin{aligned}\operatorname{Cov}[V_i, \Delta V_i] &= E[(V_i - \mu_i)(\Delta V_i - \Delta \mu_i)] \\ &= E[(V_i - \mu_i)(\Delta V_i - \rho_n \eta_i(\boldsymbol{\mu})]\end{aligned} \tag{7.43}$$

since $\Delta\mu_i$ is a deterministic function of n, and $E[(V_i - \mu_i)q_n] = 0$ whenever q_n is a deterministic function of n. Thus,

$$\begin{aligned}\operatorname{Cov}[V_i, \Delta V_i] &= E\{E[(V_i - \mu_i)(\Delta V_i - \rho_n \eta_i(\boldsymbol{\mu}))|\mathbf{V}]\} \\ &= E\{(V_i - \mu_i)E[\Delta V_i - \rho_n \eta_i(\boldsymbol{\mu})|\mathbf{V}]\} \\ &= \rho_n E\{(V_i - \mu_i)[\eta_i(\mathbf{V}) - \eta_i(\boldsymbol{\mu})]\} \\ &= \rho_n E\left\{(V_i - \mu_i)^2 \left[\frac{\eta_i(\mathbf{V}) - \eta_i(\boldsymbol{\mu})}{V_i - \mu_i}\right]\right\} \\ &\leqslant \rho_n \Sigma_{ii} \gamma_{ii}.\end{aligned} \tag{7.44}$$

Putting the results of Equations (7.40) and (7.44) into Equation (7.36), we obtain

$$\Sigma_{ii}(n+1) \leqslant (1 + \rho_n \gamma_{ii})^2 \Sigma_{ii}(n) + \rho_n^2 \alpha_{ii}. \tag{7.45}$$

Replacing ρ_n by $\rho g(n\rho)$ and recalling that $g(t)$ is nonincreasing, we have

$$\Sigma_{ii}(n+1) = l_i \Sigma_{ii}(n) + k_i, \tag{7.46}$$

where

$$l_i = (1 + \rho g(0)\gamma_{ii})^2 > 1 \tag{7.47}$$

$$k_i = \rho^2 g^2(0) \alpha_{ii} > 0. \tag{7.48}$$

Noting that $\Sigma_{ii}(0) = 0$, and using Equation (7.46) to find successively the expressions for $\Sigma_{ii}(1)$, $\Sigma_{ii}(2), \ldots$, we arrive at the following solution of the above difference equation:

$$\Sigma_{ii}(n) \leqslant \frac{l_i^n - 1}{l_i - 1} k_i. \tag{7.49}$$

Using the property

$$\begin{aligned} l_i^n = (1 + g(0)\gamma_{ii}\rho)^{2n} &= \sum_{j=0}^{2n} \binom{2n}{j} (g(0)\gamma_{ii}\rho)^j \\ &\leqslant \sum_{j=0}^{\infty} \frac{(2n)^j}{j!} (g(0)\gamma_{ii}\rho)^j = \exp 2g(0)\gamma_{ii} n\rho, \end{aligned} \tag{7.50}$$

we obtain, finally,

$$\Sigma_{ii}(n) \leqslant \left(\frac{\exp(2g(0)\gamma_{ii}t) - 1}{2\rho g(0)\gamma_{ii} + \rho^2 g(0)\gamma_{ii}}\right)\rho^2 g^2(0)\alpha_{ii} \\ = \mathcal{O}(\rho) \quad \text{for } n\rho = t < \infty. \tag{7.51}$$

7.4 The Covariance Equation

In this section we show how the continuous-state model leads to an approximate differential equation for the covariance matrix $\boldsymbol{\Sigma}(n)$. Specifically we show that, for sufficiently small values of ρ,

$$\boldsymbol{\Sigma}(n) \cong \rho g(n\rho)\mathbf{R}(n\rho), \tag{7.52}$$

where $\mathbf{R}(t)$ is a matrix satisfying

$$\frac{d\mathbf{R}}{dt} = g(t)\left[\frac{d\boldsymbol{\eta}}{d\boldsymbol{\mu}}\mathbf{R}(t) + \mathbf{R}(t)\left(\frac{d\boldsymbol{\eta}}{d\boldsymbol{\mu}}\right)^T + \boldsymbol{\Omega}(\boldsymbol{\mu}(t))\right]. \tag{7.53}$$

We also show that $\mathbf{V}(n)$ asymptotically approaches a normal distribution with mean $\boldsymbol{\mu}(n)$ and covariance matrix $\boldsymbol{\Sigma}(n)$ as $\rho \to 0$, $n\rho = t$. We refer to $\mathbf{R}(t)$ as the *relative covariance matrix.*

In Equation (7.52), as well as Equation (7.15), we see that $\boldsymbol{\Sigma}(n) \to \mathbf{0}$ as $\rho \to 0$, while in Equations (7.29) and (7.31) we see that $\boldsymbol{\mu}(n)$ can remain nonzero (and usually does so) as $\rho \to 0$. In the theory of distributions [4] it is shown that if a sequence of functions of $\boldsymbol{\xi}$ say $\{f_n(\boldsymbol{\xi})\}$, asymptotically approaches a sequence of a Gaussian function $\{g_n(\boldsymbol{\xi})\}$ about a mean $\{\boldsymbol{\mu}\}$, if $\int f_n(\boldsymbol{\xi})d\boldsymbol{\xi} = 1$ for all n, and if the covariance matrix associated with $g_n(\boldsymbol{\xi})$ approaches zero as $n \to \infty$, then $f_n(\boldsymbol{\xi})$ approaches a Dirac delta function, $\delta(\boldsymbol{\xi} - \boldsymbol{\mu})$, as $n \to \infty$; i.e.,

$$f_n(\boldsymbol{\xi}) \xrightarrow[n\to\infty]{} \delta(\boldsymbol{\xi} - \boldsymbol{\mu}), \tag{7.54}$$

where $\delta(\boldsymbol{\xi} - \boldsymbol{\mu})$ has the following properties:

(a) $\int d\boldsymbol{\xi}\, h(\boldsymbol{\xi})\,\delta(\boldsymbol{\xi} - \boldsymbol{\mu}) = h(\boldsymbol{\mu})$ for any bounded function $h(\boldsymbol{\xi})$ that is continuous at $\boldsymbol{\xi} = \boldsymbol{\mu}$, and
(b) $\delta(\boldsymbol{\xi} - \boldsymbol{\mu}) = 0$ for $\boldsymbol{\xi} \neq \boldsymbol{\mu}$.

The notation $\int d\boldsymbol{\xi}$ denotes the multiple integral $\int_{-\infty}^{\infty} d\xi_1 \cdots \int_{-\infty}^{\infty} d\xi_m$ for an m-dimensional vector $\boldsymbol{\xi}$.

Now consider the density function $p_{\mathbf{V}(n)}(\mathbf{v})$ of the random vector $\mathbf{V}(n)$. Since $\mathbf{V}(n)$ is asymptotically normal with a mean of $\boldsymbol{\mu}(n)$ (which we demonstrate below) and since $\boldsymbol{\Sigma}(n) \to 0$ as $\rho \to 0$ while $n\rho = t$, it follows that $p_{\mathbf{V}(n)}(\mathbf{v}) \to \delta(\mathbf{v} - \boldsymbol{\mu}(t))$ as $\rho \to 0$ while $n\rho = t$. We will make use of this property later in our derivation of learning curves and variance curves for small-step multiple-feature classifiers.

To show that $\mathbf{V}(n)$ is asymptotically normal, define

$$\left.\begin{aligned}\hat{\mathbf{K}}(n) &= g(n)^{-1/2}[\mathbf{V}(n) - \boldsymbol{\mu}(n)]\\ \mathbf{K}(n\rho) &= [\rho g(n\rho)]^{-1/2}[\mathbf{V}(n) - \boldsymbol{\mu}(n)].\end{aligned}\right\} \tag{7.55}$$

Let $\Phi_n(\boldsymbol{\omega})$ and $\Phi(\boldsymbol{\omega}, n\rho)$ denote the characteristic functions of $\mathbf{K}(n)$ and $\mathbf{K}(n\rho)$, respectively, i.e.,

$$\left.\begin{aligned}\Phi_n(\boldsymbol{\omega}) &= E[\exp i\boldsymbol{\omega}^T\hat{\mathbf{K}}(n)]\\ \Phi(\boldsymbol{\omega}, n\rho) &= E[\exp i\boldsymbol{\omega}^T\mathbf{K}(n\rho)],\end{aligned}\right\} \tag{7.56}$$

where E is the expectation operator, $\boldsymbol{\omega}$ is a vector of complex angular frequency, and $i = \sqrt{-1}$. Note that $\Phi_n(\boldsymbol{\omega}) = \Phi(\boldsymbol{\omega}, n)$. We will show that $\Phi(\boldsymbol{\omega}, n\rho)$ approaches the characteristic function of a normal density as $n \to \infty$, $n\rho = t$. (For a rigorous discussion of asymptotic normality of sums of dependent random vectors, see Rosen [6].) Hereafter, we will often omit the argument $n\rho$; e.g., we will use $\mathbf{K}$ in place of $\mathbf{K}(n\rho)$.

From Equation (7.56),

$$\begin{aligned}\Phi(\boldsymbol{\omega}, n\rho) &= E[\exp i\boldsymbol{\omega}^T(\mathbf{K} + \Delta\mathbf{K})]\\ &= E[(\exp i\boldsymbol{\omega}^T\mathbf{K})(\exp i\boldsymbol{\omega}^T\,\Delta\mathbf{K})]\\ &= E\{E[(\exp i\boldsymbol{\omega}^T\mathbf{K})(\exp i\boldsymbol{\omega}^T\,\Delta\mathbf{K})\,|\,\mathbf{V}]\}\\ &= E\{(\exp i\boldsymbol{\omega}^T\mathbf{K})E[\exp i\boldsymbol{\omega}^T\,\Delta\mathbf{K}\,|\,\mathbf{V}]\}.\end{aligned} \tag{7.57}$$

since $\mathbf{K}(n\rho)$ is a deterministic function of $\mathbf{V}(n\rho)$.

Expanding $\exp i\boldsymbol{\omega}^T\,\Delta\mathbf{K}$ in a Taylor series, we have

$$\exp i\boldsymbol{\omega}^T\,\Delta\mathbf{K} = 1 + i\boldsymbol{\omega}^T\,\Delta\mathbf{K} - \tfrac{1}{2}\boldsymbol{\omega}^T\,\Delta\mathbf{K}\Delta\mathbf{K}^T\boldsymbol{\omega} - \frac{i}{6}(\boldsymbol{\omega}^T\,\Delta\mathbf{K})^3 + \cdots \tag{7.58}$$

Hence

$$\begin{aligned}E\{\exp i\boldsymbol{\omega}^T\,\Delta\mathbf{K}\,|\,\mathbf{V}\} = 1 &+ i\boldsymbol{\omega}^T E(\Delta\mathbf{K}\,|\,\mathbf{V}) - \tfrac{1}{2}\boldsymbol{\omega}^T E(\Delta\mathbf{K}\,\Delta\mathbf{K}^T\,|\,\mathbf{V})\boldsymbol{\omega}\\ &- \frac{i}{6}E[(\boldsymbol{\omega}^T\,\Delta\mathbf{K})^3\,|\,\mathbf{V}] + \cdots\end{aligned} \tag{7.59}$$

$$E(\Delta\mathbf{K}\,|\,\mathbf{V}) = (\rho g)^{-1/2}E(\Delta\mathbf{V} - \Delta\boldsymbol{\mu}\,|\,\mathbf{V}) \tag{7.60}$$

$$E(\Delta\mathbf{V}\,|\,\mathbf{V}) = \rho g\boldsymbol{\eta}(\mathbf{V}) \tag{7.61}$$

$$\begin{aligned}E(\Delta\boldsymbol{\mu}\,|\,\mathbf{V}) &= E(\Delta\boldsymbol{\mu}) = \Delta\boldsymbol{\mu} = \Delta E(\mathbf{V})\\ &= E\Delta\mathbf{V} = E[E(\Delta\mathbf{V}\,|\,\mathbf{V})]\\ &= \rho g E\boldsymbol{\eta}(\mathbf{V}).\end{aligned} \tag{7.62}$$

Hence

$$E(\Delta\mathbf{K}\,|\,\mathbf{V}) = (\rho g)^{1/2}[\boldsymbol{\eta}(\mathbf{V}) - E\boldsymbol{\eta}(\mathbf{V})]. \tag{7.63}$$

By a Taylor series expansion of $\boldsymbol{\eta}(\mathbf{V})$ about $\boldsymbol{\mu}$, we obtain

$$\boldsymbol{\eta}(\mathbf{V}) = \boldsymbol{\eta}(\boldsymbol{\mu}) + \frac{d\boldsymbol{\eta}}{d\boldsymbol{\mu}}(\mathbf{V} - \boldsymbol{\mu}) + \hat{\mathcal{O}}(\rho)\mathbf{1}, \tag{7.64}$$

where $d\boldsymbol{\eta}/d\boldsymbol{\mu}$ is defined as the matrix

$$\frac{d\boldsymbol{\eta}}{d\boldsymbol{\mu}} = \left\{\frac{d\eta_i}{d\mu_j}\right\},$$

and where $\hat{\mathcal{O}}(\rho)$ = a random vector such that

$$E(\hat{\mathcal{O}}(\rho)) = \mathcal{O}(\rho). \tag{7.65}$$

The term $\hat{\mathcal{O}}(\rho)$ follows from our earlier observation that $\boldsymbol{\Sigma}(n) = \mathcal{O}(\rho)\mathbf{U}$. Taking the expectation of Equation (7.64) we get

$$E\boldsymbol{\eta}(\mathbf{V}) = \boldsymbol{\eta}(\boldsymbol{\mu}) + \mathcal{O}(\rho)\mathbf{1} \tag{7.66}$$

where $\mathbf{1} = [1, \ldots, 1]^T$. Hence

$$\begin{aligned}\boldsymbol{\eta}(\mathbf{V}) - E\boldsymbol{\eta}(\mathbf{V}) &= \frac{d\boldsymbol{\eta}}{d\boldsymbol{\mu}}(\mathbf{V} - \boldsymbol{\mu}) + \hat{\mathcal{O}}(\rho)\mathbf{1} \\ &= (\rho g)^{1/2}\frac{d\boldsymbol{\eta}}{d\boldsymbol{\mu}}\mathbf{K} + \hat{\mathcal{O}}(\rho)\mathbf{1}.\end{aligned} \tag{7.67}$$

By Equations (7.63) and (7.67),

$$E(\Delta\mathbf{K}\,|\,\mathbf{V}) = (\rho g)^{1/2}\left[(\rho g)^{1/2}\frac{d\boldsymbol{\eta}}{d\boldsymbol{\mu}}\mathbf{K} + \hat{\mathcal{O}}(\rho)\mathbf{1}\right] = (\rho g)\frac{d\boldsymbol{\eta}}{d\boldsymbol{\mu}}\mathbf{K} + \hat{\mathcal{O}}(\rho^{3/2})\mathbf{1} \tag{7.68}$$

$$\begin{aligned}\Phi(\omega, (n+1)\rho) = E\Bigg\{&(\exp i\omega^T\mathbf{K})\Bigg[1 + i\omega^T\rho g\frac{d\boldsymbol{\eta}}{d\boldsymbol{\mu}}\mathbf{K} + i\omega^T\hat{\mathcal{O}}(\rho^{3/2})\mathbf{1} \\ &- \tfrac{1}{2}\,\omega^T E(\Delta\mathbf{K}\,\Delta\mathbf{K}^T\,|\,\mathbf{V})\omega - \frac{i}{6}E[(\omega^T\Delta\mathbf{K})^3\,|\,\mathbf{V}] + \cdots\Bigg]\Bigg\}\end{aligned} \tag{7.69}$$

$$\begin{aligned}E(\Delta\mathbf{K}\,\Delta\mathbf{K}^T\,|\,\mathbf{V}) &= (\rho g)^{-1}E[(\Delta\mathbf{V} - \Delta\boldsymbol{\mu})(\Delta\mathbf{V} - \Delta\boldsymbol{\mu})^T\,|\,\mathbf{V}] \\ &= \rho g E\left[\left(\frac{\Delta\mathbf{V}}{\rho g} - \frac{\Delta\boldsymbol{\mu}}{\rho g}\right)\left(\frac{\Delta\mathbf{V}}{\rho g} - \frac{\Delta\boldsymbol{\mu}}{\rho g}\right)^T\,\middle|\,\mathbf{V}\right].\end{aligned}$$

By Equation (7.62),

$$E(\Delta\mathbf{K}\,\Delta\mathbf{K}^T\,|\,\mathbf{V}) = \rho g E\left[\left(\frac{\Delta\mathbf{V}}{\rho g} - E\boldsymbol{\eta}(\mathbf{V})\right)\left(\frac{\Delta\mathbf{V}}{\rho g} - E\boldsymbol{\eta}(\mathbf{V})\right)^T\,\middle|\,\mathbf{V}\right]$$

and by Equation (7.66),

$$\begin{aligned}E(\Delta\mathbf{K}\,\Delta\mathbf{K}^T\,|\,\mathbf{V}) &= \rho g E\left[\left(\frac{\Delta\mathbf{V}}{\rho g} - \boldsymbol{\eta}(\boldsymbol{\mu}) + \mathcal{O}(\rho)\mathbf{1}\right)\left(\frac{\Delta\mathbf{V}}{\rho g} - \boldsymbol{\eta}(\boldsymbol{\mu}) + \mathcal{O}(\rho)\mathbf{1}\right)^T\,\middle|\,\mathbf{V}\right] \\ &= \rho g[\boldsymbol{\Omega}(\boldsymbol{\mu}) + \mathcal{O}(\rho)\mathbf{U}] = \rho g\boldsymbol{\Omega}(\boldsymbol{\mu}) + \mathcal{O}(\rho^2)\mathbf{U}.\end{aligned} \tag{7.70}$$

Substituting Equation (7.70) into Equation (7.69) we obtain

$$\begin{aligned}\Phi(\omega, (n+1)\rho) = E\Bigg\{&(\exp i\omega^T\mathbf{K})\Bigg[1 + \rho g i\omega^T\frac{d\boldsymbol{\eta}}{d\boldsymbol{\mu}}\mathbf{K} - \frac{\rho g}{2}\omega^T\boldsymbol{\Omega}(\boldsymbol{\mu})\omega \\ &+ i\omega^T\hat{\mathcal{O}}(\rho^{3/2})\mathbf{1} + \omega^T\mathcal{O}(\rho^2)\mathbf{U}\omega + \cdots\Bigg]\Bigg\}.\end{aligned} \tag{7.71}$$

By Equation (7.65),

$$\Phi(\omega,(n+1)\rho) = \Phi(\omega,n\rho) + \rho g i\omega^T \frac{d\eta}{d\mu} E\Big(\mathbf{K}\exp i\omega^T\mathbf{K}\Big) - \frac{\rho g}{2}\omega^T\boldsymbol{\Omega}(\mu)\omega\Phi(\omega,n\rho) + \mathcal{O}(\rho^{3/2}). \tag{7.72}$$

Therefore,

$$[\Phi(\omega,(n+1)\rho) - \Phi(\omega,n\rho)]/\rho = gi\omega^T \frac{d\eta}{d\mu} E(\mathbf{K}\exp i\omega^T\mathbf{K}) - \frac{g}{2}\omega^T\boldsymbol{\Omega}(\mu)\omega\Phi(\omega,n\rho) + \mathcal{O}(\rho^{1/2}). \tag{7.73}$$

Let $t = n\rho$. Then Equation (7.73) becomes

$$(\Phi(\omega,t+\rho) - \Phi(\omega,t))/\rho = gi\omega^T \frac{d\eta}{d\mu} E(\mathbf{K}\exp i\omega^T\mathbf{K}) - \frac{g}{2}\omega^T\boldsymbol{\Omega}(\mu)\omega\Phi(\omega,t) + \mathcal{O}(\rho^{1/2}). \tag{7.74}$$

where g denotes $g(t)$, and $\mathbf{K}$ denotes $\mathbf{K}(t)$. Note that

$$\frac{\partial}{\partial\omega}E(\exp i\omega^T\mathbf{K}) = E\left[\frac{\partial}{\partial\omega}\exp i\omega^T\mathbf{K}\right] = iE(\mathbf{K}\exp i\omega^T\mathbf{K}).$$

Hence

$$\frac{\partial}{\partial\omega}\Phi(\omega,t) = iE(\mathbf{K}\exp i\omega^T\mathbf{K}). \tag{7.75}$$

Substituting Equation (7.75) into Equation (7.74), we have

$$(\Phi(\omega,t+\rho) - \Phi(\omega,t))/\rho = g\omega^T\frac{d\eta}{d\mu}\frac{\partial\Phi}{\partial\omega} - \frac{g}{2}\omega^T\boldsymbol{\Omega}(\mu)\omega\Phi + \mathcal{O}(\rho^{1/2}). \tag{7.76}$$

Letting $\rho \to 0$ with $n\rho = t$, Equation (7.76) becomes

$$\frac{\partial\Phi}{\partial t} = g(t)\left[\omega^T\frac{d\eta}{d\mu}\frac{\partial\Phi}{\partial\omega} - \tfrac{1}{2}\omega^T\boldsymbol{\Omega}(\mu)\omega\Phi\right]. \tag{7.77}$$

The solution to Equation (7.77) is [1]

$$\Phi(\omega,t) = \exp(-\tfrac{1}{2}\omega^T\mathbf{R}(t)\omega), \tag{7.78}$$

where $\mathbf{R}(t)$ is a matrix satisfying

$$\frac{d\mathbf{R}}{dt} = g(t)\left[\frac{d\eta}{d\mu}\mathbf{R}(t) + \mathbf{R}(t)\left(\frac{d\eta}{d\mu}\right)^T + \boldsymbol{\Omega}(\mu(t))\right]. \tag{7.79}$$

We refer to Equation (7.79) as the *covariance equation.*

Thus $\mathbf{K}(n\rho)$ is normally distributed with a mean of $\mathbf{0}$ and a covariance matrix of $\mathbf{R}(t)$ as $\rho \to 0$, $n\rho = t$. Hence, $\mathbf{V}(n)$ is normally distributed with a covariance matrix $\mathbf{\Sigma}(n)$ given by

$$\mathbf{\Sigma}(n) \cong \rho g(n\rho)\mathbf{R}(n\rho). \tag{7.80}$$

7.5 Learning Curves and Variance Curves

As indicated in Equation (7.80), small values of ρ result in small values of the covariance matrix of $\mathbf{V}(n)$. Furthermore, for small ρ, $\mathbf{V}(n)$ is approximately normally distributed with a mean of approximately $\hat{\boldsymbol{\mu}}(n\rho)$. Consequently, the probability function $p_{\mathbf{V}(n)}(\mathbf{v})$ may be approximated by a Dirac delta function, $\delta(\mathbf{v} - \hat{\boldsymbol{\mu}}(n\rho))$. This result yields relatively simple formulas for the learning curves and variance curves of the training processes.

The learning curve was defined in Section 6.5 as the curve of success probability as a function of n. For multifeature training processes, this function is given by

$$z(n) = E\{S[\mathbf{V}(n)]\} \tag{7.81}$$

where

$$S(\mathbf{v}) = P[\Omega(n) = R(n) \mid \mathbf{V}(n) = \mathbf{v}] \tag{7.82}$$

is the conditional success probability. Equations (7.81) and (7.82) are extensions of the single-feature formulas, Equations (6.5) and (6.6). It follows that

$$z(n) = \int S(\mathbf{v}) p_{\mathbf{V}(n)}(\mathbf{v})\, d\mathbf{v}, \tag{7.83}$$

where

$$\int d\mathbf{v} \quad \text{denotes} \quad \int_{-\infty}^{\infty} dv_0 \cdots \int_{-\infty}^{\infty} dv_d. \tag{7.84}$$

Let $\hat{z}(t) = \lim z(n)$ as $\rho \to 0$ with $n\rho = t$. Since $p_{\mathbf{V}(n)}(\mathbf{v}) \to \delta(\mathbf{v} - \hat{\boldsymbol{\mu}}(t))$ as $\rho \to 0$ with $n\rho = t$, it follows that

$$\hat{z}(t) = \int S(\mathbf{v})\, \delta(\mathbf{v} - \hat{\boldsymbol{\mu}}(t))\, d\mathbf{v} = S[\hat{\boldsymbol{\mu}}(t)]. \tag{7.85}$$

Hence

$$z(n) \cong \hat{z}(n\rho) = S[\hat{\boldsymbol{\mu}}(n\rho)]. \tag{7.86}$$

This equation gives us a continuous-state approximation for learning curves.

Again extending the definitions of Section 6.5 to multifeature training processes, the variance of the conditional success probability at trial n is defined by

$$\begin{aligned} q(n) &= E\{[S(\mathbf{V}(n)) - z(n)]^2\} \\ &\cong S^2[\hat{\boldsymbol{\mu}}(n\rho)] - z^2(n), \quad \text{when } q(n) \text{ is not too close to zero.} \end{aligned} \tag{7.87}$$

It can be shown that $q(n) \to 0$ as $\rho \to 0$ with $n\rho = t$. (See Exercise 7.1.) But 0 is generally not an acceptable estimate of $q(n)$, since it is useful to know how $q(n)$ varies as a function of n. Furthermore $q(n)$ may contribute significantly to a performance measure of the training process at trial n. To obtain this estimate, let

$$\boldsymbol{\delta}_n = \mathbf{V}(n) - \boldsymbol{\mu}(n) \tag{7.88}$$

and, as previously defined, let

$$\nabla = \left[\frac{\partial}{\partial v_i}\right] \tag{7.89}$$

be a column vector operator. Thus $\nabla S(\mathbf{v}) = [\partial S/\partial v_i]$ and $\nabla S(\boldsymbol{\mu}) = \{\partial S/\partial v_i | \mathbf{v} = \boldsymbol{\mu}\}$. Note that

$$S[\mathbf{V}(n)] \cong S[\boldsymbol{\mu}(n)] + \boldsymbol{\delta}_n^T \nabla S[\boldsymbol{\mu}(n)] \tag{7.90}$$

by a Taylor series expansion. Hence

$$z(n) = ES[\mathbf{V}(n)] \cong S[\boldsymbol{\mu}(n)], \tag{7.91}$$

since $E\boldsymbol{\delta}_n = \mathbf{0}$. Hence

$$S[\mathbf{V}(n)] - z(n) \cong \boldsymbol{\delta}_n^T \nabla S[\boldsymbol{\mu}(n)]. \tag{7.92}$$

Hence, by Equations (7.87) and (7.92),

$$q(n) = (\nabla S[\boldsymbol{\mu}(n)])^T E(\boldsymbol{\delta}_n^T \boldsymbol{\delta}_n) \nabla S[\boldsymbol{\mu}(n)] + \cdots. \tag{7.93}$$

Let

$$\hat{q}(t) = \lim_{\rho \to 0} \frac{q(n\rho)}{\rho}, \qquad n\rho = t. \tag{7.94}$$

Now let $\rho \to 0$ with $n\rho = t$ in Equation (7.93). By Equations (7.80) and (7.93),

$$\hat{q}(t) = (\nabla S[\hat{\boldsymbol{\mu}}(t)])^T g(t)\mathbf{R}(t)\nabla S[\hat{\boldsymbol{\mu}}(t)] \tag{7.95}$$

and

$$q(n) \cong \rho\hat{q}(n\rho) = \rho g(n\rho)(\nabla S[\hat{\boldsymbol{\mu}}(n\rho)])^T \mathbf{R}(n\rho)\nabla S[\hat{\boldsymbol{\mu}}(n\rho)]. \tag{7.96}$$

This equation gives us a continuous-state approximation for variance curves.

7.6 Normalization with Respect to t

It is possible and often convenient to simplify the description of the learning dynamics of continuous-state Markov training processes by distorting the t-axis in such a way that $g(t)$ is eliminated from the centroid and variance equations.

Let $a(t)$ denote the distortion of the t-axis obtained by

$$\tau = a(t) = \int_0^t g(u)\, du. \tag{7.97}$$

Since $g(t) > 0$ for all finite t, $a(t)$ is monotonic increasing. Hence there exists an inverse function $a^{-1}(\tau)$ which is also monotonic increasing. One can show

(Exercise 7.4) that, in terms of τ, Equations (7.31) and (7.79) are transformed into

$$\frac{d\hat{\mathbf{m}}(\tau)}{d\tau} = \boldsymbol{\eta}[\hat{\mathbf{m}}(\tau)] \tag{7.98}$$

and

$$\frac{d\hat{\boldsymbol{\rho}}(\tau)}{dt} = \boldsymbol{\Omega}[\hat{\mathbf{m}}(\tau)] + 2\frac{d\boldsymbol{\eta}[\hat{\mathbf{m}}(\tau)]}{d\mathbf{v}}\hat{\boldsymbol{\rho}}(\tau), \tag{7.99}$$

where

$$\hat{\mathbf{m}}(\tau) = \hat{\boldsymbol{\mu}}[a^{-1}(\tau)], \qquad \hat{\boldsymbol{\rho}}(\tau) = \mathbf{R}[a^{-1}(\tau)]. \tag{7.100}$$

7.7 Illustrative Examples

The continuous-state approximation is illustrated here by two simple examples.

EXAMPLE 7.1. Consider a one-feature classifier with the training procedure of Equation (7.2). In this case

$$\Delta\Theta(n)/\rho_n = -\Gamma(n), \tag{7.101}$$

where

$$\Gamma(n) = \Omega(n) - R(n), \qquad \Omega(n) = i \quad \text{when } \omega(n) = \omega_i,\ i = 1, 2. \tag{7.102}$$

First we find $\eta_\theta = -\eta_0(\mathbf{v})$:

$$\Omega(n) = 1 \text{ or } 2, \qquad R(n) = 1 \text{ or } 2.$$

$$\begin{aligned}\eta_\theta &= E[-\Omega(n) + R(n) \mid \Theta(n) = \theta]\\ &= -E(\Omega(n)\mid\theta) + E(R(n)\mid\theta) = -E(\Omega(n)) + E(R(n)\mid\theta)\\ &= -(1-\alpha) + P(X(n) > \theta),\end{aligned}$$

where $\alpha = P(\Omega(n) = 1) = \int_{-\infty}^{\infty} f_1(x)\,dx$. Hence

$$\eta_\theta = -1 + \alpha + \int_\theta^\infty [f_1(x) + f_2(x)]\,dx,$$

where $f_i(x)$, $i = 1, 2$, are the constituent densities.

Suppose $f_i(x)$ are the triangular densities shown in Figure 7.2. Then $\alpha = \frac{1}{2}$, and, since $f_1(x) + f_2(x)$ is a rectangular mixture density,

$$\eta_\theta = -\tfrac{1}{2} + \int_\theta^\infty \frac{1}{b}\left[u\left(x + \frac{b}{2}\right) - u\left(x - \frac{b}{2}\right)\right]dx,$$

where $u(y)$ is a unit step function, defined as: $u(y) = 0$ for $y < 0$, $u(y) = 1$ for $y \geqslant 0$. Hence

$$\eta_\theta = -\tfrac{1}{2} + \begin{cases} 1, & \text{if } \theta < -b/2,\\ [(b/2) - \theta](1/b), & \text{if } -b/2 \leqslant \theta \leqslant b/2,\\ 0, & \text{if } \theta > b/2,\end{cases}$$

$$= \begin{cases} \tfrac{1}{2}, & \text{if } \theta < -b/2,\\ -\theta/b, & \text{if } -b/2 \leqslant \theta \leqslant b/2,\\ -\tfrac{1}{2}, & \text{if } \theta > b/2.\end{cases} \tag{7.103}$$

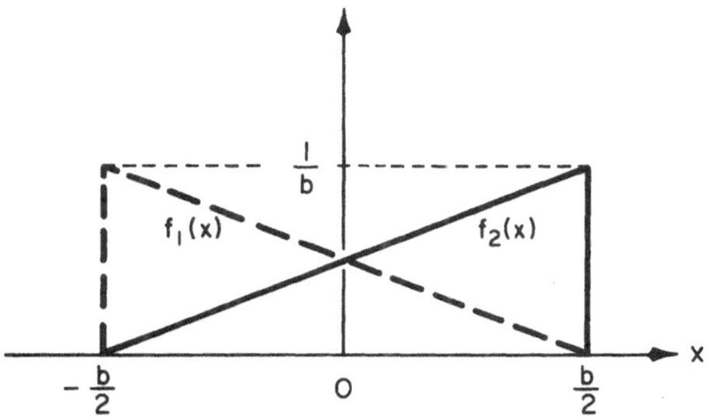

Figure 7.2. Triangular densities in Example 7.1.

Assume $g(t) = 1$, and that the initial value of θ is $-b/2$. Thus

$$p_{\Theta(0)}(\theta) = \delta\left(\theta + \frac{b}{2}\right) \tag{7.104}$$

and, by Equation (7.31),

$$\hat{\mu}'(t) = \eta_{\hat{\mu}(t)} = -(1/b)\hat{\mu}(t).$$

Hence

$$d\hat{\mu}/\hat{\mu} = -dt/b$$

from which

$$\log(\hat{\mu}(t)/\hat{\mu}(0)) = -t/b$$

so that

$$\hat{\mu}(t) = \hat{\mu}(0)\exp(-t/b) \tag{7.105}$$

and

$$\hat{\mu}(t) = -(b/2)\exp(-t/b). \tag{7.106}$$

We see from Equation (7.105) that $\hat{\mu}(t)$ approaches 0 asymptotically as $t \to \infty$, assuming $-b/2 \leqslant \hat{\mu}(0) \leqslant b/2$. Actually, it is easy to show that $\hat{\mu}(\infty) = 0$ for any real value of $\hat{\mu}(0)$. To find $\mu(n)$ in terns of $\hat{\mu}(t)$, use Equation (7.25):

$$\mu(n) \cong \hat{\mu}(n\rho) = -(b/2)\exp(-n\rho/b). \tag{7.107}$$

EXAMPLE 7.2. Consider a two-feature, two-class classifier with a proportional increment training procedure. In this case

$$\Delta\mathbf{V}(n)/g(n) = \Gamma(n)\mathbf{Y}(n), \tag{7.108}$$

where $\Gamma(n) = \Omega(n) - R(n)$; $\Omega(n)$ is an input class number; $R(n)$ is output; the alphabet of $\Omega(n)$ is $(1, 2)$; the alphabet of $R(n)$ is $(1, 2)$; and

$$\mathbf{Y}(n) = \begin{bmatrix} 1 \\ Y_1(n) \\ Y_2(n) \end{bmatrix} \tag{7.109}$$

is an augmented feature vector.

Now we show how to find $\boldsymbol{\eta}(\hat{\boldsymbol{\mu}})$ and $\boldsymbol{\Omega}(\hat{\boldsymbol{\mu}})$, since these functions are needed in the continuous-state approximations of the learning curves and variance curves. For convenience, the arguments t and n are often omitted in the discussion below when the omission causes no confusion.

Suppose the constituent densities are $f_1(\mathbf{x})$ and $f_2(\mathbf{x})$. Then

$$\begin{aligned}\boldsymbol{\eta}(\boldsymbol{\mu}) &= E(\Delta\mathbf{V}/g(n)|\mathbf{V}=\boldsymbol{\mu}) = E(\Gamma\mathbf{Y}|\mathbf{V}=\boldsymbol{\mu})\\ &= \int_{\boldsymbol{\mu}^T\mathbf{y}<0}\mathbf{y}f_2(\mathbf{x})\,d\mathbf{x} - \int_{\boldsymbol{\mu}^T\mathbf{y}>0}\mathbf{y}f_1(\mathbf{x})\,d\mathbf{x}.\end{aligned} \tag{7.110}$$

Thus

$$\begin{aligned}\eta_0(\boldsymbol{\mu}) &= \int_{\boldsymbol{\mu}^T\mathbf{y}<0}f_2(\mathbf{x})\,d\mathbf{x} - \int_{\boldsymbol{\mu}^T\mathbf{y}>0}f_1(\mathbf{x})\,d\mathbf{x}\\ \eta_i(\boldsymbol{\mu}) &= \int_{\boldsymbol{\mu}^T\mathbf{y}<0}y_if_2(\mathbf{x})\,d\mathbf{x} - \int_{\boldsymbol{\mu}^T\mathbf{y}>0}y_if_1(\mathbf{x})\,d\mathbf{x}, \quad \text{for } i=1,2\end{aligned} \tag{7.111}$$

and

$$\begin{aligned}\boldsymbol{\Omega}(\boldsymbol{\mu}) &= E\left\{\left[\frac{\Delta\mathbf{V}}{g(n)} - \boldsymbol{\eta}(\boldsymbol{\mu})\right]\left[\frac{\Delta\mathbf{V}}{g(n)} - \boldsymbol{\eta}(\boldsymbol{\mu})\right]^T\middle|\mathbf{V}=\boldsymbol{\mu}\right\}\\ &= E\left[\left(\frac{\Delta\mathbf{V}}{g(n)}\right)\left(\frac{\Delta\mathbf{V}}{g(n)}\right)^T\right] - \boldsymbol{\eta}(\boldsymbol{\mu})\boldsymbol{\eta}^T(\boldsymbol{\mu}).\end{aligned} \tag{7.112}$$

Hence

$$\begin{aligned}\Omega_{ij}(\boldsymbol{\mu}) &= E\left[\frac{\Delta V_i\Delta V_j}{g^2(n)}\middle|\mathbf{V}=\boldsymbol{\mu}\right] - \eta_i(\boldsymbol{\mu})\eta_j(\boldsymbol{\mu})\\ &= E\left[\frac{\Gamma^2(n)Y_iY_j}{g^2(n)}\middle|\mathbf{V}=\boldsymbol{\mu}\right] - \eta_i(\boldsymbol{\mu})\eta_j(\boldsymbol{\mu})\\ &= \int_{\boldsymbol{\mu}^T\mathbf{y}<0}y_iy_jf_2(\mathbf{x})\,d\mathbf{x} + \int_{\boldsymbol{\mu}^T\mathbf{y}>0}y_iy_jf_1(\mathbf{x})\,d\mathbf{x} - \eta_i(\boldsymbol{\mu})\eta_j(\boldsymbol{\mu}).\end{aligned} \tag{7.113}$$

The learning curve is given by Equation (7.86), i.e.,

$$z(n) \cong S[\hat{\boldsymbol{\mu}}(n\rho)], \tag{7.114}$$

where

$$S(\boldsymbol{\mu}) = \int_{\boldsymbol{\mu}^T\mathbf{y}<0}f_1(\mathbf{x})\,d\mathbf{x} + \int_{\boldsymbol{\mu}^T\mathbf{y}>0}f_2(\mathbf{x})\,d\mathbf{x} \tag{7.115}$$

and $\hat{\boldsymbol{\mu}}(t)$ is determined by

$$d\hat{\boldsymbol{\mu}}(t)/dt = \boldsymbol{\eta}[\hat{\boldsymbol{\mu}}(t)]. \tag{7.116}$$

(Note that in this case $g(t) =$ constant, which we have set equal to 1).

The variance curve is given by Equation (7.96), i.e.,

$$q(n) \cong \rho[\nabla S(\hat{\boldsymbol{\mu}})]^T\mathbf{R}\nabla S(\hat{\boldsymbol{\mu}})|_{t=n\rho}, \tag{7.117}$$

where $\mathbf{R}(t)$ is determined by Equation (7.79).

Because of the complexity of the integrals in Equations (7.110), (7.113), and (7.115), it is usually not possible to find closed-form solutions for $\hat{\boldsymbol{\mu}}(t)$, $\mathbf{R}(t)$, $\hat{z}(t)$, and $\hat{q}(t)$. Hence in most cases, one must resort to numerical integration to compute the continuous-state learning curves and variance curves.

To be more specific, suppose the two constituent densities are normal densities of the form

$$f_i(\mathbf{x}) = \frac{1}{4\pi|\boldsymbol{\Sigma}|^{1/2}}\exp\left[-\tfrac{1}{2}(\mathbf{x}-\boldsymbol{\mu}_i)^T\boldsymbol{\Sigma}^{-1}(\mathbf{x}-\boldsymbol{\mu}_i)\right], \qquad \text{for } i=1,2 \tag{7.118}$$

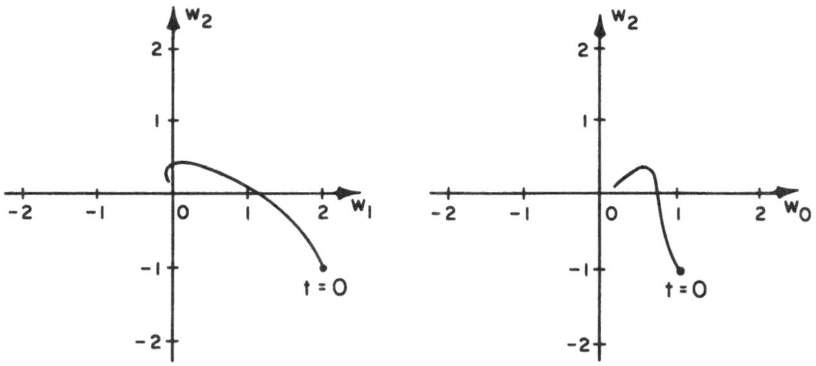

Figure 7.3. Centroid curve for Example 7.2.

where $\boldsymbol{\Sigma}$ is the covariance matrix of $\mathbf{X}$ given either class. Suppose

$$\boldsymbol{\Sigma} = \begin{bmatrix} 1 & 0 \\ 0 & 1 \end{bmatrix} \tag{7.119}$$

$$\boldsymbol{\mu}_1 = [0, 2], \qquad \boldsymbol{\mu}_2 = [1, 1], \tag{7.120}$$

so that

$$f_1(\mathbf{x}) = \frac{1}{4\pi} \exp\{-\tfrac{1}{2}[x_1^2 + (x_2 - 2)^2]\} \tag{7.121}$$

$$f_2(\mathbf{x}) = \frac{1}{4\pi} \exp\{-\tfrac{1}{2}[(x_1 - 1)^2 + (x_2 - 1)^2]\}. \tag{7.122}$$

Also suppose $\mathbf{V}(0) = [1, 2, -1]^T$. Then numerical integration yields the centroid curve shown in Figure 7.3 and the learning curve and variance curve shown superimposed in Figure 7.4. The centroid curve in Figure 7.3 is depicted by its projections on the w_0, w_2 and w_1, w_2 coordinate planes.

To evaluate the accuracy of these curves, one may use a Monte Carlo technique to generate a large number of runs of the training process on a computer, and from these

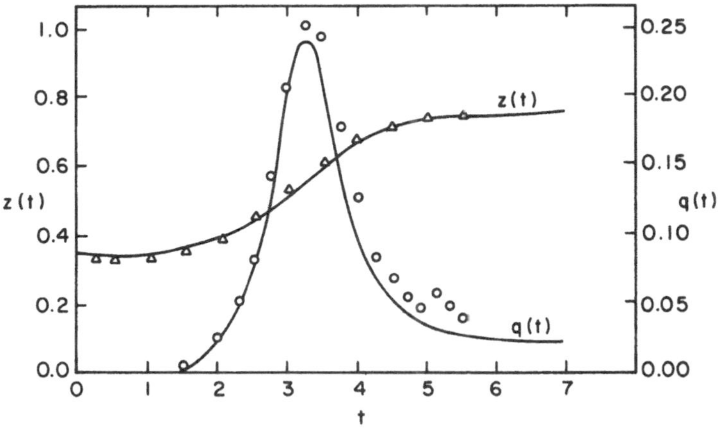

Figure 7.4. Learning curves and variance curves for Example 7.2.

runs to compute the fraction of correct guesses at trial $n (n = 0, 1, 2, \ldots)$, and the variance of the guesses about the average. The small circles and small triangles in Figure 7.4 show the resulting learning curve and variance curve obtained from 200 runs [1]. Here it is seen that the continuous-state approximation and the experimentally derived data are quite close.

In the next two sections we illustrate how the continuous-state model provides interesting insights into the learning dynamics of trainable classifiers.

7.8 Shapes of Learning Curves in Single-Feature Classifiers

In this section the continuous-state model provides the means for deriving a set of constraints on the shapes of the learning curves in constant increment training of single-feature classifiers.

The notation for single-feature constant increment training given in Chapter 6 is used here. The training procedure is

$$\begin{aligned} V_1(n) &= +1, \quad \text{with respect to } n, \\ \Theta(n) &= -V_0(n)/V_1(n) = -V_0(n), \\ \Delta\Theta(n) &= -\Delta V_0(n) = -\rho Z(n) \\ &= -\rho(\Omega(n) - R(n)), \end{aligned} \tag{7.123}$$

where for this single feature classifier, $Z(n) = \Gamma(n)$. Thus

$$\rho g(n\rho) = \rho, \qquad g(t) = 1. \tag{7.124}$$

Now define

$$\eta_\theta = -\eta_0(\mathbf{v}) = E\left[\frac{-\Delta V_0}{\rho}\middle|\mathbf{V} = \mathbf{v}\right] = E\left[\frac{\Delta\Theta}{\rho}\middle|\Theta = \theta\right]. \tag{7.125}$$

Since $\Delta V_1(n) = 0$, we need only consider $\eta_0(\mathbf{v})$ in our analysis. By Equation (7.123),

$$\begin{aligned} \eta_\theta(\mathbf{v}) &= P[\Omega(n) = 1, R(n) = 2] - P[\Omega(n) = 2, R(n) = 1] \\ &= \int_\theta^\infty f_1(x)\,dx - \int_{-\infty}^\theta f_2(x)\,dx \\ &= \alpha - \int_{-\infty}^\theta [f_1(x) + f_2(x)]\,dx, \end{aligned} \tag{7.126}$$

where

$$\alpha = \int_{-\infty}^\infty f_1(x)\,dx \equiv P[\Omega(n) = 1]. \tag{7.127}$$

Because $\Delta V_1(n) = 0$, η_θ is a function of only θ. This is shown in Equation (7.126). Let

$$F(\theta) = -\eta_\theta. \tag{7.128}$$

(Since $f_1(x) + f_2(x)$ is the probability density of $y_1 = x$, $F(x)$ is a distribution function that has been displaced vertically—i.e., along the $F(x)$ axis—by an amount α. This is illustrated in Figure 7.5 where a typical form of $F(x)$ is shown.) Then, by Equations (7.3), (7.31), (7.126), (7.128), and by defining $\hat{\mu}_\theta(t) = -\hat{\mu}_0(t)$,

$$d\hat{\mu}_\theta(t)/dt = -F[\hat{\mu}_\theta(t)] \tag{7.129}$$

and

$$\hat{\mu}_\theta(0) = \Theta(0). \tag{7.130}$$

The learning curve is determined from Equations (7.86) and (7.129):

$$z(n) \cong S[\hat{\mu}_\theta(n\rho)] \tag{7.131}$$

where

$$\begin{aligned} S(\theta) &= P[\Omega(n) = R(n) \mid \Theta(n) = \theta] \\ &= \int_{-\infty}^{\theta} f_1(x)\,dx + \int_{\theta}^{\infty} f_2(x)\,dx \\ &= 1 - \alpha + \int_{-\infty}^{\theta} [f_1(x) - f_2(x)]\,dx. \end{aligned} \tag{7.132}$$

Note from Equation (7.132) that $S(\theta)$ has a maximum or minimum at every value of x where $f_1(x)$ and $f_2(x)$ intersect—i.e., for every θ^* such that $f_1(\theta^*) = f_2(\theta^*)$. In the subsequent discussion it is assumed that $f_1(x)$ and $f_2(x)$ have a unique intersection at $x = \theta^*$, so that $S(\theta)$ has precisely one maximum, and the location of this maximum is $\theta = \theta^*$. This assumption holds quite often, in particular whenever the $f_i(x)$'s are unimodal, $E[X \mid \omega = \omega_1] < E[X \mid \omega = \omega_2]$ and the $f_i(x)$'s intersect in just one point. (See Exercise 7.5.) The quantity θ^* is referred to here as the *minimum-error point.*

The final value of $z(n)$ occurs when

$$0 = \frac{d\hat{\mu}}{dt} = \frac{d^2\hat{\mu}}{dt^2} = \cdots = \frac{d^m\hat{\mu}}{dt^m} = \cdots. \tag{7.133}$$

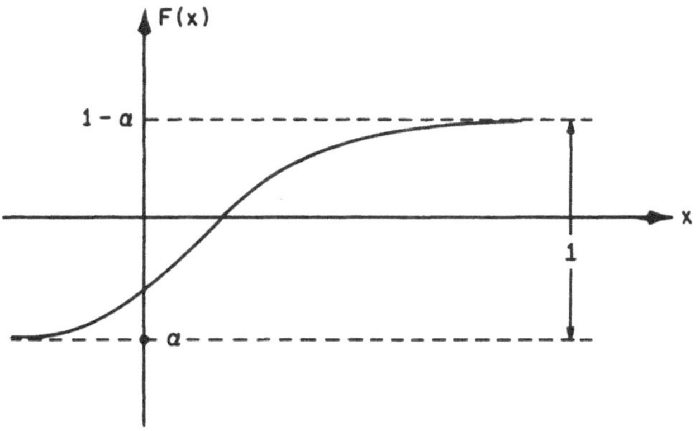

Figure 7.5. A typical form of $F(x)$.

But, by Equation (7.129),

$$d^m\hat{\mu}/dt^m \text{ has } F(\hat{\mu}) \text{ as a factor for every } m. \tag{7.134}$$

Hence the final value of $z(n)$ occurs when $F(\hat{\mu}) = 0$, i.e., when

$$\int_{\xi}^{\infty} f_1(x)\,dx = \int_{-\infty}^{\xi} f_2(x)\,dx. \tag{7.135}$$

The left- and right-hand sides of the Equation (7.135) are the probabilities of a false ω_1 and a false ω_2, respectively. Hence ξ is referred to as an *equal error point*. Usually $\xi \neq \theta^*$, although they are often near each other. The equal error point and the minimum error point are illustrated in Figure 7.6. A demonstration of the fact that ξ and θ^* are near each other for typical pairs of constituent densities is given in Section 7.9.

In Equation (7.129), note that $d\hat{\mu}/dt$ always has the opposite sign from $F(\hat{\mu})$. Since $F(\hat{\mu}) > 0$ for $\hat{\mu} > 0$ and $F(\hat{\mu}) < 0$ for $\hat{\mu} < 0$, it follows that $\hat{\mu}(t)$ progresses monotonically in one direction, approaching asymptotically to ξ, i.e., $\hat{\mu}(t) \to \xi$ as $t \to \infty$.

We are now in a position to describe the constraints on the shapes of the learning curves. Assuming that $S(\theta)$ has a single peak (a very common occurrence), there are three shapes of learning curves, depending on the relative ordering among $\Theta(0)$, ξ, and θ^*.

Case 1. $\Theta(0) < \xi \leqslant \theta^*$ or $\Theta(0) > \xi \geqslant \theta^*$. In this case $z(n)$ increases monotonically from $S[\Theta(0)]$ to $S(\xi)$. At no time does $z(n)$ achieve the optimum value of $S[\theta^*]$, except in the limit as $t \to \infty$ if $\xi = \theta^*$.

Case 2. $\Theta(0) < \theta^* \leqslant \xi$ or $\Theta(0) > \theta^* \geqslant \xi$. In this case $z(n)$ increases from $S[\Theta(0)]$ to its maximum value of $S(\theta^*)$, thereafter decreasing toward its asymptotic value of $S(\xi)$.

Case 3. $\theta^* \leqslant \Theta(0) < \xi$ or $\theta^* \geqslant \Theta(0) > \xi$. In this case $z(n)$ decreases from $S[\Theta(0)]$ toward its asymptotic value of $S(\xi)$. If $\Theta(0) = \theta^*$, then $z(n)$ equals the maximum possible value at $n = 0$, and never attains that value again.

Case 4. $\Theta(0) = \xi$. In this case $z(n) = S(\xi) =$ constant for all n.

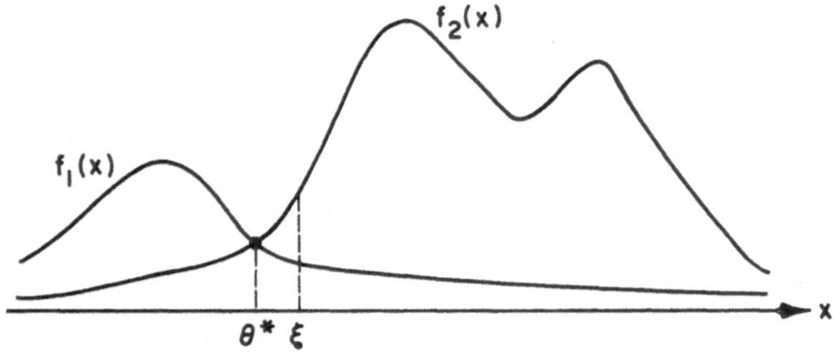

Figure 7.6. A pair of constituent densities having precisely one equal error point ξ and one minimum error point θ^*.

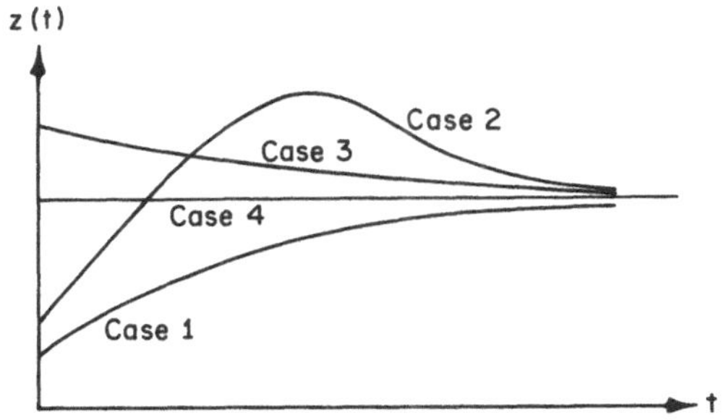

Figure 7.7. Four possible shapes of learning curves in single-feature constant increment training.

In all four cases $S(\xi) \leqslant S(\theta^*)$, and usually $S(\xi) < S(\theta^*)$. Thus the shapes of the learning curves in the single-feature constant increment training process are restricted to:

Case 1: Monotonic increasing.
Case 2: Single-peaked.
Case 3: Monotonic decreasing.
Case 4: Constant.

In all cases the asymptotic value of $z(n)$ is $z(\infty) = S(\xi)$. (These four cases are illustrated in Figure 7.7.) Thus $z(\infty)$ is independent of $\Theta(0)$.

$z(n)$ is a result of a traveling delta wave:

$$p_{\Theta(t)}(\theta) = \delta(\theta - \hat{\mu}_0(t)).$$

Hence the continuous-state Markov model of the single-feature constant increment training process is ergodic in the sense that the final value of $p_{\Theta(t)}(\theta)$ is independent of $\Theta(0)$.

These restrictions on the learning waves are independent of whether or not $g(t)$ is constant. To see this, note that the restrictions on the learning waves apply in general when they are plotted as functions of τ, where τ is defined by Equation (7.97). Then note that the properties of monotonicity, single-peakedness, and constantness of the learning waves are unaffected by the τ-to-t transformation.

7.9 How Close Are the Equal Error and Minimum Error Points?

Usually one finds that the equal error point ξ and the minimum error point θ^* in a single-feature classifier are close to each other. We demonstrate this here for the class of pairs of constituent densities illustrated in Figure 7.8.

This class of pairs of densities is defined as those having the following properties: (a) the derivatives of $f_1(x)$ and $f_2(x)$ exist for all x; (b) the tangent to $f_1(x)$ at the point of intersection $x = \theta^*$ lies entirely below $f_1(x)$ for $x \geqslant \max(\theta^*, \xi)$. (If $\xi > \theta^*$, replace x by $-x$, apply the construction described

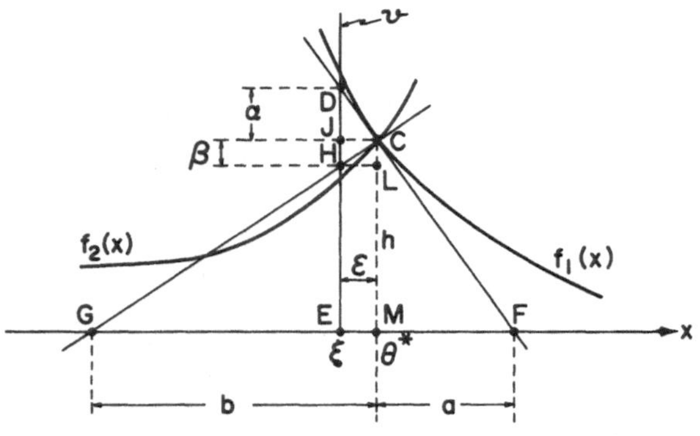

Figure 7.8. Tails of a pair of constituent densities.

below, and again change the sign of x.) For this class of constituent densities, carry out the following geometrical construction (illustrated in Figure 7.8).

Draw a vertical straight line through $(\xi, 0)$. Let υ denote this line. Let E denote the point $(\xi, 0)$. Let M denote $(\theta^*, 0)$. Assume, without loss of generality, that E lies to the left of M. Let C denote $(\theta^*, f_1(\theta^*))$. Let l denote the tangent to $f_1(x)$ at $x = \theta^*$. Construct a straight line through C that intersects υ at point D and intersects the x-axis at point F. The points D and F are determined by making the area of triangle DEF equal to the equal error false alarm probability, i.e., by making the area of triangle DEF equal to the area under $f_1(x)$ to the right of ξ. Let A denote this area. This construction is always possible for the following reason. The area of triangle DEF is less than A when DCF coincides with l, because of our constraints on the class of constituent densities, and it is infinite when DCF is horizontal. Clearly the area of the triangle is a continuous function of the angle DFE. Hence the desired construction of DCF is always possible. Now construct another straight line through C that meets the x-axis at G, the line υ at H, and such that the angle HGE is acute and the area of triangle HGE is equal to A. This is always possible since the area of triangle HGE is infinite when CHG is horizontal and zero when CHG is vertical (coinciding with υ).

Let $GM = b, MF = a, EM = HL = JC = \varepsilon, DJ = \alpha, CL = \beta$, and $CM = h$. Then

$$\alpha/\varepsilon = h/a \quad \text{or} \quad \alpha = h\varepsilon/a.$$

$$\beta/\varepsilon = h/b \quad \text{or} \quad \beta = h\varepsilon/b.$$

By the construction,

$$\tfrac{1}{2}(b - \varepsilon)(h - \beta) = \tfrac{1}{2}(a + \varepsilon)(h + \alpha) = A.$$

So,

$$\frac{b - \varepsilon}{a + \varepsilon} = \frac{h + \alpha}{h - \beta} = \frac{1 + (\varepsilon/a)}{1 - (\varepsilon/b)}$$

or

$$\frac{(1 - (\varepsilon/b))^2}{(1 + (\varepsilon/a))^2} = \frac{a}{b}$$

or

$$\frac{1 - (\varepsilon/b)}{1 + (\varepsilon/a)} = (a/b)^{1/2} = \lambda \quad \text{(say)}.$$

Thus,

$$1 - \frac{\varepsilon}{b} = \lambda\left(1 + \frac{\varepsilon}{a}\right) = \lambda\left(1 + \frac{\varepsilon}{b}\frac{b}{a}\right) = \lambda + \frac{1}{\lambda}\frac{\varepsilon}{b},$$

or

$$1 - \lambda = \frac{\varepsilon}{b}\left(1 + \frac{1}{\lambda}\right),$$

i.e.,

$$\frac{\varepsilon}{b} = \frac{1 - \lambda}{1 + (1/\lambda)} = \frac{\lambda - \lambda^2}{\lambda + 1}.$$

ε/b has an extremum when

$$\frac{d}{d\lambda}\left(\frac{\varepsilon}{b}\right) = \frac{(\lambda + 1)(1 - 2\lambda) - (\lambda - \lambda^2)}{(\lambda + 1)^2} = \frac{1 - 2\lambda - \lambda^2}{(1 + \lambda)^2} = 0,$$

i.e., when $\lambda = -1 \pm \sqrt{2}$. Now, $\lambda = -1 + \sqrt{2}$ corresponds to the maximum of ε/b. Therefore,

$$\left(\frac{\varepsilon}{b}\right)_{\max} = \frac{(\sqrt{2} - 1)(2 - \sqrt{2})}{\sqrt{2}} < 0.17.$$

Thus ξ and θ^* are near each other relative to the distance b.

If $f_1(x)$ is Gaussian and the area A is small enough for ξ to be to the right of the point at which $d^2f_1(x)/dx^2 = 0$, and if $f_2(x)$ intersects $f_1(x)$ at $(\theta^*, f_1(\theta^*))$ where $\theta^* > \xi$, then it is easy to see that $(f_1(x), f_2(x))$ belongs to the class of pairs of densities defined above. Our class is more general than this class of Gaussian pairs.

7.10 Asymptotic Stability in the Large

Another insight provided by the continuous-state model of learning is asymptotic stability in the large. In the analysis of multidimensional training processes by the continuous-state model, it has been found that in many cases more than one asymptotic value of $\hat{\boldsymbol{\mu}}(t)$ exists, and that the asymptote $\hat{\boldsymbol{\mu}}(\infty)$ depends on the initial weight vector $\mathbf{V}(0)$. Such behavior is said to be *asymptotically unstable in the large.* Asymptotically unstable in the large behavior is a clear indication that the associated Markov process is not ergodic. (See Section 6.6.)

Since the value of $\boldsymbol{\mu}(\infty)$ can contribute significantly to the error probability in the working phase of a trainable classifier, it is important, first, to have a means of detecting asymptotic instability in the large and, second, to have a means of choosing a training procedure that provides an effective tradeoff in the expected performance associated with the various possible values of

$\boldsymbol{\mu}(\infty)$. These aspects of stability in the large are discussed below from the point of view of Lyapunov's second method of analysis of stability of dynamical systems.

To start the discussion, we introduce some nomenclature from the literature on nonlinear automatic control systems.

In a dynamical system having a state space, a state $\mathbf{p}$ is an *equilibrium state* if there exists a trajectory $\mathbf{s}(t)$ and a time t_0 such that $\mathbf{s}(t) = \mathbf{p}$ for all $t \geqslant t_0$. An equilibrium state $\mathbf{p}$ is *stable* if, after $\mathbf{s}(t)$ moves directly from $\mathbf{p}$ to another point within a small neighborhood of $\mathbf{p}$, $s(t)$ remains within a bounded distance of $\mathbf{p}$. The state $\mathbf{p}$ is *asymptotically stable* if, after $\mathbf{s}(t)$ moves from $\mathbf{p}$ directly to another point within a small neighborhood of $\mathbf{p}$, $\mathbf{s}(\infty) = \mathbf{p}$. The state $\mathbf{p}$ is *asymptotically stable in the large* if the trajectory subsequent to a direct move of $s(t)$ from $\mathbf{p}$ to another nearby point approaches $\mathbf{p}$ asymptotically regardless of the initial state $\mathbf{s}(0)$.

In Markovian training processes, the weight vector $\mathbf{V}(t)$ takes the role of a state. But since the motion of $\mathbf{V}(t)$ is stochastic in nature, the definition of stability must take into account the stochastic nature of $\mathbf{V}(t)$. In this definition *expected state* replaces the role of state. We denote the trajectory of the expected state by $\boldsymbol{\mu}(t)$.

Thus an expected state $\mathbf{p}$ is *stochastically stable* if the following conditions hold for any $\varepsilon > 0$: (a) in response to the expected state $\boldsymbol{\mu}(t)$ moving directly from $\mathbf{p}$ to $\mathbf{p} + \Delta\mathbf{p}$ with $|\Delta\mathbf{p}| < \varepsilon$, the expected state $\boldsymbol{\mu}(t)$ subsequently remains within a bounded distance of $\mathbf{p}$; (b) the relative covariance matrix $\mathbf{r}(t)$ remains bounded as $t \to \infty$. State $\mathbf{p}$ is *stochastically asymptotically stable* if $\boldsymbol{\mu}(\infty) = \mathbf{p}$ whenever $\boldsymbol{\mu}(t)$ is moved directly from $\mathbf{p}$ to $\mathbf{p} + \Delta\mathbf{p}$, $|\Delta\mathbf{p}| < \varepsilon$, and $\mathbf{R}(t)$ remains bounded as $t \to \infty$. State $\mathbf{p}$ is *stochastically asymptotically stable in the large* if it is stochastically asymptotically unstable for every initial state $\mathbf{V}(t)$. (These definitions are somewhat similar, but not identical, to Aoki's [5].) Hereafter we often omit the term "stochastic" when discussing stochastic stability if the omission causes no confusion.

Lyapunov's Second Method. Lyapunov's second method is a technique for analyzing the stability of systems governed by differential equations of the form

$$d\boldsymbol{\mu}/dt = \boldsymbol{\eta}[\boldsymbol{\mu}(t)] \tag{7.136}$$

$$\boldsymbol{\eta}(\mathbf{p}) = \mathbf{0}, \tag{7.137}$$

provided $d\boldsymbol{\eta}(\boldsymbol{\mu})/d\boldsymbol{\mu}$ satisfies the Lipschitz condition

$$\sup_{\boldsymbol{\mu} \neq \mathbf{a}} \frac{|\boldsymbol{\mu}'(\boldsymbol{\mu}) - \boldsymbol{\eta}(\mathbf{a})|}{|\boldsymbol{\mu} - \mathbf{a}|} < \infty \tag{7.138}$$

where

$$\boldsymbol{\eta}'(\boldsymbol{\mu}) \equiv d\boldsymbol{\eta}/d\boldsymbol{\mu} = \{\partial\eta_i/\partial\mu_j\}.$$

(This condition implies that $\boldsymbol{\eta}'(\boldsymbol{\mu})$ is everywhere continuous in $\boldsymbol{\mu}$.)

Note that Equation (7.136) is in the form of the centroid equation (7.31) for the case $g(t) = 1$. But in Section 7.6 it was shown that this is really not a restriction at all, since Equation (7.31) can be transformed into the form of Equation (7.136) by deforming the time axis in accordance with Equation (7.97). Thus the centroid equation belongs precisely to the class of equations suitable for analysis by Lyapunov's second method, provided the proper transformation is made in t.

Referring to Equations (7.136) and (7.137), Lyapunov's second method for the analysis of stochastic stability is as follows. Search for a positive definite scalar function $T(\boldsymbol{\mu})$ such that $T(\mathbf{0}) = 0$, and $dT(\boldsymbol{\mu})/dt$ is negative semidefinite or identically zero. If such a function (a *Lyapunov function*) is found, and if $|\mathbf{R}(\infty)| < \infty$, then $\boldsymbol{\mu} = \mathbf{0}$ is necessarily a stochastically stable state. If, in addition, $T(\boldsymbol{\mu}) \to \infty$ as $\boldsymbol{\mu} \to \infty$, then $\boldsymbol{\mu} = \mathbf{0}$ is a stochastically asymptotically stable state.

Below we discuss the stability of two types of training processes: single-feature constant increment training and proportional increment training. The first training process is asymptotically stable in the large; the second is unstable in the large.

Single-Feature Constant Increment Training. We saw in Section 7.8 that the continuous-state Markov model of single-feature constant increment training is an ergodic process. Hence we should be able to demonstrate that this process is asymptotically stable in the large. To do so, choose a Lyapunov function as follows. Let

$$s = \hat{\mu}_\theta - \Theta(0), \tag{7.139}$$

so that Equation (7.129) becomes

$$ds/dt = -F(s + \Theta(0)). \tag{7.140}$$

Let

$$r(s) = sF(s + \Theta(0)) \tag{7.141}$$

be a Lyapunov function. The Lyapunov conditions for asymptotic stability in the large are checked as follows:

$$r(0) = 0$$

$$\frac{dr}{dt} = -sF\frac{dF}{ds} - F^2.$$

Since $sF > 0$ for all $s \neq 0$, and since $dF/ds = f_1(s) + f_2(s) \geqslant 0$ for all s, it follows that

$$dr/dt \leqslant 0 \quad \text{for all } s \tag{7.142}$$

Hence $s = 0$ is a stable state. Furthermore, if $F[s(t^*) + \Theta(0)] = 0$, then $ds(t^*)/dt = 0$ by Equation (7.140). Then by Equation (7.137), it follows that $s(t^*)$ is a stable point. By Equations (7.126) and (7.128), $r(s) \cong s(1 - \alpha)$ when s is sufficiently large. Hence $r(s) \to \infty$ as $s \to \infty$.

Thus by Lyapunov's second method we have shown that the continuous model of the single-feature constant increment training process is asymptotically stable in the large. This confirms our observation in Section 7.8 that this process is ergodic.

Proportional Increment Training. Example 7.2 describes a continuous-state model of proportional increment training for a two-feature, two-class classifier. The continuous-state model in this example may be extended to a d-feature classifier. For simplicity, assume $g(t) = 1$. Then the centroid equation becomes

$$d\hat{\boldsymbol{\mu}}/dt = \boldsymbol{\eta}(\hat{\boldsymbol{\mu}}), \tag{7.143}$$

where

$$\boldsymbol{\eta}(\boldsymbol{\mu}) = \int_{\boldsymbol{\mu}^T\mathbf{y}<0} \mathbf{y} f_2(\mathbf{x})\, d\mathbf{x} - \int_{\boldsymbol{\mu}^T\mathbf{y}>0} \mathbf{y} f_1(\mathbf{x})\, d\mathbf{x}. \tag{7.144}$$

We refer to the set of pairs $\{(\mathbf{x}, f_2(\mathbf{x})) \mid \boldsymbol{\mu}^T\mathbf{y} < 0\}$ as the *error tail* of $f_2(\mathbf{x})$, and $\{(\mathbf{x}, f_1(\mathbf{x})) \mid \boldsymbol{\mu}^T\mathbf{y} > 0\}$ as the error tail of $f_1(\mathbf{x})$.

Equations (7.143) and (7.144) tell us that $\boldsymbol{\eta}(\hat{\boldsymbol{\mu}}) = \mathbf{0}$ is a stable point only if each moment of the error tail of $f_2(\mathbf{x})$ equals the corresponding moment of the error tail of $f_1(\mathbf{x})$. This can be fully satisfied only if the centroids of the two error tails coincide—an impossible requirement if either $f_1(\mathbf{x})$ or $f_2(\mathbf{x})$ is nonzero for some $\mathbf{x}$ such that $\boldsymbol{\eta}^T\mathbf{y} < 0$ or $f_1(\mathbf{x})$ is nonzero for some $\mathbf{x}$ such that $\boldsymbol{\eta}^T\mathbf{y} > 0$. Hence the proportional increment training process has no stable point in weight space.

Experiments on the case where $d = 2$ and where the constituent densities are Gaussian have shown that proportional increment training generates a trajectory in weight space ($\mathbf{w}$-space) that approaches $\mathbf{0}$ asymptotically from one of two directions, depending on the initial weight vector [1]. One of these directions yields a hyperplane with weight vector $\mathbf{W}$ corresponding approximately to the situation where the areas (zeroth moments) of the two error tails are equal, and the other direction corresponding to the situation where the weight vector of the decision hyperplane is approximately the negative of $\mathbf{W}$. Thus for this case, the proportional increment training process seems to be bistable in the large, provided the training process is stopped before $\mathbf{W}(n) = \mathbf{0}$. Here the choice of the initial weight vector determines which of two approximately stable states $(\mathbf{w}_1, \mathbf{w}_2)$ is approached by $\mathbf{W}(n)$. This is a special form of stochastic instability in the large. We say that the two states $(\mathbf{w}_1, \mathbf{w}_2)$ are *approximately* stable, because as $n \to \infty$ the values of both $\mathbf{w}_1$ and $\mathbf{w}_2$ approach zero, resulting in the corruption of the asymptotic estimates of $\mathbf{w}_1$, and $\mathbf{w}_2$ by roundoff noise.

The equalized error training procedure, presented in Section 5.5, is a modification of the proportional increment training procedure which overcomes the instability of the latter procedure. In the equalized error training procedure the stable point in weight space corresponds to a coincidence of the *projections* of the centroids of the two error tails onto the decision hyperplane.

Exercises

7.1. Let $q(n)$ denote the variance of $S[V(n)]$, given by Equation (7.87).

(a) Show that $q(n) = E\{S[V(n)]\}^2 - z^2(n)$.

(b) Show that $q(n) \to 0$ as $\rho \to 0$ with $n\rho = t$.

7.2. Let $a(t) = \int_0^t g(u)\,du$. Assume $g(t) > 0$. Let $a(t) = \tau$.

(a) Show that if $g(t) > 0$ for all finite positive t, there exists an inverse function $a^{-1}(\tau)$ such that

$$a[a^{-1}(\tau)] = \tau \quad \text{for all } \tau > 0$$

and such that $a^{-1}(\tau)$ is a monotonic increasing function of τ.

(b) Demonstrate the validity of Equations (7.98) and (7.99).

7.3. Consider the following procedure for a single-feature classifier (discussed in Reference [7]):

$$\Theta(n+1) = \Theta(n) - \rho[\Omega(n) - R(n)],$$

using the nomenclature in Section 6.4. Assume ρ is constant with respect to n. Let $p_n(\theta)$ denote the probability that $\Theta(n) = \theta$.

(a) Derive a difference equation satisfied by $p_n(\theta)$ under certain boundary conditions. Describe these boundary conditions.

(b) Let $P(\omega_i)$ be the a priori probability that the observed class is ω_i. Let

$$F(\theta) = \int_{-\infty}^{\theta} [f_1(x) + f_2(x)]\,dx - P(\omega_1).$$

Let $t = \rho n$. Show that under certain conditions the function $p_n(\theta)$ approaches $p(\theta, t)\,d\theta$, where $p(\theta, t)$ is differentiable in both θ and t, and satisfies the following differential equation:

$$\partial p(\theta, t)/\partial t = g(t)\,\partial[p(\theta, t)F(\theta)]/\partial\theta,$$

where $g(t)$ is a continuous function of t. Show that under these conditions the difference equation obtained in part (a) approaches this differential equation. What are these conditions?

7.4. (a) Show that, by the use of the distorted time τ defined by Equation (7.97), $g(t)$ may be eliminated from the centroid equation and the covariance equation. Specifically, show that Equations (7.31) and (7.79) are transformed into Equations (7.98) and (7.99).

(b) Suppose $g(t) = t^{-\gamma}$, where γ is a constant satisfying $0 < \gamma < 1$. Find τ in terms of t.

(c) Suppose $g(t)$ is defined as in part (b); suppose the constituent densities are those in Figure 7.2; and suppose the training procedure is a constant increment training procedure. Find expressions for $\hat{m}(\tau)$ and $\hat{\rho}(\tau)$.

(d) Using the results of parts (b) and (c), find $\hat{\mu}(t)$ and $R(t)$.

7.5. Prove that $S(\theta)$ has precisely one maximum whenever the one-dimensional constituent densities $f_1(x)$ and $f_2(x)$ are unimodal, $E(X|\omega = \omega_1) < E(X|\omega = \omega_2)$, and $f_1(x)$ intersects $f_2(x)$ in precisely one point, θ^*. Show that $S(\theta^*)$ is the maximum value of $S(\theta)$.

References

1. P. Merryman, Dynamics of multidimensional Markov learning. Technical Report TP-72-3, Pattern Recognition Project, School of Engineering, University of California, Irvine, CA, May 1972.

2. A. Einstein, *Investigation on the Theory of the Brownian Movement*, Dover Publications, New York, 1956.

3. M. F. Norman, *Markov Processes and Learning Models*. Academic Press, New York, 1972.

4. J. Arsac, *Fourier Transforms and the Theory of Distributions*. Prentice-Hall, Englewood Cliffs, NJ, 1966.

5. M. Aoki, *Optimization of Stochastic Systems*. Academic Press, New York, 1967.

6. B. Rosén, On asymptotic normality of sums of dependent random vectors. *Z. Warsheinlichkeitstheorie Verw.*, 7: 95–102 (1967).

7. J. Sklansky and N. J. Bershad, The dynamics of time-varying threshold learning. *Information and Control*, **15**, (6): 455–486 (1969).

APPENDIX A

Vectors and Matrices

A.1 Vector Inequalities and Other Vector Notation

It is often convenient to use the symbols $<$, $\neq$, $\geqslant$, etc. to express inequalities among vectors. These inequalities cannot be interpreted unambiguously without special definitions. The definitions used in this book are given in Table A.1. Other vector notation used in this book is given in Table A.2, using m to denote the number of components in the vector.

Table A.1.

Expression	Definition
$\mathbf{a} \neq \mathbf{b}$	$a_i \neq b_i$ for some i
$\mathbf{a} \geqslant \mathbf{b}$	$a_i \geqslant b_i$ for all i
$\mathbf{a} > \mathbf{b}$	$a_i > b_i$ for all i
$\mathbf{a} \ngeqslant \mathbf{b}$	$a_i < b_i$ for some i
$\mathbf{a} \ngtr \mathbf{b}$	$a_i \leqslant b_i$ for some i
$\mathbf{a} \leqslant \mathbf{b}$	$a_i \leqslant b_i$ for all i
$\mathbf{a} < \mathbf{b}$	$a_i < b_i$ for all i
$\mathbf{a} \nleqslant \mathbf{b}$	$a_i > b_i$ for some i
$\mathbf{a} \nless \mathbf{b}$	$a_i \geqslant b_i$ for some i

Table A.2.

Expression	Definition
$\mathbf{a} = \mathbf{b}$	$a_i = b_i$ for all i
$\|\mathbf{a}\|$	$\left(\sum_i a_i^2\right)^{1/2}$
$\lvert\mathbf{a}\rvert$	$[\lvert a_1\rvert, \lvert a_2\rvert, \ldots, \lvert a_m\rvert]^T$

A.2 Permutation Matrices

An $n \times n$ matrix $\mathbf{P}$ is defined as a *postmultiplier permutation matrix* if, for any $m \times n$ matrix $\mathbf{A}$, $\mathbf{AP}$ is obtained from $\mathbf{A}$ by permuting the columns of $\mathbf{A}$, the permutation being independent of $\mathbf{A}$. $\mathbf{P}$ is a *premultiplier permutation matrix* if $\mathbf{PA}$ yields a permutation of the rows of $\mathbf{A}$, the permutation being independent of $\mathbf{A}$. Suppose we wish to find a permutation matrix $\mathbf{P}$ which will permute the columns $j = 1, \ldots, n$ to the columns $\pi_1, \ldots, \pi_n$. To do this, apply this permutation to the rows of the $n \times n$ identity matrix $\mathbf{I}$. This permutation of the rows of the identity matrix $\mathbf{I}$ is the desired $\mathbf{P}$. For example, suppose we wish to interchange columns 1 and 3 in a 4×4 matrix $\mathbf{A}$. Interchanging rows 1 and 3 in $\mathbf{I}$ yields

$$\mathbf{P} = \begin{bmatrix} 0 & 0 & 1 & 0 \\ 0 & 1 & 0 & 0 \\ 1 & 0 & 0 & 0 \\ 0 & 0 & 0 & 1 \end{bmatrix}.$$

Then $\mathbf{AP}$ yields the desired interchange for any $\mathbf{A}$. Premultiplier matrices for the rows of $\mathbf{A}$ can be obtained in an analogous manner, by permuting the columns of the $m \times m$ identity matrix $\mathbf{I}$. It can be shown that the determinant of any permutation matrix is ± 1.

The following theorem establishes the Gauss–Jordon elimination procedure for solving a set of linear equations.

Theorem. *For every matrix* $\mathbf{A} \neq \mathbf{0}$ *there exist matrices* $\mathbf{F}, \mathbf{P}, \mathbf{R}$ *such that*

$$\mathbf{F}\mathbf{A}^T\mathbf{P} = \begin{bmatrix} \mathbf{I} & \mathbf{R} \\ \mathbf{0} & \mathbf{0} \end{bmatrix} \quad \text{or} \quad \begin{bmatrix} \mathbf{I} \\ \mathbf{0} \end{bmatrix},$$

where $\mathbf{I}$ *is an identity matrix;* $\mathbf{A}^T$ *is the transpose of* $\mathbf{A}$; $\mathbf{A}^T\mathbf{P}$ *is the result of a permutation on the columns of* $\mathbf{A}^T$; $\mathbf{F}\mathbf{A}^T\mathbf{P}$ *is the result of a sequence of elementary row operations* on* $\mathbf{A}^T\mathbf{P}$; *and* $|\mathbf{F}| \neq 0$, *where* $|\mathbf{F}|$ *denotes the determinant of* $\mathbf{F}$.

* An elementary row operation on a matrix $\mathbf{Y}$ consists of replacing $\mathbf{y}_j$ by $b\mathbf{y}_j + c\mathbf{y}_i$, where $\mathbf{y}_i$, $\mathbf{y}_j$ are rows of $\mathbf{Y}$.

Note that no row operation involves the replacement of a row by a row of zeros, because $|\mathbf{F}| \neq 0$.

PROOF. Let $\mathbf{F}$ be expressed as

$$\mathbf{F} = \mathbf{F}_p \cdots \mathbf{F}_1 \mathbf{F}_0$$

where the $\mathbf{F}_i$'s are matrices performing elementary row operations in the product $\mathbf{F}\mathbf{A}^T$.

Let m, n denote the number of rows and columns, respectively, in $\mathbf{A}$. If $a_{11} = 0$, interchange column 1 with another column k, choosing k so that $a_{1k} \neq 0$. Choose the permutation matrix $\mathbf{P}$ so that $\mathbf{A}^T\mathbf{P}$ yields the desired interchange of columns. If such an interchange of columns is not possible, then the first row of $\mathbf{A}$ necessarily consists only of zeros. In that case interchange the first row of $\mathbf{A}$ with a row of $\mathbf{A}$ that doesn't consist only of zeros. Such an interchange of rows may be incorporated into $\mathbf{F}$ without affecting $|\mathbf{F}|$.

For example, suppose $n = 4$. Let $\boldsymbol{\alpha}_i$ denote the ith row of $\mathbf{A}^T$. Suppose $\boldsymbol{\alpha}_2 = \mathbf{0}$, $\boldsymbol{\alpha}_4 \neq \mathbf{0}$. To interchange $\boldsymbol{\alpha}_2$ and $\boldsymbol{\alpha}_4$, add $\boldsymbol{\alpha}_4$ to $\boldsymbol{\alpha}_2$ to form a new $\boldsymbol{\alpha}_2$. Then subtract the new $\boldsymbol{\alpha}_2$ from $\boldsymbol{\alpha}_4$ to form a new $\boldsymbol{\alpha}_4$. This operation is equivalent to the postmultiplier

$$\mathbf{F}_0 = \begin{bmatrix} 1 & 0 & 0 & 0 \\ 0 & 1 & 0 & 0 \\ 0 & 0 & 1 & 0 \\ 0 & -1 & 0 & 1 \end{bmatrix} \begin{bmatrix} 1 & 0 & 0 & 0 \\ 0 & 1 & 0 & 1 \\ 0 & 0 & 1 & 0 \\ 0 & 0 & 0 & 1 \end{bmatrix}$$

$$= \begin{bmatrix} 1 & 0 & 0 & 0 \\ 0 & 1 & 0 & 1 \\ 0 & 0 & 1 & 0 \\ 0 & -1 & 0 & 0 \end{bmatrix}.$$

Note that $|\mathbf{F}_0| = 1$.

Let $\mathbf{B} = \mathbf{A}^T\mathbf{P}$. Let

$$\mathbf{B}^{(k)} = \mathbf{F}_k \cdots \mathbf{F}_0 \mathbf{A}^T \mathbf{P}.$$

Let $\mathbf{b}_i$ denote the ith row of $\mathbf{B}$, and let $\mathbf{b}_i^{(k)}$ denote the ith row of $\mathbf{B}^{(k)}$. Define $\mathbf{F}$ as a matrix that replaces $\mathbf{b}_1$ by $b_{11}^{-1}\,\mathbf{b}_1$. Thus, if $m = n = 4$,

$$\mathbf{F}_1 = \begin{bmatrix} b_{11}^{-1} & 0 & 0 & 0 \\ 0 & 1 & 0 & 0 \\ 0 & 0 & 1 & 0 \\ 0 & 0 & 0 & 1 \end{bmatrix}.$$

Note that $|\mathbf{F}_1| = b_{11}^{-1}$.

Define $\mathbf{F}_2$ as a matrix that replaces $\mathbf{b}_i^{(1)}$ by $\mathbf{b}_i^{(1)} - b_{i1}^{(1)}\mathbf{b}_1^{(1)}$, for $i = 2, \ldots, n$. Thus, for $m = n = 4$,

$$\mathbf{F}_2 = \begin{bmatrix} 1 & 0 & 0 & 0 \\ -b_{21}^{(1)} & 1 & 0 & 0 \\ -b_{31}^{(1)} & 0 & 1 & 0 \\ -b_{41}^{(1)} & 0 & 0 & 1 \end{bmatrix}$$

Note that $|\mathbf{F}_2| = 1$.

If $b_{22}^{(2)} = 0$, interchange column 2 with a column k of $\mathbf{B}^{(2)}$ such that $k > 2$ and $b_{2k}^{(2)} \neq 0$, and incorporate this interchange in $\mathbf{P}$. If such an interchange is not possible, then $\mathbf{b}_2^{(2)}$ necessarily consists of only zeros. In that case interchange $\mathbf{b}_2^{(2)}$ with a row l of $\mathbf{B}^{(2)}$ such that $l > i$ and $b_{l2}^{(2)} \neq 0$. This interchange of rows may be incorporated into $\mathbf{F}$ without affecting $|\mathbf{F}|$. If no such row l exists, then $\mathbf{b}_j = \mathbf{0}$ for $j \geqslant 2$, and $\mathbf{FA}^T\mathbf{P}$ has the desired form.

Define $\mathbf{F}_3$ as a matrix that replaces $\mathbf{b}_2^{(2)}$ by $(b_{22}^{(2)})^{-1}\mathbf{b}_2^{(2)}$. Thus, for $m = n = 4$,

$$\mathbf{F}_3 = \begin{bmatrix} 1 & 0 & 0 & 0 \\ 0 & (b_{22}^{(2)})^{-1} & 0 & 0 \\ 0 & 0 & 1 & 0 \\ 0 & 0 & 0 & 1 \end{bmatrix}.$$

Note that $|\mathbf{F}_3| = (b_{22}^{(2)})^{-1}$

Define $\mathbf{F}_4$ as a matrix that replaces $\mathbf{b}_i^{(3)}$ by $\mathbf{b}_i^{(3)} - b_{i2}^{(3)}\mathbf{b}_2^{(3)}$. Thus, for $m = n = 4$,

$$\mathbf{F}_4 = \begin{bmatrix} 1 & -b_{12}^{(3)} & 0 & 0 \\ 0 & 1 & 0 & 0 \\ 0 & -b_{32}^{(3)} & 1 & 0 \\ 0 & -b_{42}^{(3)} & 0 & 1 \end{bmatrix}$$

Note that $|\mathbf{F}_4| = 1$.

If $b_{33}^{(4)} = 0$, go through a process of column interchange or row interchange in a manner similar to that which took place for the case $b_{22}^{(2)} = 0$.

This process is continued until $\mathbf{FA}^T\mathbf{P}$ is of the form

$$\begin{bmatrix} \mathbf{I} & \mathbf{R} \\ \mathbf{0} & \mathbf{0} \end{bmatrix}.$$

Note that $|\mathbf{F}| = \prod_k |\mathbf{F}_k| \neq 0$. □

APPENDIX B

Proof of Convergence for the Window Procedure

In this appendix we prove Theorem 5.1 of Subsection 5.3.2. First we establish a boundedness property which is used in our proof.

Assume that the $\mathbf{v}^*$'s and $\hat{\mathbf{v}}^*$'s have been normalized so that $|v_k^*| = |\hat{v}_k^*| = 1$. The bound we wish to establish is

$$(1/c)E(\|\mathbf{Z}\|^2) < \infty. \tag{B.1}$$

Letting Ψ denote $\Psi[\mathbf{V}^T\mathbf{Y}/\|\mathbf{W}\|, c]$, (5.77) yields

$$\begin{aligned}\frac{1}{2c}E\|\mathbf{Z}\|^2 &= \frac{1}{2c}E\{E[\|\mathbf{Z}\|^2|\mathbf{V}]\}\\ &= E\left\{\frac{1}{2c}\int\left\|2c\|\mathbf{W}\|\left[\tilde{\mathbf{Y}} - \frac{\mathbf{V}^T\mathbf{Y}}{\|\mathbf{W}\|^2}\left[\frac{0}{\tilde{\mathbf{W}}}\right]\right]\Psi\right\|^2[f_1(\mathbf{x}) + f_2(\mathbf{x})]\,d\mathbf{x}\right\}.\end{aligned} \tag{B.2}$$

Let $\tilde{\mathbf{U}}$ denote the vector $[\tilde{\mathbf{X}} - (\mathbf{V}^T\mathbf{Y}/\|\mathbf{W}\|^2)\tilde{\mathbf{W}}]$, i.e.,

$$\left[\tilde{\mathbf{Y}} - \frac{\mathbf{V}^T\mathbf{Y}}{\|\mathbf{W}\|^2}\left[\frac{0}{\tilde{\mathbf{W}}}\right]\right]$$

with the first component omitted. Since $Y_0 = 1$, (B.2) can be weakened to

$$\frac{1}{2c}E\|\mathbf{Z}\|^2 \leqslant 2c\left\{E\left[\|\mathbf{W}\|^2\int\Psi^2 p(\mathbf{x})\,d\mathbf{x}\right] - E(\|\mathbf{W}\|^2\|\tilde{\mathbf{U}}\|^2\Psi^2)\right\}. \tag{B.3}$$

It will next be demonstrated that

$$\int \Psi^2 p(\mathbf{x})\,d\mathbf{x} \leqslant K \int_{-\infty}^{\infty} \Psi^2(\xi, c)\,d\xi \quad \text{for all } \mathbf{V},\ 0 < K < \infty. \tag{B.4}$$

First, rotate the coordinate system prior to any integrations, so that one of the new coordinate axis lies parallel to $\mathbf{W}$. Note that

$$\frac{\mathbf{V}^T\mathbf{Y}}{\|\mathbf{W}\|} = \frac{\mathbf{W}^T\mathbf{X}}{\|\mathbf{W}\|} + \frac{W_0}{\|\mathbf{W}\|}$$

so that Ψ does not vary with motion of $\mathbf{X}$ normal to $\mathbf{W}$. Since, from constraint 5 of Theorem 5.1, $p(\mathbf{x})$ is a bounded probability density function, the function found by first integrating $p(\mathbf{x})$ over a hyperplane normal to $\mathbf{W}$ will also have an upper bound. Let this bound be denoted by K. Therefore since $\Psi^2 > 0$

$$\int \Psi^2 p(\mathbf{x})\,d\mathbf{x} \leqslant K \int_{-\infty}^{\infty} \Psi^2\left(\frac{\mathbf{W}^T\mathbf{X}}{\|\mathbf{W}\|} + \frac{W_0}{\|\mathbf{W}\|}, c\right) d\left(\frac{\mathbf{W}^T\mathbf{X}}{\|\mathbf{W}\|}\right),$$

which proves (B.4). Therefore, recalling that $\|\mathbf{W}\| < M$, (B.3) can be weakened to

$$\frac{1}{2c} E\|\mathbf{Z}\|^2 \leqslant M^2\left[2cK \int_{-\infty}^{\infty} \Psi^2(\xi, c)\,d\xi + 2cE(\|\tilde{\mathbf{U}}\|^2\Psi^2)\right]. \tag{B.5}$$

When constraints 9 and 10 of the theorem are applied, the validity of (B.1) is proved.

The following proof of Theorem 5.1 is an extension of a proof by Blum. (See Reference [6] of Chapter 5.) First we consider any one $\hat{\mathbf{v}}^*$ from the set of all $\hat{\mathbf{v}}^*$'s. Let $h(\mathbf{V})$ denote the following real valued function:

$$h(\mathbf{V}) = \|\mathbf{V} - \hat{\mathbf{v}}^*\|^2 \tag{B.6}$$

where all the $\mathbf{v}^*$'s and $\hat{\mathbf{v}}^*$'s have been normalized so that $|v_k^*| = |\hat{v}_k^*| = 1$.

From (B.6), $h(\mathbf{V} + \rho\mathbf{Z}_\mathbf{V}) = (\mathbf{V} - \hat{\mathbf{v}}^* + \rho\mathbf{Z}_\mathbf{V})^T(\mathbf{V} - \hat{\mathbf{v}}^* + \rho\mathbf{Z}_\mathbf{V})$, which when expanded becomes

$$h(\mathbf{V} + \rho\mathbf{Z}_\mathbf{V}) = h(\mathbf{V}) - 2\rho(\hat{\mathbf{v}}^* - \mathbf{V})^T\mathbf{Z}_\mathbf{V} + \rho^2\|\mathbf{Z}_\mathbf{V}\|^2,$$

where $\mathbf{Z}_\mathbf{V}$ denotes the random variable $\mathbf{Z}$ for the given observation $\mathbf{V}$. Taking expectations on both sides

$$E[h(\mathbf{V} + \rho\mathbf{Z})|\mathbf{V}] = h(\mathbf{V}) - 2\rho(\hat{\mathbf{v}}^* - \mathbf{V})^T E(\mathbf{Z}|\mathbf{V}) + \rho^2 E(\|\mathbf{Z}\|^2|\mathbf{V}). \tag{B.7}$$

Now let $H(n)$ denote $h[\mathbf{V}(n)]$. Then from (5.76),

$$H(n+1) = \begin{cases} h[\mathbf{V}(n) + \rho(n)\mathbf{Z}(n)], & \text{if } \|\mathbf{W}(n) + p(n)\mathbf{S}(n)\| \leqslant M, \\ h[\mathbf{V}(n)], & \text{otherwise.} \end{cases}$$

Utilizing (B.7),

$$E[H(n+1)|H(0),\ldots,H(n)]$$

$$= H(n) - \Big(2\rho(n)E\{[\hat{\mathbf{v}}(n)^* - \mathbf{V}(n)]^T E[\mathbf{Z}(n)|\mathbf{V}(n)]|H(0),\ldots,H(n)\}$$

$$+\rho(n)^2 E[\|\mathbf{Z}(n)\|^2|H(0),\ldots,H(n)]\Big)P[\|\mathbf{W}(n)-\rho(n)\mathbf{S}(n)\|<M|H(0),\ldots,H(n)]. \tag{B.8}$$

Now from constraint 6 of the theorem, (B.8) can be weakened to

$$E[H(n+1) - H(n)|H(0),\ldots,H(n)] \leqslant \rho(n)^2 E[\|\mathbf{Z}(n)\|^2|H(0),\ldots,H(n)]$$

whenever each $\hat{\mathbf{v}}^*(n)$ has been chosen to satisfy constraint 6. From Equation (B.1) and constraint 4 of the theorem, it follows that

$$\sum_{n=0}^{\infty} E[H(n+1) - H(n)|H(0),\ldots,H(n)] < \infty. \tag{B.9}$$

From a corollary by Blum based upon a theorem of Doob (see Reference [6] of Chapter 5), the property (B.9) together with the conditions that the $H(n)$ must be integrable and bounded from below uniformly in n, ensure the convergence of $H(n)$ with probability one. Therefore, since $H(n) = \|\mathbf{V}(n) - \hat{\mathbf{v}}^*(n)\|^2$ is integrable, is bounded from below by zero for all n, and (B.9) holds, it follows that there exists a set of random variables $H^* \in \mathscr{H}^*$ such that

$$P\Big(\lim_{n\to\infty} \|\mathbf{V}(n) - \hat{\mathbf{v}}^*(n)\|^2 = H^*\Big) = 1 \tag{B.10}$$

for some $\hat{\mathbf{v}}^*(n)$.

From constraint 2 and constraint 11, $\lim_{n\to\infty} \hat{\mathbf{v}}^*(n) = \hat{\mathbf{v}}^*$. (This was earlier expressed as Equation (5.69)). Therefore,

$$\lim_{n\to\infty} \|\mathbf{V}(n) - \hat{\mathbf{v}}^*(n)\|^2 = \Big\|\lim_{n\to\infty} (\mathbf{V}(n) - \hat{\mathbf{v}}^*(n))\Big\|^2 = \lim_{n\to\infty} \|\mathbf{V}(n) - \mathbf{v}^*\|^2, \qquad \mathbf{v}^* \in \mathscr{V}^*.$$

Now (B.10) can be written

$$P\Big(\lim_{n\to\infty} \|\mathbf{V}(n) - \mathbf{v}^*\|^2 = H^*\Big) = 1, \qquad \mathbf{v}^* \in \mathscr{V}^*, H^* \in \mathscr{H}^*. \tag{B.11}$$

Taking the unconditional expectation of both sides of (B.8) and iterating, the following expression is obtained:

$$EH(n+1) = H(0) - \sum_{j=0}^{n} 2\rho(j)E\{[\hat{\mathbf{v}}^*(j) - \mathbf{V}(j)]^T E[\mathbf{Z}(j)|\mathbf{V}(j)]\}P[\|\mathbf{W}(j)$$

$$+ \rho(j)\mathbf{S}(j)\| < M] + \sum_{j=0}^{n} \rho(j)^2 E\|\mathbf{Z}(j)\|^2 P[\|\mathbf{W}(j) + \rho(j)\mathbf{S}(j)\| < M]. \tag{B.12}$$

Taking the limit as $n \to \infty$ in (B.12), $\lim_{n\to\infty} EH(n+1)$ is finite since from (B.10) it converges to some random variable H^*. $H(0) = \|\mathbf{V}(n) - \hat{\mathbf{v}}^*(0)\|^2$ is finite since $\mathbf{V}(0)$ is an arbitrary initial finite vector, $\|\hat{\mathbf{v}}^*(0)\| < \infty$ from constraint 11, and $\|\mathbf{V}(0) - \hat{\mathbf{v}}^*(0)\| \leqslant \|\mathbf{V}(0)\|^2 + \|\hat{\mathbf{v}}^*(0)\|^2$. $\sum_{j=0}^{\infty} \rho^2(j)E\|\mathbf{Z}(j)\|^2$ is finite since $\sum_{j=0}^{\infty} \rho(j)^2 c(j) < \infty$ from constraint 4 and $(1/c(j))E\|\mathbf{Z}(j)\|^2 < \infty$ from (B.1). Therefore $\sum_{j=0}^{\infty} 2\rho(j)E\{(\hat{\mathbf{v}}^*(j) - \mathbf{V}(j))^T E[\mathbf{Z}(j)|\mathbf{V}(j)]\}$ must also be finite since $P[\|\mathbf{W}(j) + \rho(j)\mathbf{S}(j) < M|\mathbf{V}(j)] > 0$ from constraint 12. But from constraint 3, $\sum_{j=0}^{\infty} \rho(j)c(j) = \infty$ and from constraint 6,

$$\frac{1}{c(j)}[\hat{\mathbf{v}}^*(j) - \mathbf{V}(j)]^T E[\mathbf{Z}(j)|\mathbf{V}(j)] \geqslant 0, \qquad j > N$$

for at least one $\hat{\mathbf{v}}^*(j)$ so that if such a $\hat{\mathbf{v}}^*(n)$ is selected at each $n \geqslant N$, the following relationships must hold:

$$\limsup_{N\to\infty,\, n\geqslant N} \frac{1}{c(n)} E\{[\hat{\mathbf{v}}^*(n) - \mathbf{V}(n)]^T E[\mathbf{Z}(n)|\mathbf{V}(n)]\} = 0$$

$$\liminf_{N\to\infty,\, n\geqslant N} \frac{1}{c(n)} E\{[\hat{\mathbf{v}}^*(n) - \mathbf{V}(n)]^T E[\mathbf{Z}(n)|\mathbf{V}(n)]\} = 0.$$

Now let $\{n_e\}$ be an infinite sequence of integers such that

$$\lim_{n_e\to\infty} \frac{1}{c(n_e)} E|[\hat{\mathbf{v}}^*(n_e) - \mathbf{V}(n_e)]^T E[\mathbf{Z}(n_e)|\mathbf{V}(n_e)]| = 0.$$

This implies convergence in the mean which in turn implies the convergence of

$$\frac{1}{c(n_e)}[\hat{\mathbf{v}}^*(n_e) - \mathbf{V}(n_e)]^T E[\mathbf{Z}(n_e)|\mathbf{V}(n_e)]$$

to zero in probability. There therefore exists a further subsequence, say

$$\left\{\frac{1}{c(m_e)}[\hat{\mathbf{v}}^*(m_e) - \mathbf{V}(m_e)]^T E[\mathbf{Z}(m_e)|\mathbf{V}(m_e)]\right\},$$

such that

$$P\left[\lim_{m_e\to\infty} \frac{1}{c(m_e)}[\hat{\mathbf{v}}^*(m_e) - \mathbf{V}(m_e)]^T E[\mathbf{Z}(m_e|\mathbf{V}(m_e)] = 0\right] = 1. \tag{B.13}$$

From constraint 6 it follows that

$$\frac{1}{c(m_e)}[\hat{\mathbf{v}}^*(m_e) - \mathbf{V}(m_e)]^T E[\mathbf{Z}(m_e)|\mathbf{V}(m_e)] = 0, \text{ iff } \|\hat{\mathbf{v}}^*(m_e) - \mathbf{V}(m_e)\| = 0.$$

Therefore, from (5.69), $\lim_{m_e\to\infty} \|\hat{\mathbf{v}}^*(m_e) - \mathbf{V}(m_e)\| = 0$ with probability one. As before, from (5.69) and constraint 2, $\lim_{m_e\to\infty} \hat{\mathbf{v}}^*(m_e) \in \mathscr{V}^*$ so that

$$P\left[\lim_{m_e\to\infty} \|\mathbf{v}^* - \mathbf{V}(m_e)\| = 0\right] = 1, \qquad \mathbf{v}^* \in \mathscr{V}^*. \tag{B.14}$$

Relationship (B.11) states that the limit of all sequences $\{\mathbf{V}(n)\}$ lies, with probability one, on a hypersphere whose radius squared is described by H^* and whose center is located at some $\mathbf{v}^*$, $\mathbf{v}^* \in \mathscr{V}^*$. But since (B.14) states that for one of these sequences $H^* = 0$ with probability one, H^* must equal zero with probability one for all sequences $\{\mathbf{V}(n)\}$. Therefore (B.11) can be written

$$P\left(\lim_{n\to\infty} \|\mathbf{V}(n) - \mathbf{v}^*\|^2 = 0\right) = 1, \qquad \mathbf{v}^* \in \mathscr{V}^*$$

or

$$P\left(\lim_{n\to\infty} \mathbf{V}(n) = \mathbf{v}^*\right) = 1, \qquad \mathbf{v}^* \in \mathscr{V}^*,$$

which is the desired result proving Theorem 5.1.

APPENDIX C

Proof of Convergence for the Equalized Error Procedure

In this appendix we prove Theorem 5.3 of Subsection 5.5.2. Before proceeding to this proof we establish two boundedness properties of **Z**.

C.1 Proof that $E(\|\mathbf{Z}\|^2|\mathbf{V}) < \infty$ and $\|E(\mathbf{Z}|\mathbf{V})\|^2 < \infty$

From the definition of **Z** given by (5.126),

$$E(\|\mathbf{Z}\|^2|\mathbf{V}) = \|\mathbf{W}\|^2\left\{P(\omega_2, \mathbf{Y} \in \mathcal{R}_1)E\left[\left\|\tilde{\mathbf{Y}} - \frac{\mathbf{V}^T\mathbf{Y}}{\|\mathbf{W}\|^2}\begin{bmatrix} 0 \\ \tilde{\mathbf{W}} \end{bmatrix}\right\|^2 \middle| \omega_2, \mathbf{Y} \in \mathcal{R}_1\right]\right.$$
$$\left. + P(\omega_1, \mathbf{Y} \in \mathcal{R}_2)E\left[\left\|\tilde{\mathbf{Y}} - \frac{\mathbf{V}^T\mathbf{Y}}{\|\mathbf{W}\|^2}\begin{bmatrix} 0 \\ \tilde{\mathbf{W}} \end{bmatrix}\right\|^2 \middle| \omega_1, \mathbf{Y} \in \mathcal{R}_2\right]\right\}. \tag{C.1}$$

But the zeroth component of $\tilde{\mathbf{Y}}$ is always one and the X_k and W_k components of $\tilde{\mathbf{Y}}$ and $\tilde{\mathbf{W}}$ are zero. Therefore,

$$\left\|\tilde{\mathbf{Y}} - \frac{\mathbf{V}^T\mathbf{Y}}{\|\mathbf{W}\|^2}\begin{bmatrix} 0 \\ \tilde{\mathbf{W}} \end{bmatrix}\right\|^2 = 1 + \left\|\tilde{\mathbf{X}} - \frac{\mathbf{V}^T\mathbf{Y}}{\|\mathbf{W}\|^2}\tilde{\mathbf{W}}\right\|^2 \leqslant 1 + \left\|\mathbf{X} - \frac{\mathbf{V}^T\mathbf{Y}}{\|\mathbf{W}\|^2}\mathbf{W}\right\|^2 \tag{C.2}$$

Furthermore

$$\left\|\mathbf{X} - \frac{\mathbf{V}^T\mathbf{Y}}{\|\mathbf{W}\|^2}\mathbf{W}\right\|^2 = \|\mathbf{X}\|^2 - 2\frac{(\mathbf{V}^T\mathbf{Y})(\mathbf{W}^T\mathbf{X})}{\|\mathbf{W}\|^2} + \frac{(\mathbf{V}^T\mathbf{Y})^2}{\|\mathbf{W}\|^2}$$
$$= \|\mathbf{X}\|^2 + \frac{W_0^2 - (\mathbf{W}^T\mathbf{X})^2}{\|\mathbf{W}\|^2} \leqslant \|\mathbf{X}\|^2 + \frac{W_0^2}{\|\mathbf{W}\|^2}. \tag{C.3}$$

Therefore, (C.1) can be weakened to

$$E(\|\mathbf{Z}\|^2 | \mathbf{V}) \leqslant \|\mathbf{W}\|^2 \Big\{ P(\omega_2, \mathbf{Y} \in \mathscr{R}_1) + P(\omega_1, \mathbf{Y} \in \mathscr{R}_2)$$

$$+ P(\omega_2, \mathbf{Y} \in \mathscr{R}_1)\left[E(\|\mathbf{X}\|^2 | \omega_2, \mathscr{R}_1) + \frac{W_0^2}{\|\mathbf{W}\|^2}\right]$$

$$+ P(\omega_1, \mathbf{Y} \in \mathscr{R}_2)\left[E(\|\mathbf{X}\|^2 | \omega_1, \mathscr{R}_2) + \frac{W_0^2}{\|\mathbf{W}\|^2}\right]\Big\}. \tag{C.4}$$

Now since $P(\omega_2, \mathbf{Y} \in \mathscr{R}_1) + P(\omega_1, \mathbf{Y} \in \mathscr{R}_2) \leqslant 1$,

$$E(\|\mathbf{Z}\|^2 | \mathbf{V}) \leqslant \|\mathbf{W}\|^2 \left\{ 1 + \frac{W_0^2}{\|\mathbf{W}\|^2} + \int_{\mathscr{R}_1} \|\mathbf{X}\|^2 f_2(\mathbf{x})\, d\mathbf{x} + \int_{\mathscr{R}_2} \|\mathbf{X}\|^2 f_1(\mathbf{x})\, d\mathbf{x} \right\}$$

$$\leqslant \|\mathbf{W}\|^2 \left\{ 1 + \frac{W_0^2}{\|\mathbf{W}\|^2} + \int \|\mathbf{X}\|^2 [f_1(\mathbf{x}) + f_2(\mathbf{x})]\, d\mathbf{x} \right\} \tag{C.5}$$

$$= \|\mathbf{V}\|^2 + \|\mathbf{W}\|^2 E(\|\mathbf{X}\|^2) . \tag{C.6}$$

But $\|\mathbf{V}\| < M$ and $\|\mathbf{W}\|^2 \leqslant \|\mathbf{V}\|^2$, so that

$$E(\|\mathbf{Z}\|^2 | \mathbf{V}) \leqslant M^2(1 + E(\|\mathbf{X}\|^2)). \tag{C.7}$$

Finally, from contraint 4 of Theorem 5.3,

$$E(\|\mathbf{Z}\|^2 | \mathbf{V}) < \infty. \tag{C.8}$$

Next, the variance of $\mathbf{Z}$ for a given $\mathbf{V}$ is expressed as

$$E[\|\mathbf{Z} - E(\mathbf{Z}|\mathbf{V})\|^2 | \mathbf{V}] = E\{[\|\mathbf{Z}\|^2 - 2\mathbf{Z}^T E(\mathbf{Z}|\mathbf{V}) + \|E(\mathbf{Z}|\mathbf{V})\|^2] | \mathbf{V}\}, \tag{C.9}$$

which simplifies to

$$E[\|\mathbf{Z} - E(\mathbf{Z}|\mathbf{V})\|^2 | \mathbf{V}] = E(\|\mathbf{Z}\|^2 | \mathbf{V}) - \|E(\mathbf{Z}|\mathbf{V})\|^2. \tag{C.10}$$

Therefore

$$\|E(\mathbf{Z}|\mathbf{V})\|^2 \leqslant E(\|\mathbf{Z}\|^2 | \mathbf{V}) \tag{C.11}$$

so that, from (C.8),

$$\|E(\mathbf{Z}|\mathbf{V})\|^2 < \infty. \tag{C.12}$$

C.2 Proof of Theorem 5.3

The multidimensional extension of the method of Dvoretzky discussed in Subsection 4.8.2 will be the basis for this convergence proof. It will therefore be sufficient to prove that Dvoretzky's conditions 1–7 presented in Subsection 4.8.2 all hold.

Dvoretzky's condition 5 is satisfied by comparing (5.128) with this condition so that his transformation $\mathbf{t}_n$ becomes

$$\mathbf{t}_n = \mathbf{V}(n) + \rho(n) E[\mathbf{Z}(n) | \mathbf{V}(n)] P[\|\mathbf{V}(n) + \rho(n)\mathbf{Z}(n)\| < M | \mathbf{V}(n)]$$

and $\mathbf{R}_n$ becomes

$$\mathbf{R}_n = \rho(n)\left[\begin{Bmatrix}\mathbf{Z}(n), & \text{for } \|\mathbf{V}(n) + \rho(n)\mathbf{Z}(n)\| < M \\ \mathbf{0}, & \text{otherwise}\end{Bmatrix}\right.$$
$$\left. - E[\mathbf{Z}(n)|\mathbf{V}(n)]P[\|\mathbf{V}(n) + \rho(n)\mathbf{Z}(n)\| < M|\mathbf{V}(n)]\right]. \tag{C.13}$$

For the proof, we shall only be interested in those $\mathbf{v}^*$'s whose v_k^* component has a magnitude of one. Therefore, we shall assume that all $\mathbf{v}^*$'s have been so normalized.

The satisfaction of Dvoretzky's conditions 1–4 is demonstrated through the following expansion for any $\mathbf{v}^*$ at trial n:

$$\begin{aligned}\|\mathbf{t}_n - \mathbf{v}^*\|^2 = {} & \|\mathbf{V}(n) - \mathbf{v}^*\|^2 - 2\rho(n)[\mathbf{v}^* - \mathbf{V}(n)]^T E[\mathbf{Z}(n)|\mathbf{V}(n)] \\ & \times P[\|\mathbf{V}(n) + \rho(n)\mathbf{Z}(n)\| < M|\mathbf{V}(n)] \\ & + \rho(n)^2\|E[\mathbf{Z}(n)|\mathbf{V}(n)]\|^2 P^2[\|\mathbf{V}(n) + \rho(n)\mathbf{Z}(n)\| < M|\mathbf{V}(n)].\end{aligned} \tag{C.14}$$

First the case for which

$$\|\mathbf{t}_n - \mathbf{v}^*\| > \|\mathbf{V}(n) - \mathbf{v}^*\| + \varepsilon, \tag{C.15}$$

ε any positive real number, will be discussed. From constraint 6, (C.14) can be weakened to

$$\|\mathbf{t}_n - \mathbf{V}^*(n)\|^2 \leqslant \|\mathbf{V}(n) - \mathbf{V}^*(n)\|^2 + \rho(n)^2\|E[\mathbf{Z}(n)|\mathbf{V}(n)]\|^2,$$

where $\mathbf{V}^*(n)$ denotes any $\mathbf{v}^*$ which satisfies constraint 6 at trial n.

For the case under discussion, this can be written

$$\|\mathbf{t}_n - \mathbf{V}^*(n)\| \leqslant \frac{\rho(n)^2\|E[\mathbf{Z}(n)|\mathbf{V}(n)]\|^2}{\|\mathbf{t}_n - \mathbf{V}^*(n)\| - \|\mathbf{V}(n) - \mathbf{V}^*(n)\|} - \|\mathbf{V}(n) - \mathbf{V}^*(n)\|.$$

Using (C.15) this can be weakened to

$$\|\mathbf{t}_n - \mathbf{V}^*(n)\| \leqslant \rho(n)^2\frac{1}{\varepsilon}\|E[\mathbf{Z}(n)|\mathbf{V}(n)]\|^2. \tag{C.16}$$

But

$$\sup_{\mathbf{V}} \|E(\mathbf{Z}|\mathbf{V})\|^2 < \infty,$$

which is given by (C.12). Now define α_n by

$$\alpha_n = \rho(n)^2\frac{1}{\varepsilon}\sup_{\mathbf{V}}\|E[\mathbf{Z}(n)|\mathbf{V}(n)]\|^2. \tag{C.17}$$

Then, since $p(n)^2 \to 0$ from constraint 3,

$$\alpha_n \to 0,$$

which together with $\|\mathbf{t}_n - \mathbf{V}^*(n)\| \leqslant \alpha_n$ from (C.16) and (C.17) satisfies Dvoretzky's condition 1 and the first part of 4.

Now the case for which

$$\|\mathbf{t}_n - \mathbf{V}^*(n)\| \leqslant \|\mathbf{V}(n) - \mathbf{V}^*(n)\|$$

will be discussed. Define γ_n as

$$\gamma_n = \frac{1}{\|\mathbf{V}(n) - \mathbf{V}^*(n)\|} \Bigg[\rho(n) \inf_{\varepsilon \leqslant \|\mathbf{V}(n) - \mathbf{V}^*(n)\|} [\mathbf{V}^*(n) - \mathbf{V}(n)]^T E[\mathbf{Z}(n) | \mathbf{V}(n)] \\ \times P[\|\mathbf{V}(n) + \rho(n)\mathbf{Z}(n)\| < M | \mathbf{V}(n)] - \tfrac{1}{2}\rho(n)^2 \sup_{\mathbf{V}(n)} \|E[\mathbf{Z}(n)|\mathbf{V}(n)]\|^2 \Bigg] \tag{C.18}$$

so that (C.14) can be weakened to

$$\|\mathbf{t}_n - \mathbf{V}^*(n)\|^2 \leqslant \|\mathbf{V}(n) - \mathbf{V}^*(n)\|^2 - 2\|\mathbf{V}(n) - \mathbf{V}^*(n)\|\gamma_n + \gamma_n^2$$

or

$$\|\mathbf{t}_n - \mathbf{V}^*(n)\| \leqslant \|\mathbf{V}(n) - \mathbf{V}^*(n)\| - \gamma_n, \tag{C.19}$$

which is of the form of the second part of Dvoretzky's condition 4 with $\beta_n = 0$. Now from (C.18) and constraints 2, 3, 5, and 6,

$$\sum_{n=0}^{\infty} \gamma_n = \infty$$

which together with $\beta_n = 0$ for all n and (C.19) satisfies Dvoretzky's conditions 2 and 3 together with the remainder of 4.

The satisfaction of Dvoretzky's condition 6 is demonstrated by first expanding $E(\|\mathbf{R}_n\|^2)$, where $\mathbf{R}_n$ is given in (C.13). For economy of notation, let $\boldsymbol{\Gamma}_n$ denote

$$\boldsymbol{\Gamma}_n = \begin{cases} \mathbf{Z}(n), & \text{if } \|\mathbf{V}(n) + \rho(n)\mathbf{Z}(n)\| < M \\ \mathbf{0}, & \text{otherwise.} \end{cases}$$

Then

$$\begin{aligned} E\|\mathbf{R}_n\|^2 &= \rho^2(n) E(\|\boldsymbol{\Gamma}_n - E[\boldsymbol{\Gamma}_n|\mathbf{V}(n)]\|^2) \\ &= \rho^2(n) E(E\{\|\boldsymbol{\Gamma}_n - E[\boldsymbol{\Gamma}_n|\mathbf{V}(n)]\|^2|\mathbf{V}(n)\}) \\ &= \rho^2(n) E(E\{\|\boldsymbol{\Gamma}_n\|^2 - 2\boldsymbol{\Gamma}_n^T E[\boldsymbol{\Gamma}_n|\mathbf{V}(n)] + \|E[\boldsymbol{\Gamma}_n|\mathbf{V}(n)]\|^2|\mathbf{V}(n)\}) \end{aligned}$$

or, taking the inner conditional expectation,

$$E\|\mathbf{R}_n\|^2 = \rho(n)^2 E\{E[\|\boldsymbol{\Gamma}_n\|^2|\mathbf{V}(n)] - \|E[\boldsymbol{\Gamma}_n|\mathbf{V}(n)]\|^2\}.$$

But this can be weakened to

$$E\|\mathbf{R}_n\|^2 \leqslant \rho(n)^2 E\{E[\|\boldsymbol{\Gamma}_n\|^2|\mathbf{V}(n)]\} \leqslant \rho(n)^2 E\{E[\|\mathbf{Z}(n)\|^2|\mathbf{V}(n)]\}. \tag{C.20}$$

However, from (C.8),

$$E(\|\mathbf{Z}\|^2|\mathbf{V}) < \infty. \tag{C.21}$$

Now from (C.20), (C.21) and constraint 3,

$$\sum_{n=0}^{\infty} E(\|\mathbf{R}_n\|^2) < \infty,$$

so that Dvoretzky's condition 6 is satisfied. His condition 7 is trivially satisfied from (C.13). Therefore, since $\mathbf{V}^*(n)$ always equals some $\mathbf{v}^*$, the sequence $\{\mathbf{V}(n)\}$ converges to a $\mathbf{v}^*$ both in mean square and probability one and Theorem 5.3 is proved.

Index

Greek Alphabet